Advances in Intelligent Systems and Computing

Volume 1337

The series “Advances in Intelligent Systems and Computing” contains publications on theory, applications, and design methods of Intelligent Systems and Intelligent Computing. Virtually all disciplines such as engineering, natural sciences, computer and information science, ICT, economics, business, e-commerce, environment, healthcare, life science are covered. The list of topics spans all the areas of modern intelligent systems and computing such as: computational intelligence, soft computing including neural networks, fuzzy systems, evolutionary computing and the fusion of these paradigms, social intelligence, ambient intelligence, computational neuroscience, artificial life, virtual worlds and society, cognitive science and systems, Perception and Vision, DNA and immune based systems, self-organizing and adaptive systems, e-Learning and teaching, human-centered and human-centric computing, recommender systems, intelligent control, robotics and mechatronics including human-machine teaming, knowledge-based paradigms, learning paradigms, machine ethics, intelligent data analysis, knowledge management, intelligent agents, intelligent decision making and support, intelligent network security, trust management, interactive entertainment, Web intelligence and multimedia.

The publications within “Advances in Intelligent Systems and Computing” are primarily proceedings of important conferences, symposia and congresses. They cover significant recent developments in the field, both of a foundational and applicable character. An important characteristic feature of the series is the short publication time and world-wide distribution. This permits a rapid and broad dissemination of research results.

Indexed by DBLP, INSPEC, WTI Frankfurt eG, zbMATH, Japanese Science and Technology Agency (JST).

All books published in the series are submitted for consideration in Web of Science.

More information about this series at https://link.springer.com/bookseries/11156

Preface

This book contains papers from the 2020 Annual Conference of the North American Fuzzy Information Processing Society (Redmond, Washington, USA, August 20–22, 2020). These papers cover many important theoretical issues related to fuzzy techniques, as well as many successful applications of fuzzy methodology.

A special section of this book is devoted to papers from the affiliated 13th International Workshop on Constraint Programming and Decision Making CoProd'2020 (Redmond, Washington, USA, August 19, 2020).

Our many thanks to all the authors for their interesting papers, to all the anonymous referees for their important work, to NAFIPS leadership for their support, to Janusz Kacprzyk and to all the Springer staff for helping us to publish this volume, and, of course, to the readers for their interest—this is for you, the readers, that these books are published. We hope that in this book, you will find new ideas that can be directly applied, as well as ideas and challenges that will inspire you to make further progress.

Thanks again!

Redmond, Washington, USA — Barnabás Bede
El Paso, Texas, USA — Martine Ceberio
Tacoma, Washington, USA — Martine De Cock
El Paso, Texas, USA — Vladik Kreinovich

Contents

Powerset Operators in Categories with Fuzzy Relations Defined by Monads

Jiří Močkoř

Abstract Powerset operators in categories with relations defined by monads as morphisms are investigated. Such categories are, in fact, Kleisli categories of monads in clone form. Examples of such categories used in lattice-valued F-transform theory are presented. It is proven that for arbitrary powerset operator $\mathcal{P}$ in a category $\mathbf{K}$ with a monad $\mathcal{T}$ there exists a powerset operator $\widetilde{\boldsymbol{P}}$ of the Kleisli category $\mathbf{K_T}$, which extends the original powerset operator of the category $\mathbf{K}$.

1 Introduction

Fuzzy set theory was introduced by Zadeh [16] as a generalization of the classical set theory, allowing working with vagueness, as one of the basic features of real-world applications. Concurrently with its origins, the theory of fuzzy sets dealt not only with objects, i.e., with fuzzy sets, but also investigated the functional relations between these objects. This naturally led to research into the categorical aspects of fuzzy sets and, in general, to exploration of fuzzy set categories. However, it was still valid that the key role of morphisms in these new categories had two main structures, namely the mappings between underlying sets of fuzzy sets with specific properties on the one hand and the fuzzy relation (again with special properties) on the other.

Recently, the theory of fuzzy set categories was applied to, among other things, new types of objects designed in fuzzy environments, which are often used in other fields of mathematics as well as in applications, including various fuzzy algebraic and topological structures, ordered structures and various aggregation operators.

All these structures require the use of extensions of mappings between underlying sets of these structures to mapping between these structures, i.e., generalizations of the classical Zadeh's extension principle. These extensions can generally be included

J. Močkoř (✉)
Institute for Research and Applications of Fuzzy Modeling, University of Ostrava, Centre of Excellence IT4Innovations, 30. dubna 22, 701 03 Ostrava 1, Ostrava, Czech Republic
e-mail: mockor@osu.cz

B. Bede et al. (eds.), *Fuzzy Information Processing 2020*,
Advances in Intelligent Systems and Computing 1337,
https://doi.org/10.1007/978-3-030-81561-5_1

under the notion of *powerset operators*, which was in a general form introduced by Rodabaugh [12].

The powerset operators and corresponding powerset theories are widely used in algebra, logic, topology, computer science ans especailly in fuzzy set theory. The standard Zadeh's example of a powerset extension of a mapping $f : X \to Y$ to the map $Z(f) : \mathcal{L}^X \to \mathcal{L}^Y$ is widely used in almost all branches of mathematics and their applications, including computer science.

A common feature of the categories of fuzzy sets is that morphisms in these categories are generally perceived as special mappings, i.e., the second standard possibility of defining morphisms as a special fuzzy relation is neglected.

Recently, however, a number of results have emerged in the theory of fuzzy sets, which are based on the application of fuzzy relations in suitable categories. A typical example is the category of sets as objects and $\mathcal{L}$-valued fuzzy relations between sets as morphisms. This category is frequently used in constructions of approximation functors which represent various approximations of fuzzy sets, defined by fuzzy relations. This approximation was for the first time defined by Goguen [1], when he introduced the notion of the image of a fuzzy set under a fuzzy relation. Many examples using explicitly or implicitly approximation functors can be found in rough fuzzy sets theory, F-transform theory and many others (see, e.g., [9, 10, 13, 14]). The increasing importance of using categories with morphisms defined as relations also implies the increasing importance of powerset theories defined in these relational categories.

In this paper we want to show what might be powerset operators and corresponding powerset theories in the case of categories of various fuzzy objects, where morphisms are fuzzy relations. We use the idea of Manes [4], who defined the notion of a $\mathcal{T}$-relation from an object X to an objects Y in a category $\mathbf{K}$ as a morphism in the Kleisli category $\mathbf{K}_{\mathcal{T}}$, where $\mathcal{T}$ is monad in a category $\mathbf{K}$. In that way we are able to define relations in many categories of fuzzy objects, including classical fuzzy sets, fuzzy sets in sets with similarity relations (i.e., in total fuzzy sets), or in spaces with fuzzy partitions as the ground structures for F-transforms and approximation functors.

As the main result of the paper we show that any powerset theory $\mathcal{T}$ in a category $\mathbf{K}$ can be extended to a powerset theory in the Kleisli category $\mathbf{K}_{\mathcal{T}}$, where $\mathcal{T}$ is a monad in $\mathbf{K}$. This result allows to extend the powerset structures defined in categories to the powerset structures in categories whose morphisms are monadic relations. This result corresponds to current trends in the theory of fuzzy structures, where the relationships between structures are defined using fuzzy relations.

2 Preliminaries

In this section we introduce principal notions and categories based on fuzzy sets which we use in the paper. A principal lattice structure used in fuzzy set theory in the paper is a *complete residuated lattice* (see e.g. [8]), i.e. a structure $\mathcal{L} = (L, \wedge, \vee, \otimes, \to, 0_L, 1_L)$ such that $(L, \wedge, \vee)$ is a complete lattice, $(L, \otimes, 1_L)$

is a commutative monoid with operation $\otimes$ isotone in both arguments and $\rightarrow$ is a binary operation which is residuated with respect to $\otimes$, i.e.

$$\alpha \otimes \beta \leq \gamma \quad \text{iff} \quad \alpha \leq \beta \rightarrow \gamma.$$

If $\mathcal{L}$ is a complete residuated lattice, a *$\mathcal{L}$-fuzzy set* in a crisp set X is a map $f : X \rightarrow L$. f is a non-trivial $\mathcal{L}$-fuzzy set, if f is not identical to the zero function. The core of a $\mathcal{L}$-fuzzy set f in a set X is defined by $core(f) = \{x \in X : f(x) = 1_L\}$.

A set with $\mathcal{L}$-valued similarity relation (or $\mathcal{L}$-set) is a couple (A, δ), where $\delta : A \times A \rightarrow L$ is a map such that

(a) $(\forall x \in A) \quad \delta(x, x) = 1$,
(b) $(\forall x, y \in A) \quad \delta(x, y) = \delta(y, x)$,
(c) $(\forall x, y, z \in A) \quad \delta(x, y) \otimes \delta(y, z) \leq \delta(x, z)$ (generalized transitivity).

The notion of $\mathcal{L}$-sets were introduced under the name *totaly fuzzy sets* by Wyler [15].

In the paper we use some standard categories with (sometimes special) maps as morphisms. Namely,

1. Category **CSLAT** of complete $\vee$-semilattices as objects and with semilattices homomorphisms as morphisms.
2. Category **Set** of sets as objects with mappings as morphisms.
3. Category **Set**($\mathcal{L}$) with $\mathcal{L}$-sets (X, δ) as objects and with $f : (X, \delta) \rightarrow (Y, \gamma)$ as morphisms, where $f : X \rightarrow Y$ is a map and $\delta(x, y) \leq \gamma(f(x), f(y)), x, y \in X$.

Besides these well known categories we use also the category **SpaceFP** of spaces with fuzzy partitions which was introduce in [5] and which is the basic category for lattice-valued F-transforms. This category is based on $\mathcal{L}$-valued *fuzzy partitions* [11].

Definition 1 Let X be a set. A system $\mathcal{A} = \{A_\lambda : \lambda \in \Lambda\}$ of normal $\mathcal{L}$-valued fuzzy sets in X is a *fuzzy partition* of X, if $\{core(A_\lambda) : \lambda \in \Lambda\}$ is a partition of X. A pair $(X, \mathcal{A})$ is called a space with a fuzzy partition. The index set of $\mathcal{A}$ will be denoted by $|\mathcal{A}|$.

Definition 2 The category **SpaceFP** is defined by

1. Fuzzy partitions $(X, \mathcal{A})$, as objects,
2. Morphisms $(f, g) : (X, \{A_\lambda : \lambda \in |\mathcal{A}|\}) \rightarrow (Y, \{B_\omega : \omega \in |\mathcal{B}|\})$, such that
   a. $f : X \rightarrow Y$ is a map,
   b. $g : |\mathcal{A}| \rightarrow |\mathcal{B}|$ is a map,
   c. $\forall \lambda \in |\mathcal{A}|, A_\lambda(x) \leq B_{g(\lambda)}(f(x))$, for each $x \in X$.
3. The composition of morphisms in **SpaceFP** is defined by $(h, \tau) \circ (g, \sigma) = (h \circ g, \tau \circ \sigma)$.

On the underlying set L of a lattice $\mathcal{L}$ we can define a special fuzzy partition $\mathcal{Q}$, where $\mathcal{Q} = \{L_\alpha : \alpha \in L\}$ and fuzzy sets L_α in a set L are defined by $L_\alpha(\beta) = \alpha \leftrightarrow \beta$, where $\leftrightarrow$ is a bi-residuum operation in $\mathcal{L}$. Using this fuzzy partition we can define the notion of an extensional fuzzy set in a space with a fuzzy partition, which generalizes the classical notion of an extensional fuzzy set in a set with a similarity relation.

Definition 3 An $\mathcal{L}$-valued fuzzy set $f : X \to \mathcal{L}$ is *extensional with respect to* $(X, \mathcal{A})$, if there exists a map $g : |\mathcal{A}| \to \mathcal{L}$ such that $(f, g) : (X, \mathcal{A}) \to (L, \mathcal{Q})$ is a morphism in the category **SpaceFP**.

In the paper [6] we proved that for arbitrary space with a fuzzy partition $(X, \mathcal{A})$ there exists $\mathcal{L}$-valued similarity relation $\delta_{X,\mathcal{A}}$ (called *a characteristic similarity relation of* $(X, \mathcal{A})$) defined on X such that $f : X \to \mathcal{L}$ is extensional in $(X, \mathcal{A})$ if and only if $f : X \to \mathcal{L}$ is an extensional map with respect to $\delta_{X,\mathcal{A}}$.

3 Fuzzy Type Relations Defined by Monads

In categories where morphisms are defined as various types of mappings, morphisms defined as relations can be implemented in many different ways. One of the ways to define relations as morphisms in a category with fuzzy type objects is to use the approach introduced by Manes in the published review [4]. In this concept, a relation is not defined absolutely, it is tied to the existence of a monad in this category. Therefore to different monads $\mathcal{T} = (T, \Diamond, \eta)$ in a category **K**, different types of relations between objects in this category, called $\mathcal{T}$-relations, can be defined in **K**. This approach allows to link the theory of categories with relations as morphisms with more general structures, i.e., monads in categories, while at the same time allowing to justify the more general nature of the choice of the specific type of relations in the category representing morphisms.

In this section we repeat basic definitions of monads and corresponding Kleisli categories, which are necessary for the definition of $\mathcal{T}$-relations in a category. The idea of using monads for fuzzyfication is based on extension of an object X of a category **K** to another object $T(X) \in$ **K**, which may be regarded as a "cloud of fuzzy states" with a morphism $\eta : X \to T(X)$, representing "crisp" states in the object of "fuzzy states".

We present definitions of all these notions.

Definition 4 A structure $\mathcal{T} = (T, \Diamond, \eta)$ is a *monad* (in clone form) in a category **K**, if

1. $T :$ **K** $\to$ **K** is a mapping defined on objects of **K**,
2. η is a set of **K**-morphisms $\{\eta_A | \eta_A : A \to T(A), A \in \mathbf{K}\}$,
3. for each pair of **K**-morphisms $f : A \to T(B), g : B \to T(C)$, there exists a composition (called a *Kleisli composition*) $g \Diamond f : A \to T(C)$, which is associative,
4. for every **K**-morphism $f : A \to T(B)$, $\eta_B \Diamond f = f$ holds,
5. $\Diamond$ is compatible with composition of morphisms of **K**, i.e., for each **K**-morphisms $f : A \to B, g : B \to T(C)$, we have $g \Diamond (\eta_B.f) = g.f$, where "." is the composition of morphisms in **K**.

It should be noted that if $(T, \Diamond, \eta)$ is a monad in a category $\mathbf{K}$, then $T : \mathbf{K} \to \mathbf{K}$ is a functor. In fact, for each morphism $f : A \to B$, the morphism $T(f)$ is defined by

$$T(f) = (\eta_B.f)\Diamond 1_{T(A)}. \tag{1}$$

For a monad $\mathcal{T}$ in a category $\mathbf{K}$ a new category $\mathbf{K}_{\mathcal{T}}$, called the *Kleisli category* (see, e.g., [2, 3]), is defined by

1. Objects of $\mathbf{K}_{\mathcal{T}}$ are the same as objects of $\mathbf{K}$,
2. Morphisms $A \rightsquigarrow B$ in $\mathbf{K}_{\mathcal{T}}$ between objects A, B are morphisms $A \to T(B)$ in the category $\mathbf{K}$.
3. A composition of morphisms $f : A \rightsquigarrow B$, $g : B \rightsquigarrow C$ is defined by $g\Diamond f$.

It should be noted that for any morphism $R : X \rightsquigarrow Y$ in $\mathbf{K}_{\mathcal{T}}$ (i.e., a morphism $R : X \to T(Y)$ in the category $\mathbf{K}$), R can be extended to the morphism $R^{\#} : T(X) \to T(Y)$ in the category $\mathbf{K}$, defined by $R^{\#} = R\Diamond 1_{T(X)}$. Moreover, there exists an embedding functor $I_{\mathbf{K}} : \mathbf{K} \hookrightarrow \mathbf{K}_{\mathcal{T}}$ such that $I_{\mathbf{K}}(X) = X$ and $I_{\mathbf{K}}(f) = \ulcorner f \urcorner$ for morphisms $f : X \to Y$, where $\ulcorner f \urcorner$ is the *graph* of f, defined by $\ulcorner f \urcorner = \eta_Y.f : X \rightsquigarrow Y$.

Now, according to Manes [4], we can define $\mathcal{T}$-relations in a category.

Definition 5 Let $\mathbf{K}$ be a category and let $\mathcal{T} = (T, \Diamond, \eta)$ be a monad in $\mathbf{K}$. A $\mathcal{T}$-relation R from an object X to an object Y in $\mathbf{K}$ is a morphism $R : X \rightsquigarrow Y$ in the Kleisli category $\mathbf{K}_{\mathcal{T}}$, or, equivalently, the morphism $R : X \to T(Y)$ in the category $\mathbf{K}$. A composition of $\mathcal{T}$-relations $R : X \rightsquigarrow Y$ and $S : Y \rightsquigarrow Z$ is defined by as a composition of morphisms $S\Diamond R$ in $\mathbf{K}_{\mathcal{T}}$.

From the above definition it follows that the Kleisli category $\mathbf{K}_{\mathcal{T}}$ can be considered the relational variant of a category $\mathbf{K}$ where instead of morphisms of category $\mathbf{K}$, $\mathcal{T}$-relations are used.

In the next part we will show examples of categories and fuzzy type relations between objects of these categories, which are actually $\mathcal{T}$-relations for some monads defined in these categories. As this overview shows, these relations include, among others, most standard fuzzy type relations and their compositions, which are used in fuzzy theory.

Example 1 Let X, Y be objects from the category **Set**. By a *relation* R from X to Y we understand a classical relation $R \subseteq X \times Y$, with a standard composition of relations.

Example 2 Let X, Y be objects from the category **Set**. A *fuzzy relation* from X to Y in **Set** is a $\mathcal{L}$-valued fuzzy relation $R : X \times Y \to \mathcal{L}$ with a standard composition of $\mathcal{L}$-valued fuzzy relations R and $S : Y \times Z \to \mathcal{L}$ defined by

$$(S \circ R)(x, z) = \bigvee_{y \in Y} R(x, y) \otimes S(y, z).$$

Example 3 Let (X, δ), (Y, γ) be objects from the category $\mathbf{Set}(\mathcal{L})$. By a relation from (X, δ) to (Y, γ) we understand an $\mathcal{L}$-valued fuzzy relation $R : X \times Y \to \mathcal{L}$, such that

$$R(x, y) \otimes \delta(x, x') \leq R(x', y), \quad R(x, y) \otimes \gamma(y, y') \leq R(x, y'),$$

for arbitrary $x, x' \in X$, $y, y' \in Y$, with standard composition of fuzzy relations.

Example 4 Let $(X, \mathcal{A})$, $(Y, \mathcal{B})$ be objects from the category **SpaceFP**. By a relation from $(X, \mathcal{A})$ to $(Y, \mathcal{B})$ we understand a pair (R, S), such that

1. R and S are $\mathcal{L}$-valued fuzzy relations $R : X \times Y \to \mathcal{L}$, $S : |\mathcal{A}| \times |\mathcal{B}| \to \mathcal{L}$,
2. For all $x \in X, i \in |\mathcal{A}|, j, j' \in |\mathcal{B}|, y, y' \in Y, \ y_j \in core(B_j), y_{j'} \in core(B_{j'})$ hold

$$\begin{aligned} A_i(x) &\leq R(x, y) \leftrightarrow S(i, j_y), \\ R(x, y) \otimes \delta_{Y,\mathcal{B}}(y, y') &\leq R(x, y'), \\ S(i, j) \otimes \delta_{Y,\mathcal{B}}(y_j, y_{j'}) &\leq S(i, j'), \end{aligned}$$

where $j_y \in |\mathcal{B}|$ is such that $y \in core(B_{j_y})$.

Composition of relations $(R, S) : (X, \mathcal{A}) \to (Y, \mathcal{B})$ and $(R', S') : (Y, \mathcal{B}) \to (Z, \mathcal{C})$ is defined by

$$(R', S').(R, S) = (R' \circ R, S' \circ S),$$

where $\circ$ is the standard composition of $\mathcal{L}$-valued fuzzy relations.

In the paper [6] we proved the following theorem.

Theorem 1 *For arbitrary category* $\mathbf{K}$ *from Examples 1–4 there exists a monad* $\mathcal{T}$ *in* $\mathbf{K}$ *such that relations defined in these Examples are* $\mathcal{T}$*-relation in* $\mathbf{K}$.

To obtain a relational version $\mathbf{K}_{rel}$ of the category $\mathbf{K}$ from Examples 1–4, instead of classical morphisms from a category $\mathbf{K}$ we can consider relations defined in these examples. In that case the previous Theorem can be re-formulated as follows:

Theorem 2 *For arbitrary category* $\mathbf{K}$ *from Examples 1–4 there exists a monad* $\mathcal{T}$ *in* $\mathbf{K}$ *such that the relational version* $\mathbf{K}_{rel}$ *is isomorphic to the Kleisli category* $\mathbf{K}_{\mathcal{T}}$ *of* $\mathbf{K}$ *with respect to the monad* $\mathcal{T}$, $\mathbf{K}_{rel} \cong \mathbf{K}_{\mathcal{T}}$.

4 Powerset Theory of Categories with Relational Morphisms Defined by Monads

As we mentioned in the Introduction, the powerset structures are widely used in algebra, logic, topology and also in computer science. A lot of papers were published about Zadeh's extension and its generalizations, which could be considered as the first example of a *powerset operator* in the fuzzy sets theory. The theoretical justification of the Zadeh's extension principle was, for the first time, presented by Rodabaugh [12]. Whatever works of Rodabaugh gave very serious basis for further research of powerset objects and operators, only abstract theory of powerset objects based on similar principles as the theory of monads has brought another important ideas to the research of powerset objects.

However, in all these categories, the role of morphisms was played only by standard mappings. It is therefore natural to ask how the powerset theory would look like in categories where morphisms are variants of relations. In this Section, we will present one of the ways how to construct these powerset structures in categories of relations as morphisms, where relations are $\mathcal{T}$-relations defined by monads.

We introduce a generalized form of the **CSLAT**-powerset theory introduced by Rodabaugh [12].

Definition 6 Let **K** be a category. Then $\mathcal{P} = (P, V, \chi)$ is called **CSLAT** *-powerset theory in* **K**, if

1. $P : \mathbf{K} \to \mathbf{CSLAT}$ is a functor,
2. $V : \mathbf{K} \to \mathbf{Set}$ is a functor,
3. For each $X \in \mathbf{K}$, χ determines in **Set** a mapping $\chi_X : V(X) \to WP(X)$, where $W : \mathbf{CSLAT} \to \mathbf{Set}$ is a forgetful functor,
4. For each $f : X \to Y$ in **K**, $WP(f).\chi_X = \chi_Y.V(f)$.

For simplicity, instead of **CSLAT**-powerset theory we will speak only about powerset theory. Let us consider the following examples of powerset theories. For simplicity we present only simplified description of these powerset theories, for further details see [7].

Example 5 Powerset theory $\mathcal{P} = (P, 1_{\mathbf{Set}}, \chi)$ in the category **Set**, where

1. $P : \mathbf{Set} \to CSLAT$ is defined by $P(X) = (2^X, \subseteq)$, and any element S of $P(X)$ is identified with the characteristic function χ_S^X of a subset $S \subseteq X$ in X.
2. for each $f : X \to Y$ in **Set**, $P(f) : P(X) \to P(Y)$ is defined by $P(f)(\chi_S^X) = \chi_{f(S)}^Y$,
3. for each $X \in \mathbf{Set}$, $\chi_X : X \to P(X)$ is the characteristic function $\chi_{\{x\}}^X$ of a subset $\{x\}$ in X.

Example 6 Powerset theory $\mathbf{Z} = (Z, 1_{\mathbf{Set}}, \eta)$ in the category **Set**, where

1. $Z : \mathbf{Set} \to \mathbf{CSLAT}$ is defined by $Z(X) = (\mathcal{L}^X, \leq)$, ordered point-wise,
2. for each $f : X \to Y$ in **Set**, $Z(f) : \mathcal{L}^X \to \mathcal{L}^Y$ is defined by $Z(f)(s)(y) = \bigvee_{x \in X, f(x)=y} s(x)$,
3. for each $X \in \mathbf{Set}$, $\eta_X : X \to \mathcal{L}^X$ is defined by $\eta_X(a) = \chi_{\{a\}}^X$, for $a \in X$, where $\chi_{\{a\}}^X$ is the $\mathcal{L}$-valued characteristic function of $\{a\}$ in X.

Example 7 Powerset theory $\mathbf{E} = (E, V, \rho)$ in the category $\mathbf{Set}(\mathcal{L})$, where

1. $E : \mathbf{Set}(\mathcal{L}) \to \mathbf{CSLAT}$ is defined by $E(X, \delta) =$ the set of all fuzzy sets $f \in \mathcal{L}^X$ extensional with respect to the similarity relation δ, ordered point-wise,
2. for each morphism $f : (X, \delta) \to (Y, \gamma)$ in $\mathbf{Set}(\mathcal{L})$, $E(f) : E(X, \delta) \to E(Y, \gamma)$ is defined by $E(f)(s)(y) = \bigvee_{x \in X} s(x) \otimes \gamma(f(x), y)$,
3. $V : \mathbf{Set}(\mathcal{L}) \to \mathbf{Set}$ is the forgetfull functor,
4. for each $(X, \delta) \in \mathbf{Set}(\mathcal{L})$, $\rho_{(X,\delta)} : X \to E(X, \delta)$ is defined by $\rho_{(X,\delta)}(a)(x) = \delta(a, x)$, for $a, x \in X$.

Example 8 Powerset theory $\mathbf{F} = (F, U, \vartheta)$ in the category **SpaceFP** is defined by

1. $F : \mathbf{SpaceFP} \to \mathbf{CSLAT}$, defined by

$$F(X, \mathcal{A}) = (\{f | f : X \to \mathcal{L} \text{ is extensional in } (X, \mathcal{A})\}, \leq),$$

ordered point-wise.
2. For each $(f, u) : (X, \mathcal{A}) \to (Y, \mathcal{B})$ in **SpaceFP**, $F(f, u) : F(X, \mathcal{A}) \to F(Y, \mathcal{B})$ is defined by

$$g \in F(X, \mathcal{A}), y \in Y, \quad F(f, u)(g)(y) = \bigvee_{x \in X} g(x) \otimes \delta_{Y,\mathcal{B}}(f(x), y),$$

where $\delta_{Y,\mathcal{B}}$ is the characteristic similarity relation of $(Y, \mathcal{B})$.
3. $U : \mathbf{SpaceFP} \to \mathbf{Set}$ is the forgetful functor, $U(X, \mathcal{A}) = X$,
4. For each $(X, \mathcal{A})$ in **SpaceFP**, $\vartheta_{(X,\mathcal{A})} : X \to F(X, \mathcal{A})$, $\vartheta_{(X,\mathcal{A})}(a)(x) = \delta_{X,\mathcal{A}}(a, x)$, for each $a, x \in X$.

The idea how to extend a powerset theory in a category **K** to the new powerset theory in a category $\mathbf{K}_{rel}$ with relations is rather simple. As we showed in previous Theorem 2, for many standard categories **K** with fuzzy objects there exist monads $\mathcal{T}$ in **K** such that the relational versions $\mathbf{K}_{rel}$ of these categories are isomorphic to Kleisli categories $\mathbf{K}_{\mathbf{T}}$. *Hence, by a powerset theory in these relational categories* $\mathbf{K}_{rel}$ *we can understand (standard) powerset theories in Kleisli categories* $\mathbf{K}_{\mathcal{T}}$. In the following theorem we show that powerset theories in categories **K** from Examples 5 to 8 can be extended to powerset theories in the relational versions $\mathbf{K}_{rel} \cong \mathbf{K}_{\mathcal{T}}$ of these categories.

Theorem 3 *Let* $\mathcal{P} = (P, V, \xi)$ *be any of* **CSLAT**-*powerset theories from Exampled 5 to 8 in a category* **K**. *Let* $\mathcal{T}$ *be a monad which defines relations in* **K**, *introduced in corresponding Example 1–4.Then there exists a* **CSLAT**-*powerset theory* $\widetilde{\mathcal{P}} = (\widetilde{P}, V, \xi)$ *in the category* $\mathbf{K}_{\mathcal{T}}$, *which is an extension of the powerset theory* $\mathcal{P}$.

Proof The principal idea of the proof is simple. The new powerset functor $\widetilde{P} : \mathbf{K}_{\mathcal{T}} \to$ **CSLAT** has the same object function as the functor P, in a different way are defined only $\bigvee$-preserving mappings $\widetilde{P}(R)$, where $R : X \rightsquigarrow Y$ are morphisms in $\mathbf{K}_{\mathcal{T}}$, i.e., $R : X \to P(Y)$. For example, for $\mathcal{P} = \mathcal{Z}$ from the Example 6 we set

$$s \in Z(X), y \in Y, \quad \widetilde{P}(s)(y) = \bigvee_{x \in X} s(x) \otimes R(x)(y).$$

Because, by a simple calculation, $\widetilde{P}(S\Diamond R) = \widetilde{P}(S).\widetilde{P}(R)$, where $\Diamond$ is a composition of morphisms in the Kleisli category $\mathbf{Set}_{\mathcal{T}}$, where $\mathcal{T}$ is a monad which defines relations R, $\widetilde{P}$ is a functor $\mathbf{Set}_{\mathcal{T}} \to \mathbf{CSLAT}$ and it is easy to prove that $\widetilde{\mathcal{P}}$ is a powerset theory in $\mathbf{Set}_{\mathcal{T}}$, extending the powerset theory $\mathcal{P}$ in **Set**. We can proceed analogously for the other examples.

Unfortunately, as follows from the construction of extended functor $\widetilde{P}$, this procedure can be used only in some cases of powerset theories and relations defined by monads and cannot be applied to arbitrary powerset theory and monad in an arbitrary category. Our goal in this part is to prove that for arbitrary **CSLAT**-powerset theory $\mathcal{P}$ in a category **K** and arbitrary monad $\mathcal{T}$ in **K** we can construct a **CSLAT**-powerset theory of the Kleisli category $\mathbf{K}_{\mathcal{T}}$ which extends the powerset theory $\mathcal{P}$.

This represents a general solution of the problem how to extend an arbitrary powerset theory in a category to a powerset theory in a Kleisli category of this category **K**.

Theorem 4 *Let $\mathcal{P} = (P, V, \chi)$ be a* **CSLAT***-powerset theory in the category* **K** *and let $\mathcal{T} = (T, \Diamond, \xi)$ be a monad in* **K***. Then $\widetilde{\mathcal{P}} = (\widetilde{P}, U, \mu)$ is a* **CSLAT***-powerset theory in the Kleisli category $\mathbf{K}_{\mathcal{T}}$, where,*

1. *The object function of $\widetilde{P}$ is defined by $\widetilde{P} = P.T : \mathbf{K}_{\mathcal{T}} \to \mathbf{CSLAT}$,*
2. *$\widetilde{P}(R) := P(R\Diamond 1_{T(X)}) : \widetilde{P}(X) \to \widetilde{P}(Y)$ for arbitrary morphism $X \rightsquigarrow Y$ in $\mathbf{K}_{\mathcal{T}}$.*
3. *$U : \mathbf{K}_{\mathcal{T}} \to \mathbf{Set}$ is defined by $U(X) := V.T(X), U(R) := V(R\Diamond 1_{T(X)}) : U(X) \to U(Y)$,*
4. *$\mu_X := \chi_{T(X)} : U(X) \to W.\widetilde{P}(X)$.*

The powerset theory $\widetilde{\mathcal{P}}$ is an extension of a powerset theory $\mathcal{P}$, i.e., the following diagram commutes:

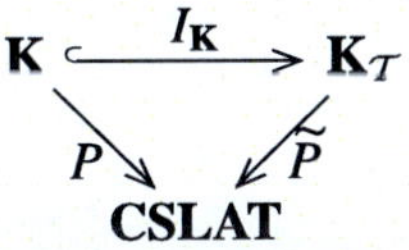

The proof of this theorem will be published elsewhere.
For illustration we present a simple example.

Example 9 Let us consider the powerset theory $\mathcal{Z}$ from Example 6 and let $\mathcal{T}$ be a monad defining relations from Example 2. According to [6], we have $\mathcal{T} = (T, \Diamond, \xi)$, where $T(X) = \mathcal{L}^X$, $\xi_X(a) = \chi^X_{\{a\}}$ and for $f : X \to \mathcal{L}^Y$ and $g : Y \to \mathcal{L}^Z$ we have $g\Diamond f(x)(z) = \bigvee_{y\in Y} f(x)(y) \otimes g(y)(z)$. Let $\widetilde{\mathcal{Z}} = (\widetilde{Z}, U, \mu)$ be the powerset theory in the Kleisli category $\mathbf{Set}_{\mathcal{T}}$ defined in Theorem 4. For illustration we show only how $\widetilde{Z}(\ulcorner f \urcorner)$ is defined for the graph $\ulcorner f \urcorner : X \rightsquigarrow Y$ of a map $f : X \to Y$. For $X, Y \in$ **Set**, $f : X \to Y$ we have

$$\widetilde{Z}(\ulcorner f\urcorner):\mathcal{L}^{\mathcal{L}^X}\to\mathcal{L}^{\mathcal{L}^Y},$$

$$S\in\mathcal{L}^{\mathcal{L}^X}, t\in\mathcal{L}^Y, \widetilde{Z}(\ulcorner f\urcorner)(S)(t)=Z(\ulcorner f\urcorner\Diamond 1_{T(X)})(S)(t)=\bigvee_{\{u\in\mathcal{L}^X:\forall y\in Y,\bigvee_{x\in X}u(x)\otimes\ulcorner f\urcorner(x)(y)=t(y)\}}S(u)=$$

$$\bigvee_{\{u\in\mathcal{L}^X:\forall y\in Y,\bigvee_{x\in X}u(x)\otimes\chi^Y_{\{f(x)\}}(y)=t(y)\}}S(u)=\bigvee_{\{u\in\mathcal{L}^X:Z(f)(u)=t\}}S(u).$$

5 Conclusions

In this paper we tried to show how we can define a variant of the category **K** of fuzzy type objects, where we use different types of fuzzy relations instead of the standard mappings as **K**-morphisms. For this purpose, we used the fact that a large part of the fuzzy relations that are normally used in fuzzy theory are in fact defined by monads in suitable categories. Since these relations defined by monads $\mathcal{T}$ in category **K** are actually morphisms in Kleisli category $\mathbf{K}_{\mathcal{T}}$ of this monad, we can consider this Kleisli category as relational variant of category a **K**. We have shown how using this principle we can extend a powerset operator $\mathcal{P}$ in a category **K** to the powerset operator $\widetilde{\mathcal{P}}$ in the category $\mathbf{K}_{\mathcal{T}}$, which in fact represents the extension of the powerset theory in category**K** to the powerset theory in relational version of this category.

References

1. J.A. Goguen, L-fuzzy sets. J. Math. Anal. Appl. **18**, 145–174 (1967)
2. H. Kleisli, Every standard construction is induced by a pair of adjoint functors. Proc. Am. Math. Soc. **16**, 544–546 (1965)
3. E.G. Manes, *Algebraic Theories* (Springer, Berlin, 1976)
4. L.A. Manes, Book review fuzzy sets and systems, theory and applications. Bull. (New Ser.) Am. Math. Soc. **7**(3), 603–612 (1982)
5. J. Močkoř, Spaces with fuzzy partitions and fuzzy transform. Soft Comput. **21**(13), 3479–3492 (2017)
6. J. Močkoř, Relational variants of categories of fuzzy sets defined by monads, in *Conference of the International Fuzzy Systems Association and the European Society for Fuzzy Logic and Technology (EUSFLAT)* (Atlantis Press, 2019), pp. 44–51
7. J. Močkoř, Some Examples, of Relations Between F-Transforms and Powerset Theories, in *Fuzzy Information Processing, NAFIPS*, ed. by G. Barreto, R. Coelho *Communications in Computer and Information Science*, vol. 831 (Springer, Cham, 2018), pp. 72–83
8. V. Novák, I. Perfilijeva, J. Močkoř, *Mathematical Principles of Fuzzy Logic* (Kluwer Academic Publishers, Boston, Dordrecht, London, 1999)
9. I. Perfilieva, A.P. Singh, S.P. Tiwari, On the relationships among F-transform, fuzzy rough sets and fuzzy topology. Soft Comput. **21**(13), 3513–3523 (2017)
10. I. Perfilieva, Fuzzy transforms: theory and applications. Fuzzy Sets Syst. **157**, 993–1023 (2006)

11. I. Perfilieva, A. P. Singh, S.P. Tiwari, On the relationship among F-transform, fuzzy rough set and fuzzy topology, in *Proceedings of IFSA-EUSFLAT* (Amsterdam, Atlantis Press, 2015), pp. 1324–1330
12. S.E. Rodabaugh, Powerset operator foundation for poslat fuzzy set theories and topologies, in *Mathematics of Fuzzy Sets: Logic, Topology and Measure Theory, The Hnadbook of Fuzzy Sets Series*, vol. 3, ed. by U. Höhle, S.E. Rodabaugh (Kluwer Academic Publishers, Boston, Dordrecht, 1999), pp. 91–116
13. C.Y. Wang, Fuzzy rough sets based on generalized residuated lattices. Inf. Sci. **248**, 31–49 (2013)
14. W.Z. Wu, Y. Leung, J.S. Mi, On characterization of $(\mathcal{J}, \mathcal{T})$-fuzzy rough approximation operators. Fuzzy Sets Syst. **154**, 76–102 (2005)
15. O. Wyler, Fuzzy logic and categories of fuzzy sets, in *Non-Classical Logics and Their Applications to Fuzzy Subsets*, ed. by U. Höhle, E.P. Klement (Kluwer Academic Publishers, 1995), pp. 235–268
16. L.A. Zadeh, Fuzzy sets. Inf. Control **8**, 338–353 (1965)

Improved Fuzzy Q-Learning with Replay Memory

Xin Li and Kelly Cohen

Abstract Artificial Intelligence (AI) is becoming more and more popular after 2000s. Many AI-related technologies are developed and applied to our daily life which brought us benefits, such as Reinforcement Learning (RL) (L. P. Kaelbling, M. L. Littman, A. W. Moore: "Reinforcement Learning: A Survey", 1996, Journal of Artificial Intelligence Research, Vol 4, (1996), 237–285; 9605103.) which allows the first computer to defeat human champion in Game Go. However, such AI technology are heavily depended on the Neural Network (NN) which is too complicated for human to understand and interpret. On the other hand, Fuzzy Logic (FL) is an AI technology that is well known for its interpretation, however, with relatively poor performance compare to Neural Network related methods. This paper is introducing an example of replacing NN with FL for optimizing RL, so that we can gain better performance and easier tuning process, and also with interpretation on the results.

Keywords Fuzzy logic · Q-Learning · Continuous input space · Replay memory

1 Introduction

1.1 Fuzzy Logic & Q-Learning

The most fundamental method of RL is Q-Learning (QL) [2]. Most of the current RL algorithms are developed upon QL and it is still a widely used RL method for Monte Carlo [3] problems due to its simplicity and solid mathematical proof. The updating equation (Richard, 2018) [4] of QL is shown as following:

X. Li (✉)
Department of Computer Science, University of Cincinnati, Cincinnati, OH, USA
e-mail: li4x7@mail.uc.edu

K. Cohen
Department of Aerospace Engineering, University of Cincinnati, Cincinnati, OH, USA
e-mail: cohenky@ucmail.uc.edu

B. Bede et al. (eds.), *Fuzzy Information Processing 2020*,
Advances in Intelligent Systems and Computing 1337,
https://doi.org/10.1007/978-3-030-81561-5_2

$$Q(s_t, a_t)^* \leftarrow (1 - \alpha) * Q(s_t, a_t) + \alpha * (r_{t+1} + \gamma * \max_a Q(s_{t+1}, a)) \quad (1)$$

Each update involves current state s_t, current action a_t, corresponding reward r_{t+1} and next state s_{t+1}. There are two hyperparameters which are learning rate α and discount rate γ. Q is called Q-value, which represents the corresponding value of taking certain actions under certain states. When the model is converged, QL results a Q-value matrix called "Q-Table" which serves as a look-up table, so that it can be used to find the best action in a given state. However, due to its weakness of handling continuous inputs and outputs, QL without NN optimization is usually used on "toy problems" which normally involved with discrete inputs and outputs in low dimensions for research purposes. In order to apply QL on realistic problems, researchers have to use complicated optimization models which sacrifice the interpretability in exchange for better performance. Fortunately, the fuzzification in Fuzzy Logic [5] (FL) is well known for converting continuous values into discrete values with probabilities of confidence. It solves the issue of handling continuous inputs and outputs for QL, so that it can take any continuous values and re-recognize them as discrete values with certain degree of confidence. Such degree of confidence can be used as interpretation to explain the performance along with Q-Table. Normally we need to implement FL as Fuzzy Rule-Based System (FRBS) with optimization algorithms such as Generic Algorithm [6] (GA) for tuning membership functions of inputs and outputs as well as optimizing the rule-based table. However, such optimization for real life problems is a difficult task due to the "curse of dimensionality". For example, if we have 3 memberships for only 1 input, we will need to tune 9 values (3 memberships * 3 coordinates of each membership). Given 4 inputs, we will need to tune at least 36 values (9 * 4) just for the input membership functions. Also, for the rule-based table, we are facing a matrix of 81 (3 * 3 * 3 * 3) cells for tuning. This is just a very simple case and in reality, we need to handle much more complicated problems which not only has large scale of input and output spaces that require large number of memberships for precision purpose, but also involves with high dimensional inputs and outputs that exponentially increase the number of memberships we need to handle.

As a result, FL is good for interpreting continuous variables as discrete variables with degree of confidence, but at the same time it is hard to be optimized for high dimensional problems. On the other hand, QL is easy for being optimized, but it is weak at handling continuous variables. Note that QL is not a good algorithm for handling high dimensional optimization, but by combining FL with QL, QL can utilized the rule-based nature of FL to avoid many wrong state-action choices which dramatically reduce the converging time.

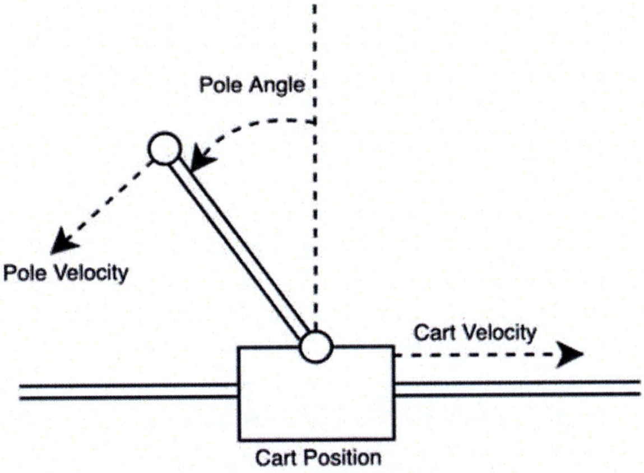

Fig. 1 A cart is attached to a pole which can rotate 180° and the cart can be only moved to the left of right

1.2 *Simulator*

A commonly used simulator for RL problem called "CartPole" [7] is used for this experiment (see Fig. 1).

There are 4 continuous inputs: (1) Cart Position [−4.8, 4.8], (2) Cart Velocity [−3, 3], (3) Pole Angle [−24, 24], and (4) Pole Velocity [−3, 3] and these input domains do not have units for the simulation purpose. There are also 2 discrete outputs which are 0 for moving cart to left and 1 for moving cart to right. The goal is to keep the pole from falling horizontally as long as possible (500 steps is the maximum score) by adjusting cart's position by moving it to left or right. The reason of using this simulator is that it is a widely used one for testing RL algorithms. Thus, the result of Fuzzy Q-Learning can be used for future comparison with other RL algorithms which are implemented under Neural Networks. This simulator is also light-weighted, so that it can return the performance faster than other complicated simulators.

2 Methodology

2.1 *Fuzzy Logic*

In this paper, we assume the membership functions for all inputs are evenly distributed which means the triangles used for membership functions are all identical and are evenly overlapping each other to cover the entire input domain. The setup for input membership functions are shown as following (see Fig. 2).

As a result, we avoid the complicated part of our tuning process by skipping the membership function tuning. Such assumption is based on the fact that in reality, most of the states are independent (cart position is independent of cart velocity). We

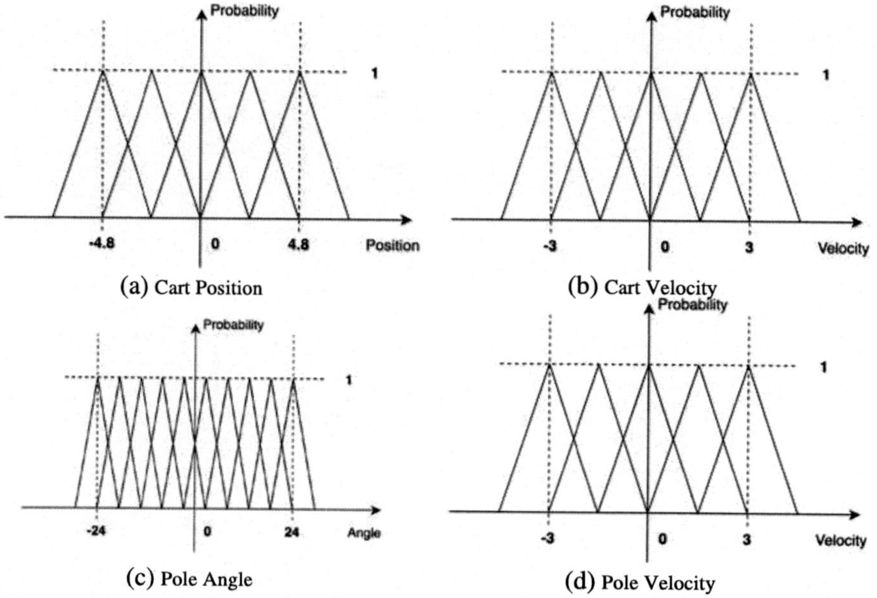

(a) Cart Position (b) Cart Velocity (c) Pole Angle (d) Pole Velocity

Fig. 2 Fuzzy membership functions: **a** 5 memberships; **b** 5 memberships; **c** 10 memberships; **d** 5 memberships

can use this setup to speed up the tuning process, unless some states are strongly dependent on others.

Next, we use Q-Table to represent the rule-based table. Since we only have 2 discrete outputs, the Q-Table is a (5*5*10*5)*2 matrix with 2500 cells. It seems to be a huge matrix for tuning, but this is where combination of FL and QL is getting interesting. The traditional way of implementing FRBS with optimization algorithms such as GA, requires multiple candidates to carry an individual rule-based table. The more candidates the GA has, the more tables that we need to optimize. On another hand, QL, by its nature, will always focus on the same table and update one cell at a time during the tuning process. Also, in practice, it is easy to identify the "ridiculous" cells which represent unrealistic state-action combinations. As the Q-Table downgrades these cells, it is harder for them to be revisited and Q-Table then focuses on optimize the "reasonable" cells. By using QL's updating equation and environment observations, we can update our Q-Table and the general setup of Fuzzy Q-Learning is shown as following (see Fig. 3).

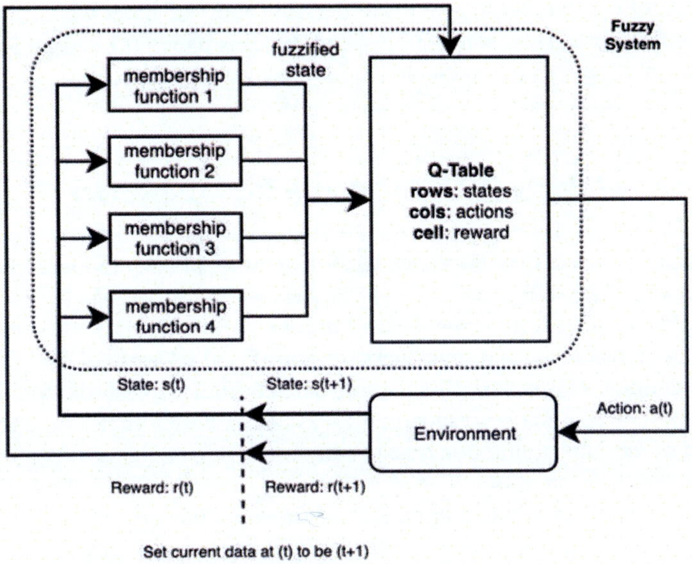

Fig. 3 Fuzzy Q-Learning diagram that illustrates how fuzzification converts continuous inputs into discrete inputs which can be used by Q-Table

2.2 *Normal Fuzzy Q-Learning*

The following is the setup of combination of fuzzification and Q-Learning (see Fig. 4).

This implementation will update one cell in the matrix at a time after each feedback of the environment. A hyperparameter epsilon (ϵ) [2] is used to demonstrate the randomness of exploration. However, due to the tradeoff of exploration vs. exploitation, it is usually hard to decide the value for epsilon. If we provide higher degree for exploration, our system will explore the searching space more aggressively to discover better solutions, however, it will learn bad solutions more easily at the same time. If we provide higher degree for exploitation, then the system will be more

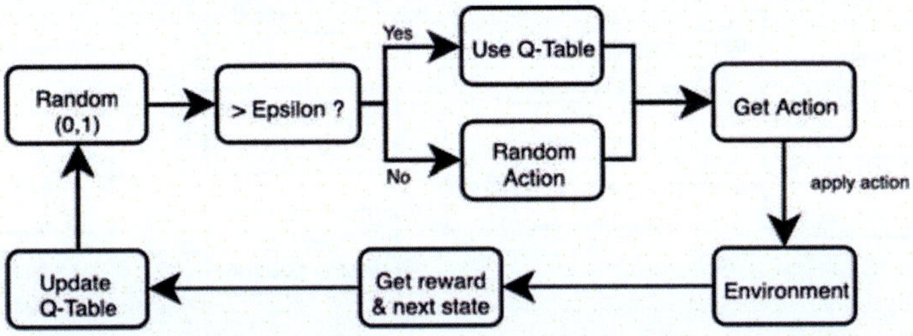

Fig. 4 Normal fuzzy Q-Learning diagram

careful and follow the Q-Table to make decision most of time, which result slow learning progress and slow converging speed. For this paper, 0.1 is used for epsilon for the reason of being a common choice for most of the cases.

2.3 *Improved Fuzzy Q-Learning with Replay Memory*

The following is the setup of improved version of normal Fuzzy Q-Learning [8] with replay memory [9] (see Fig. 5).

The replay memory is implemented as a list of solutions and it is added to enhance the speed of convergence and robustness by storing and re-learning the best current on-hand solutions. This is done by sampling a batch from the best solutions in replay memory and increase the corresponding Q-Values in the Q-Table. As a result, the side-effect of random exploration on bad solutions will be balance-off by reviewing the discovered good solutions. The size of replay memory pool is limited, and all solutions are stored in a queue in respect of time. As the system will learn better solutions during the training process, the replay memory pool will be updated with "first-in, first-out" (FIFO) rule, which allows the newest (better) solutions to be stored at the end of the queue, and the oldest solutions are removed from the head of the queue.

3 Result Analysis

By using CartPole-v0 [10] which is an open source library developed by OpenAI Gym, we can compare the performances between normal version and improved version of Fuzzy Q-Learning. Because the maximum steps of playing CartPole is

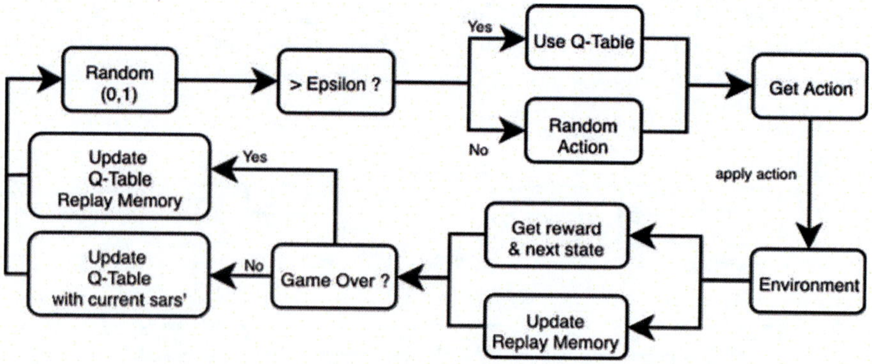

Fig. 5 Improved fuzzy Q-Learning with memory replay diagram

Table 1 Normal fuzzy Q-Learning parameters

Iteration	γ	α	ε	Memory size	Batch size	Replay
100	0.9	0.1	0.1	NA	NA	False

Table 2 Normal fuzzy Q-Learning parameters

Iteration	γ	α	ε	Memory size	Batch size	Replay
500	0.9	0.1	0.1	NA	NA	False

500, we pursue a score of 500 for each iteration as our goal which means the successful completion of each game.

3.1 *Result of Normal Fuzzy Q-Learning*

Performance of 100 iterations. This is for testing the performance of normal FQL given relatively short training iterations (Table 1).

Performance of 500 iterations. This is for testing the performance of normal FQL given relatively long training iterations (Table 2).

3.2 *Results of Fuzzy Q-Learning with Replay Memory*

Performance of 100 iterations. This is for testing the performance of improved FQL given relatively short training iterations (Table 3).

Performance of 500 iterations. This is for testing the performance of improved FQL given relatively long training iterations (Table 4).

Table 3 Improved fuzzy Q-Learning parameters

Iteration	γ	α	ε	Memory size	Batch size	Replay
100	0.9	0.1	0.1	500	25	True

Table 4 Improved fuzzy Q-Learning parameters

Iteration	γ	α	ε	Memory size	Batch size	Replay
500	0.9	0.1	0.1	1000	25	True

3.3 Robustness Comparison

Both normal Fuzzy Q-Learning and improved Fuzzy Q-Learning with Replay Memory are facing the challenge of randomness, which means there is a possibility of continuously encountering bad solutions. This usually causes the whole Q-Table to be corrupted by continuously following the wrong solutions without a chance to get recovered from later good solutions. Bellowing is the comparison of the robustness test over 25 iterations (see Fig. 10).

3.4 Result Summary

From Figs. 7 and 9, it is clear to see that given enough iterations, for example, 500 iterations of training, both normal and improved Fuzzy Q-Learning (FQL) can reach convergency with good validation results. However, with only 100 iterations given for training, as the results shown in Figs. 6 and 8, the validation of normal FQL indicates a trend of decreasing in performance while the improved FQL with Replay Memory has a stable performance with all perfect scores (500). This indicates that improved FQL can utilize the Replay Memory to further strength the Q-Table with discovered good solutions to speed up the learning process. Also, in both Figs. 6 and 7, the normal FQL's validation results have both trends of decreasing and unstable performances while the improved FQL's validation result shown in Figs. 8 and 9 are both expected and acceptable. This is because even the training result shows the signal of convergency, it does not really mean the solution space is fully covered and learned. It usually takes more than just 100 or 500 iterations to fully discover the solution space in order to guarantee a solid validation result. With Replay Memory, FQL can re-learning good solutions multiple times instead of only once to reduce the chance of forgetting them which is caused by learning new but bad solutions. It also provides a chance for FQL to repeatedly learn some potential good solutions.

In Fig. 10, with only 100 iteration training iterations are given, it is clear to see that the robustness of normal FQL suffers from the randomness much more than the improved FQL. Normal FQL has either perk or bottom performances while improved FQL has not only similar number of perk and bottom performances, but also with relatively acceptable scores. It means improved FQL can still fixed the corrupted Q-Table that is caused by randomness within short training time period.

4 Conclusion

In general, using Replay Memory can help to reduce the training time greatly. This is important for applying FRBS on real life problems which may require thousands of iterations of training due to the "curse of dimensionality". Other than that, Replay

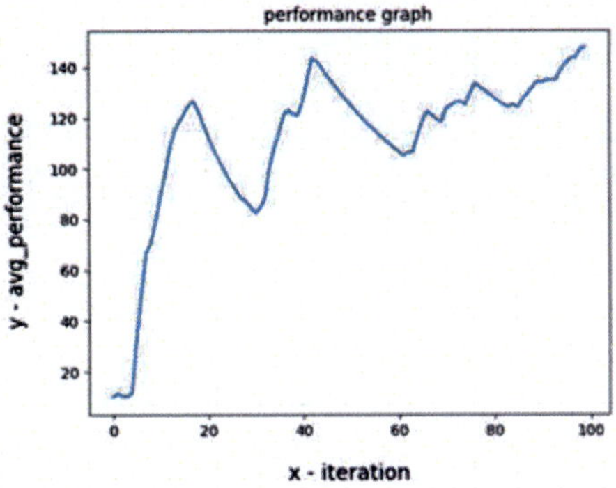

(a) Training of 100 iterations

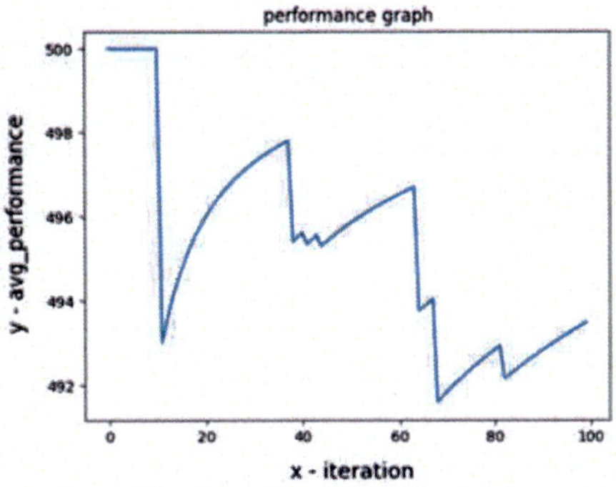

(b) Validation of 100 iterations

Fig. 6 Normal fuzzy Q-Learning performance plots (trained with 100 iterations)

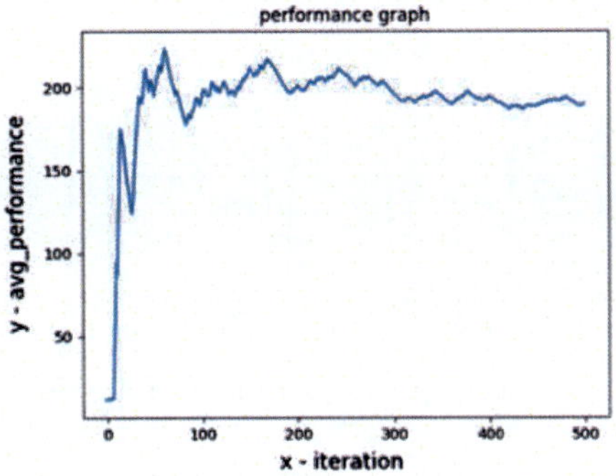

(a) Training of 500 iterations

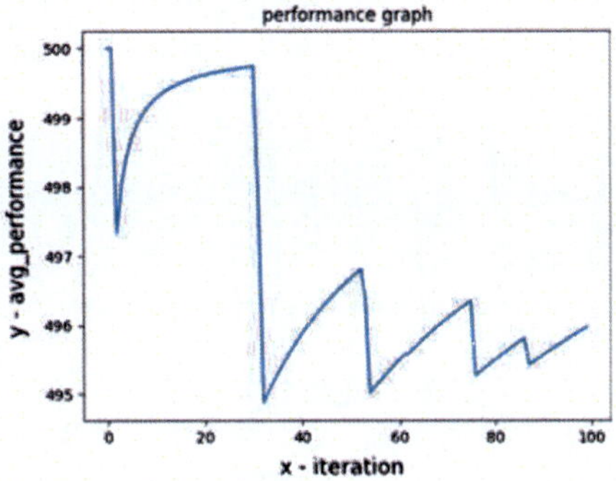

(b) Validation of 100 iterations

Fig. 7 Normal fuzzy Q-Learning performance plots (trained with 500 iterations)

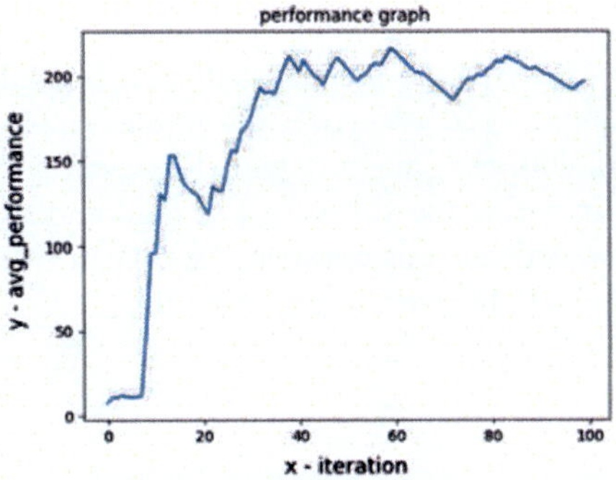

(a) Training of 100 iterations

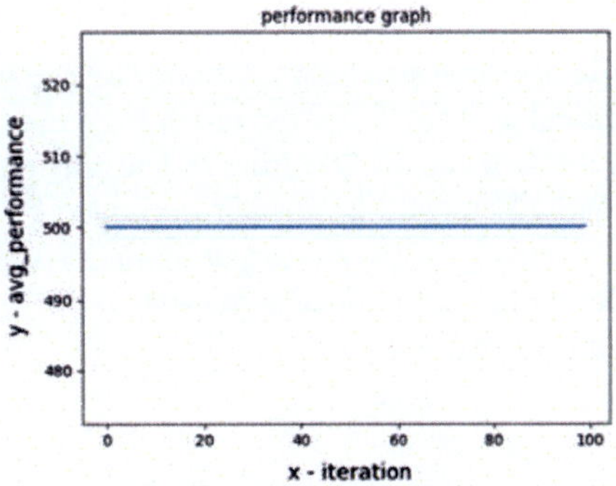

(b) Validation of 100 iterations

Fig. 8 Improved fuzzy Q-Learning with memory replay performance plots (trained with 100 iterations)

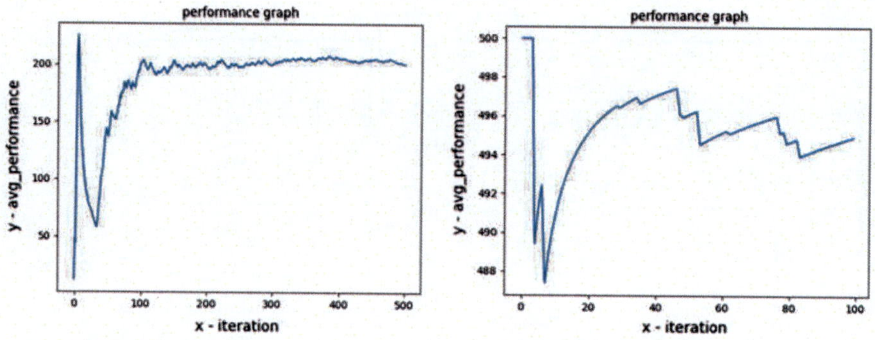

(a) Training of 500 iterations

(b) Validation of 100 iterations

Fig. 9 Improved fuzzy Q-Learning with memory replay performance plots (trained with 500 iterations)

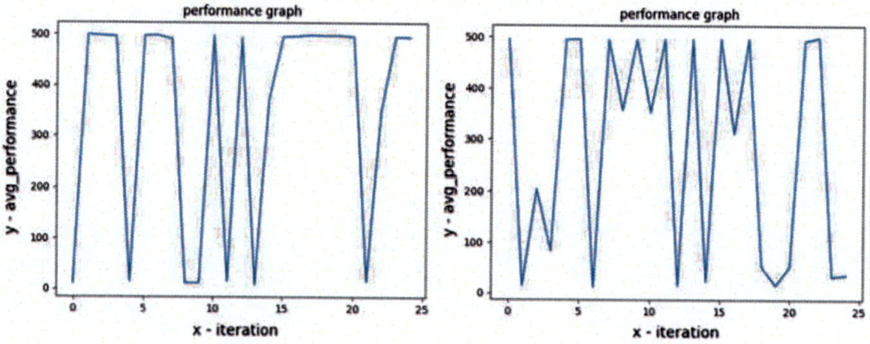

(a) 25 Validations on Normal Version

(b) 25 Validations on Improved Version

Fig. 10 Robustness validations between normal version and improved version

Memory can also help to balance off the side-effect of randomness, so that our trained model will not suffer too much from the frequently encountered bad solutions and as a result, it can recover from re-learning good solutions in memory pool. However, this method can still be further improved by eliminating the side-effect of randomness completely. Also, the output domain is discrete value in this problem, thus, it can also be improved by introducing defuzzification of FRBS into this approach to extend it versatility.

References

1. L.P. Kaelbling, M.L. Littman, A.W. Moore, Reinforcement learning: a survey. J Artif Intell Res **4**, 237–285 (1996). arXiv:cs/9605103
2. Y. Hosoya, M. Umano,Dynamic fuzzy Q-learning with facility of tuning and removing fuzzy rules, in *2012 IEEE International Conference on Fuzzy Systems (FUZZ-IEEE)* (2012), pp 1–8
3. H. Wang, M. Emmerich, A. Plaat, Monte carlo Q-learning for general game playing (2018). arXiv:1802.05944
4. R.S. Sutton, A.G. Barto, Reinforcement learning: an introduction. The MIT Press (2018)
5. L.A. Zadeh, Fuzzy logic. Computer **21**(4), 83–93 (1988). https://doi.org/10.1109/2.53.
6. B. Mayoh, Genetic algorithms in optimisation, simulation and modelling. AI Commun **8**(3–4), 196–197 (1995). https://doi.org/10.3233/aic-1995-83-404
7. G. Surma, Cartpole—Introduction to Reinforcement Learning (DQN—Deep Q-Learning) (2019, November 10). Accessed 09 June 2020. https://towardsdatascience.com/cartpole-introduction-to-reinforcement-learning-ed0eb5b58288
8. P. Glorennec, Fuzzy Q-learning and dynamical fuzzy Q-learning, in *Proceedings of 1994 IEEE 3rd international fuzzy systems conference.* (n.d.). https://doi.org/10.1109/fuzzy.1994.343739
9. R. Liu, J. Zou, The effects of memory replay in reinforcement learning, in *2018 56th annual allerton conference on communication, control, and computing (Allerton)*
10. CartPole-v0—OpenAI Gym (n.d.). Accessed 10 June 2020. https://gym.openai.com/envs/CartPole-v0/

Agnesi Quasi-fuzzy Numbers

Flaulles Bergamaschi, Natan Jesus, Regivan Santiago, and Alesxandra Oliveira

Abstract In this paper we proposed a concept of Agnesi quasi-fuzzy numbers based on Agnesi's curve. Also, in the set of all Agnesi quasi-fuzzy numbers is defined an arithmetic where field properties are satisfied.

Keywords Agnesi · Curve · Fuzzy sets · Fuzzy numbers

1 Introduction

Introduced by Zadeh [1], fuzzy numbers are a specific type of fuzzy sets used for representing values of real world parameters when the exact values are not known [2]. Thus, fuzzy numbers have been applied in many different areas, such as data analysis, engineering problems, and decision making problems [3–5]. In such applications, fuzzy arithmetic is frequently applied to solve mathematical equations with fuzzy numbers by using one of the two approaches introduced in the literature: first, the α-cut (or α-level) approach, second the extension principle approach using different t-norms. First approach is the most common approach for implementing fuzzy arithmetic in different applications, due its simplicity and to the availability of computational methods. However, the accumulation of fuzziness, a phenomenon that causes overestimation of uncertainties in the resulting fuzzy numbers is noted. On the other hand, operations on α-cuts brings similar interval

F. Bergamaschi (✉) · N. Jesus · A. Oliveira
Southwest Bahia State University, Conquista, Brazil
e-mail: flaulles@yahoo.com.br

N. Jesus
e-mail: nataniuri@outlook.com

A. Oliveira
e-mail: alexsandra@uesb.edu.br

R. Santiago
Federal University of Rio Grande do Norte, Natal, Brazil
e-mail: regivan@dimap.ufrn.br

B. Bede et al. (eds.), *Fuzzy Information Processing 2020*,
Advances in Intelligent Systems and Computing 1337,
https://doi.org/10.1007/978-3-030-81561-5_3

arithmetic issues [6]. The second approach using the min t-norm produces the same result as the first case, leading to similar overestimation of uncertainty in the resulting fuzzy numbers [7]. In [8] Seresht and Fayek introduce a computational method for implementing extended fuzzy arithmetic using product and Lukasiewicz t-norms on triangular fuzzy numbers, which reduces the uncertainty overestimation problem in the resulting fuzzy numbers, but the issue is still there.

In [9] it is shown that each fuzzy number can be written in a unique way as the sum of a symmetric and a skew fuzzy number. One of the benefits of this representation is that the algebraic order-theoretic and metric structures on the product can make the set of fuzzy numbers into a commutative semiring. But, again the overestimation problem is still there. This problems comes from interval arithmetic defined by Moore [6] as we can see in Neumaier [10]. Recently Lodwick [11] proposed a new arithmetic called constrained interval arithmetic. Unlike Moore arithmetich, constrained interval arithmetic has an additive inverse, multiplicative inverse and satisfies the distributive law. Again, Lodwick in [12] define a single level constraint interval arithmetic as a variant of the last one. This results in a computationally simple interval arithmetic that has several desirable properties not shared in general by interval operations previously defined in the literature. But, the constrained arithmetic lost the inclusion isotonicity and its arithmetic is based on complex optimization process. Thus, we would like to define a kind of fuzzy numbers with field properties and in a certain way has inclusion isotonicity. For this end, we decided to Agnesi's curve as we will see.

As it is know Agnesi's curve, or witch of Agnesi is a cubic plane curve defined from two diametrically opposite points of a circle. It gets its name from Italian mathematician Maria Gaetana Agnesi, and from a mistranslation of an Italian word for a sailing sheet. Before Agnesi, the same curve was studied by Fermat, Grandi, and Newton. The main properties of this curve can be derived from integral calculus. As a first attempt, we take the Agnesi's classical definition to create a special fuzzy set. This set is almost a fuzzy number, as we will see. Also, we define an arithmetic such that the set of all Agnesi quasi-fuzzy numbers became a field. This paper has the following structure: Sect. 2 provides the basic definitions and the Agnesi's curve; Sect. 3 provides the fuzzy concept of Agnesi's quasi-fuzzy number, the main parts of this section is Propositions 1 and 2; Sect. 4 provides the final remarks.

2 Basic Definitions

In 1965, Zadeh [13] introduced the concept of fuzzy sets. Fuzzy numbers are a kind of fuzzy sets that are of particular importance in fuzzy set theory. A fuzzy number [14] is a function $u : \mathbb{R} \longrightarrow [0, 1]$ satisfying the following conditions:

(1) u is normal, that is, there is a real number x_0 such that $u(x_0) = 1$;

(2) u is compactly supported, that is, the closure of $\{x \in \mathbb{R} : u(x) > 0\}$ is bounded;

(3) u is convex, that is, $r \leq x \leq s$ implies $\min\{u(r), u(s)\} \leq u(x)$ for all $r, s, x \in \mathbb{R}$;

(4) u is upper semi-continuous, that is, $\{x : u(x) \geq \alpha\}$ is closed for each $\alpha \in [0, 1]$.

As analyzed in [15, 16], fuzzy numbers are an extension of intervals, rather than real numbers. So, a fuzzy number is also called a fuzzy interval. The set of all fuzzy numbers is denoted by $\mathscr{F}$. The set $\mathbb{R}$ of real numbers is canonically embedded in $\mathscr{F}$, identifying each real number a with the crisp fuzzy number $\tilde{a} : \mathbb{R} \longrightarrow [0, 1]$ given by

$$\tilde{a}(x) = \begin{cases} 1, & if\ x = a; \\ 0, & otherwise. \end{cases}$$

Given a fuzzy number u and $\alpha \in (0, 1]$, the α-level or α-cut set $u_\alpha = \{x : u(x) \geq \alpha\}$ is a closed interval, hence, $u_\alpha = [u_L(\alpha), u_R(\alpha)]$ for some real numbers $u_L(\alpha)$ and $u_R(\alpha)$. Letting $u_L(0) = inf\{u_L(\alpha) : \ 0 < \alpha \leq 1\}$ and $u_R(0) = sup\{u_R(\alpha) : 0 < \alpha \leq 1\}$, we obtain an increasing function $u_L : [0, 1] \longrightarrow \mathbb{R}$ and a decreasing function $u_R : [0, 1] \longrightarrow \mathbb{R}$.

A field [17] is a set F together with two operations on F called addition and multiplication. These operations are required to satisfy the following properties, referred to as field axioms. In these axioms, a, b, and c are arbitrary elements of the set F:

(a) Associativity of addition and multiplication: $(a + b) + c = a + (b + c)$, $a \cdot (b \cdot c) = (a \cdot b) \cdot c$.

(b) Commutativity of addition and multiplication: $a + b = b + a$ and $a \cdot b = b \cdot a$.

(c) Additive and multiplicative identity: there exist two different elements 0 and 1 in F such that $a + 0 = a$ and $a \cdot 1 = a$.

(d) Additive inverses: for every a in F, there exists an element in F, denoted a, called the additive inverse of a, such that $a + (-a) = 0$.

(e) Multiplicative inverses: for every $a \neq 0$ in F, there exists an element in F, denoted by a^{-1} or $1/a$, called the multiplicative inverse of a, such that $a \cdot a^{-1} = 1$.

(f) Distributivity of multiplication over addition: $a \cdot (b + c) = a \cdot b + a \cdot c$ and $(b + c) \cdot a = b \cdot a + c \cdot a$ for $a, b, c \in R$.

Definition 1 Given $a \in \mathbb{R}$, the Agnesi Curve or Witch of Agnesi, is a cubic plane curve defined by the equation $y = \dfrac{\mathrm{a}^3}{x^2 + \mathrm{a}^2}$.

3 Agnesi Quasi-fuzzy Numbers

This section introduces the concept of Agnesi quasi-fuzzy numbers and the first investigation on the set of all Agnesi quasi-fuzzy numbers.

Consider $\mathrm{a} \in \mathbb{R}$ and $f : \mathbb{R} \longrightarrow \mathbb{R}$ such that $f(x) = \dfrac{\mathrm{a}^3}{x^2 + \mathrm{a}^2}$, its easy to see that the Agnesi curve (see Fig. 1) has the following properties:

$$\lim_{x\to\infty} f(x) = \lim_{x\to-\infty} f(x) = 0 \text{ and } \lim_{x\to0^-} f(x) = \lim_{x\to0^+} f(x) = \mathrm{a}$$

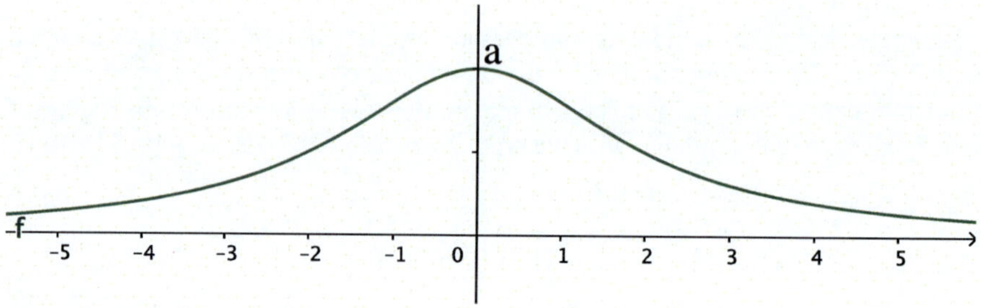

Fig. 1 Agnesi curve

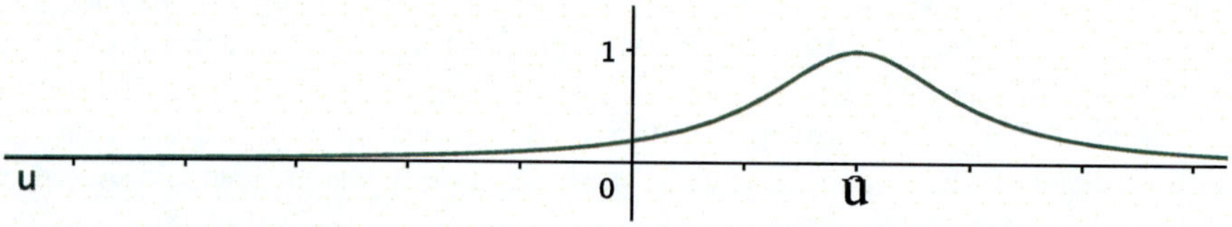

Fig. 2 Agnesi quasi-fuzzy number

Therefore, we conclude that: if x is close to 0, then $f(x)$ is close to a. Inspired on these properties to create the Agnesi quasi-fuzzy numbers, we simple modify Agnesi's curve to $u(x) = \dfrac{1}{1+(x-\mathrm{a})^2}$. Thus, if x is close to a, then $u(x)$ is close to 1 (Fig. 2).

Definition 2 Let $u \in \mathbb{R}$, the Agnesi quasi-fuzzy number u is the function $u : \mathbb{R} \longrightarrow (0, 1]$, where $u(x) = \dfrac{1}{1+(x-u)^2}$.

Note that $u(u) = 1$ and $\lim_{x\to\infty} u(x) = \lim_{x\to-\infty} u(x) = 0$, but $u(x) \neq 0$ for all $x \in \mathbb{R}$.

Agnesi quasi-fuzzy numbers defined above is normal, convex, upper semi-continuous, but the support is not a compactly set, since $\{t \in \mathbb{R} : u(t) > 0\}$ is not bounded. In this case $\{t \in \mathbb{R} : u(t) > 0\} = \mathbb{R}$. For this reason, we called quasi-fuzzy number.

Definition 3 Consider $\mathscr{F}_A$ the se of all Agnesi quasi-fuzzy numbers. In $\mathscr{F}_A$ define:

$$u(x) + v(x) = \frac{1}{1+(x-(u+v))^2}, \quad u(x)\cdot v(x) = \frac{1}{1+(x-(u\cdot v))^2},$$
$$1(x) = \frac{1}{1+(x-1)^2} \ \textit{and} \ 0(x) = \frac{1}{1+(x-0)^2}.$$

Note that operations in $\mathscr{F}_A$ depend of real numbers operations. Also, the same letter was used for the quasi-fuzzy number and the real number. Our intention, like fuzzy numbers are consider the real number with uncertainties, the same letter helps us to identify the number.

Proposition 1 $(\mathscr{F}_A, +, \cdot, 1(x), 0(x))$ *is a Field.*

Proof Note that the operations $(+, \cdot)$ in $\mathscr{F}_A$ was defined based on arithmetic of real numbers. So, the associativity, commutativity and distributivity properties are satisfy straightforward. Finally, consider the additive and multiplicative inverses:

$$-u(x) = \frac{1}{1 + (x - (-u))^2} \text{ and } u^{-1}(x) = \frac{1}{1 + (x - (u^{-1}))^2}.$$

Now, is easy to see all field properties.

Consider the α-cut set $u_\alpha = \{x : \ u(x) \geq \alpha\}$ and $u_\alpha = [u_L(\alpha), u_R(\alpha)]$ for some real numbers $u_L(\alpha)$ and $u_R(\alpha)$. Letting $u_L(0) = inf\{u_L(\alpha) : \ 0 < \alpha \leq 1\}$ and $u_R(0) = sup\{u_R(\alpha) : 0 < \alpha \leq 1\}$, we obtain an increasing function $u_L : [0, 1] \longrightarrow \mathbb{R}$ and a decreasing function $u_R : [0, 1] \longrightarrow \mathbb{R}$.

Let u in $\mathscr{F}_A$, using the inverse function for $u(x)$, the α-cut is given by $u_\alpha = [u_L(\alpha), u_R(\alpha)]$ where $u_L(\alpha) = u - \sqrt{\frac{1-\alpha}{\alpha}}$ and $u_R(\alpha) = u + \sqrt{\frac{1-\alpha}{\alpha}}$. Thus, $u_0 = [u_L(0), u_R(0)] = (-\infty, +\infty)$ and $u_1 = [u_L(1), u_R(1)] = [u, u]$.

The next proposition shows that the arithmetic on $\mathscr{F}_A$ is not compatible to the arithmetic based on α-cuts.

Proposition 2 *Let $u, v \in \mathscr{F}_A$ and $0 \leq \alpha \leq 1$, then $(u + v)_\alpha \subseteq u_\alpha + v_\alpha$ and $(u \cdot v)_\alpha \subseteq u_\alpha \cdot v_\alpha$.*

Proof Note that:

$$(u + v)_\alpha = \left[(u + v) - \sqrt{\frac{1-\alpha}{\alpha}}, (u + v) + \sqrt{\frac{1-\alpha}{\alpha}}\right] \text{ and}$$

$$u_\alpha + v_\alpha = \left[u - \sqrt{\frac{1-\alpha}{\alpha}}, u + \sqrt{\frac{1-\alpha}{\alpha}}\right] + \left[v - \sqrt{\frac{1-\alpha}{\alpha}}, v + \sqrt{\frac{1-\alpha}{\alpha}}\right]$$

Using Moore interval arithmetic:

$$u_\alpha + v_\alpha = \left[(u + v) - 2\sqrt{\frac{1-\alpha}{\alpha}}, (u + v) + 2\sqrt{\frac{1-\alpha}{\alpha}}\right].$$

Therefore $(u + v)_\alpha \subseteq u_\alpha + v_\alpha$. In similar way we can have $(u \cdot v)_\alpha \subseteq u_\alpha \cdot v_\alpha$.

As we can see, the operations $(+, \cdot)$ are not compatible with Moore [6] operations.

Note that, given an uncertainty u, for a better approximation we can change the exponent "2" on the Definition 2 for high exponent. The curve will be more close to the line $x = u$.

4 Final Remarks

This paper defines an approximation for an uncertainty called Agnesi quasi-fuzzy number. As was showed the set of this numbers has a complete arithmetic allowing to solve equations. But some questions have arisen, for instance, is there an algorithm that can reduce the cost of evaluations on real numbers? This arithmetic has or not inclusion isotonicity?

We hope to have contributed, in a certain way, to draw attention for Agnesi quasi-fuzzy numbers.

Acknowledgements The authors would like to thank UESB (Southwest Bahia State University) and UFRN (Federal University of Rio Grande do Norte) for their financial support.

References

1. L.A. Zadeh, The concept of a linguistic variable and its application to approximate reasoning. Inf. Sci. **8**(3), 199–249 (1975)
2. W. Pedrycz, F. Gomide, *Fuzzy Systems Engineering: To RD Human-Centric Computing* (Wiley, 2007)
3. J. Booker, T. Ross, An evolution of uncertainty assessment and quantification. Sci. Iran. **18**(3), 669–676 (2011)
4. M. Alavidoost, M. Tarimoradi, M.F. Zarandi, Fuzzy adaptive genetic algorithm for multi-objective assembly line balancing problems. Appl. Soft Comput. **34**, 655–677 (2015)
5. A. Blanco-Fernández, M. Casals, A. Colubi, N. Corral, M. García-Bárzana, M. Gil, G. González-Rodríguez, M. López, M. Lubiano, M. Montenegro et al., A distance-based statistical analysis of fuzzy number-valued data. Int. J. Approx. Reason. **55**(7), 1487–1501 (2014)
6. R.E. Moore, *Interval Analysis*, vol. 4 (Prentice-Hall Englewood Cliffs, NJ, 1966)
7. P. Kechagias, B.K. Papadopoulos, Computational method to evaluate fuzzy arithmetic operations. Appl. Math. Comput. **185**(1), 169–177 (2007)
8. N.G. Seresht, A.R. Fayek, Computational method for fuzzy arithmetic operations on triangular fuzzy numbers by extension principle. Int. J. Approx. Reason. **106**, 172–193 (2019)
9. D.Z. Yingming Chai, A representation of fuzzy numbers. Fuzzy Sets Syst. **295**, 1–18 (2016)
10. A. Neumaier, A. Neumaier, *Interval Methods for Systems of Equations*, vol. 37 (Cambridge University Press, 1990)
11. W.A. Lodwick, Constrained interval arithmetic. Citeseer (1999)
12. Y. Chalco-Cano, W.A. Lodwick, B. Bede, Single level constraint interval arithmetic. Fuzzy Sets Syst. **257**, 146–168 (2014)
13. L. Zadeh, Fuzzy sets, 338–353 (1965). https://doi.org/10.1016/S0019-9958(65)90241-X
14. R. Goetschel Jr., W. Voxman, Topological properties of fuzzy numbers. Fuzzy Sets Syst. **10**(1–3), 87–99 (1983)
15. D. Dubois, H. Fargier, J. Fortin, The empirical variance of a set of fuzzy intervals, in, *The 14th IEEE International Conference on Fuzzy Systems, FUZZ'05* (IEEE, 2005), pp. 885–890
16. D. Dubois, H. Prade, Gradual elements in a fuzzy set. Soft Computing **12**(2), 165–175 (2008)
17. I.N. Herstein, *Topics in Algebra* (John Wiley & Sons, 2006)
18. J.J. Buckley, E. Eslami, *An Introduction to Fuzzy Logic and Fuzzy Sets*, vol. 13 (Springer Science & Business Media, 2002)

Fuzzy Mathematical Morphology and Applications in Image Processing

Alexsandra Oliveira Andrade, Flaulles Boone Bergamaschi, Roque Mendes Prado Trindade, and Regivan Hugo Nunes Santiago

Abstract Fuzzy Mathematical Morphology extends binary morphological operators to grayscale and color images using concepts from fuzzy logic. To define the morphological operators of erosion and fuzzy dilation, the R-implications and fuzzy T-norm respectively are used. This work presents the application of the fuzzy morphological operators of Lukasiewicz, Gödel and Goguen and of the epsilon and delta functions of Weber and Fodor in the counting of mycorrhizal fungi spores.

Keywords Fuzzy mathematical morphology · Fuzzy logic · Counting · Mycorrhizal

1 Introduction

Mathematical morphology is a collection of operations that produces good results in the image processing area. Its origins are in the studies of porous media in the 1960s decade, by the group of French researchers led by Georges Matheron and Jean Serra, of the cole Superieure de Mines de Paris [21, 27].

They introduced a study in set theory that led to the analysis of binary images. Because mathematical morphology was originally defined in binary images, most theories for the extension of mathematical morphology tried to extend this fact to grayscale and colored images [28, 35].

A. Oliveira Andrade (✉) · F. Boone Bergamaschi · R. Mendes Prado Trindade
Southwest Bahia State University, Conquista, Brazil
e-mail: alexsandra@uesb.edu.br

F. Boone Bergamaschi
e-mail: flaulles@yahoo.com.br

R. Mendes Prado Trindade
e-mail: roquetrindade@uesb.edu.br

R. Hugo Nunes Santiago
Federal University of Rio Grande do Norte, Natal, Brazil
e-mail: regivan@dimap.ufrn.br

B. Bede et al. (eds.), *Fuzzy Information Processing 2020*,
Advances in Intelligent Systems and Computing 1337,
https://doi.org/10.1007/978-3-030-81561-5_4

The literature brings other methods of extending classical morphology to fuzzy morphology. The way we used it was through fuzzy logic, where the first author who used it was Goetcherian [14]. But there were others like: Sinha and Dougherty [30, 31], Bloch and Maitre [11, 12], De Beats, Nachtegael and Kerre [9, 10, 24] and Deng and Heijmans [16].

The image processing area is quite wide, with several applications and many tools. These tools are used according to the image and what we want to extract from them [15].

In this article we present an application of fuzzy morphology for the automatic counting of mycorrhizal fungus sporesan association between a group of soil fungi and most "terrestrial, epiphytic, aquatic vascular plants and also with rhizoid and stems of bryophytes and other basal vegetables" [33]. This application consists of counting these spores in an automated way, that is, without the participation of man. The first proposal for the algorithm was presented in the work [2], a second version, with uncertainties in [5] and the latest and most efficient version was developed in the article [4]. Currently, the spore count with impurities was presented by Melo et al. using neural networks and can be seen in [22, 23].

In the last decades, the studies on mycorrhizae have increased in a very intense way, both in Brazil [29] and in the world, in the search for knowledge of the effects, besides a perspective of its use in species vegetables of economic interest, mainly agricultural ones [20, 26]. These are of great importance for better absorption of water and nutrients by the plant [32], which makes this automated counting process of great relevance. There are several methods for spore extraction [13, 18, 34], at the end of the execution of these, we will have a certain volume of spores to quantify. Thus, after the extraction procedure, the spores are counted manually, but using only a stereoscopic microscope (magnifying glass) to help visualize these structures, which vary from 22 to 1050μm [33]. One of the premises in such an analysis is that it will be make by a unique person.

This work has as objective to make an overview of the works developed with the residual R-implications that formed or not adjunctions and its repercussion in the processing of gray and colored images. The article is organized as follows. Section 2 presents the fuzzy morphological operators developed [3]. Section 3 describes the experiments with the fuzzy morphological operators applied in the counting of mycorrhizal spores based on the works of [2, 4, 5] and still in Sect. 3, we take a look at the fuzzy operators and the Epsilon functions of Weber and Fodor behave in counting mycorrhizal fungi spores based on the work [6]. And finally in Sect. 4 final considerations and proposals for future work are expressed.

2 The Fuzzy Morphological Operators

Fuzzy logic is one of the ways to extend binary operators to gray and color images. According to Ronse [25], the basic morphological operators can be defined in any complete lattice opening the possibility of being binary images, gray-scale or

Table 1 Some fuzzy T-norms and fuzzy implications

Name	T-Norm	Implication
Lukasiewicz	$T_{LK}(x, y) = \max(0, x + y - 1)$	$I_{LK}(x, y) = \min(1, 1 - x + y)$
Gödel	$T_{GD}(x, y) = \min(x, y)$	$I_{GD}(x, y) = \begin{cases} 1 \; if & x \leq y \\ y \; if & x > y \end{cases}$
Goguen	$T_{GG}(x, y) = x.y$	$I_{GG}(x, y) = \begin{cases} 1 \; if & x \leq y \\ \frac{y}{x} \; if & x > y \end{cases}$
Weber	$T_{WB}(x, y) = \begin{cases} 1 & if \;\; x < 1, y < 1 \\ \min(x, y) & \text{otherwise} \end{cases}$	$I_{WB}(x, y) = \begin{cases} 1 \; if & x < 1 \\ y \; if & x = 1 \end{cases}$
Fodor	$T_{FD}(x, y) = \begin{cases} 1 & if \;\; x + y < 1 \\ \min(x, y) & \text{otherwise} \end{cases}$	$I_{FD}(x, y) = \begin{cases} 1 & if \;\; x \leq y \\ \max(1 - x, y) & if \;\; x > y \end{cases}$

Table 2 Dilation and erosion of Lukasiewicz, Gödel and Goguen

Operator	Dilation	Erosion
Lukasiewicz	$\delta_B^{LK} A(x) = \sup_{x \in A} \max[0, \widehat{B(x)} + A(x) - 1]$	$\varepsilon_B^{LK} A(x) = \inf_{x \in A} \min[1, 1 - B(x) + A(x)]$
Gödel	$\delta_B^{GD} A(x) = \sup_{x \in A}[\min[\widehat{B(x)}, A(x)]]$	$\varepsilon_B^{GD} A(x) = \inf_{x \in A} \begin{cases} 1 & if \;\; B(x) \leq A(x) \\ A(x) & if \;\; B(x) > A(x) \end{cases}$
Goguen	$\delta_B^{GG} A(x) = \sup_{x \in A}[\widehat{B(x)} A(x)]$	$\varepsilon_B^{GG} A(x) = \inf_{x \in A} \begin{cases} 1 & if \;\; B(x) \leq A(x) \\ \frac{A(x)}{B(x)} & if \;\; B(x) > A(x) \end{cases}$

coloured. However it is not any pair (I,T) which gives rise to **adjunctions**, for that it is necessary that the implication and conjunction have the following relation

$$\delta_I^T(x) = T(z, x) \leq y \leftrightarrow x \leq I(z, y) = \varepsilon_I^T(y)$$

To define these new operators it is necessary the implications and conjunctions (T-norms) fuzzy that must be analyzed if the pair of R-implication and T-norm form an adjunct. The Table 1 presents the fuzzy T-norms and the R-implications that were analyzed as an adjunct in the work of Andrade et al. [3].

In Andrade et al. [3] it was demonstrated that the T-norms T_{LK}, T_{GD}, T_{GG}, are continuous to the left, and that each pair (I_{LK}, T_{LK}), (I_{GD}, T_{GD}), (I_{GG}, T_{GG}) were an adjunct and give rise to a family of erosions and fuzzy dilations. However the T-norms T_{WB}, T_{FD} are not continuous to the left, that is, the pairs (I_{WB}, T_{WB}), (I_{FD}, T_{FD}) were not adjunctions and do not constitute expansion /erosion pairs.

Like the pairs (I_{LK}, T_{LK}), (I_{GD}, T_{GD}), (I_{GG}, T_{GG}) form dilation /erosion pairs. Operators **dilation** and **erosion** of a A image depend on a structural element B respectively is given by Table 2.

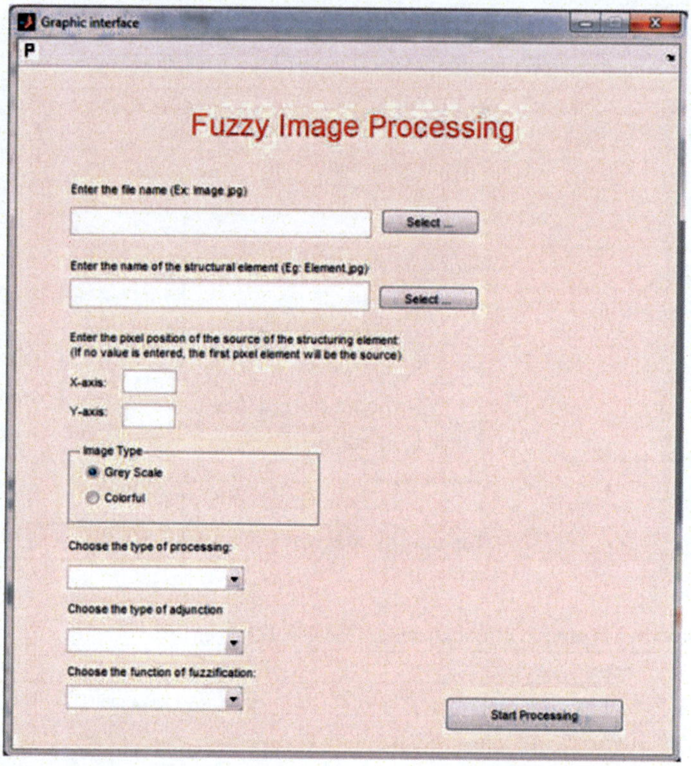

Fig. 1 Image of Matlab interface

Table 3 Functions delta and epsilon of weber and fodor

Function	Delta	Epsilon
Weber	$\Delta^T_B A(x) = \sup_{x\in A}\begin{cases} 1 \ \ if \ \ \widehat{B(x)} < 1, A(x) < 1 \\ \min[\widehat{B(x)}, A(x)] \ \ \text{otherwise} \end{cases}$	$\epsilon^I_B A(x) = \inf_{x\in A}\begin{cases} 1 \ \ if \ \ B(x) \leq 1 \\ A(x) \ \ if \ \ B(x) = 1 \end{cases}$
Fodor	$\Delta^T_B A(x) = \sup_{x\in A}\begin{cases} 1 \ \ if \ \ \widehat{B(x)} + A(x) < 1 \\ \min[\widehat{B(x)}, A(x)] \ \ \text{otherwise} \end{cases}$	$\epsilon^I_B A(x) = \inf_{x\in A}\begin{cases} 1 \ \ if \ \ B(x) \leq A(x) \\ \max(1 - B(x), A(x)), B(x) > A(x) \end{cases}$

Although they do not form an adjunct, we call **epsilon function** and **delta function** for equations of erosion and dilation operators for the purpose of experiments implemented in Matlab using the interface according Fig. 1 (Table 3).

Applying all operators erosion and dilation to images, we observed that the erosion accentuates the black pixels and the dilation softens the image as seen in Fig. 2.

Fig. 2 **a** Original image of color Lena, **b** Image of Lena with Gödel's drosion, **c** Image of Lena with Gödel's dilation

3 Application of Fuzzy Morphology to Mycorrhizal Fungi Spores

In this section we present the application of operators on mycorrhizal fungi spores, an association between a group of soil fungi and most "terrestrial, epiphytic, aquatic vascular plants and also with rhizoid and stems of bryophytes and other basal vegetables" [33].

3.1 Lukasiewicz's Fuzzy Morphology

Initially, an experiment was carried out at Matlab in which the morphological operators from Lukasiewicz's involvement in the mycorrhizal fungi spore counting process were used, according to the work of Andrade et al. in [4].

As an initial step we have the choice of the color image in [17], more specifically an image with jpg extension with 205.4 KB and 430 × 311 pixels of the mycorrhizal fungal spore species Glomus claroideum as Fig. 3a.

The structuring element, image with which the original image is processed, was chosen based on tests, Fig. 3d. Firstly, a smaller element, a 5 × 5 matrix, but more needed to be done erosions to improve the image. Then an 9 × 9 element was used, which erases some very small spores. So a 7 × 7 element was used, with which a better result was obtained, as erosion and an opening operator were needed that didn't erase the smaller spores. As for the structuring element, the one that showed the best performance was in the form of a white cross in relation to the one that had only a white square, as it separated the spores with less computational effort.

The counting process consists of three steps:

1. Use of Lukasiewicz's Morphological Operators;
2. Use of the Dark Background Spore Detection Algorithm;
3. Use of the Algorithm to quantify points.

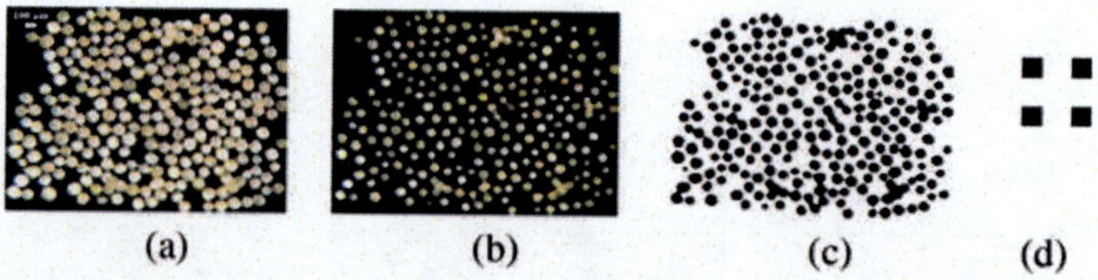

Fig. 3 **a** Original image of spores, **b** Image with erosion and opening **c** Fungi spore detection image, **d** Structuring element

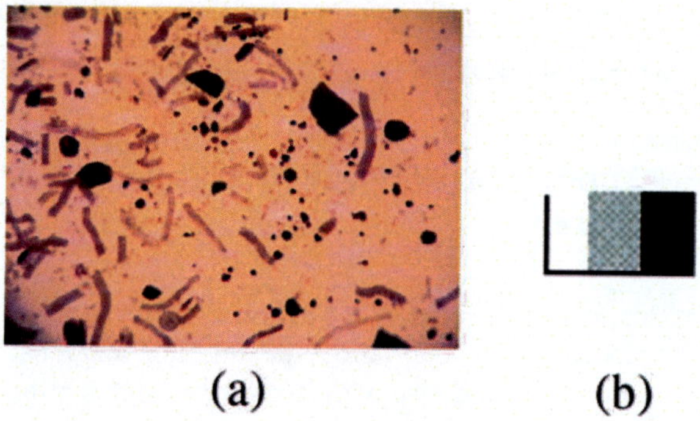

Fig. 4 **a** Image of spores with uncertainties **b** Structuring element

With images having uncertainties, which are other materials such as plant root, stone, soil, pre-processing was used, a neural network. This network was of the feedforward type with two hidden layers, and the first has five neurons and the second three neurons. He has been trained for 10 seasons using a 1116542 pixel sample taken from 7 photos of mycorrhizal fungi with impurities that were acquired in the soil microbiology laboratory at the State University of Southwest Bahia with the dimension of 461×346 pixels. Figure 4a illustrates how the spores are arranged as they are and the uncertainties.

After pre-processing, the images were ready to use the algorithm. For this, we use two structuring elements, the first of order 4 of the white color used to process the erosion operator and the second of dimension 2×1 as shown in Fig. 4b used for the opening operator. The 80% hit index on spore count on 12 out of 19 dataset images.

3.2 Gödel's Fuzzy Morphology

This section presents an account of the experiments carried out with the operators developed with the R-implications: Lukasiewicz, Gödel and Goguen found in Andrade et al. [3]. The experiments used several structuring elements to obtain an algorithm capable of counting mycorrhizal fungi spores efficiently and effectively. Where the same images found in [17] were used, according to the work of Andrade et al. in [5].

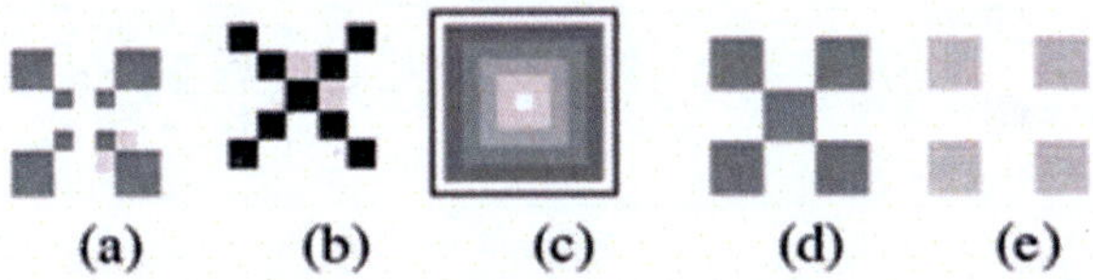

Fig. 5 Structuring elements of the five tests

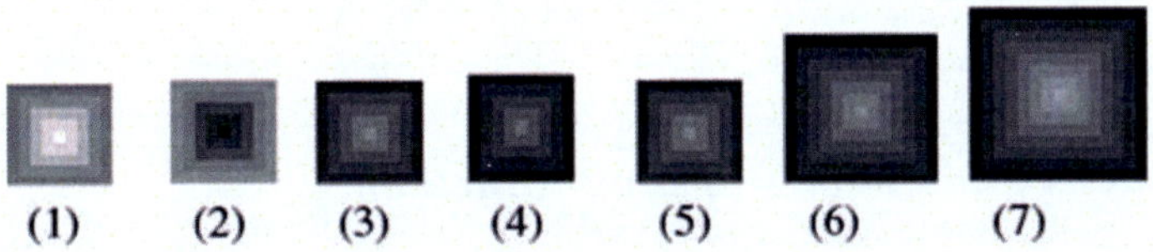

Fig. 6 Structuring elements used in tests with Gödel's operator

The experiments take place in two stages: in the first, the processing was carried out combining the possibilities of one to three of the morphological operators resulting in a total of 1741 images. Then, a visual analysis of the acquired images was made to determine which operators were more efficient in separating the spores. In the next stage, a counting process was applied to determine the morphology and the most efficient structuring element.

As a result of the first stage, the most efficient operator is erosion because it promotes good spore separation. In the second step, the number of operators and the appropriate structuring element were determined, which can be seen in Fig. 5, because five tests were carried out with different structuring elements.

With these tests, we can see that the Gödel's morphology is the morphology that shows the best performance in spore separation. And at the same time, the average calculation can better measure the amount of spores and reduce the problem of overlapping spores.

Analyzing the number of operations and the structuring element, a relationship could be observed: *the number of operations is inversely proportional to the size of a structuring element, that is, if the element used was greater then the number of operations will be less, the result was still kept close to the others* e *the lighter the color of the structuring element, the more operations will be needed to separate the spores*.

Figure 6 illustrates the structuring elements used in the tests. The structuring elements from (1) to (6) are of order 9; (7) is of order 11; (8) is of order 13; and the last and order 15.

After the tests, it was revealed that the best structuring element used to separate and count the spores is number 1. Processing time ranged from 10 to 34 s, using a 4-core processor and 2,6 GHz. The method achieved a hit rate of 85%. The counting process consists of four steps:

1. Use the Gödel Morphological Operator;
2. Use the White Background Spore Detection Algorithm;

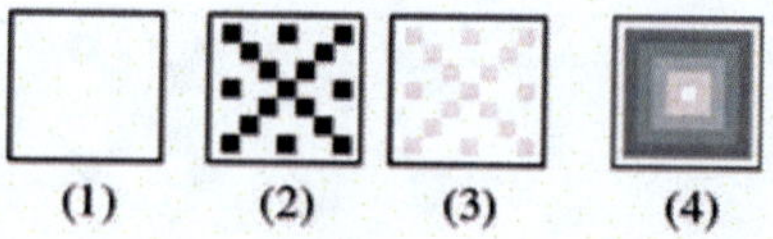

Fig. 7 Structuring elements used in tests with the functions of weber and fodor

3. Use the Mask;
4. Use the Algorithm to Quantify Spores by Average.

3.3 Weber's and Fodor's Epsilon Functions

In Andrade et al. work [3] was introduced the Weber's Epsilon and Fodor's Epsilon functions it does not form an adjunct. In this section, an analysis was made of the morphological operators including the Epsilon functions of Weber and Fodor in the counting of mycorrhizal fungi spores. For this we use the same 37 images from the site [17].

Several experiments were carried out with 4 different structuring elements illustrated in Fig. 7 using the erosion of Gödel, Lukasiewicz, the Epsilon functions of Weber and Fodor. With these experiments, a statistical study called Bland-Altman [7]. In each experiment with one of the structuring elements, the performance of each operator in counting the spores of mycorrhizal fungi was compared using this statistical method. The results of which indicated that in the first, second and third experiments, any operator: the Lukasiewicz erosion, the Gödel erosion, the Epsilon Weber function and the Epsilon Fodor function can be used in spore counts. In the case of the fourth experiment, it is concluded that only the operator: the erosion of Gödel can be used in spore counts.

4 Final Considerations

This work made an overview of the applications in the image processing of the fuzzy morphological operators developed in Andrade et al. [3] and this study reinforces that the image processing area is a vast field of performance with the presentation in counting mycorrhizal fungi spores based on [2, 4–6].

As the mycorrhizal fungi spore counting process is done manually, its automation was a scientific gain of great relevance. In the identification of license plates, the use of fuzzy morphology had a gain in processing, since it reduces the number of processes used, but we had problems to establish a pattern in the image to identify the plate with the desired success.

As future work, we would like to apply fuzzy morphology to automation processes in the biomedical or biological fields.

References

1. L.D. Amaral, A.O. Andrade, Uso da Morfologia Matemtica Fuzzy no Processamento de Imagens para Reconhecimento de Placas de Veculos. Semana de Computao da UESB (2017)
2. A.O. Andrade, R.M.P. Trindade, D.S. Maia, D.L. Miguel, R.H.N. Santiago, A.M.G. Guerreiro, Uso da Morfologia Matemtica Fuzzy na contagem Esporos de Fungos Micorrzicos. Recentes Avanos em Sistemas Fuzzy. II Congresso Brasileiro de Sistema Fuzzy (2012)
3. A.O. Andrade, R.M.P. Trindade, D.S. Maia, R.H.N. Santiago, A.M.G. Guerreiro, Analysing some R-implications and its application in fuzzy mathematical morphology. J. Intell. Fuzzy Syst. **27**(1), 201–209 (2014)
4. A.O. Andrade, R.M.P. Trindade, V.B.F. Neves, A.S. Barros, I.B. Soares, R.P. Costa, D.L. Miguel, R.H.N. Santiago, A.M.G. Guerreiro, Analysis of fuzzy morphology in spore counts of mycorrhizal fungi, in *Fuzzy Information Processing Society (NAFIPS) held jointly with 2015 5th World Conference on Soft Computing (WConSC). 2015 Annual Conference of the North American.* Proceedings of the IEEE
5. A.O. Andrade, R.M.P. Trindrade, V.B.F. Neves, D.S. Maia, D.L. Miguel, R.H.N. Santiago, A.M.G. Guerreiro, The counting of mycorrhizal fungi spores using fuzzy mathematical morphology, in *Fuzzy Information Processing Society (NAFIPS) held jointly with 2015 5th World Conference on Soft Computing (WConSC). 2015 Annual Conference of the North American.* Proceedings of the IEEE
6. A.O. Andrade, R.M.P. Trindade, F.B. Bergamasch, A.S. Barros, R.H.N. Santiago, A.M.G. Guerreiro, Analyzing the R-implications of weber and fodor in the counting of mycorrhizal fungi spores. J. Commun. Comput. **13**(3) (2016)
7. D.G. Altman, J.M. Bland, The measurement in medicine: the analysis of method comparison studies. Statistician **32**, 307–317 (1983)
8. D.L. Baggio, OpenCV 3.0 computer vision with java. Packt publishing. Birmingam-Mumbai (2015)
9. B. De Baets, Fuzzy Morphology: a logical approach, in *Uncertainty Analysis in Engineering and Science: Fuzzy Logic, Statistics, and Neural Network Approach* (Kluwer Academic Publishers, Norwell), pp. 53–67 (1997)
10. B. De Baets, E. Kerre, The fundamentals of fuzzy mathematical morphology part 1: basic concepts. Int. J. Gen. Syst. **23**, 155–171 (1995)
11. I. Bloch, H. Maitre, Fuzzy mathematical morphology. Ann. Math. Artif. Intell. **10**, 55–84 (1994)
12. I. Bloch, H. Maitre, fuzzy mathematical morphologies: a comparative study. Pattern Recognit. **28**(9), 1341–1387 (1995)
13. J.E. Gerdeman, T.H. Nicolson, Spores of mycorrhizal endogone species extracted from soil by wet sieving and decanting. Trans. Br. mycol. Soc. **46**, 235–244 (1963)
14. V. Goetcherian, Form Binary to Grey tone Image Processing using Fuzzy Logic Concept. Pattern Recognit. **12**, 7–15 (1980)
15. R.C. Gonzalez, R.E. Woods, Digital image processing. Editora Edgard Blcher Ltda. (2000)
16. H.J.A.M. Heijmans, T.Q. Deng, Grey-scale morphology based on fuzzy logic. J. Math. Imaging Vis. **16**(2) (2002)
17. Invam: international culture collection of (Vesicular) arbuscular mycorrhizal fungi. Disponvel em http://invam.caf.wvu.edu. Consulta em 29/06/2012
18. W.R. Jenkins, A rapid centrifugal-flotation technique for separating nematodes from soil. Plant Dis. Rep. **8**, 692 (1964)
19. S.H. Mohades Kasaei, S.M. Mohades Kasaei, S.A. Monadjemi, A novel morphological method for detection and recognition of vehicle license plates. Am. J. Appl. Sci. **6**(12), 2066–2070 (2009)
20. P.E. Lovato, A. Trouvelot, V. Gianinazzi-Pearson, S. Gianinazzi, Avanos em fundamentos e aplicao em micorrizas (Advances in fundamentals and application in mycorrhizae), in *Micorrizao de plantas micropropagadas(Mycorrhization of micropropagated plants)*, pp. 175–201 (1996)

21. G. Matheron, *Elements Pour Une Theorie Des Milieux Poreni* (Masson, 1967)
22. C.A.O. Melo, J.G. Lopes, A.O. Andrade, R.M.P. Trindade, R.S. Magalhes, Semi-automated counting of arbuscular mycorrhizal fungi spores using artificial neural network. IEEE Lat. Am. Trans. **15**, 1566–1573 (2017)
23. C.A.O. Melo, J.G. Lopes, A.O. Andrade, R.M.P. Trindade, R.S. Magalhes, *Semi-automated counting model for arbuscular mycorrhizal fungi spores using the circle hough transform and an artificial neural network.* ANAIS DA ACADEMIA BRASILEIRA DE CINCIAS (ONLINE), vol. 91, p. e2018016 (2019)
24. M. Nachtegael, E. Kerre, Connections between binary, grey-scale and fuzzy mathematical morphology. Fuzzy Sets Syst. **129**, 73–86 (2001)
25. C. Ronse, Why mathematical morphology needs complete lattices. Signal Process. **21** 129–154 (1990)
26. O. Saggin Jr, J.O. Siqueira, Avanos em fundamentos e aplicao em micorrizas (Advances in fundamentals and application in mycorrhizae). Micorrizas arbusculares em cafeeiro (Arbuscular mycorrhizae in coffee). Editora UFLa, pp. 203–254 (1996)
27. J. Serra, Introduction a la Morphologie Mathmatique. Book-let no 3. Cahiers du Centre de Morphologie Mathmatique, Fontainebleau (1969)
28. J. Serra, *Image Analysis and Mathematical Morphology* (Academic Press, 1982)
29. J.O. Siqueira, F.A. De Souza, E.J.N. Cardoso, S.M. Tsai, Micorrizas: 30 anos de pesquisas no Brasil(Mycorrhizae: 30 years of research in Brazil). *Histrico e evoluo da micorrizologia no Brasil em trłs dcadas(History and evolution of micorrizologia in Brazil in three decades).* Editora UFLa. vol. 55, pp. 1–14 (2010)
30. D. Sinha, E.R. Dougherty, Fuzzy mathematical morphology. J. Vis. Commun. Image Represent. **3**(3), 286–302 (1992)
31. D. Sinha, E.R. Dougherty, Fuzzification of set inclusion: theory and applications. Fuzzy Sets Syst.**55**, 15–42 (1003)
32. S.E. Smith, J.D. Read, *Mycorrhizal symbiosis.* S.l:S.n. (1997)
33. F.A. de Souza, S.L. Stmer, R. Carrenho and S.F.B. Trufem, Micorrizas: 30 anos de pesquisas no Brasil(Mycorrhizae: 30 years of research in Brazil). *Classificao e taxonomia de fungos micorrzicos arbusculares e sua diversidade e ocorrłncia no Brasil (Classification and taxonomy of mycorrhizal fungi and their diversity and occurrence in Brazil).* Editora UFLa. vol. 55, pp. 15–73 (2010)
34. J.C. Sutton, G.L. Barron, Population dynamics of Endogone spores in soil. Can. J. Bot. **50**, 1909–1914 (1972)
35. S.R. Sternberg, Grayscale morphology. Comput. Vis. Graph. Image Process. **35**(3) (1986)

Interval Fuzzy Models Based on Evolving Gaussian Clustering—eGauss+

Igor Škrjanc

Abstract In this paper a new approach to interval fuzzy model identification is given. It is based on evolving Gaussian clustering algorithm, eGauss+. This algorithm clusters the data form the stream into small clusters called granules. These granules are then merged together if they fulfill all necessary criteria. This means that the cluster partitions are learned incrementally online from data streams and after that they are merged together in bigger structures. The proposed approach is not limited to the use in the data stream clustering, but can be used also in the case of classical batch clustering methods, especially for big data problems. The idea of interval fuzzy model is to find a lower and upper bound of the data set and describe these bounds by fuzzy model. The band or confidence interval should contain the prescribed number of samples. The interval fuzzy model is described and shown on simple examples.

Keywords Data stream · Evolving clustering · Evolving cluster models · Incremental learning · Interval fuzzy model

1 Introduction

The problem of nonlinear model identification from data streams has received a great attention and has become very important in last years. This is because of an emerging number of sensors used, which produce an enormous number of data in real-time which should be processed in real time. The most frequently used techniques to achieve this are incremental, evolving or single-pass clustering methods. These methods from are able to process the data step-by-step, on-line in real-time, update the parameters, and evolve the structure of the identified model. One way to realize all these goals is to use the so-called evolving clustering algorithms, [1, 2], which are sometimes also called incremental [3–5], single-pass clustering methods [6], which

I. Škrjanc (✉)
University of Ljubljana, Faculty of Electrical Engineering, Ljubljana, Slovenia
e-mail: igor.skrjanc@fe.uni-lj.si

B. Bede et al. (eds.), *Fuzzy Information Processing 2020*,
Advances in Intelligent Systems and Computing 1337,
https://doi.org/10.1007/978-3-030-81561-5_5

means that the algorithm is working by processing the data in a step-wise way, and updating as well as evolving the structure and the parameters [7, 8].

Basically a lot of these algorithms originate from classical unsupervised fuzzy clustering algorithms, such as the evolving Gustafson-Kessel algorithm in [9, 10, 30]. The possibilistic approach to clustering proposed in [11–14] is also a great inspiration for the further development of evolving approaches based on density criteria, as proposed in [15, 16], and in [17], in which the Cauchy data distribution is assumed and in [18], in which it is shown how different inner matrix norms can be used to deal with different clustering problems. The evolving principle based on principal component analysis is presented in [19]. A generalized evolving fuzzy system in an incremental single-pass manner, in which clusters appear that are of different size and rotation, is presented in [20, 21]. A survey about evolving fuzzy and neuro-fuzzy approaches in clustering, regression, identification, and classification is given in [22].

When dealing with evolving clustering, the problem of clusters that are close and very similar frequently arises [23]. This is because the data arrive step-wise over time, and the size of the cluster is not known in advance. The two clusters can, therefore, move together and can be presented as only one. In such situations, clusters should be merged to obtain the minimum number of cluster partitions. This effect is called cluster fusion and is usually caused by samples successively filling the gaps between two or more clusters, which seem to be disjointed at a former point of time in the data stream, but later it turns out that they are not, and thus should be merged to eliminate overlapping and redundant information. The merging of clusters not only provides a more accurate representation of the local data distributions but also keeps evolving neuro-fuzzy systems more compact and thus more easy to adapt and interpret. Several different merging approaches are given in the literature, from an adaptive threshold for merging proposed in [24, 25], to semi-supervised approach given in [26, 27], merging based on correlation between the previous data and membership values is given in [28]. In [29], two merging criteria, the touching and homogeneity condition, are used to decide when two clusters should be merged, in [30], merging is based on the normalized distance between their centers.

The main contribution of this paper is development of interval fuzzy model which is defined by the lower and the upper fuzzy bounds. This is a method for approximation of nonlinear functions of a finite sets of input and output measurements. It is especially important in the case of nonlinear function families to obtain the interval which contains the whole or a part of measurement data set. The interval fuzzy model is described in [34–36], where the interval fuzzy model is obtained by linear programming approach, and in [37–39], where it is obtained from least-squares optimization and used in fault detection.

This paper is organized as follows: in the beginning, in Sect. 1, the main advantages of evolving interval fuzzy systems are discussed; in Sect. 2, the Gaussian evolving algorithm for modeling from data streams with recursive computation of cluster parameters, is given. In Sect. 3, the evolving clustering for regression with some examples is presented, and in Sect. 4 the interval fuzzy model is discussed. At the end, the conclusion is given.

2 Gaussian Probability Density Distribution for Clustering from Data Streams

The multivariate Gaussian probability density of m measured variables is defined as follows:

$$f_m(z(k), \mu, \Sigma) = \frac{1}{2\pi|\Sigma|^{\frac{1}{2}}} e^{-\frac{1}{2}(z(k)-\mu)^T \Sigma^{-1}(z(k)-\mu)}. \quad (1)$$

where, the data sample at the time instant k is written as $z(k) = [z_1(k)\ z_2 \cdots z_m(k)]^T$, the mean values of data samples inside the same cluster are written as $\mu = [\mu_1\ \mu_2 \cdots \mu_m]^T$, the covariance matrix of the corresponding data set as Σ, and the determinant of Σ is defined as $|\Sigma|$.

The covariance matrix of particular cluster Σ is, in the matrix form, defined as follows:

$$\Sigma = \frac{1}{n-1}(Z - EM)^T (Z - EM), \quad (2)$$

where Z stands for the data matrix of dimension $n \times m$, $Z^T = [z(1), \ldots, z(n)]$, M stands for the matrix $M = diag(\mu_1, \ldots, \mu_m)$ and E stands for the matrix of dimension $n \times m$ with all elements equal to 1. The diagonal elements of matrix M are equal to

$$\mu_i = \frac{1}{n} \sum_{k=1}^{n} z_i(k), \quad (3)$$

and define the components of the center of the cluster, and n defines the number of the samples in the cluster.

The covariance matrix can be written in its singular value decomposed form as

$$\Sigma = P \Lambda P^T, \quad (4)$$

where p_i stands for the ith eigenvector, $P = [p_i]$, $i = 1, \ldots, m$, and λ_i for the ith eigenvalue, $\Lambda = diag(\lambda_i)$, $i = 1, \ldots, m$, of the covariance matrix Σ. The covariance matrix defines the shape of the cluster. The volume of the hyper-ellipsoid, defined by covariance matrix in m dimensional hyper-space, is defined as follows:

$$V = \frac{2\pi^{m/2}}{m\Gamma(m/2)} \Pi_{i=1}^{m} \lambda_i, \quad (5)$$

where Γ stands for the gamma function. This means that the volume of the clusters in a certain hyper-dimensional domain depends on the product of the eigenvalues of the cluster covariance matrix. Each cluster is then fully represented with the center, the covariance matrix and the number of samples, (μ_j, Σ_j, n_j).

The multivariate Gaussian density of the current sample $z(k)$ is further called typicality, and is defined as follows

$$\gamma(k) = e^{-\frac{1}{2}d^2(k)}$$

where

$$d^2(k) = (z(k) - \mu)^T \Sigma^{-1}(z(k) - \mu) \tag{6}$$

stands for the Mahalanobis distance.

2.1 *Recursive Adaptation of Cluster Parameters*

All the elements of each cluster should be in the framework of evolving algorithms calculated in recursive manner to be used in on-line procedure. manner. The typicality should be calculated first for each sample, and the sample should be, completely assigned to the cluster where the typicality, is maximal, or a new cluster should be formed.

2.1.1 Calculation of Typicalities

For each new sample $z(k)$, the typicalities to all existing clusters are calculated, at the beginning based on the Euclidean distance,

$$\gamma_i(k) = e^{-d_i^2(k)}, \tag{7}$$

with

$$d_i^2(k) = (z(k) - \mu_i)^T (z(k) - \mu_i), \tag{8}$$

and after some time, when the the number of samples in the set n_i is bigger than N_{max}, the typicality is calculated based on Mahalanobis distance

$$d_i^2(k) = \frac{1}{2}(z(k) - \mu_i)^T \Sigma_i^{-1}(z(k) - \mu_i) . \tag{9}$$

This is because of the singularity problems which arise with bad conditioning of covariance matrix. Using Mahalanobis distance the clusters become more flexible and have much higher approximative ability. N_{max} is user defined parameter.

2.1.2 Adapting Clusters with a New Sample

If the maximal typicality of the sample overcome the user defined threshold Γ_{max}, this means that this sample is assigned to that particular cluster. For example, if the sample $z(k)$ belongs to the ith cluster, then the number of samples in the cluster is incremented, the cluster mean and the covariance matrix of the cluster are adapted recursively, by taking into account the previous values as follows:

$$e_j(k) = z(k) - \mu_j^{n_j}. \tag{10}$$

Next, the mean is updated

$$\mu_j^{n_j+1} = \mu_j{}^{n^j} + \frac{1}{n_j+1} e_j(k). \tag{11}$$

After that, the states of the unnormalized covariance matrix are computed as:

$$S_j^{n_j+1} = S_j^{n_j} + e_j(k)\left(z(k) - \mu_j^{n_j+1}\right)^T \tag{12}$$

and the covariance matrix is then obtained as:

$$\Sigma_j^{n_j+1} = \frac{1}{n_j} S_j^{n_j+1}. \tag{13}$$

2.2 *Adding New Clusters*

If the typicalities to all existing clusters are less then the predefined maximal threshold, then a new cluster is formed and initialized with this sample, the current number of clusters c is incremented, the number of elements in the cluster is initialized to $n_c = 1$, and the center and the covariance matrix of the cluster are initialized to $\mu_c = z(k)$ and $\Sigma_c = \mathbf{0}$, respectively.

The complete evolving Gaussian clustering algorithm is shown in Algorithm 1.

2.3 *Merging Clusters*

In each evolving clustering algorithm, a merging mechanism is of essential meaning. This is the most important when the samples come to the learning algorithm randomly from different classes and are very dispersed, and the clusters are big and of different sizes. In such cases, the algorithm usually creates more clusters than needed. The basic evolving concept, without merging, results in clusters which have similar constraint volume. Therefore, at the beginning, the bigger cluster will be partitioned into more small clusters. After using a merging approach, some of them, these which are similar and close together, will be merged together. The number of clusters decreases, and the model structure is simplified which is important, because of the transparency.

Algorithm 1 Algorithm of Evolving Gaussian Clustering.

Choice of Γ_{max}, N_{max}
Initialization:
$c \leftarrow 1$, $\mu_c \leftarrow z(1)$, $k \leftarrow 1$
repeat $k \leftarrow k+1$
  for i = 1 : c
    if $n_i < N_{max}$
      $d_i^2(k)$ is the Euclidean distance
    else
      $d_i^2(k)$ is Mahalanobis distance
    end
    Calculation of typicality $\gamma_i(k) = e^{-d_i^2(k)}$
  end
  Choice of maximal typicality $j = arg\max_i \gamma_i(k)$
  if $\gamma_j(k) \geq \Gamma_{max}$
    Update of cluster j with new sample
    $e_j(k) \leftarrow z(k) - \mu_j$
    $\mu_j \leftarrow \mu_j + \frac{1}{n_j+1} e_j$
    $S_j = S_j + e_j(k)(z(k) - \mu_j)^T$
    $n_j \leftarrow n_j + 1$
  else
    Add and initialize new cluster
    $c \leftarrow c+1$
    $n_c \leftarrow 1$
    $\mu_c \leftarrow z(k)$
  end
  Merging clusters
until $k > N$

To start the merging mechanism, for every cluster pair (i, j) with a triplet (μ_i, Σ_i, n_i) and (μ_j, Σ_j, n_j), the estimated joined triplet $(\mu_{ij}, \Sigma_{ij}, n_{ij})$ is calculated. The number of data samples in the joined cluster and the joined cluster center are respectively equal to

$$n_{ij} = n_i + n_j, \tag{14}$$

and

$$\mu_{ij} = \frac{n_i \mu_i + n_j \mu_j}{n_{ij}}. \tag{15}$$

The matrix of data samples for the joint cluster is now $Z_{ij}^T = \left[Z_i^T \; Z_j^T \right]$. The calculation of the covariance matrix for the joint cluster is then written as follows

$$\Sigma_{ij} = \frac{1}{n_{ij} - 1} \left(Z_i^T Z_i + Z_j^T Z_j - M_{ij}^T E_{ij}^T E_{ij} M_{ij} \right). \tag{16}$$

where

$$Z_i^T Z_i = (n_i - 1)\Sigma_i + M_i^T E_i^T E_i M_i, \tag{17}$$

and

$$Z_j^T Z_j = (n_j - 1)\Sigma_j + M_j^T E_j^T E_j M_j. \tag{18}$$

This means that the covariance matrix of the joint cluster is calculated using (14, 15, 16, 17, and 18). The proposed calculation enables the exact calculation of the joint cluster covariance matrix without storing the old data sample. The only information needed is the triples of cluster parameters.

The compactness of two clusters (i, j) can be measured by calculating the ratio between the joined volume and the sum of both cluster volumes as follows

$$\kappa_{ij} = \frac{V_{ij}}{V_i + V_j}. \tag{19}$$

This ratio is calculated for all possible pairs of clusters that are activated with a typicality that is higher than the required activation threshold Γ_{ACT}.

These clusters with minimal ratio, which is defined as follows

$$\kappa_{i^* j^*} = \min_{i,j} \kappa_{ij}, \;\; i, j \;\in \mathscr{A}, \tag{20}$$

where

$$(i^*, j^*) = arg \min_{i,j} \kappa_{ij}, \tag{21}$$

are than merged together, if the ratio is less than the predefined joint threshold value κ_{join}. After merging the number of clusters is decreased, $c \leftarrow c - 1$, and the corresponding clusters triplets are updated.

The complete merging algorithm is described in Algorithm 2.

3 Evolving Clustering for Regression

Dealing with regression problems, the set of input samples is mapped into the output samples. In our case the approximation with Takagi-Sugeno model is used, where the approximation is given by overlapping linear models. In our approach the structure is changed and the parameters are estimated from the data stream. The structure change was already discussed, the estimation of model parameters θ_j can be obtained from clusters, i.e. clustering of input-output data space and singular value decomposition of the covariance matrices of the clusters lead to the model parameters in the case of regression problems. This will be discussed in more details.

Talking about regression problems the measured data vectors z should be treated as the input vector u, which could be of an arbitrary dimension, and the corresponding output y. This means that the problem can be seen as the influence of the input data vector u to the output of the process y. All needed information lies in the covariance matrices of the corresponding clusters formed in the input-output data space. In

Algorithm 2 Merging clusters.

Choice of κ_{join}
Initialization: $merge = 1$
while $merge == 1$
  Computation of $V_i = \frac{2\pi^{m/2}}{m\Gamma(m/2)} \Pi_{j=1}^{m} \lambda_j^i, \; i \in \mathcal{A}$,
  where λ_j^i stands for the jth eigenvalue of Σ_i
  For every $i, j \in \mathcal{A}, \; i \neq j$, compute
  $\Sigma_{ij} = \frac{1}{n_{ij}-1}\left(Z_i^T Z_i + Z_j^T Z_j - M_{ij}^T E_{ij}^T E_{ij} M_{ij}\right)$,
  the volume V_{ij}, and overlapping $\kappa_{ij} = \frac{V_{ij}}{V_i+V_j}$.
  Find minimal ratio $\kappa_{i^*j^*} = arg\,\min_{i,j} \kappa_{ij}$,
  if $\kappa_{i^*j^*} < \kappa_{join}$
    Merge clusters i^* and j^* in $(\Sigma_{new}, \mu_{new}, n_{new})$
    $c \leftarrow c - 1$
    $merge = 1$
  else
    $merge = 0$
  end
end

the case of regression problems the input-output data should lie along the hyper-surface which represent the input-output mapping. In the real data the disturbances, measurement noise, parasitic disturbances and other sources of errors are always present and therefore the data do not lie exactly on the surface, but close.

3.1 Estimation of the Local Model Parameters

The local linear model of the data is a hyper-plane in the input-output data space, and defined by the normal vector to the model hyper-plane. The normal vector is the eigenvector of the jth covariance matrix which fulfill 0.95 of the whole variance. Let's say this vector is defined as $p_{j,i}$. This normal vector is orthogonal to the model hyper-plane which goes through the cluster center μ_j. This tangential hyper-plane represents the local linear model and can be obtained in the implicit equation as follows

$$\left(z - \mu_j\right)^T p_{j,i} = 0 \tag{22}$$

In the case of regression problems, the regressors are usually very carefully chosen, meaning that they are linearly independent, and the excitation of the process is adequate. In this way, the rank of the current covariance matrix is $q - 1$, which means that only one eigenvalue of the matrix is close to zero, i.e. close to noise variance.

The nonlinear nature of the data results in a number of different clusters with different normal vectors and different centers. In the context of fuzzy approximators the observed point in the data space will be modelled by weighted linear combination of a number of hyper-planes. The weights in this case are given as the typicalities. This leads to the following estimation of the model output:

$$y_m(k) = \frac{\sum_{j=1}^{c} \gamma_j(k) y_{m,j}(k)}{\sum_{j=1}^{c} \gamma_j(k)} \tag{23}$$

where c stands for the number of clusters. The typicality is calculated with projection principle. The projection of the cluster to the model hyper-plane is called the membership function and is called $\mathcal{Z}_j$.

The fuzzy model in this case is defined as:

$$R_j : \ if \ u(k) \ is \ \mathcal{Z}_j \ then \ y_{m,j}(k) = \theta_j^T u(k) \tag{24}$$

In the example of using the proposed algorithm in regression problems, the data stream is formed as follows: first, the independent variable $z_1(k) = \mathcal{N}(0, 1)$ of $n = 400$ samples is generated. The value of the dependent variable is a nonlinear function, $z_2(k) = 0.4z_1^3(k) + \mathcal{N}(0, 0.2)$.

The algorithm is initialized with the following tuning parameters: the parameter $\kappa_{join} = 1.25$, the maximal number to use the Euclidean distance measure is $N_{max} = 20$, the maximal typicality is defined as $\gamma_{Max} = 0.85$, and the minimal number of samples in a cluster to remove it is defined as $N_{min} = 4$. The data samples in the final situation and final clusters with centers and 3σ contour are shown in Fig. 1. The removed clusters are shown only with centers of clusters.

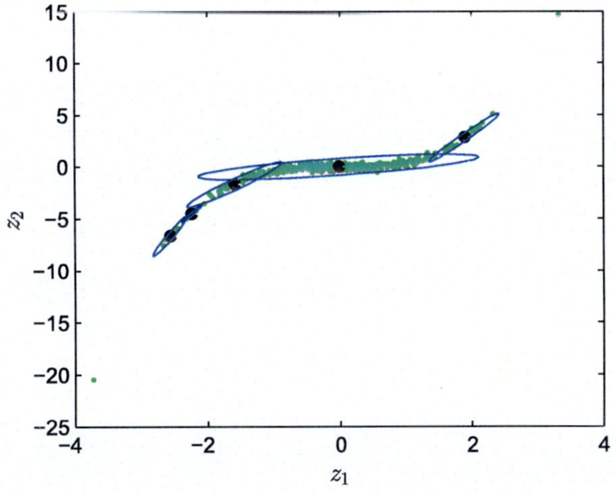

Fig. 1 The data and final clusters with centers and 3σ contour, and $\kappa_{join} = 1.25$

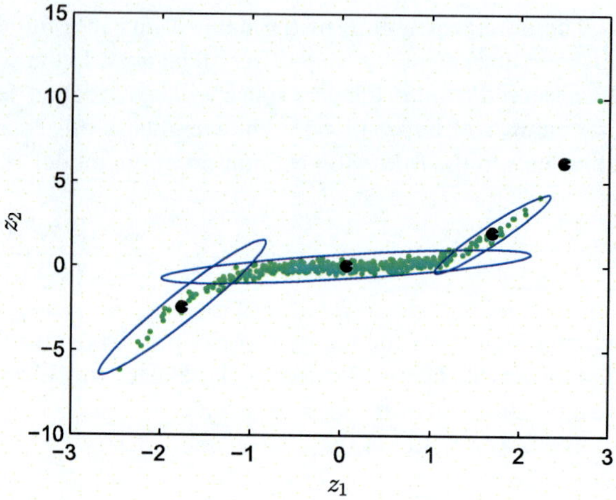

Fig. 2 The data and final clusters with centers, 3σ contour, and $\kappa_{join} = 1.8$

The merging parameter $\kappa_{join} = 1.8$ merge also the clusters which are more different and this results in fewer final clusters, which is shown in Fig. 2.

The final fuzzy model is than defined in Takagi-Sugeno form, for example, from results in Fig. 2 as follows:

$$R_1 : \; if \; z_1 \; is \; \mathcal{Z}_1 \;\; then \; z_2 = 4.47z_1 - 10.33 \tag{25}$$

$$R_2 : \; if \; z_1 \; is \; \mathcal{Z}_2 \;\; then \; z_2 = 0.41z_1 + 0.05 \tag{26}$$

$$R_3 : \; if \; z_1 \; is \; \mathcal{Z}_3 \;\; then \; z_2 = 3.76z_1 - 8.36 \tag{27}$$

In this example the variable z_1 acts as input u, and the variable z_2 as output y.

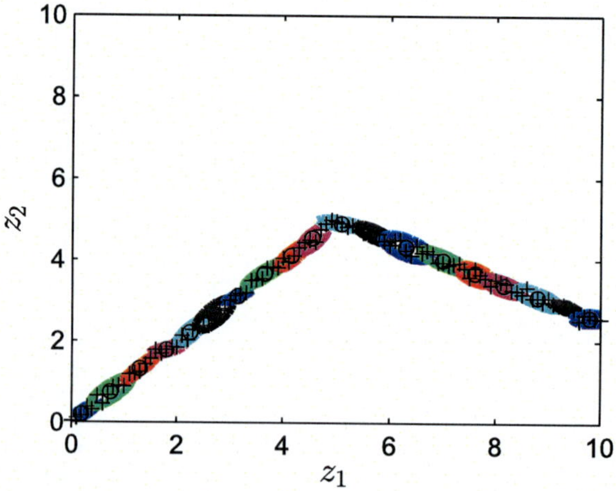

Fig. 3 Clusters with 3σ contour before merging, $\gamma_{Max} = 0.85$, and $\kappa_{join} = 1.25$

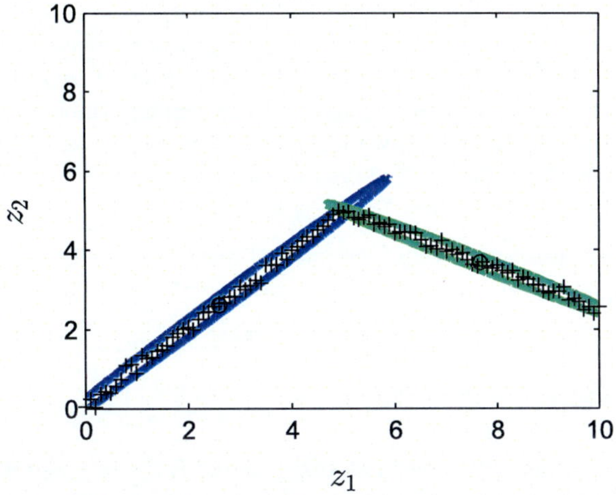

Fig. 4 Clusters with 3σ contour after merging, $\gamma_{Max} = 0.85$, and $\kappa_{join} = 1.25$

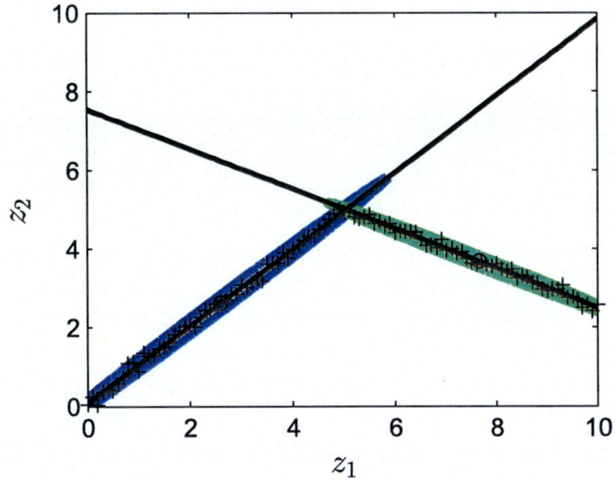

Fig. 5 Clusters with 3σ contour after merging and local linear models, $\gamma_{Max} = 0.85$, and $\kappa_{join} = 1.25$

In the second example the proposed algorithm was used to identify the parameters of piecewise linear function which is given as follows: first, the independent variable $z_1(k)$ goes from 0 to 10 with step 0.1 is generated. The dependent variable is piecewise linear and defined as $z_2(k) = z_1(k) + \mathcal{N}(0, 0.08)$, if $z_1(k) \leq 5$, and $z_2(k) = -0.5z_1(k) - 7.5 + \mathcal{N}(0, 0.08)$, if $z_1(k) \geq 5$.

The results of clustering are shown in Fig. 3, where the clusters are shown when no merging is used, and $\gamma_{Max} = 0.85$, and $\kappa_{join} = 1.25$, and $N_{max} = 20$.

After merging only two clusters are obtained as shown in Fig. 4.

In Fig. 5 the clusters and the local linear model are given.

4 Interval Fuzzy Model

The idea of interval fuzzy model is to find a lower and upper bound of the data set and describe these bounds by fuzzy model. The band or confidence interval should contain the prescribed number of samples. The lower and upper hyper-plane are parallel to the model hyper-plane. The distance is defined with the eigenvalue of the latent eigenvector of covariance matrix, or more exactly with $\sigma_{j,q}$ which is defined as square root, $\lambda_{j,q}^{\frac{1}{2}}$. The interval model of jth fuzzy model is than defined as

$$\left(z - \mu_j + \delta\sigma_{j,q} p_{j,q}\right)^T p_{j,q} = 0 \tag{28}$$

where δ stands for the number of multiplication for the interval, i.e. it could ba defined as 1σ, 2σ or 3σ.

The resulting interval fuzzy model is than defined by two different fuzzy models. The lower fuzzy model is defined as: The fuzzy model in this case is defined as:

$$R_j : if\ u(k)\ is\ \mathcal{Z}_j\ then\ \underline{y}_{m,j} = \underline{\theta}_j^T u(k) \tag{29}$$

and the upper fuzzy model as:

$$R_j : if\ u(k)\ is\ \mathcal{Z}_j\ then\ \overline{y}_{m,j} = \overline{\theta}_j^T u(k) \tag{30}$$

This means that each measurement of the output variable y lies in the interval $\underline{y}(k) \leq y(k) \leq \overline{y},\ k = 1, \ldots, N$.

The example of interval fuzzy model is given as follows: first, the independent variable $z_1(k)$ goes from 0 to 10 with step 0.05 is generated. The dependent variable is defined as $z_2(k) = \exp(-z_1(k)/10)\sin(0.5z_1(k)) + \mathcal{N}(0, 0.05)$, if $z_1(k)$.

In Fig. 6 the data and the clusters with 3σ contour after merging and the centers are shown in the case where $\gamma_{Max} = 0.85$, and $\kappa_{join} = 1.02$.

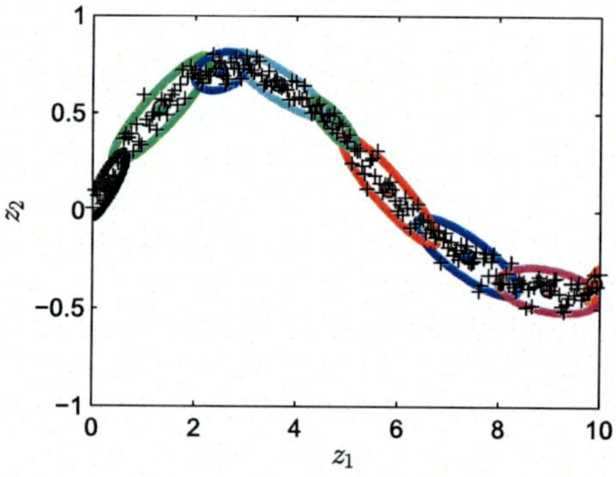

Fig. 6 Clusters with 3σ contour after merging, $\gamma_{Max} = 0.85$, and $\kappa_{join} = 1.02$

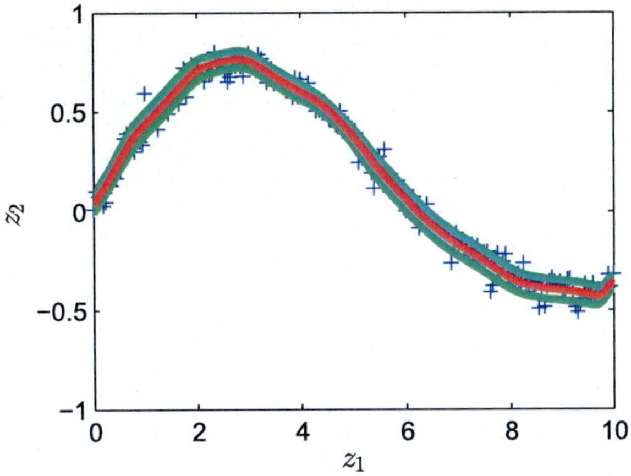

Fig. 7 Confidence interval with 1σ, $\gamma_{Max} = 0.85$, and $\kappa_{join} = 1.02$

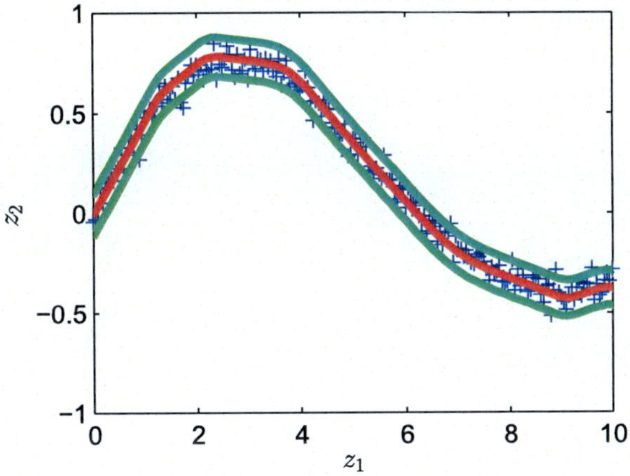

Fig. 8 Confidence interval with 2σ, $\gamma_{Max} = 0.85$, and $\kappa_{join} = 1.02$

The lower and the upper bound, and the output model are shown in Fig. 7, where for confidence interval the parameter δ was chosen as 1, i.e. 1σ confidence interval.

The same as in Fig. 7 was also shown in Fig. 8, but here 2σ confidence interval is shown. This means that around 95 percent of samples lies inside the confidence interval.

In Fig. 9 the data and the clusters with 3σ contour after merging and the centers are shown in the case where $\gamma_{Max} = 0.85$, and $\kappa_{join} = 1.2$. By choosing bigger κ_{join}, as previously, less clusters are obtained.

The lower and the upper bound, and the output model are shown in Fig. 10, where 3σ confidence interval is defined. This means that in general only 2% of sample can lie outside the confidence interval.

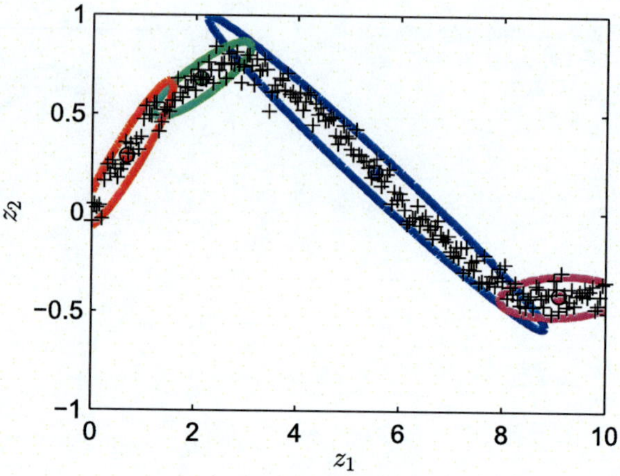

Fig. 9 Clusters with 3σ contour after merging, $\gamma_{Max} = 0.85$, and $\kappa_{join} = 1.2$

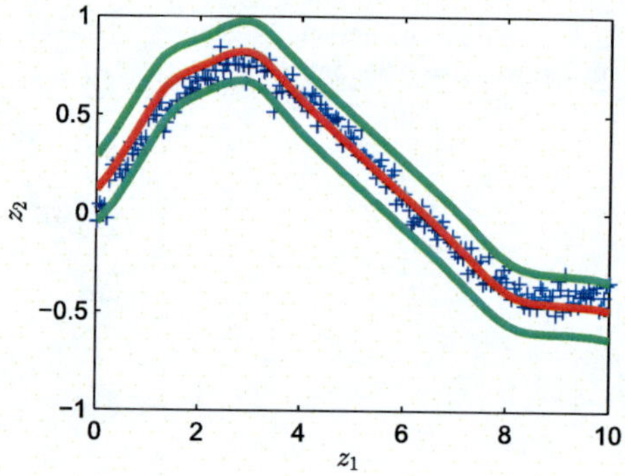

Fig. 10 Confidence interval with 3σ, $\gamma_{Max} = 0.85$, and $\kappa_{join} = 1.2$

5 Conclusion

A new evolving clustering eGauss+ approach can be very usefully applied for regression problems. The evolving algorithm is proposed for achieving more efficient data stream clustering. It is basically based on incremental evolving principal together with dynamic cluster merging which is based on the cluster volume and the cluster covariance matrix. The approach is very much suitable for regression problems, where the fuzzy models are used. Here the extension from, fuzzy model was done to the interval fuzzy models which can be very useful in many different real problems when dealing with uncertain processes which generate the streaming data.

Acknowledgements This work has been supported by the Slovenian Research Agency with the Research Program P2-0219.

References

1. P.P. Angelov, X. Zhou, Evolving fuzzy-rule-based classifiers from data streams. IEEE Trans. Fuzzy Syst. **16**(6), 1462–1475 (2008)
2. P.P. Angelov, E. Lughofer, X. Zhou, Evolving fuzzy classifiers using different model architectures. Fuzzy Sets Syst. **159**(23), 3160–3182 (2008)
3. Y. Wang, L. Chen, J.P. Mei, Incremental fuzzy clustering with multiple medoids for large data. IEEE Trans., Fuzzy Syst. **22**(6), 1557–1568 (2014)
4. T.C. Havens, J. Bezdek, C. Leckie, L. Hall, M. Palaniswami, Fuzzy c-means algorithms for very large data. IEEE Trans. Fuzzy Syst. **20**(6), 1130–1146 (2012)
5. E. Lughofer, FLEXFIS: a robust incremental learning approach for evolving Takagi-Sugeno fuzzy models. IEEE Trans. Fuzzy Syst. **16**(6), 1393–1410 (2008)
6. L. Hall, D.B. Goldof, Convergence of the single-pass and online fuzzy C-means algorithms. IEEE Trans., Fuzzy Syst. **19**(4), 792–794 (2011)
7. P. Angelov, An approach for fuzzy rule-base adaptation using on-line clustering. Int. J. Approx. Reason. **35**(3), 275–289 (2004)
8. P. Angelov, D. Filev, N. Kasabov, *Evolving Intelligent Systems-Methodology and Applications* (Wiley, New York, 2010)
9. D. Dovžan, I. Škrjanc, Recursive clustering based on a Gustafson-Kessel algorithm. Evol. Syst. J. **2**, 15–24 (2011)
10. D. Filev, O. Georgieva, An extended version of the Gustafson-Kessel algorithm for evolving data stream clustering, in *Evolving Intelligent Systems: Methodology and Applications*, ed. by P. Angelov, D. Filev, A. Kasabov (John Willey and Sons, IEEE Press Series on Computational Intellegence, 2010), pp. 273–300
11. R. Krishnapuram, J.M. Keller, Possibilistic approach to clustering. IEEE Trans. Fuzzy Syst. **1**(2), 98–100 (1993)
12. B. Ojeda-Magana, R. Ruelas, M.A. Corona-Nakamura, D. Andina, An improvement to the possibilistic fuzzy c-means clustering algorithm. Intell. Autom. Soft Comput. **20**(1), 585–592 (2006)
13. N.R. Pal, K. Pal, J.M. Keller, J.C. Bezdek, A possibilistic fuzzy c-means clustering algorithm. IEEE Trans. Fuzzy Syst. **13**(4), 517–530 (2005)
14. H. Timm, C. Borgelt, C. Doering, R. Kruse, An extension to possibilistic fuzzy cluster analysis. Fuzzy Sets Syst. **147**(1), 3–16 (2004)
15. R.J. Hathaway, Y. Hu, Density-weighted fuzzy c-means clustering. IEEE Trans. Fuzzy Syst. **17**(1), 243–252 (2009)
16. P.P. Angelov, R. Yager, Simplified fuzzy rule-based systems using non-parametric antecedents and relative data density, pp. 62–69 (2011)
17. I. Škrjanc, S. Ozawa, T. Ban, D. Dovžan, Large-scale cyber attacks monitoring using evolving cauchy possibilistic clustering. Appl. Soft Comput. **62**, 2833–2839 (2017)
18. I. Škrjanc, S. Blažič, E. Lughofer, D. Dovžan, Inner matrix norms in evolving cauchy possibilistic clustering for classification and regression from data streams. Inf. Sci. **478**, 2018 (2018)
19. G. Klančar, I. Škrjanc, Evolving principal component clustering with a low run-time complexity for LRF data mapping. Appl. Soft Comput. **35**, 349–358 (2015)
20. E. Lughofer, M. Pratama, I. Škrjanc, Incremental rule splitting in generalized evolving fuzzy systems for autonomous drift compensation. IEEE Trans. Fuzzy Syst. **26**(4), 1854–1865 (2018)
21. I. Škrjanc, Cluster-volume-based merging approach for incrementally evolving fuzzy Gaussian clustering—eGAUSS+. IEEE Trans. Fuzzy Syst. 1–11 (2019). ISSN 1063-6706. [Print ed.]
22. I. Škrjanc, J.A. Iglesias, A. Sanchis, D. Leite, E. Lughofer, F. Gomide, Evolving fuzzy and neuro-fuzzy approaches in clustering, regression, identification, and classification : a survey. Inf. Sci. (2019). ISSN 0020-0255. [Print ed.] In Press
23. E. Lughofer, M. Sayed-Mouchaweh, Autonomous data stream clustering implementing incremental split-and-merge techniques—towards a plug-and-play approach. Info. Sci. **204**, 54–79 (2015)

24. U. Kaymak, M. Setnes, Fuzzy clustering with volume prototypes and adaptive cluster merging. IEEE Trans. Fuzzy Syst. **10**(6), 705–712 (2002)
25. U. Kaymak, R. Babuska, Compatible cluster merging for fuzzy modelling, in *Proceedings of 1995 IEEE International Conference on Fuzzy Systems*, vol. 2 (Yokohama, Japan, 1995), pp. 897–904
26. L. Hartert, M. Sayed-Mouchaweh, P. Billaudel, A semi-supervised dynamic version of fuzzy 1172 k-nearest neighbours to monitor evolving systems. Evol. Syst. **1**, 3–15 (2010)
27. J. Beringer, E. Huellermeier, Online clustering of parallel data streams. Data Knowl. Eng. **58**(2), 180–204 (2007)
28. N. Kasabov, Evolving fuzzy neural networks for supervised/unsupervised online knowledge-based learning. IEEE Trans. Syst. Man Cybern.—Part B **31**(6), 902–918 (2001)
29. E. Lughofer, C. Cernuda, S. Kindermann, M. Pratama, Generalized smart evolving fuzzy systems. Evol. Syst. **6**(4), 269–292 (2015)
30. D. Dovžan, V. Logar, I. Škrjanc, Implementation of an evolving fuzzy model (eFuMo) in a monitoring system for a waste water treatment process. IEEE Trans. Fuzzy Syst. **23**(5), 1761–1776 (2015)
31. E. Lughofer, A dynamic split-and-merge approach for evolving cluster models. Evol. Syst. **3**(3), 135–151 (2012)
32. L. Teslić, B. Hartmann, O. Nelles, I. Škrjanc, Nonlinear system identification by Gustafson-Kessel fuzzy clustering and supervised local model network learning for the drug absorption spectra process. IEEE Trans. Neural Netw. **22**(12), 1941–1951 (2011)
33. I. Škrjanc, Evolving fuzzy-model-based design of experiments with supervised hierarchical clustering. IEEE Trans. Fuzzy Syst. **23**(4), 861–871 (2014)
34. I. Škrjanc, S. Blažič, O. Agamennoni, Identification of dynamical systems with a robust interval fuzzy model. Automatica **41**, 327–332 (2005). ISSN 0005-1098. [Print ed.]
35. I. Škrjanc, S. Blažič, O. Agamennoni, Interval fuzzy model identification using l_∞-norm. IEEE Trans. Fuzzy Syst. **13**(5), 561–568 (2005). ISSN 1063-6706. [Print ed.]
36. I. Škrjanc, S. Blažič, O. Agamennoni, Interval fuzzy modeling applied to Wiener models with uncertainties. IEEE Trans. Syst., Man, Cybernet ICS. Part B, Cybern. **35**(5), 1092–1095. ISSN 1083-4419. [Print ed.]
37. S. Oblak, I. Škrjanc, S. Blažič, Fault detection for nonlinear systems with uncertain parameters based on the interval fuzzy model. Eng. Appl. Artif. Intell. **20**(4), 503–510. ISSN 0952-1976. [Print ed.]
38. I. Škrjanc, Confidence interval of fuzzy models : an example using a waste-water treatment plant. Chemom. Intell. Lab. Syst. **96**(2), 182–187. ISSN 0169-7439. [Print ed.]
39. I. Škrjanc, Fuzzy confidence interval for pH titration curve. Appl. Math. Model. **35**(8), 4083–4090. ISSN 0307-904X. [Print ed.]

Construction of T-Vague Groups for Real-Valued Interference Channel

Cibele Cristina Trinca, Ricardo Augusto Watanabe, and Estevão Esmi

Abstract Interference is an obstacle to communication in wireless communication networks. One of the most employed methods to efficiently reduce interference and enhance capacity of a wireless network is interference alignment. A necessary step for the interference alignment methods is the quantization of the channel coefficients. In the literature, the use of fuzzy logic in coding theory is a flourishing research topic. The present contribution intends to present a classic construction to quantize real-valued channel coefficients and include such a theory in the framework of fuzzy theory, more precisely, in the context of T-fuzzy subgroups, T-indistinguishability operator and T-vague groups.

1 Introduction

Interference is an obstacle to communication in wireless communication networks. One of the most employed methods to efficiently reduce interference and enhance capacity of a wireless network is interference alignment [7, 13]. A necessary step for interference alignment methods is the quantization of the channel coefficients [7, 14, 15].

After the introduction of fuzzy subgroups by A. Rosenfeld [10], one of the pioneering works in the use of fuzzy theory in coding theory is [12]. Recently, the authors of [1] proposed a RSA cryptosystem scheme where the representative inte-

C. Cristina Trinca (✉)
Departamento de Engenharia de Bioprocessos e Biotecnologia, Universidade Federal do Tocantins (UFT), Palmas, Tocantins 77402-970, Brazil
e-mail: cibtrinca@yahoo.com.br

R. Augusto Watanabe · E. Esmi
Instituto de Matemática, Estatística e de Computação Científica (IMECC), Universidade Estadual de Campinas (UNICAMP), Campinas, São Paulo 13083-872, Brazil
e-mail: ricardoaw18@gmail.com

E. Esmi
e-mail: eelaureano@gmail.com

B. Bede et al. (eds.), *Fuzzy Information Processing 2020*,
Advances in Intelligent Systems and Computing 1337,
https://doi.org/10.1007/978-3-030-81561-5_6

ger is converted into a triangular fuzzy number. In [11] is introduced the evaluation of RSA cryptosystem by using adaptive neural fuzzy inference system techniques and a representation of the security level of RSA algorithm by using a hybrid technique of artificial neural network and fuzzy logic.

The presented authors also have proposed a quantization method for real and complex-valued channel coefficients [16, 17] to realize interference alignment, in both cases by using support functions based on Steiner points.

The present contribution intends to present a classic construction in order to quantize real-valued channel coefficients in Sect. 2 and, in Sect. 3, explore the basic results and include such a theory in the framework of fuzzy theory, more precisely, in the context of T-fuzzy subgroups, T-indistinguishability operator and T-vague groups.

2 Interference Alignment onto a Real Ideal Lattice

Lattices have been very useful in applications in communication theory. In this section we use real ideal lattices in order to realize interference alignment onto these lattices. For formalizing this section we make use of the references [6, 15].

Definition 1 Let $v_1, v_2, \ldots, v_m$ be a set of linearly independent vectors in $\mathbb{R}^n$ such that $m \leq n$. The set of the points $\Lambda = \left\{x = \sum_{i=1}^{m} \lambda_i v_i, \text{ where } \lambda_i \in \mathbb{Z}\right\}$ is called a *lattice* of rank m and $\{v_1, v_2, \ldots, v_m\}$ is called a basis of the lattice.

Lattices have only two principal structural characteristics [7]. Algebraically, a lattice is a group. Geometrically, a lattice is endowed with the properties of the space in which it is embedded.

The definition of an ideal of a commutative ring and a principal ideal can be found in [14].

The theory of *ideal lattices* gives a general framework for algebraic lattice constructions. We recall this notion in the case of totally real algebraic number fields.

Definition 2 [14] Let K and O_K be a totally real number field of degree n and the corresponding ring of integers of K, respectively. An **ideal lattice** is a lattice $\Lambda = (I, q_\alpha)$, where I is an ideal of O_K and

$$q_\alpha : I \times I \to \mathbb{Z}, \text{ where } q_\alpha(x, y) = Tr_{K/\mathbb{Q}}(\alpha x y), \ \forall x, y \in I, \tag{1}$$

where $\alpha \in K$ is totally positive (i.e., $\sigma_i(\alpha) > 0$, for all i).

If $\{w_1, w_2, \ldots, w_n\}$ is a $\mathbb{Z}$-basis of I, the generator matrix R of an **ideal lattice** $\sigma(I) = \Lambda = \{x = R\lambda \mid \lambda \in \mathbb{Z}^n\}$ is given by

$$R = \begin{pmatrix} \sqrt{\alpha_1}\sigma_1(w_1) & \cdots & \sqrt{\alpha_1}\sigma_1(w_n) \\ \vdots & \ddots & \vdots \\ \sqrt{\alpha_n}\sigma_n(w_1) & \cdots & \sqrt{\alpha_n}\sigma_n(w_n) \end{pmatrix}, \tag{2}$$

where $\alpha_i = \sigma_i(\alpha), i = 1, \ldots, n$. One easily verifies that the Gram matrix $R^t R$ coincides with the trace form $(Tr(\alpha w_i w_j))_{i,j=1}^n$, where t denotes the transpose of the matrix R. For the $\mathbb{Z}^n$-lattice, the corresponding lattice generator matrix given in (2) becomes an orthogonal matrix ($R^{-1} = R^t$) and we talk about "rotated" $\mathbb{Z}^n$-lattices.

In order to realize interference alignment onto an *ideal lattice*, we need to quantize the channel coefficients h_{ml}. Hence, in this section, we describe a way to find a doubly infinite nested *ideal lattice* partition chain for any dimension $n = 2^{(r-2)}$, with $r \geq 3$, in order to quantize the channel coefficients.

2.1 Quantization of Real-Valued Channels onto a Lattice

In a wireless network, a transmission from a single node is heard not only by the intended receiver, but also by all other nearby nodes and, consequently, the resulting interference is usually viewed as highly undesirable.

Each node, indexed by $m = 1, 2, \ldots, M$, observes a noisy linear combination of the transmitted signals through the channel

$$y_m = \sum_{l=1}^{L} h_{ml} x_l + z_m, \tag{3}$$

where $h_{ml} \in \mathbb{R}$ are real-valued channel coefficients, x_l is a real lattice point whose message space presents a uniform distribution and z_m is an i.i.d. circularly symmetric real Gaussian noise [14].

Consider the following Galois extensions, where $r \geq 3$, ξ_{2^r} denotes the 2^r-th root of unity and $\mathbb{K} = \mathbb{Q}(\xi_{2^r} + \xi_{2^r}^{-1})$ is the totally real number field of the binary cyclotomic field $\mathbb{Q}(\xi_{2^r})$ [2, 14]:

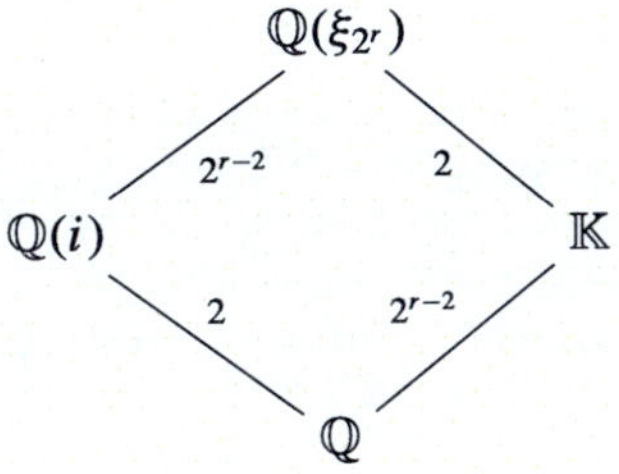

(4)

In [2], by Theorem 10, we have that $\mathbb{Z}[\xi_{2^r} + \xi_{2^r}^{-1}]$ is the ring of integers of $\mathbb{Q}(\xi_{2^r} + \xi_{2^r}^{-1})$ and

$$\{1, \xi_{2^r} + \xi_{2^r}^{-1}, \ldots, \xi_{2^r}^{n-1} + \xi_{2^r}^{-(n-1)}\} \tag{5}$$

is an integral basis of $\mathbb{Z}[\xi_{2^r} + \xi_{2^r}^{-1}]$.

Let $Gal(\mathbb{Q}(\xi_{2^r}+\xi_{2^r}^{-1})/\mathbb{Q}) = \{\sigma_1, \ldots, \sigma_n\}$ be the Galois group of $\mathbb{Q}(\xi_{2^r}+\xi_{2^r}^{-1})$ over $\mathbb{Q}$ and $\mathfrak{I} = \langle 2+\xi_{2^r}+\xi_{2^r}^{-1}\rangle = (2+\xi_{2^r}+\xi_{2^r}^{-1})\mathbb{Z}[\xi_{2^r}+\xi_{2^r}^{-1}]$ be a principal ideal of $\mathbb{Z}[\xi_{2^r}+\xi_{2^r}^{-1}]$ with norm equal to 2.

By [2], page 7, the generator matrix of the rotated $\mathbb{Z}^n$-lattice is given by

$$M_0 = \frac{1}{\sqrt{2^{r-1}}} AM^t T \text{ and } M_0^t M_0 = I, \tag{6}$$

where t denotes the transposition, I is the $n \times n$ identity matrix, M is given by

$$\begin{pmatrix} \sigma_1(1) & \sigma_2(1) & \cdots & \sigma_n(1) \\ \sigma_1(\xi_{2^r}+\xi_{2^r}^{-1}) & \sigma_2(\xi_{2^r}+\xi_{2^r}^{-1}) & \cdots & \sigma_n(\xi_{2^r}+\xi_{2^r}^{-1}) \\ \vdots & \vdots & \ddots & \vdots \\ \sigma_1(\xi_{2^r}^{n-1}+\xi_{2^r}^{-(n-1)}) & \sigma_2(\xi_{2^r}^{n-1}+\xi_{2^r}^{-(n-1)}) & \cdots & \sigma_n(\xi_{2^r}^{n-1}+\xi_{2^r}^{-(n-1)}) \end{pmatrix}, \tag{7}$$

$$A = diag\left(\sqrt{\sigma_i(2-\xi_{2^r}+\xi_{2^r}^{-1})}\right)_{i=1}^{n}, \tag{8}$$

and

$$T = \begin{pmatrix} -1 & -1 & \cdots & -1 & -1 \\ -1 & -1 & \cdots & -1 & 0 \\ \vdots & \vdots & \ddots & \vdots & \vdots \\ -1 & 0 & \cdots & 0 & 0 \end{pmatrix}. \tag{9}$$

At the receiver, we suppose that we apply M_0 to the received vector (3) to obtain

$$\bar{y}_m = M_0 y_m = \sum_{l=1}^{L} h_{ml} M_0 x_l + M_0 z_m. \tag{10}$$

Since $\mathfrak{I} = \langle 2+\xi_{2^r}+\xi_{2^r}^{-1}\rangle = (2+\xi_{2^r}+\xi_{2^r}^{-1})\mathbb{Z}[\xi_{2^r}+\xi_{2^r}^{-1}]$ is a principal ideal of $\mathbb{Z}[\xi_{2^r}+\xi_{2^r}^{-1}]$, we have that $\mathfrak{I}^k$, where $k \in \mathbb{Z}$, is an ideal of $\mathbb{Z}[\xi_{2^r}+\xi_{2^r}^{-1}]$ generated by $(2+\xi_{2^r}+\xi_{2^r}^{-1})^k$, that is, $\mathfrak{I}^k = (2+\xi_{2^r}+\xi_{2^r}^{-1})^k\mathbb{Z}[\xi_{2^r}+\xi_{2^r}^{-1}]$. Besides, since M_0 and AM^tT generate the same lattice, the $\mathbb{Z}^n$-lattice, then the real-valued channel h_{ml} can be approximated by

$$A'_k = \begin{pmatrix} (2+\theta)^k & 0 & \cdots & 0 \\ 0 & \sigma_2((2+\theta)^k) & \cdots & 0 \\ \vdots & \vdots & \ddots & \vdots \\ 0 & 0 & \cdots & \sigma_n((2+\theta)^k) \end{pmatrix}, \tag{11}$$

where $\theta = \xi_{2^r}+\xi_{2^r}^{-1}$.

Consequently the real-valued channel h_{ml} is quantized by the diagonal matrix A'_k whose elements are components of the canonical embedding of the power (positive or negative) of an element of $\mathbb{Z}[\xi_{2^r}+\xi_{2^r}^{-1}]$ with absolute algebraic norm equal to 2.

Now in the following section we present a method that describes for any dimension $n = 2^{r-2}$, with $r \geq 3$, a doubly infinite nested ideal lattice partition chain in order to quantize real-valued channels onto an ideal lattice to realize interference alignment onto an ideal lattice.

2.2 *Construction of Real Nested Ideal Lattices from the Channel Quantization*

Consider the following equation ($k \in \mathbb{Z}$):

$$A'_k(AM^tT) = (AM^tT)M_{(2+\xi_{2^r}+\xi_{2^r}^{-1})^k}. \tag{12}$$

Algorithm 1 in [6] calculates in an equivalent way Eq. (12). Such an algorithm furnishes us the generator matrix $M_k = M_{(2+\xi_{2^r}+\xi_{2^r}^{-1})^k}$ of the *real ideal lattice* Λ_k related to k and the corresponding Gram matrix G_k of such a lattice.

In this algorithm, for each r and k, we find the generator and Gram matrices of a real ideal lattice which is isomorphic to the canonical embedding of the ideal $\mathfrak{I}^k$, where we know that $\mathfrak{I}^k = (2 + \xi_{2^r} + \xi_{2^r}^{-1})^k\mathbb{Z}[\xi_{2^r} + \xi_{2^r}^{-1}]$ is an ideal of $\mathbb{Z}[\xi_{2^r} + \xi_{2^r}^{-1}]$ generated by $(2 + \xi_{2^r} + \xi_{2^r}^{-1})^k$. The position (as sublattice of the $\mathbb{Z}^n$-lattice in the nested ideal lattice partition chain) of this *real ideal lattice* Λ_k compared to the $\mathbb{Z}^n$-lattice in the nested ideal lattice partition chain related to r is exactly k.

Let $M_{(2+\xi_{2^r}+\xi_{2^r}^{-1})}$ represent the generator matrix of the real ideal lattice related to the position $k = 1$. Hence the following theorem gives us the extension by periodicity of the nested ideal lattice partition chain for the positive positions, that is, $k \geq 1$.

Theorem 1 *[6] For $k = n\beta + j$, where $\beta \in \mathbb{N}$ and $0 \leq j \leq n-1$, we have that $M_{(2+\xi_{2^r}+\xi_{2^r}^{-1})^{(n\beta+j)}} = (M_{(2+\xi_{2^r}+\xi_{2^r}^{-1})})^{k=(n\beta+j)}$ is a generator matrix of the real ideal lattice $2^\beta\Lambda_j$ seen as a $\mathbb{Z}$-lattice, where Λ_j is the real ideal lattice featured by Algorithm 1 related to the position k compared to the $\mathbb{Z}^n$-lattice.*

Thus, by Theorem 1, we can conclude that the periodicity of the nested ideal lattice partition chain for the positive positions is equal to $k = n$ because $\sigma(\mathfrak{I}^n) = 2\mathbb{Z}^n$, that is, $\sigma(\mathfrak{I}^n)$ is a scaled version of the $\mathbb{Z}^n$-lattice.

Therefore, we have the interference alignment onto a real ideal lattice for $k \geq 0$. Now the following theorem furnishes us the extension by periodicity of the nested ideal lattice partition chain for the negative positions, that is, $k \leq -1$.

Theorem 2 *[6] For all $k \in \mathbb{N}^*$, we have $\sigma(\mathfrak{I}^{-k}) = \sigma(\mathfrak{I}^k)^*$, where $\sigma(\mathfrak{I}^k)^*$ indicates the dual lattice of $\sigma(\mathfrak{I}^k)$.*

By using Theorems 1 and 2 we can conclude that we have n ($k = 0, 1, 2, \ldots, n-1$) different real ideal lattices in the doubly infinite nested ideal lattice partition chain.

Thus, for the real case, we have a generalization to obtain a doubly infinite nested ideal lattice partition chain in order to quantize the channel coefficients in order to realize interference alignment onto a real ideal lattice.

The corresponding doubly infinite nested ideal lattice partition chain is given as it follows:

$$\cdots \supset (2\mathbb{Z}^n)^* \supset (\Lambda_{n-1})^* \supset \cdots \supset (\Lambda_1)^* \supset$$

$$\supset \Lambda_0 = \mathbb{Z}^n \supset \Lambda_1 \supset \cdots \supset \Lambda_{n-1} \supset 2\mathbb{Z}^n \supset \cdots \qquad (13)$$

Besides, consequently, we have constructed nested lattice codes [7] (nested coset codes) with $2\mathbb{Z}^n$ being the corresponding sublattice.

3 T-Fuzzy Subgroups, T-Indistinguishability Operator and T-Vague Groups Based on Algebraic Lattices

The goal of this section is to present and formalize the basic concepts of T-fuzzy subgroups, T-indistinguishability operators and T-vague groups as well as connect such a theory with the *ideal lattices* from the channel quantization presented in Sect. 2.2.

Through this section a **t-norm** is a function $T : [0, 1] \times [0, 1] \rightarrow [0, 1]$ that satisfies the following properties for all $x, y, z \in [0, 1]$: 1) (commutativity) $T(x, y) = T(y, x)$; 2) (associativity) $T(x, T(y, z)) = T(T(x, y), z)$; 3)(monotocity) $T(x, y) \leq T(x, z)$, if $y \leq z$; 4) (identity) $T(x, 1) = x$. Through this section T stands as a reference for a t-norm.

Let $(G, \circ)$ be a group with a binary operation and identity element denoted by e_G.

3.1 T-Fuzzy Subgroups

Definition 3 [5, 9] Let $(G, \circ)$, μ and T be a group, a fuzzy subset of G and a t-norm, respectively. We say that μ is a T - **fuzzy subgroup** of G if and only if:

$$\mu(x \circ y^{-1}) \geq T(\mu(x), \mu(y)), \text{ for all } x, y \in G. \qquad (14)$$

Proposition 1 *[5, 9] Let $(G, \circ)$ be a group and μ a fuzzy subset of G such that $\mu(e_G) = 1$. Then μ is a T-fuzzy subgroup of G if and only if $\forall x, y \in G$, the following properties hold:*

1) $\mu(x \circ y) \geq T(\mu(x), \mu(y))$;
2) $\mu(x^{-1}) \geq \mu(x)$.

Additionaly, if the corresponding t-norm is the minimum t-norm ($T = \wedge$), we denote it only by **fuzzy subgroup**. The **core** of a fuzzy subset of a set G is defined as $core_\mu(G) := \{a \in G | \mu(a) = 1\}$.

Proposition 2 *[5, 9] Let $(G, \circ)$ be a group and μ a fuzzy subset of G such that $\mu(e_G) = 1$. Then the core $core_{e_\mu}(G)$ of μ is a subgroup of G.*

A **normal fuzzy subgroup** of $(G, \circ)$ is defined as the fuzzy subgroup μ satisfying

$$\mu(x \circ y) = \mu(y \circ x), \text{ for all } x, y \in G. \tag{15}$$

Proposition 3 *[9] The core of a normal T-fuzzy subgroup μ of $(G, \circ)$ is a normal subgroup of G.*

(Adapted from Example 10.4.3 of [8]) Let $r \in \mathbb{N}$ fixed and m be a positive integer, G be an additive abelian group and also define $nG = \{nx \mid x \in G\}$, where n is a positive integer. Define μ for all $x \in G$ as it follows:

$$\mu(x) = \vee\{1 - r^{-j} \mid j \in \mathbb{N} \text{ and } x \in m^j G\}. \tag{16}$$

One possible interpretation is that μ measures the membership grade of x in μ by the degree to which x is divisible by m. The higher the power of m dividing x, the greater the degree of the membership of x.

Consider $m = 2$ and the ideal lattices Λ_k constructed in Sect. 2.2 viewed as additive Abelian groups. Therefore we obtain the following:

Theorem 3 *μ_k as in equation (16) is a normal $\wedge$-fuzzy subgroup of $G = \Lambda_k$.*

Proof Consider

$$\mu_k(x) = \vee\{1 - r^{-j} \mid j \in \mathbb{N} \text{ and } x \in 2^j \Lambda_k\}. \tag{17}$$

Clearly $\mu_k(0) = 1$. Suppose that $x \in m^p G$ and $y \in m^q G$, then $(x + y) \in m^{p \wedge q} G$. Since $m^{p \wedge q} G$ is an additive abelian group,

$$\mu_k(x) \wedge \mu_k(-y) \leq \mu_k(x + (-y)) \Longleftrightarrow \mu_k(x) \wedge \mu_k(y) \leq \mu_k(x + (-y)). \tag{18}$$

Furthermore, $\mu_k(x + y) \geq 1 - r^{p \wedge q} = (1 - r^{-p}) \wedge (1 - r^{-q})$. Hence

$$\mu_k(x + y) \geq \vee\{(1 - r^{-p}) \wedge (1 - r^{-q}) \mid p, q \in \mathbb{N}\} = \mu_k(x) \wedge \mu_k(y). \tag{19}$$

Since $m^j G$ is a group, for all positive integers m and $k \in \mathbb{N}$, it follows that $x \in m^j G$ if, and only if, $x^{-1} \in m^j G$. Thus it follows that $\mu_k(x) = \mu_k(x^{-1})$.

As Λ_k is an abelian group, then μ_k is a normal $\wedge$-fuzzy subgroup of $G = \Lambda_k$. □

Theorem 1 implies that the periodicity of the nested lattice partition chain for the positive positions ($k \in \mathbb{N}$) is equal to $k = n$. Consequently, we have n different fuzzy subgroups μ_k and each one of them is related to the following nested lattices

$$\Lambda_k \supset 2\Lambda_k \supset 4\Lambda_k \supset 8\Lambda_k \supset \cdots \tag{20}$$

Let $k_1, k_2 \in \{0, 1, 2, \ldots, n-1\}$ such that, without loss of generality, $k_1 < k_2$. Let $x \in \Lambda_{k_2}$. Since $\Lambda_{k_2} \subset \Lambda_{k_1}$, then $2^j\Lambda_{k_2} \subset 2^j\Lambda_{k_1}$. Consequently, $x \in 2^j\Lambda_{k_1}$. Hence $\mu_{k_2}(x) = \mu_{k_1}(x)$ and, with abuse of notation, $\mu_{k_1}(\Lambda_{k_1}) \supset \mu_{k_2}(\Lambda_{k_2})$. Therefore we have showed the following:

Theorem 4 *Let $\Lambda_0, \Lambda_1, \Lambda_2, \ldots, \Lambda_{n-1}$ the lattices from the nested lattice partition chain in (13) such that $\Lambda_0 \supset \Lambda_1 \supset \Lambda_2 \supset \cdots \supset \Lambda_{n-1}$. Let $k_1, k_2 \in \{0, 1, 2, \ldots, n-1\}$ such that, without loss of generality, $k_1 < k_2$. Then $\mu_{k_1}(\Lambda_{k_1}) \supset \mu_{k_2}(\Lambda_{k_2})$.*

In other words, the corresponding $\wedge$-fuzzy subgroups are also nested and Theorem 4 provides the following chain of nested $\wedge$-fuzzy subgroups:

$$\mu_0 \supset \mu_1 \supset \mu_2 \supset \cdots \supset \mu_{n-1}. \tag{21}$$

In the next section we examine the association between the chain of nested $(T = \wedge)$-fuzzy subgroups (21) and T - indistinguishability operators.

3.2 *T*-Indistinguishability Operator

The reader might find in the literature different terminologies such as, for instance, fuzzy equivalence, T-indistinguishability operators, fuzzy bi-implication and fuzzy similarity by referring to the approach of extending the concept of equivalence relation and equality to the framework of fuzzy theory.

Definition 4 [5, 9] Let X be a universe and T a t-norm. A **T-indistinguishability operator** E on X is a fuzzy relation $E : X \times X \to [0, 1]$ on X satisfying, for all $x, y, z \in X$, the following properties:

1. **[Reflexivity]** $E(x, x) = 1$;
2. **[Symmetry]** $E(x, y) = E(y, x)$;
3. **[T-Transitivity]** $T(E(x, y), E(y, z)) \leq E(x, z)$.

Additionally, the T-indistinguishability operator E is **left invariant** if $E(x, y) = E(z \circ x, z \circ y)$; **right invariant** if $E(x, y) = E(x \circ z, y \circ z)$ and **invariant** if is left and right invariant for all $x, y, z \in G$.

Let E be a T-indistinguishability operator on X. A fuzzy subset μ of X is **extensional** with respect to E if and only if $T(\mu(y), E(x, y)) \geq \mu(x)$, $\forall x, y \in X$. Extensional fuzzy sets are also called observable fuzzy sets [9].

It is possible to construct a new T-indistinguishability operator from previous ones: given T-indistinguishability operators E on X and F on Y, the map $H(E, F) : (X \times Y) \times (X \times Y) \to [0, 1]$ defined by $H(E, F)((x_1, y_1), (x_2, y_2)) = T(E(x_1, x_2), F(y_1, y_2))$ is a T-indistinguishability operator on $X \times Y$.

Lemma 1 *[9] Let $(G, \circ)$ be a group and E a T-indistinguishability operator on G. Then $\circ$ is extensional w.r.t $H(E, E)$ on $G \times G$ and E on G if and only if E is invariant under translations w.r.t to $\circ$.*

We say that E **separates points** if and only if $E(x, y) = 1$ implies $x = y$. The operator $E(x, y)$ is usually interpreted as the degree of indistinguishability (or similarity) between x and y [9].

There are three important relations between T-fuzzy groups and T-indistinguishability operators to the applications in the construction of real nested ideal lattices, the first one being:

Definition 5 [5, 9] Let μ be a T-fuzzy subgroup of $(G, \circ)$, with $\mu(e_G) = 1$. The fuzzy relation $E_{(\mu)}$ on G defined by

$$E_{(\mu)}(x, y) := \mu(x \circ y^{-1}), \quad \forall x, y \in G \tag{22}$$

is the T-indistinguishability operator associated to μ.

Consequently, by Theorem 4 and Definition 5, the chain of nested $\wedge$-fuzzy subgroups (21) induces a sequence of $(T = \wedge)$-indistinguishability operators.

The second relation between T-fuzzy groups and T-indistinguishability operators is the following theorem:

Theorem 5 *[9] Let $(G, \circ)$ be a group, μ a T-fuzzy subgroup of G with $\mu(e_G) = 1$ and $E_{(\mu)}$ the associated T-indistinguishability operator. Then $E_{(\mu)}$ is invariant if and only if μ is a normal T-fuzzy subgroup of G.*

Since Λ_k, where $k = 0, 1, \ldots, n-1$, is an abelian additive group, consequently, μ_k is a normal $\wedge$-fuzzy subgroup of Λ_k, for all $k = 0, 1, \ldots, n-1$. Hence, in (21), we have a chain of nested *normal* $\wedge$-fuzzy subgroups.

Therefore, by Theorem 5, $E_{(\mu_k)}$ is invariant, where $E_{(\mu_k)}$ is the $\wedge$-indistinguishability operator associated to μ_k, for all $k = 0, 1, \ldots, n-1$.

3.3 *T*-Vague Groups

T-vague algebras were introduced by Demirci in [3] by considering fuzzy operations compatible with given T-fuzzy equalities. An extensive study of vague operations and T-vague groups can be found in [3, 4].

A **fuzzy binary operation** on a set G is a map $\tilde{\circ} : G \times G \times G \to [0, 1]$. The interpretation of $\tilde{\circ}(x, y, z)$ is the degree in which z is $x \circ y$. A T**-vague binary operation** on G is a function $\tilde{\circ} : G \times G \times G \to [0, 1]$ that satisfies w.r.t the T-indistinguishability operator $H(E, E)$ the following properties for all $x, y, z \in G$:

1. $T(\tilde{\circ}(x, y, z), E(x, x^{'}), E(y, y^{'}, E(z, z^{'}) \leq T(\tilde{\circ}(x^{'}, y^{'}, z^{'}))$;
2. $T(\tilde{\circ}(x, y, z), \tilde{\circ}(x, y, z^{'})) \leq E(z, z^{'})$;
3. For all $x, y \in G$, there exists $z \in G$ such that $T(\tilde{\circ}(x, y, z)) = 1$.

Definition 6 [4, 5, 9] Let $\tilde{\circ}$ be a T-vague binary operation on G with respect to a T-indistinguishability operator E on G. Then $(G, \tilde{\circ})$ is a T**-vague group** if and only if it satisfies the following properties for all $a, b, c, d, m, q, w \in G$:

1. Associativity:

$$T(\tilde{\circ}(b, c, d), \tilde{\circ}(a, d, m), \tilde{\circ}(a, b, q), \tilde{\circ}(q, c, w)) \leq E(m, w); \tag{23}$$

2. *Identity:* there exists a two sided identity element $e \in G$ such that

$$T(\tilde{\circ}(e, a, a), \tilde{\circ}(a, e, a)) = 1, \tag{24}$$

for each $a \in G$;
3. *Inverse:* for each $a \in G$ there exists a two-sided inverse element $a^{-1} \in G$ such that

$$T(\tilde{\circ}(a^{-1}, a, e), \tilde{\circ}(a, a^{-1}, e)) = 1. \tag{25}$$

We say that a T-vague group is **Abelian** (or commutative) if and only if

$$T(\tilde{\circ}(a, b, m), \tilde{\circ}(b, a, w)) \leq E(m, w), \tag{26}$$

for all $a, b, m, w \in G$.

Notice that for a given T-vague group $(G, \tilde{\circ})$ the elements identity and inverse might not be unique in general. This affirmation holds only if E separates points (either Proposition 12.8 from [9] or Proposition 2.23 from [5]).

A crisp group $(G, \circ)$ is a T-vague group by defining $\tilde{\circ}(a, b, c) = 1$, if $a \circ b = c$, and $\tilde{\circ}(a, b, c) = 0$, otherwise, and by considering the crisp equality as the T-indistinguishability operator.

Proposition 4 *[5, 9] Let $(G, \circ)$ be a group and E be a T-indistinguishability operator on G invariant w.r.t a binary operation $\circ$ on G. Let a fuzzy relation $\tilde{\circ} : G \times G \times G \to [0, 1]$ defined by $\tilde{\circ}(x, y, z) = E(x \circ y, z)$, for all $x, y, z \in G$, then $(G, \tilde{\circ})$ is a T-vague group with respect to E.*

The third relation between T-fuzzy groups, T-indistinguishability operators and T-vague groups is the following:

Proposition 5 *[5, 9] Let $(G, \circ)$ be a group, μ a normal T-fuzzy subgroup of $(G, \circ)$ with $\mu(e_G) = 1$ and E_μ its associated T-indistinguishability operator on G. If $\tilde{\circ} : G \times G \times G \to [0, 1]$ is defined by $\tilde{\circ}(x, y, z) = E_{(\mu)}(x \circ y, z) = \mu(x \circ y \circ z^{-1})$, for all $x, y, z \in G$, then $(G, \tilde{\circ})$ is a T-vague group.*

Since in (21) we have a chain of nested *normal* $\wedge$-fuzzy subgroups and $E_{(\mu_k)}$ is the invariant $\wedge$-indistinguishability operator associated to μ_k, for all $k = 0, 1, \ldots, n-1$, consequently, by Proposition 5 and the fact that $\Lambda_0 \supset \Lambda_1 \supset \Lambda_2 \supset \cdots \supset \Lambda_{n-1}$, we have the following concatenated $(T = \wedge)$-vague groups for all $n = 2^{r-2}$, with $r \geq 3$:

$$(\Lambda_0, \mu_0) \supset (\Lambda_1, \mu_1) \supset \cdots \supset (\Lambda_{n-1}, \mu_{n-1}) \Longleftrightarrow$$

$$(\Lambda_0, \tilde{\circ}_0) \supset (\Lambda_1, \tilde{\circ}_1) \supset \cdots \supset (\Lambda_{n-1}, \tilde{\circ}_{n-1}), \tag{27}$$

where $\tilde{\circ}_k : \Lambda_k \times \Lambda_k \times \Lambda_k \to [0, 1]$ is defined by $\tilde{\circ}_k(x, y, z) = E_{(\mu_k)}(x \circ y, z) = \mu_k(x \circ y \circ z^{-1})$, for all $x, y, z \in \Lambda_k$ and $k = 0, 1, \ldots, n-1$, and $\circ$ is the addition operation on Λ_k.

Now, given a real-valued channel coefficient, it is important to discuss how such a coefficient must be approximated to a $\wedge$-vague group in (27). For that, observe that every $\wedge$-vague group $(\Lambda_k, \tilde{\circ}_k)$ in (27) is associated with a generator of the ideal $\Im^k$, since Λ_k is associated with a generator of the ideal $\Im^k$.

Therefore, through [6], all the possible generators of the ideals from Sect. 2 are the elements $(\xi_{2^r}^{k'} + \xi_{2^r}^{-k'})(2 + \xi_{2^r} + \xi_{2^r}^{-1})^k$, where $k, k' \in \mathbb{Z}$.

Let $h_{ml} \in \mathbb{R}$ a real-valued channel coefficient. From Sect. 2 it is natural the approximation

$$|h_{ml}| \to |2 + \xi_{2^r} + \xi_{2^r}^{-1}|^k, \tag{28}$$

where $k \in \mathbb{Z}$. Consequently, to find the appropriate k, we have

$$\frac{log|h_{ml}|}{log|2 + \xi_{2^r} + \xi_{2^r}^{-1}|} \to k \in \mathbb{Z}, \tag{29}$$

that is, we choose k as being the closest integer value to the value $\frac{log|h_{ml}|}{log|2+\xi_{2^r}+\xi_{2^r}^{-1}|}$.

Now, after finding k, finally we can find k'. In fact, we have that $h_{ml} \to (\xi_{2^r}^{k'} + \xi_{2^r}^{-k'})(2 + \xi_{2^r} + \xi_{2^r}^{-1})^k$ (note that we already know k), then, to find k', we have $arccos\left(\frac{h_{ml}}{2(2+\xi_{2^r}+\xi_{2^r}^{-1})^k}\right)\frac{2^{(r-1)}}{\pi} \to k' \in \mathbb{Z}$, that is, we choose k' as being the closest integer value to the value $arccos\left(\frac{h_{ml}}{2(2+\xi_{2^r}+\xi_{2^r}^{-1})^k}\right)\frac{2^{(r-1)}}{\pi}$.

Consequently, by knowing k and k', it is possible to realize for dimension $n = 2^{r-2}$, where $r \geq 3$, the corresponding real-valued channel quantization onto the *real ideal lattices* Λ_k [6] and, consequently, the corresponding real-valued channel quantization onto the $\wedge$-vague groups $(\Lambda_k, \tilde{\circ}_k)$ in (27).

4 Conclusion and Final Remarks

In this work we present a method that describes for any dimension $n = 2^{r-2}$, with $r \geq 3$, concatenated $\wedge$-fuzzy subgroups and $\wedge$-vague groups in order to quantize real-valued channels, that is, in order to realize interference alignment onto a $\wedge$-fuzzy subgroup and, consequently, onto a $\wedge$-vague group. The performance analysis of the quantization method proposed depends intrinsically on the membership function defined *a priori* and their properties such as: normality of the T-fuzzy subgroup, as in Corollary 3, to ensure that the T-indistinguishability operator and the T-vague group will be invariant as well as the operation $\circ$ will be extensional; the T-indistinguishability operator should separate points to ensure that the elements identity and inverse are unique. Consequently, the performance analysis must be conducted in a wide variety of membership functions and families of t-norms and that is a subject of future works as well as a comparison of the classic quantization. It remains an open question if it is possible to obtain, instead of a crisp criteria, a criteria based only on the membership functions to obtain the closest value.

Acknowledgements The authors are greatly thankful to the anonymous referee for the valuable suggestions. Also, the authors would like to thank CAPES (Coordenação de Aperfeiçoamento de Pessoal de Nível Superior) and CNPq (Conselho Nacional de Desenvolvimento Científico e Tecnológico) under grant 141348/2019-4 for the financial support.

References

1. K. Abdullah, S.A. Bakar, N.H. Kamis, H. Aliamis, Rsa cryptosystem with fuzzy set theory for encryption and decryption, in *AIP Conference Proceedings* **1905**(1), 030001 (2017)
2. A.A. Andrade, C. Alves, T. Carlos, Rotated lattices via the cyclotomic field $\mathbb{Q}(\xi_{2^r})$. IJAM Int. J. Appl. Math. **19**(01) (2006)
3. M. Demirci, Fuzzy functions and their fundamental properties. Fuzzy Sets Syst. **106**(2), 239–246 (1999)
4. M. Demirci, Vague groups. J. Math. Anal. Appl. **230**(1), 142–156 (1999)
5. M. Demirci, J. Recasens, Fuzzy groups, fuzzy functions and fuzzy equivalence relations. Fuzzy Sets Syst. **144**(3), 441–458 (2004)
6. C.C. Trinca et al., Construction of nested real ideal lattices for interference channel coding. Int. J. Apllied Math. **32**(2) (2019)
7. G.D. Forney, Coset codes. i. introduction and geometrical classification. IEEE Trans. Inf. Theory **34**(5), 1123–1151 (1988)
8. J.N. Mordeson, K.R. Bhutani, A. Rosenfeld, *Fuzzy Group Theory* (Springer, Berlin Heidelberg, 2005)
9. J. Recasens, Indistinguishability operators: modelling fuzzy equalities and fuzzy equivalence relations. Stud. Fuzziness Soft Comput. (2010) (Springer)
10. A. Rosenfeld, Fuzzy groups. J. Math. Anal. Appl. **35**(3), 512–517 (1971)
11. S.B. Sadkhan, F.H. Abdulraheem, A proposed ANFIS evaluator for RSA cryptosystem used in cloud networking, in *Proceedings of the 2017 International Conference on Current Research in Computer Science and Information Technology (ICCIT)* (IEEE, 2017)
12. B. Šešelja, A. Tepavčević, G. Vojvodić, L-fuzzy sets and codes. Fuzzy Sets Syst. **53**(2), 217–222 (1993)
13. J. Tang, S. Lambotharan, Interference alignment techniques for MIMO multi-cell interfering broadcast channels. IEEE Trans. Commun. **61**(1), 164–175 (2013)
14. C.C. Trinca, A contribution to the study of channel coding in wireless communication systems. Ph.D. thesis, Universidade Estadual Paulista "Júlio de Mesquita Filho" (UNESP), Ilha Solteira—São Paulo, Brazil, vol. 7 (2013). An optional note
15. C.C. Trinca, J.-C. Belfiore, E.D. de Carvalho, J. Vieira Filho, Estimation with mean square error for real-valued channel quantization, in *Proceedings of the 2014 IEEE Globecom Workshops (GC Wkshps)* (IEEE, 2014)
16. R.A. Watanabe, C.C. Trinca, Application of the support function and the steiner point on the study of interference alignment channel, in *Proceedings of the 2017 IEEE International Conference on Fuzzy Systems (FUZZ-IEEE)* (IEEE, 2017)
17. R.A. Watanabe, C.C. Trinca, E. Esmi, A proposal by using fuzzy vectors on the study of complex-valued channel quantization, in *Proceedings of the 2018 IEEE International Conference on Fuzzy Systems (FUZZ-IEEE)* (IEEE, 2018)

Adaptive Interval Fuzzy Modeling from Stream Data and Application in Cryptocurrencies Forecasting

Leandro Maciel, Rosangela Ballini, and Fernando Gomide

Abstract This paper develops an adaptive interval fuzzy modeling method using participatory learning and interval-valued stream data. The model is a collection of fuzzy functional rules whose structure and parameters evolve simultaneously as data are input. The evolving nature of the method allows continuous model update using stream interval data. The method employs participatory learning to cluster interval input data, assigns to each cluster a fuzzy rule, uses the weighted recursive least squares to update the parameters of the rule consequent intervals, and returns an interval-valued output. The efficacy of the method is evaluated in modeling and forecasting daily low and high prices of the two most traded cryptocurrencies, BitCoin and Ethereum. The forecast performance of the adaptive interval fuzzy modeling method is evaluated against classic autorregressive moving average, exponential smoothing state model, and the naïve random walk. Results indicate that, similarly to with exchange rates, no model outperforms random walk in predicting prices in digital coin markets. However, when a measure of directional accuracy is accounted for, adaptive interval fuzzy modeling outperforms the remaining alternatives.

Keywords Adaptive machine learning · Fuzzy modeling · Interval-valued stream data · Forecasting · Cryptocurrencies

---

L. Maciel (✉)
Department of Business, Faculty of Economics, Business and Accounting, University of São Paulo, São Paulo, Brazil
e-mail: leandromaciel@usp.br

R. Ballini
Department of Economics Theory, Institute of Economics, University of Campinas, Campinas, Brazil
e-mail: ballini@unicamp.br

F. Gomide
Department of Computer Engineering and Automation School of Electrical and Computer Engineering, University of Campinas, Campinas, Brazil
e-mail: gomide@dca.fee.unicamp.br

B. Bede et al. (eds.), *Fuzzy Information Processing 2020*,
Advances in Intelligent Systems and Computing 1337,
https://doi.org/10.1007/978-3-030-81561-5_7

1 Introduction

Classic time series analysis uses a sequence of single-valued data samples. This may be restrictive in situations in which complex data are needed to comprehend inherent variability and/or uncertainty of a phenomenon. An example is the daily stock price of a corporation expressed as the daily minimum and maximum trading prices. If only the lowest (or the highest, or closing) price of each day is considered, then the resulting time series is single-valued and neglects intra-daily price variability. Otherwise, if lower and higher prices are accounted for daily, then an interval-valued time series (ITS) is formed with trend (or level) and volatility information (the range between the boundaries) contained in data [20]. ITS are encountered in many fields, typically in finance [2], energy [21], environment and climate [25], and agriculture [23].

Different approaches have been developed for modeling and forecasting interval-valued data. For instance [3] introduced the first linear regression model for interval-valued prediction based on the center method. A constrained center and range method adding a non-negative constraint on the coefficients of range regression model is addressed in [9]. Implementation of ARIMA and neural networks models to forecast the center and radii of intervals is given in [13]. An approach known as the interval autoregressive time series modeling is addressed in [24], and threshold autoregressive interval modeling in [20]. Nonparametric approaches such as the interval kernel regression method [6], and a nonparametric additive approach for analyzing interval-valued data that allows a nonlinear pattern [8] have also been developed. Besides nonparametric methods, machine learning has also been used in interval-valued data modeling and forecasting. Examples include interval multilayer perceptron (iMLP) model [17], multilayer perceptron (MLP) neural network and Holt's exponential smoothing [12], multiple-output support vector regression method for interval-valued stock price index forecasting [26], and possibilistic fuzzy, evolving intelligent system for interval time series modeling and forecasting [11].

This paper introduces a novel adaptive interval fuzzy method to model and forecast interval-valued time series. The method processes stream interval-valued input data, employs participatory learning to cluster interval-valued data, develops a fuzzy rule for each cluster, uses weighted recursive least squares to update the parameters of the rule consequents, and outputs interval-valued forecasts. The adaptive interval fuzzy method is evaluated in modeling and forecasting the daily low and high prices of the two most traded cryptocurrencies, respectively BitCoin and Ethereum. Its performance is compared against the autoregressive moving average (ARIMA), the exponential smoothing state model (ETS), and the naïve random walk benchmarks.

The rest of this paper proceeds as follows. Section 2 details the nature and structure of interval fuzzy models, and suggests a procedure to develop adaptive interval fuzzy models from stream data. Application in cryptocurrencies modeling and forecasting using actual daily low and high prices data is addressed in Sect. 3. Section 4 concludes the paper and suggests issues for future investigation.

2 Adaptive Interval Fuzzy Modeling

This section addresses the nature and structure of interval fuzzy models, and shows how to construct them in the realm of adaptive, evolving participatory learning from interval-valued stream data [1]. After a brief remind of interval-valued time series and interval arithmetic, the detailed steps of adaptive interval fuzzy procedure are given.

2.1 Interval Time Series and Interval Arithmetic

An interval-valued time series (ITS) is a sequence of interval-valued data indexed by successive time steps $t = 1, 2, \ldots, N$. In this paper an interval datum is $[x] = [x^L, x^U] \in \mathcal{K}_c(\mathbb{R})$ where $\mathcal{K}_c(\mathbb{R}) = \{[x^L, x^U] : x^L, x^U \in \mathbb{R}, x^L \leq x^U\}$ is the set of closed intervals of the real line $\mathbb{R}$, and x^L and x^U are the lower and upper bounds of $[x]$. When convenient, the interval $[x]$ is expressed by a two-dimensional vector $[x] = [x^L, x^U]^T$.

Interval arithmetic extends traditional arithmetic to operate on intervals. The arithmetic operations introduced by Moore [14] are:

$$
\begin{aligned}
[x] + [y] &= [x^L + y^L, x^U + y^U] \\
[x] - [y] &= [x^L - y^U, x^U - y^L] \\
[x] \cdot [y] &= \left[\min\{x^L y^L, x^L y^U, x^U y^L, x^U y^U\}, \max\{x^L y^L, x^L y^U, x^U y^L, x^U y^U\}\right] \\
[x] / [y] &= [x] (1/[y]) \text{ with } 1/[y] = [1/y^U, 1/y^L]
\end{aligned} \quad (1)
$$

It is well known that $[x] - [x] \neq [0]$ for Moore difference, $[0] = [0, 0]$. An alternative for which $[x] - [x] = [0]$ is the generalized Hukuhara difference [19]:

$$[x] - [y] = [\min\{x^L - y^L, x^U - y^U\}, \max\{x^L - y^L, x^U - y^U\}]. \quad (2)$$

This paper considers both, Moore and generalized Hukuhara differences, and evaluates their impact in the time series modeling and forecasting context. The midpoint $\bar{x} \in \mathbb{R}$, and width $w([x]) \in \mathbb{R}$ of an interval $[x]$ are:

$$\bar{x} = \frac{x^L + x^U}{2}, \quad w([x]) = x^U - x^L \quad (3)$$

The union and the intersection of intervals $[x]$ and $[y]$ are, respectively:

$$[x] \cup [y] = [\min\{x^L, y^L\}, \max\{x^U, y^U\}], \quad (4)$$

$$[x] \cap [y] = [\max\{x^L, y^L\}, \min\{x^U, y^U\}]. \quad (5)$$

The intersection of $[x]$ and $[y]$ is empty if $\max\{x^L, y^L\} > \min\{x^U, y^U\}$. Real numbers are considered as intervals of zero length.

Procedures of the adaptive interval fuzzy modeling method require a measure of distance between two intervals. Here we use the Hausdorff distance. The Hausdorff distance between $[x]$ and $[y]$, denoted $dH([x], [y])$, is:

$$dH([x],[y]) = \max\left\{|x^L - y^L|, |x^U - y^U|\right\}. \tag{6}$$

2.2 *Adaptive Interval Fuzzy Modeling*

An adaptive interval fuzzy model is a collection of fuzzy rules, a fuzzy rule base in which each rule has two parts: an antecedent part stating conditions on the input variable; and a consequent part specifying the corresponding value for the output variable. Adaptive interval fuzzy modeling (ePL-I) uses interval functional fuzzy rules, an interval extension of functional fuzzy rules of the form introduced in [22]. Interval functional fuzzy rules considered here are of fuzzy rules of the form:

$$\mathcal{R}_i : \text{ IF } [\mathbf{x}] \text{ is } \mathcal{M}_i \text{ THEN } [y_i] = \beta_{i,0} + \beta_{i,1}[x_1] + \ldots + \beta_{i,p}[x_p] \tag{7}$$

where $\mathcal{R}_i$ is the i-th fuzzy rule, $i = 1, 2, \ldots, c$, c is the number of fuzzy rules in the rule base, $[\mathbf{x}] = \left([x_1], [x_2], \ldots, [x_p]\right)^T$ the vector of inputs, $[x_j] = [x_j^L, x_j^U] \in \mathcal{K}_c(\mathbb{R})$, $j = 1, \ldots, p$, and $\{\beta_{i,0}, \ldots, \beta_{i,p}\}$ are real-valued parameters of the rule consequent. $\mathcal{M}_i$ is the fuzzy set of the antecedent whose membership function is $\mu_i([\mathbf{x}]) : \mathcal{K}_c(\mathbb{R}^p) \to [0, 1]$, and $[y_i] = [y_i^L, y_i^U] \in \mathcal{K}_c(\mathbb{R})$ is the output of the i-th rule. Fuzzy inference with interval functional rules (7) is the weighted average:

$$[y] = \sum_{i=1}^{c} \left(\frac{\mu_i([\mathbf{x}])[y_i]}{\sum_{j=1}^{c} \mu_j([\mathbf{x}])} \right) = \sum_{i=1}^{c} \lambda_i [y_i], \tag{8}$$

where $\lambda_i = \dfrac{\mu_i([\mathbf{x}])}{\sum_{j=1}^{c} \mu_j([\mathbf{x}])}$ is the normalized degree of activation of the i-th rule. The membership degree $\mu_i([\mathbf{x}])$ of datum $[\mathbf{x}]$ is given by:

$$\mu_i([\mathbf{x}]) = \left[\sum_{h=1}^{c} \left(\frac{\sum_{j=1}^{p} \left(\max\left\{ |x_j^L - v_{i,j}^L|, |x_j^U - v_{i,j}^U| \right\} \right)}{\sum_{j=1}^{p} \left(\max\left\{ |x_j^L - v_{h,j}^L|, |x_j^U - v_{h,j}^U| \right\} \right)} \right)^{\frac{2}{(m-1)}} \right]^{-1}, \tag{9}$$

where m is a fuzzification parameter (usually $m = 2$), and $[v_i] = [v^L_{i,j}, v^U_{i,j}] \in \mathcal{K}_c(\mathbb{R})$, $j = 1, \dots, p$ and $i = 1, \dots, c$ is the cluster center of the i-th cluster.

Interval functional fuzzy modeling uses parameterized fuzzy regions and associates each region with a local affine interval model. The nature of the rule-based model emerges from the fuzzy weighted combination of the collection of the multiple local interval models. The contribution of a local model to the model output is proportional to the normalized activation degree of the corresponding rule. The construction of interval-valued fuzzy models needs: (i) learning the antecedent part using fuzzy clustering to granulate the input interval data space into parameterized regions; and (ii) estimating the real-valued parameters of the rules consequents.

2.3 *Learning the Antecedent of Interval Fuzzy Rules*

Identification of the fuzzy rules of an interval-valued fuzzy model is done using the participatory learning fuzzy clustering algorithm extended to handle interval-valued data. The aim is to partition an interval data set $[\mathbf{X}] = \{[\mathbf{x}_1], \dots, [\mathbf{x}_N]\}$ in c, $2 \leq c \leq N$ fuzzy subsets. Intervals $[\mathbf{x}_j]$ bounds are assumed to be normalized:

$$x^B_{norm} = \frac{x^B - \min\{x^L, x^U\}}{\max\{x^L, x^U\} - \min\{x^L, x^U\}}, \tag{10}$$

where $B = \{L, U\}$ denotes either the lower, or the upper bound of the normalized interval.

Participatory learning clustering is based on the idea that the current cluster structure influences the cluster learning process itself. The cluster structure influences learning via the compatibility of the input data with the current cluster structure [18]. The cluster structure is defined by the cluster centers. Thus, if $[\mathbf{V}] = \{[\mathbf{v}_1], \dots, [\mathbf{v}_c]\}, [\mathbf{v}_i] = \big([v_{i,1}], \dots, [v_{i,p}]\big), [v_{i,j}] = [v^L_{i,j}, v^U_{i,j}] \in \mathcal{K}_c([0, 1])$, $j = 1, \dots, p$ and $i = 1, \dots, c$, then $[\mathbf{V}]$ represents the cluster structure. The aim of the antecedent learning is to determine $[\mathbf{V}]$ from inputs $[\mathbf{x}]^t \in [0, 1]^p$ for $t = 1, \dots$

The compatibility $\rho^t_i \in [0, 1]$ of input $[\mathbf{x}]^t$ with cluster centers $[\mathbf{v}_i]^t$ of $[\mathbf{V}]^t$, $i = 1, \dots, c$ is:

$$\rho^t_i = 1 - \frac{1}{p}\sum_{j=1}^{p}\left(\max\left\{|x^{L,t}_j - v^{L,t-1}_{i,j}|, |x^{U,t}_j - v^{U,t-1}_{i,j}|\right\}\right). \tag{11}$$

Partitipatory clustering updates only the cluster center whose compatibility with input $[\mathbf{x}]^t$ is the highest. Thus, if cluster center $[\mathbf{v}_i]^{t-1}$ is the most compatible with $[\mathbf{x}]^t$, then it is updated as follows:

$$[\mathbf{v}_i]^t = [\mathbf{v}_i]^{t-1} + G_i^t([\mathbf{x}]^t - [\mathbf{v}_i]^{t-1}), \quad G_i^t = \alpha\rho_i^t \tag{12}$$

where $\alpha \in [0, 1]$ is the basic learning rate. Notice that (12) needs the value of $([\mathbf{x}]^t - [\mathbf{v}_i]^{t-1})$. Both, Moore's difference (1), and generalized Hukuhara difference (2) are used in (12) to verify if they make any difference in time series modeling and forecasting.

Adaptive modeling is particularly relevant for time-varying domains. A sequence of input data with low compatibility with the current cluster structure indicates that the current model should be revised in front of new information. Participatory learning uses an arousal mechanism to monitor how compatibility measure values progress. Higher arousal values indicate less confidence in how the current model fits newly incoming data. The arousal mechanism can be seen as the complement of the confidence of the current cluster structure with new data. A way to express the arousal $a_i^t \in [0, 1]$ at step t is $a_i^t = a_i^{t-1} + \beta(1 - \rho_i^t - a_i^{t-1})$ where $\beta \in [0, 1]$ controls the rate of change of arousal. The closer β is to one, the faster the learning process senses compatibility variations. If the values of arousal a_i^t are greater than or equal a threshold $\tau \in [0, 1]$ for all i, then a new cluster should be created, assigning the current data as its cluster center, that is $[\mathbf{v}_{c+1}]^t = [\mathbf{x}]^t$. Otherwise, the center with the highest compatibility is updated to accommodate input data using (12). The arousal mechanism becomes part of the learning process by converting G_i^t into an effective learning rate $G_i^t = \alpha(\rho_i^t)^{1-a_i^t}$.

Updates of cluster centers (12) can be viewed as a form of exponential smoothing modulated by the compatibility of data with the model structure, the very nature of participatory learning. Participatory clustering also accounts for redundant clusters. A cluster i is redundant if its similarity $\rho_{i,h}^t$ with any other cluster h is greater than or equal to a threshold $\lambda \in [0, 1]$. Similarity between cluster centers i and h is found using:

$$\rho_{i,h}^t = 1 - \frac{1}{p} dH([\mathbf{v}_i]^t, [\mathbf{v}_h]^t) = 1 - \frac{1}{p}\sum_{j=1}^{p}\left(\max\left\{|v_{i,j}^{L,t} - v_{h,j}^{L,t}|, |v_{i,j}^{U,t} - v_{h,j}^{U,t}|\right\}\right). \tag{13}$$

If clusters i and h are found to be redundant, then they and replaced by a cluster whose center is the average of their respective centers.

2.4 Estimating the Parameters of Interval Fuzzy Rule Consequent

To complete the interval rule-based mode construction, the real-valued parameters $\beta_{i,0}, \ldots, \beta_{i,p}$ of the rule consequent are estimated using the weighted recursive least squares algorithm by taking advantage of the standard form of the least squares algorithm, and the midpoint characterization of intervals. To do this, expression (8) is expressed in terms of input interval data midpoints as $\bar{y} = \Lambda^T \Theta$ where $\Lambda = \left[\lambda_1 \bar{\mathbf{x}}_e^T, \lambda_2 \bar{\mathbf{x}}_e^T, \ldots, \lambda_c \bar{\mathbf{x}}_e^T\right]^T$ is the fuzzily weighted extended input data midpoints, $\bar{\mathbf{x}}_e = \left[1, \bar{x}_1, \bar{x}_2, \ldots, \bar{x}_p\right]^T$ is the extended input data midpoints, $\Theta = \left[\boldsymbol{\beta}_1^T, \boldsymbol{\beta}_2^T, \ldots, \boldsymbol{\beta}_c^T\right]^T$, and $\boldsymbol{\beta}_i = [\beta_{i,0}, \beta_{i,1}, \ldots, \beta_{i,p}]^T$. The weighted recursive least squares considers locally optimal error criterion:

$$\min E_i^t = \min \sum_{k=1}^{t} \lambda_i \left(\bar{y}^k - (\bar{\mathbf{x}}_e^k)^T \bar{\boldsymbol{\beta}}_i^k\right)^2, \tag{14}$$

whose solution can be expressed recursively as [10]:

$$\boldsymbol{\beta}_i^t = \boldsymbol{\beta}_i^{t-1} + \Sigma_i^{t-1} \bar{\mathbf{x}}_e^{t-1} \lambda_i^{t-1} \left(\bar{y}^{t-1} (\bar{\mathbf{x}}_e^{t-1})^T \boldsymbol{\beta}_i^{t-1}\right), \ \ \boldsymbol{\beta}_i^0 = 0, \tag{15}$$

$$\Sigma_i^t = \Sigma_i^{t-1} - \frac{\lambda_i^{t-1} \Sigma_i^{t-1} \bar{\mathbf{x}}_e^{t-1} (\bar{\mathbf{x}}_e^{t-1})^T \Sigma_i^{t-1}}{1 + \lambda_i^{t-1} (\bar{\mathbf{x}}_e^{t-1})^T \Sigma_i^{t-1} \bar{\mathbf{x}}_e^{t-1}}, \ \ \Sigma_i^0 = \Omega I, \tag{16}$$

where Ω is a large number (usually $\Omega = 1,000$), and Σ is the dispersion matrix.

Therefore, the output of the i-th fuzzy rule (7) is:

$$[y_i]^{t+1} = \beta_{i,0}^t + \beta_{i,1}^t [x_1]^t + \cdots + \beta_{i,p}^t [x_p]^t. \tag{17}$$

The model output is the weighted average (8).

2.5 Adaptive Interval Fuzzy Modeling Procedure

Adaptive interval fuzzy modeling is inherently recursive, which means that it has memory efficient form appropriate for continuous, endless learning and model adaptation from stream data input. This is of major importance especially in online and real-time application environments. The procedure to construct interval fuzzy models is as follows.

Adaptive interval fuzzy modeling

Start cluster structure with the first input datum: $c = 1$, $[\mathbf{v}_1]^1 = [\mathbf{x}]^1$
Choose control parameters $\alpha, \beta, \tau, \lambda$ and set $a_1^1 = 0,\ \Sigma_i^0 = \Omega I$
for $t = 2, 3, \ldots$ **do**
  read $[\mathbf{x}]^t$
  for $i = 1, \ldots, c$
    compute $\rho_i^t = 1 - \frac{1}{p}\sum_{j=1}^{p}\left(\max\left\{|x_j^{L,t} - v_{i,j}^{L,t-1}|, |x_j^{U,t} - v_{i,j}^{U,t-1}|\right\}\right)$
    update $a_i^t = a_i^{t-1} + \beta(1 - \rho_i^t - a_i^{t-1})$
  end for
  if $a_i^t \geq \tau$ for $i = 1, \ldots, c$
    create a new cluster: $c = c + 1$
    $[\mathbf{v}_c]^t = [\mathbf{x}]^t$
    reset $a_i^t = 0$
  else
    find the most compatible cluster $s = \arg\max_{j=1,\ldots,c}\{\rho_j^t\}$
    update $[\mathbf{v}_s]^t = [\mathbf{v}_s]^{t-1} + G_s^t([\mathbf{x}]^t - [\mathbf{v}_s]^{t-1})$,
  end if
  for $i = 1, \ldots, c - 1$ and $h = i + 1, \ldots, c$
    compute clusters compatibilities $\rho_{i,h}^t = 1 - \frac{1}{p} dH([\mathbf{v}_i]^t, [\mathbf{v}_h]^t)$
    if $\rho_{i,h}^t \geq \lambda$
      redefine $[\mathbf{v}_i]^t$ using centers i and h, and delete $[\mathbf{v}_h]^t$
      $c = c - 1$
    end if
  end for
  compute rule consequent parameters using the wRLS
  compute model output $[y]^{t+1}$
end for

3 Computational Experiments

3.1 Cryptocurrency Trade

Current interest in cryptocurrencies has offered to investors and speculators a diversity of electronic crypto assets to trade due to benefits such as anonymity, transparency, lower transaction costs, and diversification [5]. Because of their high volatile dynamics, risk management of cryptocurrencies investment positions based on volatility modeling and forecasting is a key in currency trade. Risk management methods built on intervals are more realistic in modeling economic phenomena. Intervals of daily highest and lowest currency prices convey a better idea of portfolio risk perception at a time period than a single value. Interval-valued modeling and forecasting produce more realistic price estimates because they inherently embed price variability information contained in the data samples. This section evaluate the performance of adaptive interval fuzzy modeling (ePL-I) using real-world data from the cryptocurrency market.

3.2 Data

Adaptive interval fuzzy modeling and forecasting uses a dataset with the daily interval data from high and low prices of the two leading cryptocurrencies, BitCoin and Ethereum, respectively. Data cover January 1, 2016 to December 31, 2019, totaling 1,461 samples.[1] The data set was divided in in-sample and out-of-sample sets. The in-sample set contains data from January 1, 2016 to December 31, 2018, while the out-of-sample set contains the 2019 data. Notice that ePL-I processes data as a stream, which means that there is no need to split data in training, validation and testing data as usual in machine learning. In spite of that, data set split was done to keep evaluations and comparisons of the competing methods fair. Forecasting is one-step-ahead with an iterated strategy. The forecasting techniques considered for comparison are the autoregressive moving average (ARIMA), the exponential smoothing state space model (ETS) [7], and the naïve random walk (RW). These methods produce interval forecasts from the individual lower and upper daily currency forecasts.

3.3 Performance Measures

Performance evaluation of forecasting methods with interval time series (ITS) data uses the normalized symmetric difference of intervals (NSD) [16]:

$$\mathrm{NSD}([y]^t, [\hat{y}]^t) = \frac{1}{N}\sum_{t=1}^{N} \frac{w([y]^t \cup [\hat{y}]^t) - w([y]^t \cap [\hat{y}]^t)}{w([y]^t \cup [\hat{y}]^t)}, \tag{18}$$

as well as average coverage (R^C) and efficiency rates (R^E) [16]:

$$\mathrm{R}^{\mathrm{C}} = \frac{1}{N}\sum_{t=1}^{N} \frac{w([y]^t \cap [\hat{y}]^t)}{w([y]^t)}, \quad \mathrm{R}^{\mathrm{E}} = \frac{1}{N}\sum_{t=N}^{n} \frac{w([y]^t \cap [\hat{y}]^t)}{w([\hat{y}]^t)}, \tag{19}$$

where $[y]^t = [y^{L,t}, y^{U,t}]$ is the actual price interval, and $[\hat{y}]^t = [\hat{y}^{L,t}, \hat{y}^{U,t}]$ is the forecasted price interval at t, and N is the sample size of the out-of-sample data set.

R^C and R^E give information on what part of ITS data are covered by their forecasts (coverage), and what part of the forecasts cover ITS data (efficiency). If the observed intervals are fully enclosed in the predicted intervals, then the coverage rate is 100%,

---

[1]Selection done choosing cryptocurrencies with the highest liquidity and market capitalization. Data source: https://coinmarketcap.com/.

but the efficiency could be less than 100% and reveal the fact that the forecasted ITS is in average wider than the actual ITS data samples. Hence, R^C and R^E values must be considered jointly. Good forecast performance is expected when the average coverage and efficiency rates are reasonably high, and the difference between them is small [16].

Additionally, in practice the direction of price change is as important as, sometimes even more important than the magnitude of the forecasting error [4]. A measure of forecast direction is:

$$\mathrm{DA}^B = \frac{1}{N}\sum_{t=1}^{N} Z^{B,t}, \quad Z^{B,t} = \begin{cases} 1, & \text{if } \left(\hat{y}^{B,t+1} - y^{B,t}\right)\left(y^{B,t+1} - y^{B,t}\right) > 0, \\ 0, & \text{otherwise.} \end{cases} \tag{20}$$

where B is either L or U, the lower or the upper bonds of the interval prices.

3.4 *Simulation Results*

The ePL-I is compared against ARIMA, ETS and RW in cryptocurrencies lows and highs price forecast. ARIMA, ETS and RW are univariate techniques, and their forecasts are developed for the interval bounds individually. Contrary, ePL-I as an interval-based method, produces interval-valued prices. ePL-I assumes that forecasts are produced using p lagged values of the interval time series as input.

Performance evaluation was done for out-of-sample data, i.e. from January 1, 2019 to December 31, 2019. ePL-I was implemented in MATLAB, while ARIMA and ETS were developed using R. Table 1 shows the values of the parameters of each method. Structures of ARIMA and ETS are different for lower (L) and upper (U) interval bounds. In ARIMA(p,d,q), p, d and q are the number of autoregressive, difference, and moving average terms, respectively. In ETS(er,tr,sea), er, tr, and sea mean error, trend, and season type, respectively. A, M and N denote additive, multiplicative, and none, respectively. The structures of ARIMA and ETS were selected automatically by R to achieve the highest accuracy. ePL-I parameters were chosen doing simulations to select the ones that give the best MAPE and RMSSE values. We recall that ePL-I was implemented using Moore's (ePL-$\mathrm{I_M}$) and generalized Hukuhara (ePL-$\mathrm{I_H}$) differences.

Forecasting performance considering the interval nature of the data, measured using normalized symmetric divergence (NSD), coverage ($\mathrm{R^C}$), and the efficiency ($\mathrm{R^E}$), are reported in Table 2. The naïve random walk outperforms the remaining methods. Table 2 also indicates a superior performance by using the generalized Hukuhara difference (ePL-$\mathrm{I_H}$).

Dynamic forecasting models can outperform the random walk in out-of-sample forecasting if performance is measured by direction of change and profitability [15]. Indeed, this is also the case in criptocurrency price forecasting. Table 3 shows that ePL-I outperforms random walk, ARIMA and ETS from the point of view of the

Table 1 Structures of the forecasting models and respective control parameters

Cryptocorrency	ARIMA	ETS	ePL-I_M	ePL-I_H
BitCoin	L(2,1,1) U(2,1,2)	L(M,A,N) U(M,A,N)	$p=6, \beta=0.12,$ $\tau=0.49,$ $\alpha=0.01,$ $\lambda=0.86$	$p=8, \beta=0.12,$ $\tau=0.53,$ $\alpha=0.02,$ $\lambda=0.85$
Ethereum	L(2,1,2) U(0,1,1)	L(M,A,N) U(M,A,N)	$p=6,$ $\beta=0.115,$ $\tau=0.49,$ $\alpha=0.01,$ $\lambda=0.86$	$p=9, \beta=0.07,$ $\tau=0.46,$ $\alpha=0.05,$ $\lambda=0.87$

Table 2 NSD, R^C and R^E for one-step-ahead forecast of cryptocurrencies low and high prices

Error measure	ARIMA	ETS	RW	ePL-I_M	ePL-I_H
Panel A: BitCoin					
NSD	0.6507	0.6114	**0.5926**	0.7579	0.7516
R^C	0.5418	0.5677	**0.6092**	0.2833	0.3103
R^E	**0.6812**	0.5971	0.5805	0.2102	0.5143
Panel B: Ethereum					
NSD	0.6086	0.6029	**0.5840**	0.7594	0.7341
R^C	0.5945	0.5899	**0.6165**	0.2909	0.3871
R^E	0.5761	**0.5808**	0.5781	0.4065	0.5239

Table 3 DA for one-step-ahead forecast of cryptocurrencies low and high prices

Error measure	ARIMA	ETS	RW	ePL-I_M	ePL-I_H
Panel A: BitCoin					
DA^L	0.5110	0.5852	–	**0.5137**	0.4863
DA^U	0.5385	0.5440	–	**0.5495**	0.5330
Panel B: Ethereum					
DA^L	0.5000	0.5247	–	0.5192	**0.5577**
DA^U	0.5719	0.5440	–	0.5440	**0.5742**

direction accuracy (DA). These results are in line with those of [4, 15], in which the random walk is beaten decisively when the direction is used for comparison. The high performance that ePL-I achieves in predicting directions of price changes is related with its evolving, continuous adaptation ability to capture prices changes more accurately in nonstationary environments such as digital currency markets. When trading strategies use direction, the potential to anticipate price change is crucial.

4 Conclusion

This work has introduced a novel adaptive interval fuzzy modeling method using participatory learning. The method develops fuzzy rule-based models, and continuously update their structure and parameters using stream of interval-valued data. Computational experiments considered one-step-ahead forecasting of interval-valued prices of BitCoin and Etherem, for the period from January 2016 to December 2019. Comparison of the adaptive interval fuzzy modeling and forecasting methods was evaluated against ARIMA, ETS and naïve random walk methods. Results indicate that random walk outperforms all the methods addressed in this paper. However, when performance is measured by the direction of price change, adaptive interval fuzzy modeling performance is the highest. This is a key feature when trading with price and direction-based strategies. Future work shall consider automatic mechanisms to select and adjust parameters such as thresholds, proceed with evaluations of strategies, and interval time series with distinct dynamics than cryptocurrency market.

Acknowledgements The authors are grateful to the Brazilian National Council for Scientific and Technological Development (CNPq) for grants 302467/2019-0 and 304274/2019-4, and the São Paulo Research Foundation (Fapesp) for their support

References

1. P. Angelov, D. Filev, An approach to online identification of Takagi–Sugeno fuzzy models. IEEE Trans. Syst. Man Cybern.—Part B **34**(1), 484–498 (2004)
2. J. Arroyo, R. Espínola, C. Maté, Different approaches to forecast interval time series: a comparison in finance. Comput. Econ. **27**(2), 169–191 (2011)
3. L. Billard, E. Diday, Regression analysis for interval-valued data, in (2000), pp. 369–374
4. K. Burns, I. Moosa, Enhancing the forecasting power of exchange rate models by introducing nonlinearity: does it work? Econ. Model. **50**, 27–39 (2015)
5. S. Corbet, A. Meegan, C. Larkin, B. Lucey, L. Yarovaya, Exploring the dynamic relationships between cryptocurrencies and other financial assets. Econ. Lett. **165**, 28–34 (2018)
6. R. Fagundes, R. De Souza, F. Cysneiros, Interval kernel regression. Neurocomputing **128**, 371–388 (2014)
7. R. Hyndman, A. Koehler, J. Ord, R. Snyder, *Forecasting with Exponential Smoothing: The State Space Approach* (Springer, Berlin, 2008)
8. C. Lim, Interval-valued data regression using nonparametric additive models. J. Korean Stat. Soc. **45**(3), 358–370 (2016)
9. E. Lima Neto, F. de Carvalho, Constrained linear regression models for symbolic interval-valued variables. Comput. Stat. Data Anal. **54**(2), 333–347 (2010)
10. L. Ljung, *System Identification: Theory for the User* (Prentice-Hall, Englewood Cliffs, NJ, 1988)
11. L. Maciel, F. Gomide, R. Ballini, Evolving possibilistic fuzzy modeling for financial interval time series forecasting, in Processing of the Annual Conference of the North American Fuzzy Information Processing Society (NAFIPS) Held Jointly 5th World Conference on Soft Computing (WConSC) (2015), pp. 1–6
12. A. Maia, F. de Carvalho, Holt's exponential smoothing and neural network models for forecasting interval-valued time series. Int. J. Forecast. **27**(3), 740–759 (2011)

13. A. Maia, F. de Carvalho, T. Ludermir, Forecasting models for interval-valued time series. Neurocomputing **71**(16–18), 3344–3352 (2008)
14. R. Moore, R. Kearfott, M. Cloud, *Introduction to Interval Analysis* (SIAM Press, Philadelphia, 2009)
15. I. Moosa, K. Burns, The unbeatable random walk in exchange rate forecasting: reality or myth? J. Macroecon. **40**, 69–81 (2014)
16. P. Rodrigues, N. Salish, Modeling and forecasting interval time series with threshold models. Adv. Data Anal. Classification **9**(1), 41–57 (2015)
17. A. Roque, C. Maté, J. Arroyo, A. Sarabia, iMLP: applying multi-layer perceptrons to interval-valued data. Neural Process. Lett. **25**(2), 157–169 (2007)
18. L. Silva, F. Gomide, R. Yager, in *Participatory learning in fuzzy clustering*. (Reno, Nevada, USA, 2005), pp. 857–861
19. L. Stefanini, A generalization of hukuhara difference and division for interval and fuzzy arithmetic. Fuzzy Sets Syst. **161**, 1564–1584 (2010)
20. Y. Sun, A. Han, Y. Hong, S. Wang, Threshold autoregressive models for interval-valued time series data. J. Econ. **206**, 414–446 (2018)
21. Y. Sun, X. Zhang, Y. Hong, S. Wang, Asymmetric pass-through of oil prices to gasoline prices with interval time series modelling. Energy Econ. **78**, 165–173 (2018)
22. T. Takagi, M. Sugeno, Fuzzy identification of systems and its applications to modeling and control. IEEE Trans. Syst., Man, Cybern. **SMC–15(1)**, 116–132 (1985)
23. X. Tao, C. Li, Y. Bao, Z. Hu, Z. Lu, A combination method for interval forecasting of agricultural commodity futures prices. Knowl.-Based Syst. **77**, 92–102 (2015)
24. X. Wang, S. Li, The interval autoregressive time series model, in (2011), pp. 2528–2533
25. F. Wojciech, J. Salmeron, Evolutionary learning of fuzzy grey cognitive maps for the forecasting of multivariate, interval-valued time series. Int. J. Approx. Reason. **55**, 1319–1335 (2014)
26. T. Xiong, Y. Bao, Z. Hu, Multiple-output support vector regression with a firefly algorithm for interval-valued stock price index forecasting. Knowl.-Based Syst. **55**, 87–100 (2014)

Solving Capacitated Vehicle Routing Problems with Fuzzy Delivery Costs and Fuzzy Demands

Juan Carlos Figueroa García and Jhoan Sebastian Tenjo-García

Abstract This paper shows a method for solving capacitated vehicle routing problems where its delivery costs and constraints are defined as fuzzy sets (information coming from experts). To do so, we use fuzzy numbers to represent delivery costs and constraints which is solved by an iterative algorithm based on the cumulative membership function of a fuzzy set. An equilibrium point among fuzzy costs and constraints is obtained as an overall solution of the problem.

Keywords CVRP · Fuzzy numbers · Cumulative membership function

1 Introduction and Motivation

The Capacitated Vehicle Routing Problem (CVRP) is popular in logistics since it involves not only customers and suppliers but vehicles to perform transportation tasks (see Dantzig and Ramser [3], Christofides and Eilon [2] and Borčinová [1]) where their costs/demands are defined as deterministic.

Several CVRPs have delivering costs/demands which cannot be clearly defined due to a lack of enough statistical data, so a practical approach is to use third party information i.e. experts opinions and perceptions. This way, the experts of the system can provide reliable information which can be summarized as fuzzy sets.

This paper focuses to obtain an optimal solution for a CVRP affected by uncertainty represented with fuzzy delivering costs/demands. To do so, we use the fuzzy optimization method proposed by Figueroa-García [6], and Figueroa-García and López-Bello [5, 8] to solve CVRPs with fuzzy delivering costs/demands.

J. C. Figueroa García (✉)
Universidad Distrital Francisco José de Caldas, Bogotá, Colombia
e-mail: jcfigueroag@udistrital.edu.co

J. S. Tenjo-García
Universidad Nacional de Colombia, Bogotá, Colombia
e-mail: jtenjo@unal.edu.co

B. Bede et al. (eds.), *Fuzzy Information Processing 2020*,
Advances in Intelligent Systems and Computing 1337,
https://doi.org/10.1007/978-3-030-81561-5_8

The paper is divided into five sections. Section 1 shows an introduction. In Sect. 2, the CVRP is presented; Sect. 3 presents some basics on fuzzy numbers. In Sect. 4, the Fuzzy Capacitated Vehicle Routing Problem (FCVRP) and a method to solve it are presented. Section 5 shows an example, and finally some concluding remarks are presented in Sect. 6.

2 The Crisp CVRP

A CVRP is a transportation problem where customers require goods to be sent from a depot using a transportation mean (car, aircraft, train, ship, etc.) often referred as *vehicle*. Every vehicle has a capacity and a single vehicle cannot cover all customers in a row, so multiple vehicles are required to cover customers in different routes (every customer is visited by a single vehicle).

Thus, there is a time/cost associated to cover every customer (often defined as a constant or the distance among customers/depot). In addition, every vehicle starts/ends from/to the depot which is defined as node 1. The mathematical model of a CVRP (minimizing the total delivering cost) is as follows:

$$z = \text{Min} \sum_{i} \sum_{j,i \neq j} c_{ij} x_{ij} \tag{1}$$

$$s.t.$$

$$\sum_{i,i \neq j} x_{ij} = 1 \;\forall\; j = 2, 3, \ldots, m \tag{2}$$

$$\sum_{j,i \neq j} x_{ij} = 1 \;\forall\; i = 2, 3, \ldots, m \tag{3}$$

$$\sum_{i} x_{i1} \leqslant K \tag{4}$$

$$\sum_{j} x_{1j} \leqslant K \tag{5}$$

$$\sum_{i} x_{i1} - \sum_{j,i \neq j} x_{1j} = 0 \tag{6}$$

$$\sum_{i} y_{ij} - \sum_{j,i \neq j} y_{ji} = d_i \;\forall\; i = 2, 3, \ldots, m; \tag{7}$$

$$y_{ij} \leqslant c \cdot x_{ij} \;\forall\; \{i, j\} \in \mathbb{N}_m; i \neq j \tag{8}$$

$$x_{ij} \in \{0, 1\}; \;\; y_{ij} \geqslant 0 \tag{9}$$

Index sets:

$i, j \in \mathbb{N}_m$ is the set of origin–destination nodes

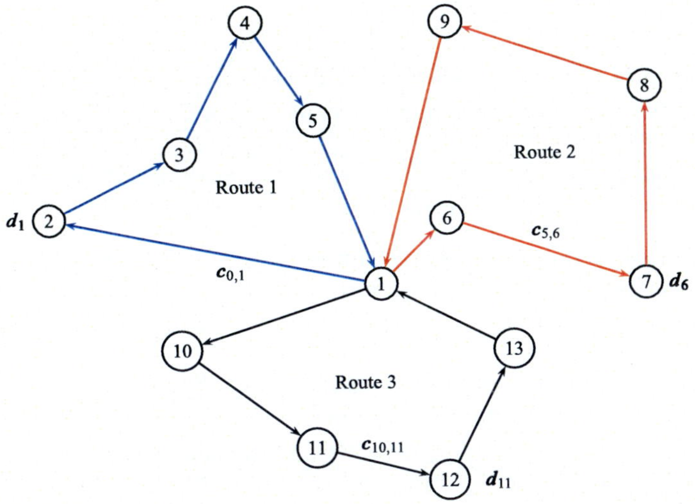

Fig. 1 CVRP for three routes

Parameters:

$c_{ij} \in \mathbb{R}$ is the delivering cost from the i_{th} node to the j_{th} node
$d_i \in \mathbb{R}$ is the demand coming from the i_{th} origin node
$K \in \mathbb{N}$ is the available amount of homogeneous vehicles
$c \in \mathbb{N}$ is the capacity of every vehicle

Decision variables:

$x_{ij} \in \mathbb{Z}$ is the amount of vehicles to come from the i_{th} node to the j_{th} node
$y_{ij} \in \mathbb{Z}$ is demand to supply from the i_{th} node to the j_{th} node

Equations (2) and (3) ensure all destination nodes j to be covered by only one vehicle coming from the origin i. Equations (4) and (5) ensure not to exceed the K available vehicles. Equation (6) prevents from any node to covered by more than one vehicle, Eq. (7) ensures to deliver the demand d_i, and Eq. (8) ensures to send the required quantities y_{ij} of all nodes of the route. All vehicles have the same capacity c and the node $(i, j) = 1$ is the depot (predetermined initial node) of the problem.

In general, what we want is to satisfy the required demands d_i from the i origins to the j destinations via y_{ij}, using a single vehicle x_{ij} per route (given by Eqs. (2) and (3)). If a vehicle x cannot supply all demands of its route (see Eq. (8)) then the model shall create a new route to cover the unsatisfied demands.

The main goal is to minimize the total transportation cost to supply all demands d_i using routes departing from a depot (node 1) and covered by a single vehicle x each. This cost includes all operational and inherent costs to the transportation task to send the required demand d_i using a vehicle x. This way, the main idea of a CVRP is to minimize the total cost z as defined in Eq. (1). Figure 1 shows how the CVRP works. Each vehicle x covers some nodes from the depot and come back to the depot with a limited amount of vehicles (routes) to cover all nodes.

3 Fuzzy Sets/Numbers

A fuzzy set $\tilde{A} = \{(x, \mu_{\tilde{A}}(x)) \mid x \in X\}$ is defined by a membership function $\mu_{\tilde{A}}(x)$, $x \in X$ which measures the membership of a value x regarding the concept/word/label A. $\mathcal{P}(X)$ is the class of all crisp sets, $\mathcal{F}(X)$ is the class of all fuzzy sets, $\mathcal{F}(\mathbb{R})$ is the class of all real-valued fuzzy sets, and $\mathcal{F}_1(\mathbb{R})$ is the class of all fuzzy numbers. A fuzzy number $\tilde{A} \in \mathcal{F}_1(\mathbb{R})$ is then defined as follows.

Definition 1 Let $\tilde{A} : \mathbb{R} \to [0, 1]$ be a fuzzy subset of the reals. Then $\tilde{A} \in \mathcal{F}_1(\mathbb{R})$ is a Fuzzy Number (FN) iff there exists a closed interval $[x_l, x_r] \neq \emptyset$ with a membership function $\mu_{\tilde{A}}(x)$ such that:

$$\mu_{\tilde{A}}(x) = \begin{cases} c(x) & \text{for } x \in [c_l, c_r], \\ l(x) & \text{for } x \in [-\infty, x_l], \\ r(x) & \text{for } x \in [x_r, \infty], \end{cases} \tag{10}$$

where $c(x) = 1$ for $x \in [c_l, c_r]$, $l : (-\infty, x_l) \to [0, 1]$ is monotonic non-decreasing, continuous from the right, i.e. $l(x) = 0$ for $x < x_l$; $l : (x_r, \infty) \to [0, 1]$ is monotonic non-increasing, continuous from the left, i.e. $r(x) = 0$ for $x > x_r$.

The α*-cut* of a fuzzy number $\tilde{A} \in \mathcal{F}_1(\mathbb{R})$ namely ${}^{\alpha}\tilde{A} \triangleq \{x \mid \mu_{\tilde{A}}(x) \geqslant \alpha\}\ \forall\, x \in X$ can be also defined as follows:

$$^{\alpha}\tilde{A} = \left[\inf_x {}^{\alpha}\mu_{\tilde{A}}(x),\ \sup_x {}^{\alpha}\mu_{\tilde{A}}(x)\right] = [\check{A}_{\alpha}, \hat{A}_{\alpha}]. \tag{11}$$

The cumulative function $F(x)$ of a probability distribution $f(x)$ is:

$$F(x) \triangleq \int_{-\infty}^{x} f(t)\, dt, \tag{12}$$

where $x \in \mathbb{R}$. Its fuzzy version is as follows (see Figueroa-García and López-Bello [5, 8], Figueroa-García [6], Pulido-López et al. [13]).

Definition 2 (*Cumulative Membership Function*) Let $\tilde{A} \in \mathcal{F}(\mathbb{R})$ be a fuzzy set and $X \subseteq \mathbb{R}$. The Cumulative Membership Function (CMF) of $\tilde{A}$, $\psi_{\tilde{A}}(x)$ is:

$$\psi_{\tilde{A}}(x) \triangleq Ps_{\tilde{A}}(X \leqslant x). \tag{13}$$

Equation (13) is the cumulative possibility of all $X \leqslant x$ i.e. $Ps(X \leqslant x)$. Then $\psi_{\tilde{A}}(x)$ is normalized by the cardinality (or total area) of $\tilde{A}$ namely Λ, as follows:

$$\psi_{\tilde{A}}(x) \triangleq \frac{1}{\Lambda_{\tilde{A}}} \int_{-\infty}^{x} \mu_{\tilde{A}}(t)\, dt = \frac{\displaystyle\int_{-\infty}^{x} \mu_{\tilde{A}}(t)}{\displaystyle\int_{-\infty}^{\infty} \mu_{\tilde{A}}(t)\, dt}. \tag{14}$$

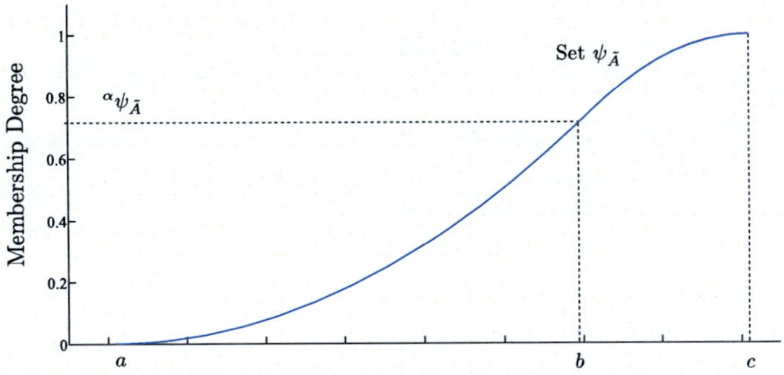

Fig. 2 Cumulative membership function $\psi_{\tilde{A}}$ of a triangular fuzzy set

Figure 2 presents the CMF $\psi_{\tilde{A}}$ of a triangular fuzzy number.

4 The Proposed FCVRP Optimization Method

The mathematical programming form of the CVRP with fuzzy delivering costs, fuzzy demands and positive integer amount of vehicles (FCVRP) is as follows:

$$z = \text{Min} \sum_{i} \sum_{j,i \neq j} \tilde{c}_{ij} x_{ij} \tag{15}$$

$$s.t.$$

$$\sum_{i,i \neq j} x_{ij} = 1 \; \forall \; j = 2, 3, \ldots, m \tag{16}$$

$$\sum_{j,i \neq j} x_{ij} = 1 \; \forall \; i = 2, 3, \ldots, m \tag{17}$$

$$\sum_{i} x_{i1} \leqslant K \tag{18}$$

$$\sum_{j} x_{1j} \leqslant K \tag{19}$$

$$\sum_{i} x_{i1} - \sum_{j,i \neq j} x_{1j} = 0 \tag{20}$$

$$\sum_{j} y_{ij} - \sum_{j,i \neq j} y_{ji} \gtrsim \tilde{d}_i \; \forall \; i = 2, 3, \ldots, m \tag{21}$$

$$y_{ij} \leqslant c \cdot x_{ij} \; \forall \; \{i, j\} \in \mathbb{N}_m; i \neq j \tag{22}$$

$$x_{ij} \in \{0, 1\}; \;\; y_{ij} \geqslant 0 \tag{23}$$

Index sets

$i, j \in \mathbb{N}_m$ is the set of origin–destination nodes

Parameters:

$\tilde{c}_{ij} \in \mathscr{F}_1(\mathbb{R})$ is the fuzzy delivering cost from the i_{th} node to the j_{th} node
$\tilde{d}_i \in \mathscr{F}_1(\mathbb{R})$ is the fuzzy demand coming from the i_{th} origin node
$K \in \mathbb{N}$ is the available amount of homogeneous vehicles
$c \in \mathbb{N}$ is the capacity of every vehicle

Decision variables:

$x_{ij} \in \mathbb{Z}$ is the amount of vehicles to come from the i_{th} node to the j_{th} node
$y_{ij} \in \mathbb{Z}$ is demand to supply from the i_{th} node to the j_{th} node

The binary relation $\lesssim$ for fuzzy sets has been proposed by Ramík and Řimánek [14] as shown as follows:

Definition 3 Let $A, B \in \mathscr{F}_1(\mathbb{R})$ be two fuzzy numbers. Then $A \lesssim B$ if and only if $\sup {}^{\alpha}A \leqslant \sup {}^{\alpha}B$ and $\inf {}^{\alpha}A \leqslant \inf {}^{\alpha}B$ for each $\alpha \in [0, 1]$, where ${}^{\alpha}A$ and ${}^{\alpha}B$ are their α-cuts ${}^{\alpha}A := [\inf {}^{\alpha}A, \sup {}^{\alpha}A]$ and ${}^{\alpha}B := [\inf {}^{\alpha}B, \sup {}^{\alpha}B]$. This binary relation is called the fuzzy max order.

Equation (15) is the total delivering cost where the demands are uncertain (see Eq. (16) for the outgoing means from the i_{th} route and Eq. (17) for the incoming means to the i_{th} route), so they are defined by the experts of the system as fuzzy numbers (see Definition 1). Basically, this model considers uncertainty over delivering costs since there are several sources that can affect them such as climate, transportation times, road conditions, etc. and uncertain amount of demands since the exact amount of goods to be sent are uncertain itself (customers demands, clients requirements, variability of the demand, etc.).

Figure 3 shows an FCVRP in which $\tilde{c}_{ij}$ is a fuzzy number (see Eq. (10)), and every demand $\tilde{d}_i$ (see Eq. (24)) is a linear fuzzy number assuming customers having non–deterministic requirements due to market conditions, fluctuations of the usage of resources, etc.:

$$\mu_{\tilde{d}_i}(f(\boldsymbol{x}), \check{d}_i, \hat{d}_i) = \begin{cases} 1, & f(\boldsymbol{x}) \geqslant \hat{d}_i \\ \dfrac{f(\boldsymbol{x}) - \check{d}_i}{\hat{d}_i - \check{d}_i}, & \check{d}_i \leqslant f(\boldsymbol{x}) \leqslant \hat{d}_i \\ 0, & f(\boldsymbol{x}) \leqslant \check{d}_i \end{cases} \tag{24}$$

where $\check{d}_i \in \mathbf{Z}$ and $\hat{d}_i \in \mathbf{Z}$ are the lower/upper bounds of $\tilde{d}_i$.

The soft constraints model defines $\mu_{\tilde{d}_i}(f(\boldsymbol{x}))$ as a linear fuzzy set that represents a soft inequality namely $\gtrsim$ where $f(\boldsymbol{x})$ is the left side of every constraint i.e. $\sum_j y_{ij} - \sum_{j,i \neq j} y_{ji}$ seen as the universe of discourse of $\tilde{d}_i$ and for which every $f(\boldsymbol{x})$ returns a membership degree (see Eq. (24)). This way, we use an iterative version of the Zimmermann soft constraints model to solve the FCVRP.

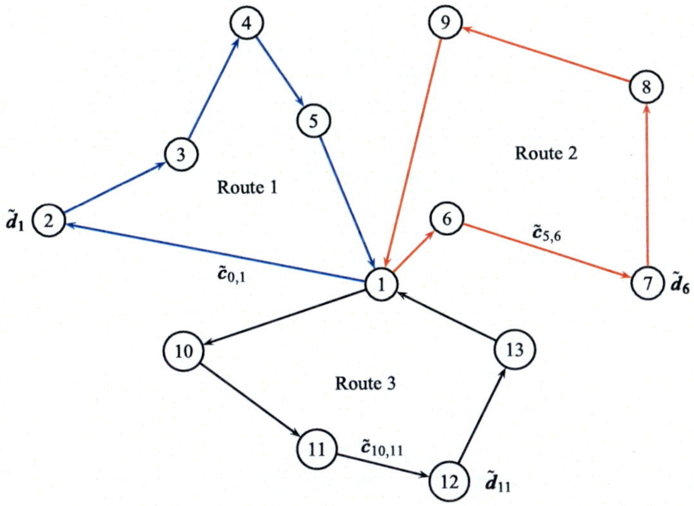

Fig. 3 Fuzzy CVRP for three routes

4.1 The Proposed Method

The main goal is to minimize $\sum_i \sum_{j,i\neq j} \tilde{c}_{ij}x_{ij}$ based on concepts introduced by Rommelfanger [16], Ramík [15], Inuiguchi and Ramík [11], Fiedler et al. [4], Ramík and Řimánek [14], Inuiguchi et al. [12], we apply the proposal of Figueroa-García and Tenjo [9], Figueroa-García [6] and Figueroa-García and López-Bello [5, 8]. Let us define the set of all crisp constraints of the FCVRP i.e. Equations (16), (17), (18), (19), (20) and (22) as $g(\boldsymbol{x}, \boldsymbol{b})$, then the proposed method is as follows:

1- Setup:

- Set $\alpha \in [0, 1]$,
- Compute $\psi'_{\tilde{c}_{ij}} = 1 - \psi_{\tilde{c}_{ij}}$ and ${}^{\alpha}\psi_{\tilde{c}_{ij}}$ $\forall\,(i, j)$,

2- Soft constraints method:

- Compute $\check{z} = \text{Min}\{z = \sum_i \sum_{j,i\neq j} {}^{\alpha}\psi'_{\tilde{c}_{ij}} x_{ij} : g(\boldsymbol{x}, \boldsymbol{b}); \sum_j y_{ij} - \sum_{j,i\neq j} y_{ji} \geqslant \check{d}_i\}$
- Compute $\hat{z} = \text{Min}\{z = \sum_i \sum_{j,i\neq j} {}^{\alpha}\psi'_{\tilde{c}_{ij}} x_{ij} : g(\boldsymbol{x}, \boldsymbol{b}); \sum_j y_{ij} - \sum_{j,i\neq j} y_{ji} \geqslant \hat{d}_i\}$
- Define the set $\tilde{z}$ with the following membership function:

$$\mu_{\tilde{z}}(z, \check{z}, \hat{z}) = \begin{cases} 1, & z \leqslant \check{z} \\ \dfrac{\hat{z} - z}{\hat{z} - \check{z}}, & \check{z} \leqslant z \leqslant \hat{z} \\ 0, & z \geqslant \hat{z} \end{cases} \tag{25}$$

– Thus, solve the following LP model:

$$\text{Max } \{\lambda\},$$
$$s.t.$$
$$\sum_i \sum_{j,i\neq j} {}^{\alpha}\psi'_{\tilde{c}_{ij}} x_{ij} + \lambda(\hat{z} - \check{z}) = \hat{z}, \tag{26}$$
$$g(\boldsymbol{x}, \boldsymbol{b}), \tag{27}$$
$$\sum_j y_{ij} - \sum_{j,i\neq j} y_{ji} - \lambda(\hat{d}_i - \check{d}_i) = \check{d}_i \; \forall\, i \in \mathbb{N}_m, \tag{28}$$
$$x_{ij} \in \{0, 1\};\; y_{ij} \tag{29}$$

3- Convergence:

– If $\lambda^* = \alpha$ then stop and return λ^* as the overall satisfaction degree of $\tilde{z}$, $\tilde{c}_{ij}$, and $\tilde{d}_i$; if $\lambda^* \neq \alpha$ then go to Step 1 and update $\alpha = \lambda^*$.

Here, we use the complement of $\psi_{\tilde{c}_{ij}}$ i.e. $\psi'_{\tilde{c}_{ij}} = 1 - \psi_{\tilde{c}_{ij}}$ because the main idea is to maximize the membership degree of the costs (as lower as better), and the constraints (as bigger as better). This way, the set $\tilde{z}$ is defined over a universe of discourse $z = \sum_i \sum_{j,i\neq j} c_{ij} x_{ij}$ and it is intended to represent *minimum delivering cost* so its highest membership degree is given by $\check{z}$ and its lowest membership degree is given by $\hat{z}$ (the same way than $\psi'_{\tilde{c}_{ij}}$).

Also note that the Zimmermann's method maximizes the overall satisfaction degree of all soft constraints namely λ (via $\boldsymbol{x}$) and its objective function z.

5 Application Example

The selected instance to solve a FCVRP was the subset `E-n13-4k` of the Set E proposed by Christofides and Eilon see [2] problem available at

`http://www.vrp-rep.org/datasets/item/2014-0010.html`

composed by $(i, j) = 1, 2, \ldots, 13$ nodes to cover using vehicles with capacity $c = 6000$. We use triangular fuzzy costs $t = (a, b, c)$ and requirements as defined in Eq. (24) with parameters $\check{d}_i$, $\hat{d}_i$ (see Appendix for its full description). We solve the problem using two approaches: the classical crisp solution (see Borčinová [1]) and the fuzzy proposed solution shown in Sect. 1.

5.1 Crisp Solution

The crisp solution of the `E-n13-4k` problem proposed by Christofides and Eilon (see [2]) is composed by the following costs/demands:

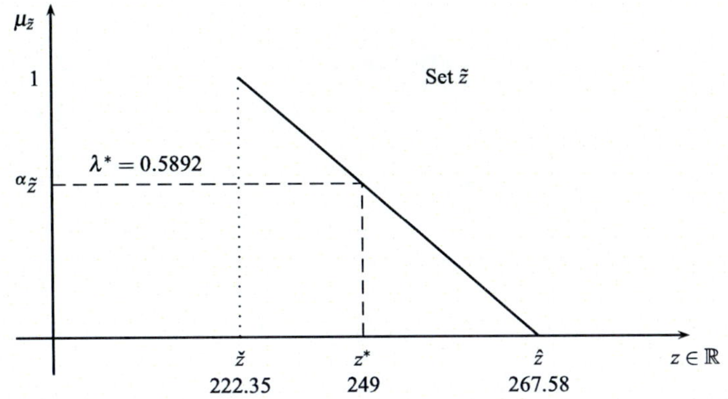

Fig. 4 Optimal solution of the problem

$$
\boldsymbol{c} = \begin{bmatrix}
0 & 9 & 14 & 21 & 23 & 22 & 25 & 32 & 36 & 38 & 42 & 50 & 52 \\
 & 0 & 5 & 12 & 22 & 21 & 24 & 31 & 35 & 37 & 41 & 49 & 51 \\
 & & 0 & 7 & 17 & 16 & 23 & 26 & 30 & 36 & 36 & 44 & 46 \\
 & & & 0 & 10 & 21 & 30 & 27 & 37 & 43 & 31 & 37 & 39 \\
 & & & & 0 & 19 & 28 & 25 & 35 & 41 & 29 & 31 & 29 \\
 & & & & & 0 & 9 & 10 & 16 & 22 & 20 & 28 & 30 \\
 & & & & & & 0 & 7 & 11 & 13 & 17 & 25 & 27 \\
 & & & & & & & 0 & 10 & 16 & 10 & 18 & 20 \\
 & & & & & & & & 0 & 6 & 6 & 14 & 16 \\
 & & & & & & & & & 0 & 12 & 12 & 20 \\
 & & & & & & & & & & 0 & 8 & 10 \\
 & & & & & & & & & & & 0 & 10 \\
 & & & & & & & & & & & & 0
\end{bmatrix} ; \boldsymbol{d} = \begin{bmatrix} 0 \\ 1200 \\ 1700 \\ 1500 \\ 1400 \\ 1700 \\ 1400 \\ 1200 \\ 1900 \\ 1800 \\ 1600 \\ 1700 \\ 1100 \end{bmatrix}
$$

In this example, delivering costs (assumed as flat distances) have no uncertainty and every vehicle has a capacity of $c = 6000$ units. The obtained solution (see Borčinová [1]) is $z^* = 247$ composed by 4 vehicles covering four routes:

R1: 1 - 2 - 1
R2: 1 - 4 - 12 - 5 - 8 - 1
R3: 1 - 7 - 11 - 13 - 10 - 1
R4: 1 - 9 - 6 - 3 - 1

5.2 *Fuzzy Solution*

We started the algorithm with $\alpha = 0.2$ to then compute $\psi'_{\tilde{t}_{ij}}$ and ${}^{\alpha}\psi'_{\tilde{t}_{ij}}$ $\forall (i, j)$ as shown in (14). After 5 iterations of the algorithm, an optimal $\lambda^* = 0.58924$ led to $z^* = 249$, $\check{z} = 222.35$, $\hat{z} = 267.58$ which is a little more expensive than the crisp solution. The obtained 4 optimal routes that cover all nodes are as follows:

R1: 1 - 4 - 6 - 9 - 1
R2: 1 - 2 - 1
R3: 1 - 7 - 11 - 8 - 3 - 1
R4: 1 - 10 - 13 - 5 - 12 - 1

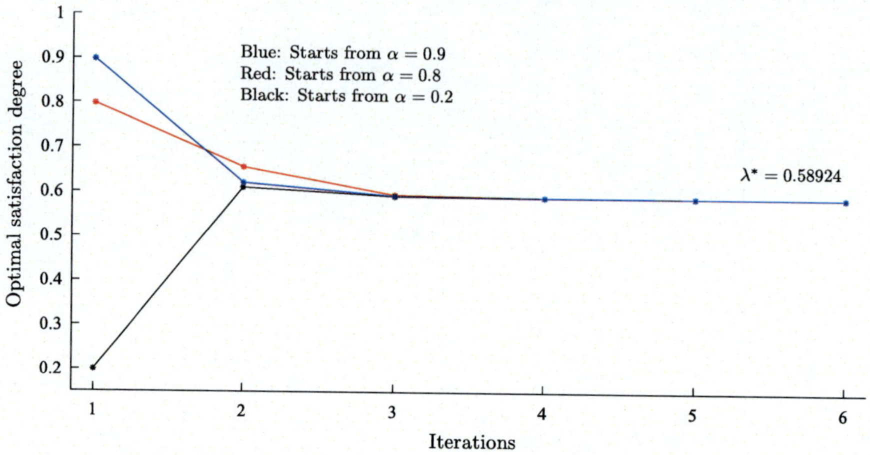

Fig. 5 Convergence of the proposed method for $\alpha = \{0.2, 0.8, 0.9\}$

where the defuzzified values of c_{ij} and d_i are:

$$c = \begin{bmatrix} 0 & 7.11 & 13.49 & 22.34 & 24.24 & 21.66 & 22.03 & 29.03 & 35.49 & 35.48 & 41.74 & 47.88 & 47.94 \\ & 0 & 5.57 & 10.94 & 19.03 & 21.67 & 21.81 & 29.57 & 32.48 & 35.85 & 43.01 & 46.81 & 50.66 \\ & & 0 & 6.74 & 14.14 & 13.81 & 22.66 & 21.94 & 27.48 & 32.22 & 35.66 & 44.31 & 44.94 \\ & & & 0 & 8.76 & 22.15 & 22.44 & 28.91 & 36.83 & 45.28 & 29.91 & 39.93 & 36.03 \\ & & & & 0 & 17.85 & 25.88 & 23.11 & 32.81 & 41.48 & 25.72 & 30.66 & 29.39 \\ & & & & & 0 & 7.85 & 10.31 & 11.63 & 19.48 & 15.40 & 25.81 & 26.72 \\ & & & & & & 0 & 6.83 & 9.57 & 13.67 & 15.57 & 25.57 & 28.05 \\ & & & & & & & 0 & 10.57 & 13.48 & 8.11 & 19.91 & 20.13 \\ & & & & & & & & 0 & 4.91 & 4.57 & 9.40 & 15.74 \\ & & & & & & & & & 0 & 13.91 & 15.97 & 16.22 \\ & & & & & & & & & & 0 & 5.03 & 12.28 \\ & & & & & & & & & & & 0 & 12.68 \\ & & & & & & & & & & & & 0 \end{bmatrix}; \boldsymbol{d} = \begin{bmatrix} 0 \\ 1353.54 \\ 1812.47 \\ 1371.39 \\ 1435.70 \\ 1694.62 \\ 1335.70 \\ 1194.62 \\ 1889.24 \\ 1630.32 \\ 1471.39 \\ 1730.32 \\ 1171.39 \end{bmatrix}$$

It is interesting to see that the obtained optimal $\lambda^* = 0.5892$ is the same no matter where it starts from ($\alpha = \{0.2, 0.8, 0.9\}$). Figure 4 shows the set $\tilde{z}$ of optimal delivering costs for a global optimal degree $\lambda^* = 0.5892$.

Note that $\lambda^* = 0.5892$ is the optimal satisfaction degree of $\tilde{z}$, $\tilde{c}_{ij}$ and $\tilde{d}_i$ found by the algorithm. Figure 5 shows λ^* per iteration for 3 different starting values $\alpha = \{0.2, 0.8, 0.9\}$ where the same optimal value $\lambda^* = 0.5892$ is reached.

6 Concluding Remarks

An FCVRP is solved by the algorithm proposed by Figueroa-García [6] and Figueroa-García and López-Bello [5, 8] with satisfactory results. The obtained optimal overall satisfaction degree e.g. $\lambda^* = 0.5892$ in the example, operates as a defuzzifier for all fuzzy parameters $\tilde{z}$, $\tilde{c}_{ij}$ and $\tilde{d}_i$.

The proposed method can deal with nonlinear fuzzy delivering costs and linear fuzzy demands. It iterates the classical Zimmermann method starting from any value $\alpha \in [0, 1]$ to finally reach a stable optimal λ^*.

The defuzzified parameters provide the optimal amount of vehicles/routes and crisp values for $c_{ij} \in \tilde{c}_{ij}$ and $d_i \in \tilde{d}_i$. Those values lead to an equilibrium between delivery costs and satisfied demands which is very useful in decision making. This way, managers can handle uncertain information and provide an expected route/cost to customers.

7 Further Topics

VRPs involving Type-2 fuzzy sets (see Figueroa-García [7]) are a natural extension of this model and the applied algorithm.

Appendix

Table 1 Fuzzy delivering costs and demands

(i, j)	$t(a, b, c)$	(i, j)	$t(a, b, c)$	(i, j)	$t(a, b, c)$	Demand	Value
(2,1)	$t(6, 9, 10)$	(8,6)	$t(19, 23, 27)$	(11,8)	$t(8, 16, 20)$	$\check{d}_1$	1000
(3,1)	$t(8, 14, 20)$	(8,7)	$t(19, 26, 29)$	(11,9)	$t(17, 22, 25)$	$\check{d}_2$	1400
(3,2)	$t(19, 21, 25)$	(9,1)	$t(25, 30, 33)$	(11,10)	$t(13, 20, 22)$	$\check{d}_3$	900
(4,1)	$t(20, 23, 28)$	(9,2)	$t(30, 36, 38)$	(12,1)	$t(24, 28, 30)$	$\check{d}_4$	1200
(4,2)	$t(18, 22, 26)$	(9,3)	$t(32, 36, 40)$	(12,2)	$t(24, 30, 33)$	$\check{d}_5$	1400
(4,3)	$t(20, 25, 27)$	(9,4)	$t(39, 44, 50)$	(12,3)	$t(5, 7, 9)$	$\check{d}_6$	1100
(5,1)	$t(27, 32, 34)$	(9,5)	$t(42, 46, 49)$	(12,4)	$t(8, 11, 13)$	$\check{d}_7$	900
(5,2)	$t(30, 36, 42)$	(9,6)	$t(4, 10, 15)$	(12,5)	$t(12, 13, 15)$	$\check{d}_8$	1300
(5,3)	$t(33, 38, 41)$	(9,7)	$t(17, 21, 27)$	(12,6)	$t(14, 17, 19)$	$\check{d}_9$	1100
(5,4)	$t(39, 42, 45)$	(9,8)	$t(18, 30, 34)$	(12,7)	$t(23, 25, 28)$	$\check{d}_{10}$	1000
(6,1)	$t(42, 50, 56)$	(10,1)	$t(23, 27, 34)$	(12,8)	$t(22, 27, 34)$	$\check{d}_{11}$	1200
(6,2)	$t(45, 52, 55)$	(10,2)	$t(35, 37, 39)$	(12,9)	$t(8, 10, 13)$	$\check{d}_{12}$	700
(6,3)	$t(3, 5, 8)$	(10,3)	$t(42, 43, 47)$	(12,10)	$t(11, 16, 19)$	$\hat{d}_1$	1600
(6,4)	$t(8, 12, 15)$	(10,4)	$t(29, 31, 32)$	(12,11)	$t(7, 10, 11)$	$\hat{d}_2$	2100
(6,5)	$t(17, 22, 24)$	(10,5)	$t(35, 37, 43)$	(13,1)	$t(14, 18, 25)$	$\hat{d}_3$	1700
(7,1)	$t(20, 21, 23)$	(10,6)	$t(34, 39, 41)$	(13,2)	$t(13, 20, 28)$	$\hat{d}_4$	1600
(7,2)	$t(20, 24, 26)$	(10,7)	$t(14, 19, 23)$	(13,3)	$t(4, 6, 7)$	$\hat{d}_5$	1900
(7,3)	$t(28, 31, 33)$	(10,8)	$t(20, 28, 34)$	(13,4)	$t(3, 6, 8)$	$\hat{d}_6$	1500
(7,4)	$t(30, 35, 38)$	(10,9)	$t(22, 25, 26)$	(13,5)	$t(7, 14, 16)$	$\hat{d}_7$	1400
(7,5)	$t(32, 37, 41)$	(11,1)	$t(31, 35, 37)$	(13,6)	$t(13, 16, 19)$	$\hat{d}_8$	2300
(7,6)	$t(38, 41, 47)$	(11,2)	$t(38, 41, 45)$	(13,7)	$t(8, 12, 19)$	$\hat{d}_9$	2000
(8,1)	$t(45, 49, 51)$	(11,3)	$t(23, 29, 32)$	(13,8)	$t(11, 12, 18)$	$\hat{d}_{10}$	1800
(8,2)	$t(47, 51, 55)$	(11,4)	$t(27, 31, 35)$	(13,9)	$t(14, 20, 22)$	$\hat{d}_{11}$	2100
(8,3)	$t(4, 7, 10)$	(11,5)	$t(25, 29, 34)$	(13,10)	$t(3, 8, 10)$	$\hat{d}_{12}$	1500
(8,4)	$t(11, 17, 21)$	(11,6)	$t(4, 9, 13)$	(13,11)	$t(9, 10, 14)$		
(8,5)	$t(12, 16, 18)$	(11,7)	$t(5, 10, 16)$	(13,12)	$t(6, 10, 18)$		

References

1. Z. Borčinová, Two models of the capacitated vehicle routing problem. Croat. Oper. Res. Rev. **8**(1), 463–469 (2017)
2. N. Christofides, S. Eilon, An algorithm for the vehicle routing dispatching problem. Oper. Res. Quaterly **20**(1), 309–918 (1969)
3. G. Dantzig, J. Ramser, The truck dispatching problem. Manag. Sci. **6**(1), 80–91 (1959)
4. M. Fiedler, J. Nedoma, J. Ramík, J. Rohn, K. Zimmermann, Linear optimization problems with inexact data (Springer, Berlin, 2006). https://doi.org/10.1007/0-387-32698-7
5. J.C. Figueroa, C.A. López, Linear programming with fuzzy joint parameters: a cumulative membership function approach, in (2008)
6. J.C. Figueroa-García, Mixed production planning under fuzzy uncertainty: a cumulative membership function approach, in *Workshop on Engineering Applications (WEA)*, vol. 1, ed. by in IEEE, (IEEE, 2012), pp. 1–6. https://doi.org/10.1109/WEA.2012.6220081
7. J.C. Figueroa-García, G. Hernández, A transportation model with interval type-2 fuzzy demands and supplies. Lecture Notes in Computer Science **7389**(1), 610–617 (2012). https://doi.org/10.1007/978-3-642-31588-6_78
8. J.C. Figueroa-García, C.A. López, Pseudo-optimal solutions of flp problems by using the cumulative membership function, in *Annual Meeting of the North American Fuzzy Information Processing Society (NAFIPS)*, vol. 28 (IEEE, 2009), pp. 1–6. https://doi.org/10.1109/NAFIPS.2009.5156464
9. J.C. Figueroa-García, J.S. Tenjo, Solving transhipment problems with fuzzy delivery costs and fuzzy constraints. Commun. Comput. Inf. Sci. **831**(1), 538–550 (2018). https://doi.org/10.1007/978-3-319-95312-0_47
10. D.P. Heyman, M.J. Sobel, *Stochastic Models in Operations Research*, vol. II: Stochastic Optimization (Dover Publishers, 2003)
11. M. Inuiguchi, J. Ramík, Possibilistic linear programming: a brief review of fuzzy mathematical programming and a comparison with stochastic programming in portfolio selection problem. Fuzzy Sets Syst. **111**, 3–28 (2000)
12. M. Inuiguchi, J. Ramik, T. Tanino, M. Vlach, Satisficing solutions and duality in interval and fuzzy linear programming. Fuzzy Sets Syst. **135**(1), 151–177 (2003). https://doi.org/10.1016/S0165-0114(02)00253-1
13. D.G. Pulido-López, M. García, J.C. Figueroa-García, Fuzzy uncertainty in random variable generation: a cumulative membership function approach. Commun. Comput. Inf. Sci. **742**(1), 398–407 (2017). https://doi.org/10.1007/978-3-319-66963-2_36
14. J. Ramík, J. ímánek, Inequality relation between fuzzy numbers and its use in fuzzy optimization. Fuzzy Sets Syst. **16**, 123–138 (1985). https://doi.org/10.1016/S0165-0114(85)80013-0
15. J. Ramík, Optimal solutions in optimization problem with objective function depending on fuzzy parameters. Fuzzy Sets Syst. **158**(17), 1873–1881 (2007)
16. H. Rommelfanger, A general concept for solving linear multicriteria programming problems with crisp, fuzzy or stochastic values. Fuzzy Sets Syst. **158**(17), 1892–1904 (2007)

Sugeno Integral over Generalized Semi-quantales

Jan Paseka, Sergejs Solovjovs, and Milan Stehlík

Abstract This paper introduces an extension of the Sugeno integral over generalized semi-quantales. We prove that our generalized Sugeno integral is a scale invariant compatible aggregation function. Moreover, we show that the generalized Sugeno integral provides a one-to-one correspondence between the set of capacities and the set of aggregation functions, which additionally are quantale module homomorphisms.

1 Introduction

In this paper, we propose a new approach to information aggregation with the help of a generalized form of the Sugeno integral [7]. The motivation for such information aggregation is enormous need to develop a predictive analytics for processing of the sets of images of the biopsies or PET/CT scans of investigated mammary tissue. This can be of tremendous help for improving probability of the correct classification of mammary cancer and/or mastopaty. For fractal based methods for the readable introduction into this topic see [6]. However, complex structure of individual images and sampling errors needs to be properly integrated, that means in the first step

J. Paseka (✉)
Department of Mathematics and Statistics, Faculty of Science, Masaryk University, Kotlářská 2, 611 37, Brno, Czech Republic
e-mail: paseka@math.muni.cz

S. Solovjovs
Department of Mathematics, Faculty of Engineering, Czech University of Life Sciences Prague, Kamýcká 129,165 00 Prague, Czech Republic
e-mail: solovjovs@tf.czu.cz

M. Stehlík
Institute of Applied Statistics and Linz Institute of Technology, Johannes Kepler University Linz, Altenberger Street 69, 4040 Linz, Austria
e-mail: Milan.Stehlik@jku.at

Institute of Statistics, Universidad de Valparaíso, Valparaíso, Chile

B. Bede et al. (eds.), *Fuzzy Information Processing 2020*,
Advances in Intelligent Systems and Computing 1337,
https://doi.org/10.1007/978-3-030-81561-5_9

construction of an appropriate aggregation technique. An attractive surrogate from perspective of current oncology can be application of the wavelet-based multifractal spectra and random carpet model to digital mammary images representing both the mastopathic and cancerous tissues in order to compute values of capacity $\mu(A)$ for a given set A from some universe. The clinical application of the method on real data images will be considered as valuable future work, which will be requiring as prerequisites both the ethical committee assessment of such analytical method and proper sharing data for cancer research protocols. From this perspective, the construction of the Sugeno integral over generalized semi-quantales is the first step, developing aggregation technique as a benchmark for further investigation.

We define thus an extension of the Sugeno integral for generalized semi-quantales (inspired by, e.g., [1, 3], which consider variants of the Sugeno integrals on lattices; these variants are different from those used in the case of the unit interval [0, 1]). In particular, we replace the meet operation in the classical definition of the Sugeno integral with a family of quantale-like tensor multiplications with convenient properties, and then single out the most suitable (for our needs) such multiplications.

Our introduced generalized Sugeno integral deserves a particular theoretical study on its own, which is *not* the topic of this paper. We do, however, provide an analogue of certain results (e.g., related to the underlying lattice congruences) of [3] in the third section of this paper. In particular, we prove that our generalized Sugeno integral is a scale invariant compatible aggregation function. Moreover, we show that the generalized Sugeno integral provides a one-to-one correspondence between the set of capacities and the set of aggregation functions, which additionally are quantale module homomorphisms.

2 Motivating Preliminaries

In multiple-criteria decision making, one often relies on a suitable aggregation function to aggregate the available information and make the final decision. The main idea of aggregation is that the output value, computed by the aggregation function, should represent or synthesize in some sense all individual inputs [2]. This paper concentrates on a particular aggregation function in the form of a modification of the Sugeno integral of [7], which has already gained popularity in the literature.

We begin with some preliminaries, where [0, 1] stands for the standard unit interval of the real line. We also assume that the set $\mathbb{N}$ of natural numbers has the form $\{1, 2, 3, \ldots\}$, namely, has no zero element. Given a set X, $\mathcal{P}(X)$ (resp. $\mathcal{P}_{fin}(X)$) stands for the set of (resp. finite) subsets of X. Additionally, $\mathcal{P}^{\varnothing}_{fin}(X)$ will denote the set of finite non-empty subsets of X. Given a complete lattice L and a subset $S \subseteq L$ (possibly empty), $\bigvee S$ (resp. $\bigwedge S$) stands for the join (resp. meet) of S.

Definition 1 Given $n \in \mathbb{N}$, we denote by $[n]$ the set $\{1, \ldots, n\}$. A map $\mu : \mathcal{P}([n]) \to [0, 1]$ is said to be a *capacity* (also known as *fuzzy measure*) provided that $\mu(\varnothing) = 0$, $\mu([n]) = 1$, and, for every $A, B \in \mathcal{P}([n])$, $A \subseteq B$ implies $\mu(A) \leqslant \mu(B)$. Given a capacity μ, the *Sugeno integral* w.r.t. μ is a map $\mathsf{Su}_\mu : [0,1]^{[n]} \to [0,1]$ defined by

$$\mathsf{Su}_\mu(\mathbf{x}) = \bigvee_{A \subseteq [n]} \left(\mu(A) \wedge \bigwedge_{i \in A} x_i \right) \tag{SI1}$$

for every $\mathbf{x} = (x_1, \ldots, x_n) \in [0,1]^{[n]}$.

We notice first that in practical examples, one can encounter maps $\mu : \mathcal{P}([n]) \to [0,1]$, which are not order-preserving, and, therefore, are not capacities. Given such a map μ, however, one can define a new map $\mu^* : \mathcal{P}([n]) \to [0,1]$ by $\mu^*(A) = \bigvee_{B \subseteq A} \mu(B)$ for $A \notin \{\varnothing, [n]\}$, $\mu^*(\varnothing) = 0$, and $\mu^*([n]) = 1$. The new map is always order-preserving (easy exercise), namely, gives a capacity. Moreover, if μ is a capacity, then $\mu = \mu^*$; otherwise, $\mu \leqslant \mu^*$, except maybe at $\varnothing$. Thus, for a map $\mu : \mathcal{P}([n]) \to [0,1]$, we can rewrite (SI1) as

$$\mathsf{Su}_{\mu^*}(\mathbf{x}) = \bigvee_{A \subseteq [n]} \left(\mu^*(A) \wedge \bigwedge_{i \in A} x_i \right). \tag{SI1a}$$

Alternatively, we can employ the order-preserving map $\mu_* : \mathcal{P}([n]) \to [0,1]$, $\mu_*(A) = \bigwedge_{A \subseteq B} \mu(B)$ for $A \notin \{\varnothing, [n]\}$, $\mu_*(\varnothing) = 0$, and $\mu_*([n]) = 1$ (it follows then that $\mu_* \leqslant \mu$, except maybe at $[n]$), getting

$$\mathsf{Su}_{\mu_*}(\mathbf{x}) = \bigvee_{A \subseteq [n]} \left(\mu_*(A) \wedge \bigwedge_{i \in A} x_i \right). \tag{SI1b}$$

In particular, $\mathsf{Su}_{\mu_*}(\mathbf{x}) \leqslant \mathsf{Su}_\mu(\mathbf{x}) \leqslant \mathsf{Su}_{\mu^*}(\mathbf{x})$ if $\mu_* \leqslant \mu \leqslant \mu^*$ (including $\varnothing$ and $[n]$).

We notice second that some practical examples can provide partial maps $\mu : \mathcal{P}([n]) \rightsquigarrow [0,1]$, i.e., maps, which are not defined for every subset $A \subseteq [n]$. To overcome such an inconvenience, we will proceed as follows. Given a subset $\mathcal{S} \subseteq \mathcal{P}([n])$ and a map $\mu : \mathcal{S} \to [0,1]$, we will define a new map $\bar{\mu}^* : \mathcal{P}([n]) \to [0,1]$ by $\bar{\mu}^*(A) = \bigvee_{B \subseteq A,\, B \in \mathcal{S}} \mu(B)$ for $A \notin \{\varnothing, [n]\}$, $\bar{\mu}^*(\varnothing) = 0$, and $\bar{\mu}^*([n]) = 1$. One obtains thus a capacity $\bar{\mu}^*$, which allows us to rewrite (SI1) in the form of

$$\mathsf{Su}_{\bar{\mu}^*}(\mathbf{x}) = \bigvee_{A \subseteq [n]} \left(\bar{\mu}^*(A) \wedge \bigwedge_{i \in A} x_i \right). \tag{SI1c}$$

Alternatively, we can use the order-preserving map $\bar{\mu}_* : \mathcal{P}([n]) \to [0,1]$, $\bar{\mu}_*(A) = \bigwedge_{A \subseteq B,\, B \in \mathcal{S}} \mu(B)$ for $A \notin \{\varnothing, [n]\}$, $\bar{\mu}_*(\varnothing) = 0$, and $\bar{\mu}_*([n]) = 1$, getting

$$\mathsf{Su}_{\bar{\mu}_*}(\mathbf{x}) = \bigvee_{A \subseteq [n]} \left(\bar{\mu}_*(A) \wedge \bigwedge_{i \in A} x_i \right). \tag{SI1d}$$

We would like to introduce two additional generalizations, namely, first, to make the unit interval [0, 1] into a convenient complete lattice L, and, second, to substitute the meet operation in the definition of Sugeno integral (SI1) with a family of suitable quantale-like tensor multiplications on L.

It is important to emphasize that there already exist several versions of the Sugeno integral over lattices, different from the unit interval [0, 1], in which the meet operation "$\wedge$" in (SI1) is replaced with a suitable residuated multiplication (see, e.g., [1] and the references within). Up to our knowledge, however, none of them replaces "$\bigwedge_{i\in A} x_i$" in (SI1) with a suitable operation "$\bigodot_{i\in A} x_i$".

The following section provides the abstract theoretical setting for our proposed generalized Sugeno integral.

3 Sugeno Integral over Generalized Semi-quantales

Given two sets X and I, we introduce the following notational conventions.

(1) Elements of the set X^I (the set of all maps from X to I) will be denoted by bold symbols $\mathbf{x}$, $\mathbf{y}$, $\mathbf{z}$, whose values $\mathbf{x}(i)$ will be denoted x_i.
(2) For every subset $K \subseteq I$ and every $\mathbf{x} \in X^I$, $\mathbf{x}|_K$ will stand for the restriction of $\mathbf{x}$ to the subset K. If K has exactly m elements $(k_1, \ldots, k_m)$, then $\mathbf{x}|_K$ is the m-tuple, obtained by taking those coordinates of $\mathbf{x}$, which belong to K, in the increasing order, i.e., $(x_{k_1}, \ldots, x_{k_m})$. For singletons, we write $\mathbf{x}|_{\{k\}} = x_k$.
(3) For every subset $K \subseteq I$ and every $\mathbf{x}, \mathbf{y} \in X^I$, $\mathbf{x}_K\mathbf{y}$ stands for the I-tuple, whose i-th coordinate is x_i, if $i \in K$, and y_i, otherwise. In particular, we write $x_K\mathbf{y}$, $\mathbf{x}_K y$, or $x_K y$, depending on whether $\mathbf{x} = (x, \ldots, x)$ (i.e., $x_i = x$ for all $i \in I$), or $\mathbf{y} = (y, \ldots y)$, or both. Notice that if $K = \varnothing$, then, e.g., $x_K x = x_\varnothing x = (x, \ldots, x)$.
(4) A map $d : X^I \to X$ is said to *have an identity element* e ($e \in X$) provided that for every $i \in I$ and every $x \in X$, it follows that $d(x_{\{i\}}e) = x$.
(5) A map $d : X^I \to X$ is said to *have a zero element* 0 ($0 \in X$) provided that for every $i \in I$ and every $\mathbf{x} \in X^I$, it follows that $d(0_{\{i\}}\mathbf{x}) = 0$.
(6) Given a map $d : X^I \to X$, an element $x \in X$ is said to be *idempotent* w.r.t. d provided that $d(x_\varnothing x) = x$. We say that the map d is *idempotent* provided that every $x \in X$ is idempotent w.r.t. d. $\mathsf{Idem}(X^I, X)$ denotes the set of all idempotent maps from X^I to X.

Example 1 Every n-copula $C : [0, 1]^n \to [0, 1]$ (cf., e.g., [2]) is an n-ary map with the zero element 0 and the unit 1 which is order-preserving in each variable.

3.1 *Generalized Semi-quantales*

Following the idea of semi-quantale of [5], we introduce its generalization, suitable for our setting.

Definition 2 (1) A *generalized semi-quantale* $(L, \leqslant, \otimes, (\gamma_n)_{n\in\mathbb{N}})$ (*gs-quantale* for short) is a complete lattice $(L, \leqslant)$, equipped with a binary operation $\otimes : L \times L \to L$ and a collection of n-ary operations $\gamma_n : L^n \to L$ for every $n \in \mathbb{N}$ with no additional assumptions (like, e.g., associativity or distributivity w.r.t. $\otimes$). We call $\otimes$ a *tensor product* on L. Morphisms of gs-quantales are maps preserving arbitrary joins, $\otimes$, and γ_n for every $n \in \mathbb{N}$. The smallest (resp. greatest) element of L will be denoted $\bot$ (resp. $\top$).

(2) A gs-quantale $(L, \leqslant, \otimes, (\gamma_n)_{n\in\mathbb{N}})$ is *multisided* provided that $\gamma_n(x_1, \ldots, x_n) \leqslant x_i$ for every $n \in \mathbb{N}$, $i \in \{1, \ldots, n\}$, and every $x_1, \ldots, x_n \in L$.

(3) An *ordered gs-quantale* $(L, \leqslant, \otimes, (\gamma_n)_{n\in\mathbb{N}})$ (*ogs-quantale* for short) is a gs-quantale such that $\otimes$ and γ_n for every $n \in \mathbb{N}$ are order-preserving in all variables.

(4) A *weak sup-gs-quantale* $(L, \leqslant, \otimes, (\gamma_n)_{n\in\mathbb{N}})$ is a gs-quantale such that $a \otimes \bigvee S = \bigvee\{a \otimes s \mid s \in S\}$ and $\bigvee S \otimes b = \bigvee\{s \otimes b \mid s \in S\}$ for every $a, b \in L$, $S \subseteq L$. A *sup-gs-quantale* $(L, \leqslant, \otimes, (\gamma_n)_{n\in\mathbb{N}})$ is a weak sup-gs-quantale with $\gamma_n(\bigvee S_{\{i\}}\mathbf{x}) = \bigvee\{\gamma_n(s_{\{i\}}\mathbf{x}) \mid s \in S\}$ for all $n \in \mathbb{N}$, $i \in \{1, \ldots, n\}$, $\mathbf{x} \in L^n$, $S \subseteq L$.

(5) A *unital gs-quantale* $(L, \leqslant, \otimes, (\gamma_n)_{n\in\mathbb{N}}, e, (e_n)_{n\in\mathbb{N}})$ (*ugs-quantale* for short) is a gs-quantale such that $\otimes$ has an identity element $e \in L$, and γ_n has an identity element $e_n \in L$ for every $n \in \mathbb{N}$ called the *units* (e is necessarily unique). Morphisms of ugs-quantales are unit-preserving gs-quantale morphisms.

(6) A *strongly ugs-quantale* $(L, \leqslant, \otimes, (\gamma_n)_{n\in\mathbb{N}}, e)$ (*sugs-quantale* for short) is an ugs-quantale such that $e_n = e$ for every $n \in \mathbb{N}$.

(7) A *gs-quantale* $(L, \leqslant, \otimes, (\gamma_n)_{n\in\mathbb{N}}, e, (e_n)_{n\in\mathbb{N}})$ *with zero* is a gs-quantale such that $\otimes$ and γ_n for every $n \in \mathbb{N}$ have the smallest element $\bot$ as a zero element (called the *zero element of* L).

(8) A *quantale* [4] $(L, \leqslant, \otimes)$ is a complete lattice $(L, \leqslant)$, equipped with an associative binary operation $\otimes$ on L distributing across arbitrary joins from both sides ($\bot$ is a two-sided zero then). Every quantale $(L, \leqslant, \otimes)$ can be identified with a sup-gs-quantale $(L, \leqslant, \otimes, (\gamma_n)_{n\in\mathbb{N}})$ where $\gamma_n(x_1, \ldots, x_n) = \bigotimes_{i=1}^{n} x_i = x_1 \otimes \cdots \otimes x_n$ for every $x_1, \ldots, x_n \in L$. A *frame* is a quantale with $\otimes = \wedge$ (binary meets).

(9) Let $(L, \leqslant, \otimes)$ be a quantale. A *left (right) L-module* [4] is a complete lattice $(M, \leqslant)$ equipped with an action $\cdot : L \times M \to M$ ($M \times L \to M$) such that

$$\begin{aligned} a \cdot \left(\bigvee m_i\right) &= \bigvee (a \cdot m_i), \\ \left(\bigvee a_i\right) \cdot m &= \bigvee (a_i \cdot m), \\ a \cdot (b \cdot m) &= (a \otimes b) \cdot m \end{aligned} \qquad \left(\begin{aligned} \left(\bigvee m_i\right) \cdot a &= \bigvee (m_i \cdot a), \\ m \cdot \left(\bigvee a_i\right) &= \bigvee (m \cdot a_i), \\ (m \cdot b) \cdot a &= m \cdot (b \otimes a) \end{aligned}\right)$$

for every $a, b, a_i \in L$ and every $m, m_i \in M$. A map $f : M \to N$ of two left (right) L-modules M, N is said to be an *L-module morphism* provided that f preserves arbitrary joins and $f(a \cdot m) = a \cdot f(m)$ ($f(m \cdot a) = f(m) \cdot a$) for every $a \in L$ and every $m \in M$.

We notice that unlike [5], gs-quantale not only strips $\otimes$ of associativity, but also assumes the existence of additional operations of finite arity. We continue by introducing (in a standard way) congruences on gs-quantales. Recall that a congruence on an algebraic structure is an equivalence relation, which preserves all the primitive operations of this structure (in case of gs-quantales these are $\bigvee$, $\otimes$, and γ_n, $n \in \mathbb{N}$).

Definition 3 Let $(L, \leqslant, \otimes, (\gamma_n)_{n\in\mathbb{N}})$ be a gs-quantale.

(1) A subset $R \subseteq L^2$ is called *r-compatible* on L provided that it is a reflexive relation preserving the joins $\bigvee$, the multiplication $\otimes$, and the operations γ_n for every $n \in \mathbb{N}$.
(2) A subset $R \subseteq L^2$ is a *congruence* on L provided that it is an r-compatible equivalence relation.

The subsequent developments of this paper will need the concept of aggregation function. We should, however, emphasize that our employed notion of aggregation function differs slightly from the standard one of, e.g., [2] (no preservation of the greatest element of the underlying complete lattice). Notice that given a gs-quantale L and a set X, the set L^X is a gs-quantale with the obvious point-wise structure.

Definition 4 Let $(L, \leqslant, \otimes, (\gamma_n)_{n\in\mathbb{N}})$ be a gs-quantale, and let I be a set.

(1) A map $A : L^I \to L$ is called an *aggregation function* provided that it is order-preserving (namely, if $\mathbf{x}, \mathbf{y} \in L^I$ are such that $\mathbf{x} \leqslant \mathbf{y}$, then $A(\mathbf{x}) \leqslant A(\mathbf{y})$), and satisfies the boundary condition $A(\bot) = A(\bot, \ldots, \bot) = \bot$ (we notice that one *does not* assume that $A(\top) = A(\top, \ldots, \top) = \top$). $\mathsf{Agg}(L^I, L)$ denotes the set of all aggregation functions from L^I to L.
(2) An *aggregation function* $A : L^I \to L$ is called *r-compatible* provided that it preserves every r-compatible relation R on $(L, \leqslant, \otimes, (\gamma_n)_{n\in\mathbb{N}})$, i.e., for every $\mathbf{x}, \mathbf{y} \in L^I$ with $(x_i, y_i) \in R$ for every $i \in I$, it follows that $(A(\mathbf{x}), A(\mathbf{y})) \in R$. $\mathsf{Agg}_\mathsf{r}(L^I, L)$ is the set of all r-compatible aggregation functions from L^I to L.
(3) An *aggregation function* $A : L^I \to L$ is called *compatible* provided that it preserves every congruence R on $(L, \leqslant, \otimes, (\gamma_n)_{n\in\mathbb{N}})$. $\mathsf{Agg}_\mathsf{c}(L^I, L)$ denotes the set of all compatible aggregation functions from L^I to L.
(4) An *aggregation function* $A : L^I \to L$ is called *$\otimes$-homogeneous* provided that $a \otimes A(\mathbf{x}) = A(a \otimes \mathbf{x})$ for every $a \in L$, $\mathbf{x} \in L^I$. $\mathsf{Agg}_\mathsf{h}(L^I, L)$ is the set of all $\otimes$-homogeneous aggregation functions from L^I to L.
(5) If $(L, \leqslant, \otimes)$ is a quantale, then both L and L^I have the obvious L-module structure. Moreover, every L-module homomorphism $L^I \to L$ is join-preserving, and, therefore, is an aggregation function. $\mathsf{Agg}_\mathsf{m}(L^I, L)$ denotes the set of all L-module homomorphisms from L^I to L.

For a thorough discussion on general aggregation of an infinite number of inputs we refer to [2]. We just notice that $\mathsf{Agg}_\mathsf{r}(L^I, L) \subseteq \mathsf{Agg}_\mathsf{c}(L^I, L) \subseteq \mathsf{Agg}(L^I, L)$. Moreover, since $\mathsf{Agg}_k(L^I, L) \subseteq L^{L^I}$, $k \in \{-, \mathsf{r}, \mathsf{c}, \mathsf{h}\}$, and L^{L^I} is an L-module (with the standard point-wise operations), one could ask whether $\mathsf{Agg}_k(L^I, L)$ is a submodule of L^{L^I}. The next two results provide the positive answer to the question.

Lemma 1 *Let $(L, \leqslant, \otimes, (\gamma_n)_{n\in\mathbb{N}})$ be a gs-quantale, and let I be a set.*

(1) $\mathsf{Agg}(L^I, L)$*,* $\mathsf{Agg_r}(L^I, L)$ *and* $\mathsf{Agg_c}(L^I, L)$ *are closed under joins in* L^{L^I}*.*
(2) If $(L, \otimes, \leqslant)$ is a quantale, then both $\mathsf{Agg_h}(L^I, L)$ *and* $\mathsf{Agg_m}(L^I, L)$ *are closed under joins in* L^{L^I}*.*

Proof For (1): Given $S \subseteq \mathsf{Agg}(L^I, L)$, we show that $\bigvee S \in \mathsf{Agg}(L^I, L)$. If $S = \varnothing$, then $\bigvee S$ is the constant map $\bot$ defined by $\bot(\mathbf{x}) = \bot$, and it is clear that $\bot \in \mathsf{Agg_r}(L^I, L) \subseteq \mathsf{Agg_c}(L^I, L) \subseteq \mathsf{Agg}(L^I, L)$. Suppose that $S \neq \varnothing$. Clearly, $\bigvee S$ is order-preserving, and $(\bigvee S)(\bot) = \bigvee_{A\in S} A(\bot) = \bigvee_{A\in S} \bot = \bot$. The remaining cases follow by the same arguments.

For (2): It is easy to see that $a \otimes \bot(\mathbf{x}) = \bot = \bot(a \otimes \mathbf{x})$ for every $a \in L$, $\mathbf{x} \in L^I$. It follows that $\bot = \bigvee \varnothing \in \mathsf{Agg_h}(L^I, L)$. Suppose we have a non-empty $S \subseteq \mathsf{Agg_h}(L^I, L)$. Since $\bigvee S \in \mathsf{Agg}(L^I, L)$, we have to show that $(a \otimes \bigvee S)(\mathbf{x}) = \bigvee\{A(a \otimes \mathbf{x}) \mid A \in S\}$ for every $a \in L, \mathbf{x} \in L^I$. One can easily see that $(a \otimes \bigvee S)(\mathbf{x}) = a \otimes (\bigvee S(\mathbf{x})) = a \otimes (\bigvee\{A(\mathbf{x}) \mid A \in S\}) = \bigvee\{a \otimes A(\mathbf{x}) \mid A \in S\} = \bigvee\{A(a \otimes \mathbf{x}) \mid A \in S\}$. The case of $\mathsf{Agg_m}(L^I, L)$ is similar and, therefore, is left to the reader.

Proposition 1 *Let $(L, \leqslant, \otimes, (\gamma_n)_{n\in\mathbb{N}})$ be a gs-quantale such that $(L, \leqslant, \otimes)$ is a quantale, let I be a set.*

(1) $\mathsf{Agg}(L^I, L)$*,* $\mathsf{Agg_r}(L^I, L)$ *and* $\mathsf{Agg_c}(L^I, L)$ *are L-modules.*
(2) If $(L, \leqslant, \otimes)$ is a commutative quantale, then both $\mathsf{Agg_h}(L^I, L)$ *and* $\mathsf{Agg_m}(L^I, L)$ *are L-modules.*

Proof For (1): From Lemma 1 get that $\mathsf{Agg}(L^I, L)$, $\mathsf{Agg_r}(L^I, L)$, $\mathsf{Agg_c}(L^I, L)$, and $\mathsf{Agg_h}(L^I, L)$ are complete lattices with the smallest element $\bot$. We check that for every $a \in L$ and every aggregation function A, the map $a \otimes A : L^I \to L$ defined by $(a \otimes A)(\mathbf{x}) = a \otimes A(\mathbf{x})$ is an aggregation function. Since $\otimes$ is order-preserving in both variables, we get that $a \otimes A$ is order-preserving as well. Moreover, $(a \otimes A)(\bot) = a \otimes A(\bot) = a \otimes \bot = \bot$. In view of the point-wise operations of the L-module L^{L^I}, $\mathsf{Agg}(L^I, L)$ is a submodule of L^{L^I}. The remaining cases use the same arguments.

For (2): Let us verify that for every $a \in L$ and every $A \in \mathsf{Agg_h}(L^I, L)$, $a \otimes A \in \mathsf{Agg_h}(L^I, L)$. Given $b \in L$ and $\mathbf{x} \in L^I$, one has $(a \otimes A)(b \otimes \mathbf{x}) = a \otimes (A(b \otimes \mathbf{x})) = a \otimes (b \otimes A(\mathbf{x})) = b \otimes (a \otimes A(\mathbf{x})) = b \otimes ((a \otimes A)(\mathbf{x}))$. The case of $\mathsf{Agg_m}(L^I, L)$ is similar and, therefore, is left to the reader.

The next result describes idempotency properties of our introduced aggregation functions. Their very definition (Definition 4) implies that the bottom element $\bot$ is always idempotent. The obvious question on the existence of other idempotent elements is answered in the next proposition (recall from item (3) at the beginning of Sect. 3 that $x_{\varnothing}x = (x, \ldots, x)$).

Proposition 2 *Let $(L, \leqslant, \otimes)$ be a commutative quantale with the unit element $\top$, and let I be a set. Put $\gamma_n(x_1, \ldots, x_n) = \bigotimes_{i=1}^{n} x_i = x_1 \otimes \cdots \otimes x_n$ for every $n \in \mathbb{N}$ and every $x_1, \ldots, x_n \in L$. Then $(L, \leqslant, \otimes, (\gamma_n)_{n\in\mathbb{N}})$ is a sup-gs-quantale such that $A \in \mathsf{Agg_c}(L^I, L)$ implies $A(u_{\varnothing}u) \leqslant u$ for every $u \in L$. Moreover, if $A(\top_{\varnothing}\top) = \top$ and $u^k = u^{k+1}$ for some $k \in \mathbb{N}$, then $A(u_{\varnothing}u) \geqslant u^k$. If $u = u^2$, then $A(u_{\varnothing}u) = u$.*

Proof Let $A \in \mathsf{Agg_c}(L^I, L)$ and $u \in L$. Define a relation $R_u \subseteq L \times L$ as follows: $(x, y) \in R_u$ if and only if $x \vee u = y \vee u$. Then R_u is clearly a congruence on $(L, \leqslant, \otimes, (\gamma_n)_{n\in\mathbb{N}})$. We are now going to compute the value of $A(u_\varnothing u)$. Since $(\bot, u) \in R_u$, $(A(\bot_\varnothing\bot), A(u_\varnothing u)) \in R_u$, and thus, $u = \bot \vee u = A(\bot_\varnothing\bot) \vee u = A(u_\varnothing u) \vee u$. The desired conclusion $A(u_\varnothing u) \leqslant u$ follows.

Suppose now that $u \in L$ is such that $u^k = u^{k+1}$ for some $k \in \mathbb{N}$. Define a relation $R^u \subseteq L \times L$ as follows: $(x, y) \in R^u$ if and only if $u^k \otimes x = u^k \otimes y$. Again, R^u is a congruence on $(L, \leqslant, \otimes, (\gamma_n)_{n\in\mathbb{N}})$. Finally, we have to compute the value of $A(u_\varnothing u)$. It is clear that $(u, \top) \in R^u$. Then $(A(u_\varnothing u), A(\top_\varnothing\top)) \in R^u$, and thus, $A(u_\varnothing u) \geq u^k \otimes A(u_\varnothing u) = u^k \otimes A(\top_\varnothing\top) = u^k \otimes \top = u^k$.

Corollary 1 *Let $(L, \leqslant, \wedge)$ be a frame with the top element $\top$, and let I be a set. Let $\gamma_n(x_1, \ldots, x_n) = \bigwedge_{i=1}^n x_i = x_1 \wedge \cdots \wedge x_n$ for every $n \in \mathbb{N}$ and every $x_1, \ldots, x_n \in L$. Then $(L, \leqslant, \wedge, (\gamma_n)_{n\in\mathbb{N}})$ is a sup-gs-quantale such that $\{A \in \mathsf{Agg_c}(L^I, L) \mid A(\top_\varnothing\top) = \top\} \subseteq \mathsf{Idem}(L^I, L)$.*

3.2 Generalized Sugeno Integral

We begin with some notational conventions. Given a finite non-empty set A, $\mathcal{G}_A$ stands for the set of all permutations on A, the elements of which are denoted σ. Given $\sigma \in \mathcal{G}_A$, $\sigma(A)$ stands for the output of the permutation σ. For example, if $A = (a_1, a_2, a_3)$, then $\sigma(A)$ could be (a_3, a_1, a_2) (the order is important).

Definition 5 (1) Let C be a set, and let $(L, \leqslant)$ be a complete lattice. A map $\mu : \mathcal{P}_{fin}(C) \to L$ is said to be an *L-capacity* on C provided that $\mu(\varnothing) = \bot$, and, for every $A, B \in \mathcal{P}_{fin}(C)$, $A \subseteq B$ implies $\mu(A) \leqslant \mu(B)$. $\mathsf{Cap}(C, L)$ denotes the set of all L-capacities on C.

(2) Let C be a set, and let $(L, \leqslant, \otimes, (\gamma_n)_{n\in\mathbb{N}})$ be a gs-quantale. Given $\mu \in \mathsf{Cap}(C, L)$, the *Sugeno integral* w.r.t. μ is a map $\mathsf{Su}_\mu : L^C \to L$ given by

$$\mathsf{Su}_\mu(\mathbf{x}) := \bigvee_{A \in \mathcal{P}^{\varnothing}_{fin}(C)} \bigvee_{\sigma \in \mathcal{G}_A} \mu(A) \otimes \gamma_{|A|}(\mathbf{x}|_{\sigma(A)}),$$

where $|A|$ stands for the number of elements in the set A. $\mathsf{Sug}(C, L)$ denotes the set of all Sugeno integrals $\{\mathsf{Su}_\mu : L^C \to L \mid \mu \in \mathsf{Cap}(C, L)\}$.

Similar to the case of aggregation functions in the previous subsection, we are going to show that the sets $\mathsf{Cap}(C, L)$ and $\mathsf{Sug}(C, L)$ have the structure of an L-module w.r.t. the point-wise operations on L^{L^C} (the proof is a direct verification).

Proposition 3 *Let C be a set, and let $(L, \leqslant)$ be a complete lattice.*

(1) The set $\mathsf{Cap}(C, L)$ of all L-capacities on C is a complete lattice.

(2) *If* $(L, \leqslant, \otimes)$ *is a quantale, then* $\mathsf{Cap}(C, L)$ *and* $\mathsf{Sug}(C, L)$ *are* L*-modules. Moreover, if* $\{\mu_j \mid j \in J\} \subseteq \mathsf{Cap}(C, L)$*, and* $\{a_j \mid j \in J\} \subseteq L$*, then* $\bigvee_{j\in J} a_j \otimes \mathsf{Su}_{\mu_j} = \mathsf{Su}_{\bigvee_{j\in J} a_j \otimes \mu_j}$.

In what follows we provide more properties of our introduced Sugeno integral.

Lemma 2 *(1) Let* $(L, \leqslant, \otimes, (\gamma_n)_{n\in\mathbb{N}})$ *be a strongly unital gs-quantale with unit* e *and zero* $\bot$*, and let* C *be a set. For every* $A \in \mathcal{P}_{fin}(C)$ *and every* $\mu \in \mathsf{Cap}(C, L)$*, it follows that* $\mathsf{Su}_\mu(e_A\bot) = \mu(A)$.

(2) Let $(L, \leqslant, \otimes, (\gamma_n)_{n\in\mathbb{N}})$ *be a gs-quantale with order-preserving multiplication* $\otimes$*, let* C *be a set, and let* $\mu, \nu \in \mathsf{Cap}(C, L)$*. Then* $\mu \leqslant \nu$ *implies* $\mathsf{Su}_\mu \leqslant \mathsf{Su}_\nu$*. Moreover, if* $(L, \leqslant, \otimes, (\gamma_n)_{n\in\mathbb{N}})$ *is a strongly unital ogs-quantale with unit* e *and zero* $\bot$*, then* $\mu \leqslant \nu$ *if and only if* $\mathsf{Su}_\mu \leqslant \mathsf{Su}_\nu$.

Proof For (1): Given $A \in \mathcal{P}_{fin}(C)$ and $\mu \in \mathsf{Cap}(C, L)$,

$$\begin{aligned}\mathsf{Su}_\mu(e_A\bot) &= \textstyle\bigvee_{B\in\mathcal{P}^{\varnothing}_{fin}(C)} \bigvee_{\sigma\in\mathcal{G}_B} \mu(B) \otimes \gamma_{|B|}(e_A\bot|_{\sigma(B)})\\ &= \textstyle\bigvee_{B\in\mathcal{P}^{\varnothing}_{fin}(A)} \bigvee_{\sigma\in\mathcal{G}_B} \mu(B) \otimes e = \bigvee_{B\in\mathcal{P}^{\varnothing}_{fin}(A)} \bigvee_{\sigma\in\mathcal{G}_B} \mu(B)\\ &\leqslant \textstyle\bigvee_{B\in\mathcal{P}^{\varnothing}_{fin}(A)} \bigvee_{\sigma\in\mathcal{G}_B} \mu(A) = \mu(A).\end{aligned}$$

Moreover, if $A = \varnothing$, then $\mu(A) = \bot \leqslant \mathsf{Su}_\mu(e_A\bot)$; otherwise, $\mu(A) = \mu(A) \otimes e = \mu(A) \otimes \gamma_{|A|}(e_A\bot|_A) \leqslant \mathsf{Su}_\mu(e_A\bot)$. Thus, it follows that $\mathsf{Su}_\mu(e_A\bot) = \mu(A)$.

For (2): For every $\mathbf{x} \in L^C$, we get

$$\begin{aligned}\mathsf{Su}_\mu(\mathbf{x}) &= \textstyle\bigvee_{A\in\mathcal{P}^{\varnothing}_{fin}(C)} \bigvee_{\sigma\in\mathcal{G}_A} \mu(A) \otimes \gamma_{|A|}(\mathbf{x}|_{\sigma(A)})\\ &\leqslant \textstyle\bigvee_{A\in\mathcal{P}^{\varnothing}_{fin}(C)} \bigvee_{\sigma\in\mathcal{G}_A} \nu(A) \otimes \gamma_{|A|}(\mathbf{x}|_{\sigma(A)}) = \mathsf{Su}_\nu(\mathbf{x}).\end{aligned}$$

If $(L, \leqslant, \otimes, (\gamma_n)_{n\in\mathbb{N}})$ is a strongly unital ogs-quantale with unit e and zero $\bot$, then, for every $A \in \mathcal{P}_{fin}(C)$, we have $\mu(A) = \mathsf{Su}_\mu(e_A\bot) \leqslant \mathsf{Su}_\nu(e_A\bot) = \nu(A)$.

Theorem 1 *Let* C *be a set, and let* $(L, \leqslant, \otimes, (\gamma_n)_{n\in\mathbb{N}})$ *be an ogs-quantale with zero. Given* $\mu \in \mathsf{Cap}(C, L)$*, the Sugeno integral* Su_μ *w.r.t.* μ *is an r-compatible aggregation function. Moreover, if* $(L, \leqslant, \otimes)$ *is a commutative quantale, and every* γ_n *is an* $\otimes$*-homogeneous map, then* $\mathsf{Su}_\mu : L^C \to L$ *is an* $\otimes$*-homogeneous map. If, additionally,* $(L, \leqslant, \otimes, (\gamma_n)_{n\in\mathbb{N}})$ *is a sup-gs-quantale, then* Su_μ *is an* L*-module morphism.*

Proof It is easy to see that if $\mathbf{x}, \mathbf{y} \in L^C$ are such that $\mathbf{x} \leqslant \mathbf{y}$, then, for every $A \in \mathcal{P}^{\varnothing}_{fin}(C)$ and every $\sigma \in \mathcal{G}_A$, we have $\gamma_{|A|}(\mathbf{x}|_{\sigma(A)}) \leqslant \gamma_{|A|}(\mathbf{y}|_{\sigma(A)})$, which yields $\mu(A) \otimes \gamma_{|A|}(\mathbf{x}|_{\sigma(A)}) \leqslant \mu(A) \otimes \gamma_{|A|}(\mathbf{y}|_{\sigma(A)})$. Thus, $\mathsf{Su}_\mu(\mathbf{x}) \leqslant \mathsf{Su}_\mu(\mathbf{y})$. Moreover, for every $A \in \mathcal{P}^{\varnothing}_{fin}(C)$ and every $\sigma \in \mathcal{G}_A$, we have that $\gamma_{|A|}(\bot|_{\sigma(A)})) = \bot$. Hence, $\mu(A) \otimes \gamma_{|A|}(\bot|_{\sigma(A)}) = \bot$, i.e., $\mathsf{Su}_\mu(\bot) = \bot$. Thus, Su_μ is an aggregation function.

Let $R \subseteq L^2$ be an r-compatible relation on L. Suppose $\mathbf{x}, \mathbf{y} \in L^I$ are such that $(x_i, y_i) \in R$ for every $i \in I$. Then, for every $A \in \mathcal{P}^{\varnothing}_{fin}(C)$ and every $\sigma \in \mathcal{G}_A$, we have that $(\gamma_{|A|}(\mathbf{x}|_{\sigma(A)}), \gamma_{|A|}(\mathbf{y}|_{\sigma(A)})) \in R$. This yields that $(\mu(A) \otimes \gamma_{|A|}(\mathbf{x}|_{\sigma(A)}), \mu(A) \otimes$

$\gamma_{|A|}(\mathbf{y}|_{\sigma(A)})) \in R$, and, therefore, $(\bigvee_{\sigma \in \mathcal{G}_A} \mu(A) \otimes \gamma_{|A|}(\mathbf{x}|_{\sigma(A)}), \bigvee_{\sigma \in \mathcal{G}_A} \mu(A) \otimes \gamma_{|A|}(\mathbf{y}|_{\sigma(A)})) \in R$. It follows then that $(\mathsf{Su}_\mu(\mathbf{x}), \mathsf{Su}_\mu(\mathbf{y})) \in R$. The remaining part of the proof is a direct verification.

Corollary 2 *Let C be a set, and let $(L, \leqslant, \otimes, (\gamma_n)_{n\in\mathbb{N}})$ be an ogs-quantale with zero. Given $\mu \in \mathsf{Cap}(C, L)$, the Sugeno integral Su_μ w.r.t. μ is a compatible aggregation function.*

Theorem 2 *Let C be a set, and let $(L, \leqslant, \otimes, (\gamma_n)_{n\in\mathbb{N}})$ be an ogs-quantale with zero and such that $(L, \leqslant, \otimes)$ is a quantale.*

(1) The map $\mathsf{Su}\colon \mathsf{Cap}(C, L) \to \mathsf{Agg}(L^C, L)$, $\mathsf{Su}(\mu) = \mathsf{Su}_\mu$ is a homomorphism of L-modules. Moreover, if $(L, \leqslant, \otimes, (\gamma_n)_{n\in\mathbb{N}})$ is strongly unital with unit e, then the map Su is one-to-one.
(2) If $(L, \leqslant, \otimes, (\gamma_n)_{n\in\mathbb{N}})$ is unital, then the map $\mathsf{Cp}\colon \mathsf{Agg}(L^C, L) \to \mathsf{Cap}(C, L)$, $\mathsf{Cp}(f)(A) = f(e_A\bot)$ is a homomorphism of L-modules.
(3) If $(L, \leqslant, \otimes, (\gamma_n)_{n\in\mathbb{N}})$ is a strongly unital ogs-quantale with unit e, then $\mathsf{Cp} \circ \mathsf{Su} = \mathsf{Id}_{\mathsf{Cap}(C,L)}$.
(4) Let $(L, \leqslant, \otimes, (\gamma_n)_{n\in\mathbb{N}})$ be unital and such that $(L, \leqslant, \otimes)$ is a commutative quantale. Then the map $\mathsf{Cp}_\mathsf{h}\colon \mathsf{Agg}_\mathsf{h}(L^C, L) \to \mathsf{Cap}(C, L)$, $\mathsf{Cp}_\mathsf{h}(f) = \mathsf{Cp}(f)$ is a homomorphism of L-modules. Additionally, if $(L, \leqslant, \otimes, (\gamma_n)_{n\in\mathbb{N}})$ is multisided, then $\mathsf{Su} \circ \mathsf{Cp}_\mathsf{h} \leqslant \mathsf{Id}_{\mathsf{Agg}_\mathsf{h}(L^C,L)}$.
(5) Let $(L, \leqslant, \otimes, (\gamma_n)_{n\in\mathbb{N}})$ be a unital multisided sup-gs-quantale such that $(L, \leqslant, \otimes)$ is a commutative quantale and $\gamma_1(a) = a$ for every $a \in L$. Then the L-module homomorphism $\mathsf{Cp}_\mathsf{m}\colon \mathsf{Agg}_\mathsf{m}(L^C, L) \to \mathsf{Cap}(C, L)$, $\mathsf{Cp}_\mathsf{m}(f) = \mathsf{Cp}(f)$ has the property $\mathsf{Su} \circ \mathsf{Cp}_\mathsf{m} = \mathsf{Id}_{\mathsf{Agg}_\mathsf{m}(L^C,L)}$.

Proof For (1): The fact that Su is a homomorphism of L-modules follows from Theorem 1 and Proposition 3 (4). The injectivity of Su follows from Lemma 2 (2).

For (2): Take $f \in \mathsf{Agg}(L^C, L)$ and $A, B \in \mathcal{P}_{fin}(C)$ such that $A \subseteq B$. Then $e_A\bot \leqslant e_B\bot$. It follows that $\mathsf{Cp}(f)(A) = f(e_A\bot) \leqslant f(e_B\bot) = \mathsf{Cp}(f)(B)$, i.e., $\mathsf{Cp}(f)$ is an order-preserving map. We notice that $\bot = e_\varnothing\bot$. It follows that $\mathsf{Cp}(f)(\varnothing) = f(\bot) = \bot$, i.e., $\mathsf{Cp}(f) \in \mathsf{Cap}(C, L)$.

Given $g \in \mathsf{Agg}(L^C, L)$, $S \subseteq \mathsf{Agg}(L^C, L)$, and $a \in L$, for every $A \in \mathcal{P}_{fin}(C)$,

$$\mathsf{Cp}(\bigvee S)(A) = (\bigvee_{f\in S} f)(e_A\bot) = \bigvee_{f\in S} f(e_A\bot) = \bigvee_{f\in S} \mathsf{Cp}(f)(A)$$

and

$$\mathsf{Cp}(a \otimes g)(A) = (a \otimes g)(e_A\bot) = a \otimes g(e_A\bot) = (a \otimes \mathsf{Cp}(g))(A).$$

For (3): Given $\mu \in \mathsf{Cap}(C, L)$, for every $A \in \mathcal{P}_{fin}(C)$, it follows that $(\mathsf{Cp} \circ \mathsf{Su}(\mu))(A) = (\mathsf{Cp}(\mathsf{Su}_\mu))(A) = \mathsf{Su}_\mu(e_A\bot) \overset{\text{Lemma 2 (1)}}{=} \mu(A)$.

For (4): The first part follows from item (2) and Proposition 1 (2). It remains to check that, for every $f \in \mathsf{Agg}_\mathsf{h}(L^C, L)$, $\mathbf{x} \in L^C$, we have $(\mathsf{Su} \circ \mathsf{Cp}_\mathsf{h}(f))(\mathbf{x}) = \mathsf{Su}_{\mathsf{Cp}_\mathsf{h}(f)}(\mathbf{x}) \leqslant f(\mathbf{x})$, which can be done as follows:

$$\begin{aligned}\mathsf{Su}_{\mathsf{Cp_h}(f)}(\mathbf{x}) &= \bigvee_{A\in\mathcal{P}^{\varnothing}_{fin}(C)} \bigvee_{\sigma\in\mathcal{G}_A} (\mathsf{Cp_h}(f))(A) \otimes \gamma_{|A|}(\mathbf{x}|_{\sigma(A)})\\ &= \bigvee_{A\in\mathcal{P}^{\varnothing}_{fin}(C)} \bigvee_{\sigma\in\mathcal{G}_A} f(e_A\bot) \otimes \gamma_{|A|}(\mathbf{x}|_{\sigma(A)})\\ &\leqslant \bigvee_{A\in\mathcal{P}^{\varnothing}_{fin}(C)} \bigvee_{\sigma\in\mathcal{G}_A} \bigwedge(\mathbf{x}|_A) \otimes f(e_A\bot)\\ &\leqslant \bigvee_{A\in\mathcal{P}^{\varnothing}_{fin}(C)} \bigvee_{\sigma\in\mathcal{G}_A} f(\bigwedge(\mathbf{x}|_A) \otimes e_A\bot) \leqslant f(\mathbf{x}).\end{aligned}$$

For (5): Correctness of the map $\mathsf{Su}\colon \mathsf{Cap}(C, L) \to \mathsf{Agg_m}(L^C, L), \mathsf{Su}(\mu) = \mathsf{Su}_\mu$ follows from Theorem 1. By the previous item, $\mathsf{Su} \circ \mathsf{Cp_m} \leqslant \mathsf{Id}_{\mathsf{Agg_m}(L^C,L)}$. For the opposite inequality, notice that given $f \in \mathsf{Agg_m}(L^C, L)$ and $\mathbf{x} \in L^C$, it follows that

$$\begin{aligned}f(\mathbf{x}) &= f(\bigvee_{c\in C}(x_c \otimes e_{\{c\}}\bot)) = \bigvee_{c\in C} f(x_c \otimes e_{\{c\}}\bot) = \bigvee_{c\in C} x_c \otimes f(e_{\{c\}}\bot)\\ &= \bigvee_{c\in C} f(e_{\{c\}}\bot) \otimes x_c = \bigvee_{c\in C}(\mathsf{Cp_m}(f))(\{c\}) \otimes x_c\\ &= \bigvee_{c\in C}(\mathsf{Cp_m}(f))(\{c\}) \otimes \gamma_1(x_c) = \bigvee_{c\in C}(\mathsf{Cp_m}(f))(\{c\}) \otimes \gamma_1(\mathbf{x}|_{\{c\}})\\ &\leqslant \bigvee_{A\in\mathcal{P}^{\varnothing}_{fin}(C)} \bigvee_{\sigma\in\mathcal{G}_A} (\mathsf{Cp_m}(f))(A) \otimes \gamma_{|A|}(\mathbf{x}|_{\sigma(A)}) = \mathsf{Su}_{\mathsf{Cp_m}(f)}(\mathbf{x}).\end{aligned}$$

Thus, we get that $f(\mathbf{x}) \leqslant (\mathsf{Su} \circ \mathsf{Cp_m}(f))(\mathbf{x})$, i.e., $\mathsf{Id}_{\mathsf{Agg_m}(L^C,L)} \leqslant \mathsf{Su} \circ \mathsf{Cp_m}$.

The next result provides an important corollary of Theorem 2.

Corollary 3 *Let $(L, \leqslant, \otimes, (\gamma_n)_{n\in\mathbb{N}})$ be a strongly unital multisided sup-gs-quantale such that $(L, \leqslant, \otimes)$ is a commutative quantale and, moreover, $\gamma_1(a) = a$ for every $a \in L$. Then the maps $\mathsf{Cp_m}\colon \mathsf{Agg_m}(L^C, L) \to \mathsf{Cap}(C, L)$, $(\mathsf{Cp_m}(f))(A) = f(e_A\bot)$ and $\mathsf{Su}\colon \mathsf{Cap}(C, L) \to \mathsf{Agg_m}(L^C, L)$, $\mathsf{Su}(\mu) = \mathsf{Su}_\mu$ are L-module homomorphisms which are inverse to each other.*

Definition 6 Let $(L, \leqslant, \otimes, (\gamma_n)_{n\in\mathbb{N}})$ be a gs-quantale, and let I be a set. Let $A\colon L^I \to L$ be an aggregation function on L. We say that A is *scale invariant* provided that for every gs-quantale $(L_1, \leqslant', \otimes', (\gamma'_n)_{n\in\mathbb{N}})$ and every surjective homomorphism $\varphi\colon L \to L_1$ of gs-quantales, there is an aggregation function $B : L_1^I \to L_1$ on L_1 such that for every $\mathbf{x} \in L^I$, it follows that $\varphi(A(\mathbf{x})) = B((\varphi(x_i))_{i\in I})$.

Theorem 3 *Let $(L, \leqslant, \otimes, (\gamma_n)_{n\in\mathbb{N}})$ be a gs-quantale, let C be a set, and let $\mu \in \mathsf{Cap}(C, L)$. Then the Sugeno integral Su_μ w.r.t. μ is scale invariant.*

Proof Given a surjective homomorphism of gs-quantales $\varphi\colon L \to L_1$, let us define a relation $R_\varphi \subseteq L \times L$ as follows: $(x, y) \in R$ if and only if $\varphi(x) = \varphi(y)$. Clearly, R_φ is a congruence on L. Let $\overline{\mathbf{x}} \in L_1^I$. Then there is an element $\mathbf{x} \in L^I$ such that $\varphi(x_i) = \overline{x}_i$ for every $i \in I$. We put $B(\overline{\mathbf{x}}) = \varphi(\mathsf{Su}_\mu(\mathbf{x}))$. Since Su_μ is r-compatible by Theorem 1, $B(\overline{\mathbf{x}})$ does not depend on the choice of $\mathbf{x} \in L^I$ such that $\varphi(x_i) = \overline{x}_i$. It is easy to see that B is the Sugeno integral $\mathsf{Su}_{\mu'}$ w.r.t. $\mu' = \varphi \circ \mu$.

Example 2 Let $(L, \leqslant)$ be a complete lattice equipped with two binary operations $\otimes$ and $\odot$, having a common zero element $\bot$ and being order-preserving in all variables. Fix $m \in \mathbb{N}$ and suppose that $\odot$ is both associative and commutative, with $x \odot y \leqslant x$ for every $x \in L$. For every $n \in \mathbb{N}$, define an n-ary operation $d_n : L^n \to L$ by:

(i) $d_1(x_1) = x_1 = \bigodot_{i=1}^{1} x_i$,
(ii) $d_{n+1}(x_1, \ldots, x_{n+1}) = d_n(x_1, \ldots, x_n) \odot x_{n+1} = \bigodot_{i=1}^{n+1} x_i$ if $n + 1 \leqslant m$,
(iii) $d_{n+1}(x_1, \ldots, x_{n+1}) = \bot$ if $n \geq m$

for every $x_1, \ldots, x_{n+1} \in L$. Then $(L, \leqslant, \otimes, (d_n)_{n\in\mathbb{N}})$ is an ogs-quantale with zero. Let μ be an L-capacity on $[m]$. The Sugeno integral $\mathsf{Su}_\mu : L^m \to L$ w.r.t. μ is then given by $\mathsf{Su}_\mu(\mathbf{x}) = \bigvee_{\varnothing\neq I\subseteq[m]} \left(\mu(I) \otimes \bigodot_{i\in I} x_i\right)$. Putting $\otimes = \odot = \wedge$ provides the usual definition of the Sugeno integral, studied in, e.g., [3].

Remark 1 Our approach covers the definition of the Sugeno integral for arbitrary lattices. For every lattice L, we have its MacNeille completion MC(L), and, moreover, the operations $\wedge$ and $\vee$ in L and MC(L) coincide. The restriction to L of our Sugeno integral computed in MC(L) coincides with the usual Sugeno integral on L.

Example 3 Let [0, 1] be the real unit interval, and let $\otimes\colon [0, 1]^2 \to [0, 1]$ be a binary operation on [0, 1] order-preserving in both variables with zero 0 and unit 1. For every $n \in \mathbb{N}$, let $\gamma_n\colon [0, 1]^n \to [0, 1]$ be an n-copula. Let C be a set such that μ is a [0, 1]-capacity on C. Then $([0, 1], \leqslant, \otimes, (\gamma_n)_{n\in\mathbb{N}})$ is a strongly unital ogs-quantale with zero. Moreover, the Sugeno integral w.r.t. μ satisfies due to [2, Proposition 3.62]

$$\bigvee_{A\in\mathcal{P}^{\varnothing}_{fin}(C)} \bigvee_{\sigma\in\mathcal{G}_A} \mu(A) \otimes max(0, \textstyle\sum_{a\in A} x_{\sigma(a)} - |A| + 1) \leqslant \mathsf{Su}_\mu(\mathbf{x}) \leqslant \bigvee_{A\in\mathcal{P}^{\varnothing}_{fin}(C)} \bigvee_{\sigma\in\mathcal{G}_A} \mu(A) \otimes min_{a\in A}(x_{\sigma(a)}).$$

In particular, if, for every $c \in C$, E_c is an event of the form $\{Z_c \leqslant z\}$ with probability $P(E_c)$, then from Sklar's theorem (see [2, Theorem 3.65]), $P(\bigcap_{i=1}^{n} E_{c_i}) = \gamma_n(P(E_{c_1}), \ldots, P(E_{c_n}))$, and the Sugeno integral w.r.t. μ is the weighted join of μ.

4 Conclusion and Future Work

In this paper we have introduced a new type of algebraic structures – generalized semi-quantales (Definition 2) and then considered a variant of the Sugeno integral as an aggregation method over them (Definition 5). We believe it is worthwhile to undertake further theoretical studies on the generalized Sugeno integral of this paper. In particular, one could provide a complete characterization of compatible aggregation functions acting on a generalized semi-quantale.

Acknowledgements Jan Paseka was supported by the Czech Science Foundation through the project No. 18-06915S. Milan Stehlík was supported by WTZ Project No. HU 11/2016.

References

1. D. Dubois, H. Prade, A. Rico, Residuated variants of Sugeno integrals: towards new weighting schemes for qualitative aggregation methods. Inf. Sci. **329**, 765–781 (2016)
2. M. Grabisch, J.-L. Marichal, R. Mesiar, E. Pap, *Aggregation Functions* (Cambridge University Press, Cambridge, 2009)
3. R. Halaš, R. Mesiar, J. Pócs, A new characterization of the discrete Sugeno integral. Inf. Fusion **29**, 84–86 (2016)
4. D. Kruml, J. Paseka, Algebraic and categorical aspects of quantales, in *Handbook of Algebra 5*, ed. by M. Hazewinkel (Elsevier/North-Holland, Amsterdam, 2008), pp. 323–362
5. S.E. Rodabaugh, Relationship of algebraic theories to powerset theories and fuzzy topological theories for lattice-valued mathematics. Int. J. Math. Math. Sci. **2007**, 1–71 (2007)
6. P. Hermann, T. Mrkvička, T. Mattfeldt, M. Minárová, K. Helisová, O. Nicolis, F. Wartner, M. Stehlík, Fractal and stochastic geometry inference for breast cancer: a case study with random fractal models and Quermass-interaction process. Stat. Med. **34**, 2636–2661 (2015)
7. M. Sugeno, *Theory of Fuzzy Integrals and Its Applications* (Tokyo Institute of Technology, Tokyo, 1974). Ph.D. thesis

Numerical Solution for Reversible Chemical Reaction Models with Interactive Fuzzy Initial Conditions

Vinícius F. Wasques, Estevão Esmi, Laécio Carvalho de Barros, and Francielle Santo Pedro

Abstract This manuscript studies the reversible chemical reactions described by a system of fuzzy differential equations, with initial conditions given by interactive fuzzy numbers. The fuzzy solution is given by the Euler method based on the arithmetic for interactive fuzzy numbers. An example is presented in order to illustrate the different types of interactivity that produce the solutions for the system.

Keywords Fuzzy initial value problem · Sup-extension principle · Interactive fuzzy numbers · Chemical reactions

1 Introduction

Chemical reactions are transformations that involve changes in the bonds of the particles of matter, resulting in the formation of a new substance with different properties than the previous one [1]. A chemical decay of a reagent is an example of a chemical reaction:

$$A \xrightarrow{k} B, \quad \text{with reaction rate k,} \tag{1}$$

where A is the initial reagent and B is the final product.

V. F. Wasques (✉)
São Paulo State University, Rio Claro, São Paulo, Brazil
e-mail: vwasques@outlook.com

E. Esmi · L. C. de Barros
University of Campinas, Campinas, São Paulo, Brazil
e-mail: eelaureano@gmail.com

L. C. de Barros
e-mail: laeciocb@ime.unicamp.br

F. S. Pedro
Federal University of São Paulo, Osasco, São Paulo, Brazil
e-mail: fran.stopedro@gmail.com

B. Bede et al. (eds.), *Fuzzy Information Processing 2020*,
Advances in Intelligent Systems and Computing 1337,
https://doi.org/10.1007/978-3-030-81561-5_10

A reaction is said to be reversible if the reagents transform into a product and the product transform into a reagent, simultaneously attained an equilibrium [2]. Otherwise, the reaction is said to be irreversible. The decay given by (1) is an example of an irreversible reaction.

The kinetic chemistry is a field of science that studies the velocity and the factors that influence the changes of chemical reactions, for instance, concentration, temperature, pressure and so on [1]. The kinetic law allows to determinate the velocity v of a reaction in terms of the concentration of the reagents, by the following equation:

$$v = k[X]^m[Y]^n,$$

where k is the reaction rate, $[X]$ and $[Y]$ are the concentrations of the reagents, and m and n are the orders of the reaction, which are experimentally determined. Thus, there may be an imprecision (or uncertain) in obtaining these parameters. Classical models do not take this fact into account [3]. On the other hand, the Fuzzy Set Theory can be used in order to describe this uncertainty.

This paper focuses on the reversible chemical reaction problem, considering that the initial quantities of molecule concentrations are uncertain and modeled by fuzzy numbers. More precisely, this manuscript proposes numerical solutions to the reaction given by [2]

$$AB \underset{k_2}{\overset{k_1}{\rightleftarrows}} A + B, \tag{2}$$

where k_1 and k_2 are distinct reaction rates such that the reagents A and B are not in equilibrium. One example of this decomposition is the carbon acid (H_2CO_3), which transform into water (H_2O) and carbon gas (CO_2):

$$H_2CO_3(aq) \rightleftarrows H_2O(l) + CO_2(g),$$

where aq, l and g represent the aqueous, liquid and gaseous states, respectively.

From the fundamental principle of kinetic modeling, which is based on the conversion of chemical reaction mechanisms into differential equations, we apply the action mass law. Hence, we obtain the following system of differential equations

$$\begin{cases} \dfrac{d[A]}{dt} = -k_1[A] + k_2[B], & [A(0)] = [A_0] \\ \\ \dfrac{d[B]}{dt} = k_1[A] - k_2[B], & [B(0)] = [B_0] \end{cases} . \tag{3}$$

Note that $[A(t)] + [B(t)] = k, \forall t \in \mathbb{R}$ and for some $k \in \mathbb{R}$, since $\dfrac{d[A]}{dt} + \dfrac{d[B]}{dt} = 0$. In particular

$$[A_0] + [B_0] = k. \tag{4}$$

The initial conditions and/or parameters of the system (3) may be uncertain [3]. If $[A_0]$ and $[B_0]$ are given by fuzzy numbers, then $[A_0]$ and $[B_0]$ must be interactive in order to guarantee that the total quantity k be a real number [4].

One could try to solve this problem by taking $[A_0] = k - [B_0]$, however $[A_0] = k - [B_0]$ must solve the original Eq. (4), which does not always occur, since the standard sum between two fuzzy numbers does not produce real numbers as a result [5]. In fact, $[A_0] + [B_0] = k \Leftrightarrow [A_0] = k - [B_0]$ if, and only if the arithmetic operations "+" and "−" are interactive [4].

Interactivity is a relationship between fuzzy numbers that resembles the notion of dependence in the case of random variables. We use this concept to guarantee the veracity of Eq. (4). In addition, we use this relationship to intrinsically model the dependence that there may be between the reagents and their concentrations.

We use the method proposed by [6] to provide a numerical solution for the system given as in (3). This method consists in adapting the arithmetic operations of the classical Euler method, via sup-J extension principle, for arithmetic between interactive fuzzy numbers. This method can be used for any system of n-dimensional differential equations [7].

Finally, we present an example to show that different types of interactivity result in different solutions for the system of differential equations (3).

2 Preliminaries

This section presents the Euler method and some basic concepts of the Fuzzy Set Theory.

2.1 Euler Method

Let $y_i : \mathbb{R} \to \mathbb{R}^n$ be functions that depend on time t, with $i = 1, \ldots, n$. Consider the Initial Value Problem (IVP) given by (5)

$$\begin{cases} \frac{dy_i}{dt} = f_i(t, y_1, y_2, \ldots, y_n) \\ y(t_0) = y_0 \in \mathbb{R}^n \end{cases}, \quad i = 1, \ldots, n, \tag{5}$$

where f_i depends on $y_1, y_2, \ldots, y_n$ and t.

The Euler method, or also called the first order Runge-Kutta method, consists in determining numerical solutions to the ordinary differential equations (ODEs) described by (5). The algorithm of this method is given by (6)

$$y_i^{k+1} = y_i^k + hf_i(t_k, y_1^k, \ldots, y_n^k), \tag{6}$$

where $0 \leq k \leq N-1$, N is the number of partitions which the time interval is divided, h is the size of the subintervals $[t_k, t_{k+1}]$ and the initial condition (t_0, y_i^0).

2.2 *Fuzzy Set Theory*

A fuzzy set A of a universe X is characterized by a function $\mu_A : X \to [0, 1]$ called membership function, where $\mu_A(x)$ represents the membership degree of x in A for all $x \in X$ [8]. For notational convenience, we may simply use the symbol $A(x)$ instead of $\mu_A(x)$. The class of fuzzy subsets of X is denoted by $\mathcal{F}(X)$. Note that each classical subset of X can be uniquely identified with the fuzzy set whose membership function is given by its characteristic function.

The α-cuts of a fuzzy set $A \subseteq X$, denoted by $[A]^\alpha$, are defined as $[A]^\alpha = \{x \in X : A(x) \geq \alpha\}$, $\forall \alpha \in (0, 1]$ [9]. In addition, if X is also a topological space, then the 0-cut of A is defined by $[A]^0 = cl\{x \in X : A(x) > 0\}$, where cl Y, $Y \subseteq X$, denotes the closure of Y. An important subclass of $\mathcal{F}(\mathbb{R})$, denoted by $\mathbb{R}_\mathcal{F}$, is the class of fuzzy numbers which includes the sets of the real numbers as well as the set of the bounded closed intervals of $\mathbb{R}$.

A fuzzy set A of $\mathbb{R}$ is said to be a fuzzy number if all α-cuts are bounded, closed and non-empty nested intervals for all $\alpha \in [0, 1]$. The α-cuts of a fuzzy number A are denoted by $[A]^\alpha = [a_\alpha^-, a_\alpha^+]$. The class of fuzzy numbers such that a_α^- and a_α^+ are continuous function with respect to α, is denoted by $\mathbb{R}_{\mathcal{F}C}$.

Note that every triangular fuzzy number is an example contained in $\mathbb{R}_{\mathcal{F}C}$. Recall that a triangular fuzzy number A is denoted by the triple $(a; b; c)$ for some $a \leq b \leq c$. By means of α-cuts we have $[A]^\alpha = [a + \alpha(b - a), c - \alpha(c - b)]$, $\forall \alpha \in [0, 1]$. Recall that for all fuzzy numbers we have $A \subseteq B \Leftrightarrow [A]^\alpha \subseteq [B]^\alpha$, for all $\alpha \in [0, 1]$.

The *Pompeiu-Hausdorff* norm and width of fuzzy numbers are given by the following definitions [10].

Definition 1 Let A and B be fuzzy numbers. The Pompeiu-Hausdorff distance $d_\infty : \mathbb{R}_\mathcal{F} \times \mathbb{R}_\mathcal{F} \to [0, +\infty)$ is given by

$$d_\infty(A, B) = \bigvee_{\alpha \in [0,1]} \max\{|a_\alpha^- - b_\alpha^-|, |a_\alpha^+ - b_\alpha^+|\}.$$

where the symbol $\bigvee$ represents the supremum operator. Moreover, the Pompeiu-Hausdorff norm of a fuzzy number $A \in \mathbb{R}_\mathcal{F}$ is defined by

$$||A||_\mathcal{F} = d_\infty(A, 0), \tag{7}$$

where the symbol 0 stands for the characteristic function of the real number 0.

Definition 2 The *width* (or *diameter*) of a fuzzy number $A \in \mathbb{R}_\mathcal{F}$ is defined by

$$width(A) = a_0^+ - a_0^-. \tag{8}$$

We say that A is more specific than B, if $width(A) \leq width(B)$. In particular, if $A \subseteq B$, then A is more specific than B.

Next section presents the definition of joint possibility distribution, which gives raise to the notion of interactivity.

3 Joint Possibility Distribution

A fuzzy relation R between two universes X and Y is given by the mapping $R : X \times Y \to [0, 1]$, where $R(x, y) \in [0, 1]$ is the degree of relationship between $x \in X$ and $y \in Y$. An n-ary relation on $X = X_1 \times ... \times X_n$ is nothing else than a fuzzy (sub)set of X.

A fuzzy relation $J \in \mathcal{F}(\mathbb{R}^n)$ is said to be a joint possibility distribution (JPD) among the fuzzy numbers $A_1, \ldots, A_n \in \mathbb{R}_{\mathcal{F}}$ if

$$A_i(y) = \Pi_J^i(y) = \bigvee_{x:x_i=y} J(x), \ \forall y \in \mathbb{R}, \tag{9}$$

for all $i = 1, \ldots, n$.

One example of JPD is the t-norm-based joint possibility distribution whose definition is given as follows. Let t be a t-norm, that is, an associative, commutative and increasing operator $t : [0, 1]^2 \to [0, 1]$ that satisfies $t(x, 1) = x$ for all $x \in [0, 1]$. The fuzzy relation J_t given by

$$J_t(x_1, \ldots, x_n) = A_1(x_1)\ t\ ...\ t\ A_n(x_n) \tag{10}$$

is called the t-norm-based joint possibility distribution of $A_1, \ldots, A_n \in \mathbb{R}_{\mathcal{F}}$.

If the t-norm is given by the minimum operator ($t = \wedge$), then the fuzzy numbers $A_1, \ldots, A_n$ are said to be non-interactive. Otherwise, that is, if J satisfies (9) and $J \neq J_\wedge$, then $A_1, \ldots, A_n$ are called interactive. Thus, the interactivity of the fuzzy numbers $A_1, \ldots, A_n$ arises from a given joint possibility distribution. The concept of interactivity resembles the relation of dependence in the case of random variables.

The sup-J extension principle is a mathematical tool that extends classical functions to fuzzy functions. Moreover, it takes the interactivity among fuzzy numbers into account. The definition is given as follows [11, 12].

Definition 3 Let $J \in \mathcal{F}(\mathbb{R}^n)$ be a joint possibility distribution of $(A_1, \ldots, A_n) \in \mathbb{R}^n_{\mathcal{F}}$ and $f : \mathbb{R}^n \to \mathbb{R}$. The sup-J extension of f at $(A_1, \ldots, A_n) \in \mathbb{R}^n_{\mathcal{F}}$, denoted $f_J(A_1, \ldots, A_n)$, is the fuzzy set defined by:

$$f_J(A_1, \ldots, A_n)(y) = \bigvee_{(x_1,\ldots,x_n)\in f^{-1}(y)} J(x_1, \ldots, x_n), \tag{11}$$

where $f^{-1}(y) = \{(x_1, \ldots, x_n) \in \mathbb{R}^n : f(x_1, \ldots, x_n) = y\}$.

Note that the sup-J extension principle generalizes the Zadeh's extension principle, since Eq. (11) boils down to the Zadeh extension, if $J = J_{\wedge}$ [13]. In this case, the fuzzy numbers $A_1, \ldots, A_n$ are non-interactive. The sup-J extension gives rise to the arithmetic on interactive fuzzy numbers, considering f as an arithmetic operator.

Other type of interactivity, which is not based on t-norm, is the one obtained by the concept of linearly interactivity, or also called completely correlation. This concept was introduced by Fullér et al. [11] but only for two fuzzy numbers. Subsequently, the authors of [14] proposed a generalization of this notion for n fuzzy numbers, $n > 2$.

The fuzzy numbers $A_1, \ldots, A_n$ are said to be linearly interactive if there exist $q = (q_1, \ldots, q_n)$, $r = (r_1, \ldots, r_n) \in \mathbb{R}^n$ with $q_1 q_2 \ldots q_n \neq 0$ such that the corresponding joint possibility distribution $J = J_{\{q,r\}}$ is given by

$$J_{\{q,r\}}(x_1, \ldots, x_n) = A_i(x_i)\chi_U(x_1, \ldots, x_n), \quad \forall (x_1, \ldots, x_n) \in \mathbb{R}^n \tag{12}$$

for all $i = 1, \ldots, n$, where χ_U stands for the characteristic function of the set $U = \{(u, q_1 u + r_1, \ldots, q_n u + r_n) : \forall u \in \mathbb{R}\}$.

The JPD given by (12) can be used to provide solutions to the fuzzy differential equations (FDEs) that consider interactivity [4, 14, 15]. However $J_{\{q,r\}}$ can only be applied to fuzzy numbers that have a co-linear relationship among their membership functions, which means that it can not be used to fuzzy numbers that do not have the same shape, triangular and trapezoidal fuzzy numbers for example [6].

Alternatively, Esmi et al. [16] employed a parametrized family of joint possibility distributions $\mathcal{J} = \{J_\gamma : \gamma \in [0, 1]\}$ to define interactive additions of fuzzy numbers. The authors of [6] used this family of JPDs to produce a numerical solution of a fuzzy initial value problem and verified that $\mathcal{J}$ is more embracing than $J_{\{q,r\}}$.

Here we focus on distribution J_γ whose definition is given as follows. Let A_1 and A_2 be two fuzzy numbers and the functions $g^i_{\wedge}$, $g^i_{\vee}$ and v^i defined in [16]

$$g^i_{\wedge}(z, \alpha) = \bigwedge_{w \in [A_{3-i}]^\alpha} |w + z|, \quad , \quad g^i_{\vee}(z, \alpha) = \bigvee_{w \in [A_{3-i}]^\alpha} |w + z|$$

and

$$v^i(z, \alpha, \gamma) = (1 - \gamma)g^i_{\wedge}(z, \alpha) + \gamma g^i_{\vee}(z, \alpha),$$

for all $z \in \mathbb{R}$, $\alpha \in [0, 1]$, $\gamma \in [0, 1]$ and $i \in \{1, 2\}$.

Also, consider the sets R^i_α and $L^i(z, \alpha, \gamma)$,

$$R^i_\alpha = \begin{cases} \{a^-_{i_\alpha}, a^+_{i_\alpha}\} & \text{if} \alpha \in [0, 1) \\ [A_i]^1 & \text{if} \alpha = 1 \end{cases}$$

$$L^i(z, \alpha, \gamma) = [A_{3-i}]^\alpha \cap [-v^i(z, \alpha, \gamma) - z, v^i(z, \alpha, \gamma) - z].$$

Finally, J_γ is defined by

$$J_\gamma(x_1, x_2) = \begin{cases} A_1(x_1) \wedge A_2(x_2), & \text{if}(x_1, x_2) \in P(\gamma) \\ 0, & \text{otherwise} \end{cases} \tag{13}$$

with $P(\gamma) = \bigcup_{i=1}^{2} \bigcup_{\alpha \in [0,1]} \{(x_1, x_2) : x_i \in R^i_\alpha \text{ and } x_{3-i} \in L^i(x_i, \alpha, \gamma)\}$.

Esmi et al. proved that the fuzzy relation J_γ, given by (13), is a joint possibility distribution of A_1 and A_2. The parameter γ intrinsically models the "level" of the interactivity between the fuzzy numbers A_1 and A_2 in the following sense, the greater the value of γ, the lower the interactivity. It is important to observe that for $\gamma = 1$ one obtains $J_1 = J_\wedge$ [16]. This means that if the JPD is given by J_1, then A_1 and A_2 are non-interactive. On the other hand, the JPD J_0 resembles the JPD $J_{q,r}$ [6].

One can observe that the Pompeiu-Hausdorff norm and the width are not equivalent, that is, $||A||_\mathcal{F} \leq ||B||_\mathcal{F}$ does not imply that $width(A) \leq width(B)$. For example, for $A = (-2; 0; 2)$ and $B = (1; 2; 3)$ we have that $||A||_\mathcal{F} = 2 \leq 3 = ||B||_\mathcal{F}$ but $width(A) = 4 > 2 = width(B)$. Sussner et al. [17] employed shifts in order to define a new family of parametrized joint possibility distributions that can be used to control the width of the corresponding interactive addition.

Definition 4 Let $A \in \mathbb{R}_\mathcal{F}$. The translation of A by $k \in \mathbb{R}$ is defined as the following fuzzy number $\tilde{A}$:

$$\tilde{A}(x) = A(x + k), \ \forall x \in \mathbb{R}. \tag{14}$$

Next, we present the distribution using the concept of Definition 4 [17].

Theorem 1 *Given $A_1, A_2 \in \mathbb{R}_\mathcal{F}$ and $c = (c_1, c_2) \in \mathbb{R}^2$. Let $\tilde{A}_i \in \mathbb{R}_\mathcal{F}$ be such that $\tilde{A}_i(x) = A_i(x + c_i)$, $\forall x \in \mathbb{R}$ and $i = 1, 2$. Let $\tilde{J}_\gamma$ be the joint possibility distribution of fuzzy numbers $\tilde{A}_1, \tilde{A}_2 \in \mathbb{R}_\mathcal{F}$ defined as Eq. (13). The fuzzy relation J^c_γ given by*

$$J^c_\gamma(x_1, x_2) = \tilde{J}_\gamma(x_1 - c_1, x_2 - c_2), \ \forall (x_1, x_2) \in \mathbb{R}^2, \tag{15}$$

is a joint possibility distribution of A_1 and A_2.

From now on, we only use the joint possibility distribution provided by Theorem 1. For simplicity of notation, we denote $J_\gamma = J^c_\gamma$.

Next section presents the arithmetic for interactive fuzzy numbers. This arithmetic is obtained from the sup-J extension principle for some $J \neq J_\wedge$. In particular we focus on the arithmetic raised by the joint possibility distribution J_γ.

4 Arithmetic for Interactive Fuzzy Numbers

Introduction section highlighted that Eq. (4) cannot be solved by taking $[A_0] = k - [B_0]$. Example 1 illustrates this fact.

Example 1 Let $X \in \mathbb{R}_\mathcal{F}$ and $B = (1; 2; 3)$. Consider the following equation

$$X + B = 10,$$

where the symbol $+$ stands for the fuzzy standard sum.

Let $A = 10 - B = 10 - (1; 2; 3) = (7; 8; 9)$. Note that A does not solve the above equation, since

$$A + B = (7; 8; 9) + (1; 2; 3) = (8; 10; 12) \neq 10.$$

The above example shows that the standard sum does not produce real numbers as result. Therefore, other types of sum must be introduced in order to solve Eq. (4).

From the definition of the sup-J extension principle, it is possible to establish an arithmetic for interactive fuzzy numbers. For example, the interactive sum and difference are given by

$$(A_1 +_J A_2)(y) = \bigvee_{x_1 + x_2 = y} J(x_1, x_2) \tag{16}$$

and

$$(A_1 -_J A_2)(y) = \bigvee_{x_1 - x_2 = y} J(x_1, x_2) \tag{17}$$

where J is an arbitrary JPD of A_1 and A_2.

Here we focus on the joint possibility distribution $J = J_\gamma$. In this case, we denote the operations defined as in (16) and (17) by $A_1 +_\gamma A_2$ and $A_1 -_\gamma A_2$, respectively. The next examples illustrate these arithmetic operations, considering $J = J_\gamma$.

Example 2 Let $A_1 = (1; 2; 3)$ and $A_2 = (2; 3; 4)$. For $\gamma \in \{0, 0.5, 0.75, 1\}$, we have

$$\begin{aligned} A_1 +_0 A_2 &= 5 \\ A_1 +_{0.5} A_2 &= (4; 5; 6) \\ A_1 +_{0.75} A_2 &= (3.5; 5; 6.5) \\ A_1 +_1 A_2 &= (3; 5; 7). \end{aligned}$$

Note that in Example 2 we obtain $A_1 +_0 A_2 = 5$, where 5 stands for the crisp number 5 whose membership function is given by the characteristic function $\chi_{\{5\}}$. This result implies that from the interactive sum it is possible to obtain a real number as a sum of two fuzzy numbers, in contrast to the standard arithmetic sum.

In addition, for $\gamma = 1$ it follows that $A_1 +_1 A_2 = (3; 5; 7) = A_1 + A_2$, corroborating the previous observation.

Example 3 Let $A_1 = (4; 5; 6)$ and $A_2 = (1; 2; 3)$. For $\gamma \in \{0, 0.5, 0.75, 1\}$, we have

$$A_1 -_0 A_2 = 3$$
$$A_1 -_{0.5} A_2 = (2; 3; 4)$$
$$A_1 -_{0.75} A_2 = (1.5; 3; 4.5)$$
$$A_1 -_1 A_2 = (1; 3; 5).$$

Note that

$$A_1 -_0 A_2 = 3 = A_1 -_g A_2 = A_1 -_{gH} A_2 = A_1 -_H A_2, \tag{18}$$

where $-_H$, $-_{gH}$ and $-_g$ represent the Hukuhara, generalized Hukuhara and generalize differences, respectively [18].

In general, the equality given by (18) always holds true, more precisely, $A_1 -_0 A_2 = A_1 -_g A_2$, for all $A, B \in \mathbb{R}_{\mathcal{F}_C}$ [19]. This means that the Hukuhara difference and its generalizations are particular types of interactive arithmetic operations.

Also, for $\gamma = 1$ it follows that $A_1 -_1 A_2 = (1; 3; 5) = A_1 - A_2$, where the symbol "−" represents the standard difference for fuzzy numbers.

The next section presents the numerical solution for (3) obtained from the method proposed by [6].

5 Fuzzy Numerical Solution to the Reversible Chemical Reactions

This paper considers that the initial concentrations $[A_0]$ and $[B_0]$ are given by interactive fuzzy numbers. Hence we adopt the method proposed by [6], which consists in extending the classical arithmetic operations presented in the Euler method (see Eq. (6)) to arithmetic for interactive fuzzy numbers.

The numerical solution based on the arithmetic provided in the previous section is given by

$$\begin{cases} [A]^{k+1} = [A]^k +_\gamma h(-k_1[A]^k +_\gamma k_2[B]^k), & [A]^0 = [A_0] \\ [B]^{k+1} = [B]^k +_\gamma h(k_1[A]^k -_\gamma k_2[B]^k), & [B]^0 = [B_0] \end{cases}, \tag{19}$$

where $[A_0], [B_0] \in \mathbb{R}_{\mathcal{F}_C}$.

According to Esmi et al. [16] the interactive sum ($+_\gamma$) and difference ($-_\gamma$) are increasing with respect to γ. This means that

$$A_1 +_0 A_2 \subseteq A_1 +_{\gamma_1} A_2 \subseteq A_1 +_{\gamma_2} A_2 \subseteq A_1 +_1 A_2, \tag{20}$$

and

$$A_1 -_0 A_2 \subseteq A_1 -_{\gamma_1} A_2 \subseteq A_1 -_{\gamma_2} A_2 \subseteq A_1 -_1 A_2, \tag{21}$$

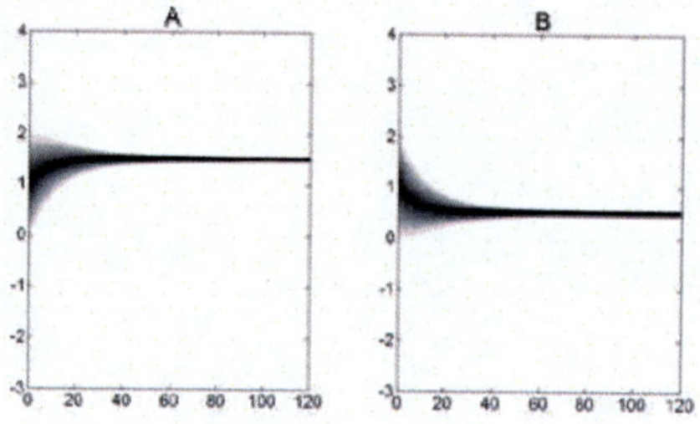

(a) Numerical solution for $\gamma = 0$.

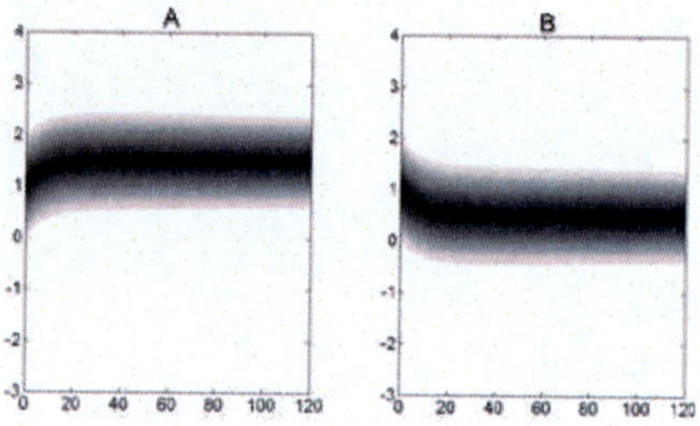

(b) Numerical solution for $\gamma = 0, 5$.

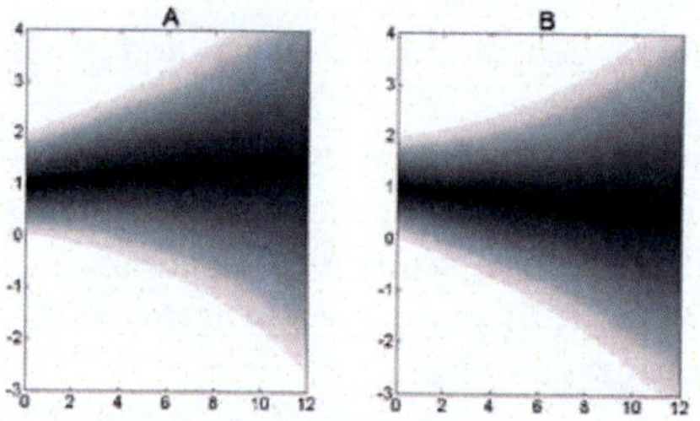

(c) Numerical solution for $\gamma = 1$.

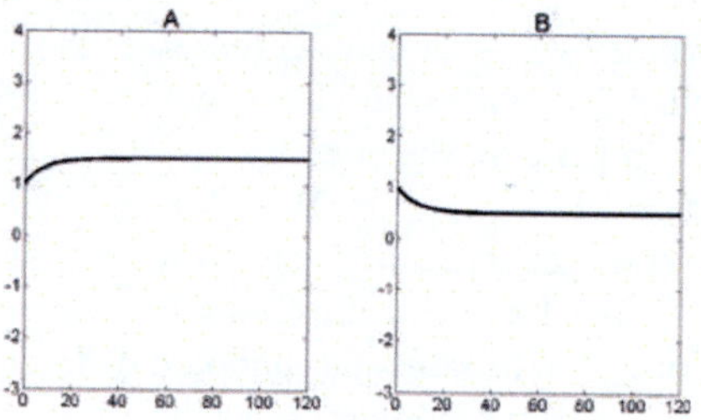

(d) Deterministic solution.

Fig. 1 The gray lines represent the α-cuts of the fuzzy solutions, where their endpoints for α varying from 0 to 1 are represented respectively from the gray-scale lines varying from white to black. The parameters are given by $h = 0, 125$, $k_1 = 0, 03$, $k_2 = 0, 09$ e $[A_0] = [B_0] = (0, 1, 2)$

for all $\gamma_1, \gamma_2 \in [0, 1]$ such that $\gamma_1 \leq \gamma_2$.

From Eqs. (20) and (21), one expects that the numerical solution produced by (19), with γ_1, is more specific than the solution with γ_2 such that $\gamma_1 \leq \gamma_2$. This fact is illustrated in Fig. 1, which depicts the numerical solutions to the system given by (3), considering three levels of interactivity: $\gamma = 0$, $\gamma = 0, 5$ and $\gamma = 1$.

Note that different values of γ imply in different numerical solutions to the problem. This fact is associated with the arithmetic for interactive fuzzy numbers that depends on the family of joint possibility distribution J_γ.

We also present the deterministic solution to the system (3) considering the initial conditions given by $[A_0] = [B_0] = 1$, in order to compare the fuzzy numerical solution qualitatively. Note that the solutions produced by $\gamma = 0$ (see Fig. 1a) and $\gamma = 0.5$ (see Fig. 1b) have a similar behaviour as the deterministic solution, in contrast to the solution produced by $\gamma = 1$ (see Fig. 1c), which is the case where the initial concentrations are considered as non-interactive.

It is important to observe that the numerical solution given by $\gamma = 0$ has a decreasing width with respect to time. Moreover, it reaches a stability value. This is a typical behaviour of reversible chemical reactions and also occurs in the deterministic solution, as Fig. 1d illustrates.

The joint possibility distribution $J_{0.5}$ allows us to control the width of the solution, which remains constant with respect to time. However, the solution for B assumes negative values which is not consistent with the nature of the phenomenon.

The joint possibility distribution J_1 produces a solution with increasing width, and since the width of the fuzzy solution is associated with the uncertainty that it models, the numerical solution for $\gamma = 1$ propagates uncertainty. This fact is already expected, since the initial conditions $[A_0]$ and $[B_0]$ are considered as non-interactive and therefore the arithmetic operations for them are the standard ones.

6 Final Remarks

This paper presented a study of reversible chemical reaction, from fuzzy differential equation point of view. Here the initial concentrations of the reagents were considered uncertainty and modeled by fuzzy numbers. The notion of interactivity was used in order to satisfy the conditions imposed by the system of differential equations (see (4)). Moreover, this relationship was used to intrinsically model the dependence that there may be between the reagents.

Numerical solutions to this FIVP were proposed by extending the arithmetic operations in the classical Euler method to the arithmetic for interactive fuzzy numbers, according to [6]. This manuscript considered the interactivity raised by the joint possibility distribution J_γ, proposed by [16]. We exhibited the solutions for $\gamma = 0$, $\gamma = 0, 5$ and $\gamma = 1$.

The numerical solution given by $\gamma = 0$ has decreasing width with respect to time, producing a more specific solution than the others one. Moreover, the numerical solution stabilizes assuming a constant real value. This stability is a typical behaviour of reversible chemical reactions, as Fig. 1d depicts.

The solution given by $\gamma = 0.5$ has constant width with respect to time, which means that the uncertainty remains constant. However, the concentration of the reagent B assumes negative values, which is not coherent with the phenomenon.

Finally, the solution given by $\gamma = 1$ has increasing width with respect to time, which means that the uncertainty increases. This behaviour was already expected since the arithmetic operations in the method, for $\gamma = 1$, boil down to the standard arithmetic and consequently it propagates uncertainty [5].

The choice of the parameter γ depends of each model and its parameters and conditions. In paritcular, for this phenomenon, the joint possibility distribution J_0 provided a solution with more similar behavior with the deterministic solution than the other JPDs. However, for other chemical reaction problems, a different JPD from J_0 can be more suitable [7].

Acknowledgements The authors would like to thank the support of CNPq under grant no. 306546/2017-5 and FAPESP under grant no. 2016/26040-7.

References

1. P. Flowers, K. Theopold, R. Langley, W.R. Robinson, Chemistry, OpenStax, Texas (2015)
2. G.F. Froment, K.B. Bischoff, J. Wilde, *Chemical Reactor Analysis and Design* (Random House, New York, 1988)
3. K. Ghosh, J. Schlipf, Formal modeling of a system of chemical reactions under uncertainty. J. Bioinform. Comput. Biol. **12**, 1–15 (2014)
4. L.C. Barros, F.S. Pedro, Fuzzy differential equations with interactive derivative. Fuzzy Sets Syst. **309**, 64–80 (2017)
5. V.F. Wasques, E. Esmi, L.C. Barros, B. Bede, Comparison between numerical solutions of fuzzy initial-value problems via interactive and standard arithmetics, *Proceedings in Fuzzy Techniques: Theory and Applications* (Springer International Publishing, Cham, 2019), pp. 704–715
6. V.F. Wasques, E. Esmi, L.C. Barros, P. Sussner, *Numerical Solutions for Bidimensional Initial Value Problem with Interactive Fuzzy Numbers* in *Proceedings in Fuzzy Information Processing*. (Springer International Publishing, Cham, 2018), pp. 84–95
7. V.F. Wasques, E. Esmi, L.C. Barros, P. Sussner, *Numerical Solution for Lotka-Volterra Model of Oscillating Chemical Reactions with Interactive Fuzzy Initial Conditions*, in *2019 Conference of the International Fuzzy Systems Association and the European Society forFuzzy Logic and Technology (EUSFLAT 2019 Proceedings)* (Atlantis Press, 2019), pp. 544–549
8. L.A. Zadeh, Fuzzy sets. Inf. Control **8**, 338–353 (1965)
9. L.C. Barros, R.C. Bassanezi, W.A. Lodwick, A first course in fuzzy logic, fuzzy dynamical systems, and biomathematics, *Studies in Fuzziness and Soft Computing* (Springer, Berlin Heidelberg, 2017)
10. P. Diamond, P. Kloeden, *Fundamentals of Fuzzy Sets*, Metric Topology of Fuzzy Numbers and Fuzzy Analysis (Springer, US, Boston, 2000), pp. 583–641
11. R. Fullér, P. Majlender, On interactive fuzzy numbers. Fuzzy Sets Syst. **143**, 355–369 (2004)
12. C. Carlsson, R. Fullér, P. Majlender, Additions of completely correlated fuzzy numbers, in *Fuzzy Systems Proceedings IEEE International Conference* (Budapest, Hungary, 2004)
13. L.A. Zadeh, Concept of a linguistic variable and its application to approximate reasoning—I. Inf. Sci. **8**, 199–249 (1975)
14. V.F. Wasques, E. Esmi, L.C. Barros, F.S. Pedro, P. Sussner, Higher order initial value problem with interactive fuzzy conditions, in *IEEE International Conference on Fuzzy Systems (FUZZ-IEEE Proceedings)* (2018), pp. 1–7
15. D.E. Sánchez, V.F. Wasques, E. Esmi, L.C. Barros, P. Sussner, Fuzzy initial value problems for fuzzy hypocycloid curves, in *IEEE International Conference on Fuzzy Systems (FUZZ-IEEE Proceedings)* (2019), pp. 1–5
16. E. Esmi, P. Sussner, G.B.D. Ignácio, L.C. Barros, A parametrized sum of fuzzy numbers with applications to fuzzy initial value problems. Fuzzy Sets Syst. **331**, 85–104 (2018)
17. P. Sussner, E. Esmi, L.C. Barros, Controling the width of the sum of interactive fuzzy numbers with applications to fuzzy initial value problems, in *Proceedings In IEEE International Conference on Fuzzy Systems* (2016), pp. 85–104
18. B. Bede, *Mathematics of Fuzzy Sets and Fuzzy Logic* (Springer, Berlin Heidelberg, Berlin, 2013)
19. V.F. Wasques, E. Esmi, L.C. Barros, P. Sussner, The generalized fuzzy derivative is interactive. Inf. Sci. **519**, 93–109 (2020)

Predictive Maintenance of Aircraft Engines Using Fuzzy Bolt©

Bhavya Mayadevi, Dino Martis, Anoop Sathyan, and Kelly Cohen

Abstract The Remaining Useful Life (RUL) of engines is a very important prognostic parameter that can be used to make a decision on when an aircraft engine needs to be sent for maintenance or repair. Today, there is no way to accurately estimate the RUL of an engine. Access to various sensor readings could provide more insights into RUL degradation. However, the relationship between these sensor readings obtained from flight data and the RUL of an engine is not well understood. In this paper, we attempt to provide an estimation of the engine RUL based on the time history data obtained from different sensors. A Genetic Fuzzy System, trained using Fuzzy Bolt©, is used to make useful estimations of RUL, which could in turn help with providing a marker for when an engine needs to be sent for maintenance. The models are trained on the NASA C-MAPSS dataset available for turbofan engines. We also compare our methodology with a similarity based model that has been proven to be one of the best models in predicting RUL on this dataset.

1 Introduction

The Remaining Useful Life (RUL) of an aircraft engine refers to the number of cycles the engine can go through before failure happens. This is an important metric that helps determine the best time to release it for maintenance or repair. The importance comes from both a cost as well as safety stand-point. Ideally, it is best to have engine be sent for maintenance before it reaches the stage of steep degradation. At the same time, it is also important to not send the engine too soon. A tool for modeling the RUL degradation will also help with gaining a better insight into how the different factors play a role. Consequently, these insights could help in making operational changes

---

B. Mayadevi · D. Martis (✉) · A. Sathyan · K. Cohen
Genexia, LLC., 2900 Reading Rd, Cincinnati, OH 45206, USA
e-mail: dmartis@genexia.co

B. Mayadevi
e-mail: bmayadevi@genexia.co

B. Bede et al. (eds.), *Fuzzy Information Processing 2020*,
Advances in Intelligent Systems and Computing 1337,
https://doi.org/10.1007/978-3-030-81561-5_11

to the engine to prolong its life. Although a lot of sensor data is available from flight systems today that could affect engine degradation, the relationship between the RUL and these sensor data remains unknown. In this paper, we try to identify the relationship through our proprietary machine learning tool called Fuzzy Bolt©, that provides the ability to train a fuzzy logic based intelligent system in an efficient manner.

Fuzzy Inference System (FIS) can be used to model the engine degradation after training using appropriate data. The FIS model uses fuzzification, rule-inference and defuzzification to make decisions. Designing a FIS involves developing the set of membership functions for each input and output variable, and defining the rules within the rulebase to define the relationship between the inputs and outputs. Along with these variables, certain other settings such as defuzzification method, conjunction, disjunction, implication methods, etc can also be decided by the designer.

Although expert knowledge can be used to build FISs and this capability is appealing to a lot of applications, it makes sense to have a mechanism to tune the parameters of the FIS automatically. Self-tuning FISs are very useful especially when there are many inputs and outputs and their relationships are not that straightforward or well known. FISs can be given such self-tuning capability by using different optimization algorithms. These include ANFIS (Adaptive Network based Fuzzy Inference System) [1], Simulated Annealing [2], Genetic Algorithm (GA) [3], etc. The process of choosing all these parameters can be automated by using an optimization approach, such as GA, that can choose a near-optimal set of parameters to minimize some pre-defined cost function. A FIS that is tuned using GA is called a Genetic Fuzzy System (GFS). Such GFSs have been developed with much success for athlete motion prediction [4], collaborative robotics [5–7], simulated air-to-air combat [8] etc.

In this paper, we discuss the applicability of Fuzzy Bolt© to engine prognostics. We train a FIS model using Fuzzy Bolt© to make accurate estimations of the RUL. The models are trained and tested on the publicly available NASA C-MAPSS data [9].

2 NASA C-MAPSS Dataset

The dataset was created using C-MAPSS (Commercial Modular AeroPropulsion System Simulation) tool that can simulate the performance of a realistic commercial turbofan engine. C-MAPSS was used to simulate an engine model of 90,000 lb thrust class for different values for three variables: (i) altitudes ranging from sea level to 40,000 ft, (ii) Mach numbers from 0 to 0.90, and (iii) sea-level temperatures from −60°F to 103°F [9]. For each run of the engine, a fault is introduced at some arbitrary point that affects the health of engine in an exponential manner.

The data has 4 sub-datasets with different number of operating conditions and fault conditions, and each sub-data set is further divided into training and test subsets, as shown in Table 1. In each of the four datasets, the rows provide information about one operational cycle of the engine. Each row provides values for the following 26 variables: (1) Engine ID, (2) Current operational cycle number of the engine,

Table 1 NASA C-MAPSS data description

Data sets	FD001	FD002	FD003	FD004
Train trajectories	100	260	100	249
Test trajectories	100	259	100	248
Number of data points in the training set	20,631	53,759	24,720	61,249
Number of data points in the test set	13,096	33,991	16,596	41,214
Operating conditions	1	6	1	6
Fault conditions	1	1	2	2

(3) Altitude, (4) Mach number, (5) Sea-level temperature, (6) Total temperature at fan inlet, (7) Total temperature at LPC outlet, (8) Total temperature at HPC outlet, (9) Total temperature at LPT outlet, (10) Pressure at fan inlet, (11) Total pressure in bypass-duct, (12) Total pressure at HPC outlet, (13) Physical fan speed, (14) Physical core speed, (15) Engine pressure ratio, (16) Static pressure at HPC outlet, (17) Ratio of fuel flow, (18) Corrected fan speed, (19) Corrected core speed, (20) Bypass Ratio, (21) Burner fuel-air ratio, (22) Bleed Enthalpy, (23) Demanded fan speed, (24) Demanded corrected fan speed, (25) HPT coolant bleed, (26) LPT coolant bleed. The engine is operating normally at the start of each time series, and develops a fault at some point in time which is unknown. In the training set, the fault grows in magnitude until a system failure. In the test set, data is provided up to some time prior to system failure. The goal is to estimate the number of remaining operational cycles remaining till failure on the test data.

Piecewise-Linear RUL Degradation

A simplified piecewise linear function [10], similar to the one shown in Fig. 1, is commonly used to model RUL degradation. Although this is an approximation, this is a commonly used assumption where the health of the engine stays constant initially for an arbitrary number of cycles after which it degrades slowly till failure. Previous works on the NASA C-MAPSS dataset use such a piecewise linear degradation model for the actual RUL function [11–14]. This is then used to evaluate the RMSE of their predictive models. For the sake of uniformity, we also use the same piecewise function in this work. In this paper, the constant part is set at an RUL value of 125. Thus, the RUL piecewise function is defined as follows:

$$RUL(x) = \begin{cases} x & x \leq 125 \\ 125 & x > 125 \end{cases}. \tag{1}$$

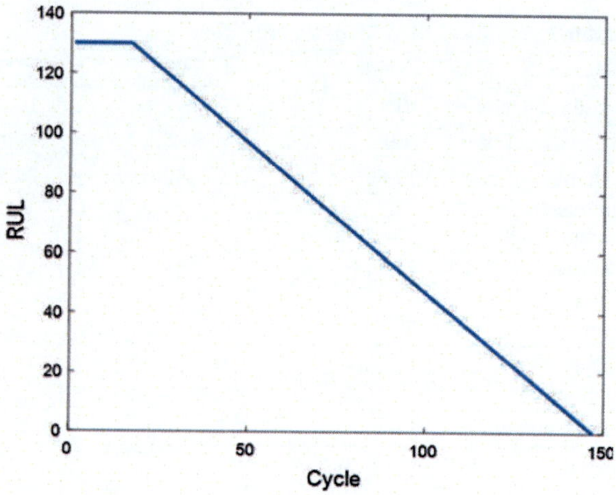

Fig. 1 Piecewise linear degradation of RUL

3 Methodology

3.1 Fuzzy Bolt©

Fuzzy Bolt© is an efficient supervised learning methodology for training FISs with many inputs. It is able to perform a targeted search while also tackling the issue of curse of dimensionality. Curse of dimensionality refers to the explosion the size of the search space caused by the exponential increase in the number of parameters to be tuned with respect to the number of input variables.

Earlier approaches of GFS used GA directly to tune the parameters of a FIS. Ideally, in a FIS, the rulebase should include all possible combinations of input membership functions in their antecedents. Therefore, if a FIS is defined using n inputs with each input defined by m membership functions, the rulebase should have m^n rules to include all possible combinations. This means that GA has to tune m^n consequent parameters along with the membership function boundaries. So, as the number of inputs increase, the number of parameters that need to be tuned using GA increases exponentially, thus increasing the overall computational complexity of the search process.

Later, Genetic Fuzzy Trees (GFTs) were developed to mitigate this problem to some extent by dividing the computations between several smaller FISs, each of which only take in two or three inputs [8]. These smaller FISs are connected together in a tree-like architecture that outputs the desired variable(s) at the last layer. This divide and conquer approach reduces the number of parameters that are tuned by GA, thus reducing the complexity of search. But, the GFT architecture needs to be defined beforehand and there is also the possibility of missing some essential connections between some input variables.

Fig. 2 An n-input FIS that can be trained by Fuzzy Bolt©

Fuzzy Bolt© provides a very efficient way to tune a FIS model. By intelligently reducing the search space, it is able to define the relationship between inputs and outputs without the need to breakdown the system to a GFT format. It works on a standard single n-input-p-output system, as shown in Fig. 2. Apart from performing a targeted search of the parameter space, Fuzzy Bolt© is also able to control the number of parameters that need to be tuned to train the system. This means the designer will have more control over the training process to prevent overfitting. The number of parameters does not increase substantially even when the number of membership functions for the inputs and outputs are increased. This provides more flexibility for the designer as they can make changes to design parameters with minimal effect on the overall training time.

4 Results

Fuzzy Bolt© was ran on all four training datasets, FD001 through FD004, separately. The best models obtained from the training were tested on the respective test sets. The RMSE values are evaluated assuming the piecewise linear degradation to be the true RUL degradation function. It is to be noted that RMSE purely compares the predicted values against the actual RUL values. In reality, it is better to use a function that penalizes predicted values that are higher than the actual RUL values. This is because for aircraft engines it is important to have a model that predicts failure before it actually happens. For example, if the actual RUL is 200, it is better to predict RUL of 197 than an RUL of 203.

For the RMSE evaluation, the piecewise linear curve has the constant part set at 125. The results obtained are shown in Table 2 along with a comparison with a similarity-based approach [15]. The outputs predicted by the model are smoothed using an averaging filter that takes the mean of the RUL predictions over the previous 10 timesteps to smoothen the RUL degradation curves. The RMSE values listed in Table 2 are obtained after this filtering process. This filtering process improves the RMSE slightly. It can be seen from the table that our approach is able to produce much better model in terms of the RMSE values.

The FIS model had the following settings:

1. Product AndMethod is used. This means that the firing strength of each rule is evaluated by multiplying the membership values of the inputs.
2. Product implication is used. So, the consequent membership function of each rule is multiplied by the firing strength value of that rule.

Table 2 RMSE values obtained on the test data

	FD001	FD002	FD003	FD004
Similarity-based method [15]	16.43	23.36	17.43	23.36
Fuzzy Bolt©	15.63	20.67	13.2	18.67

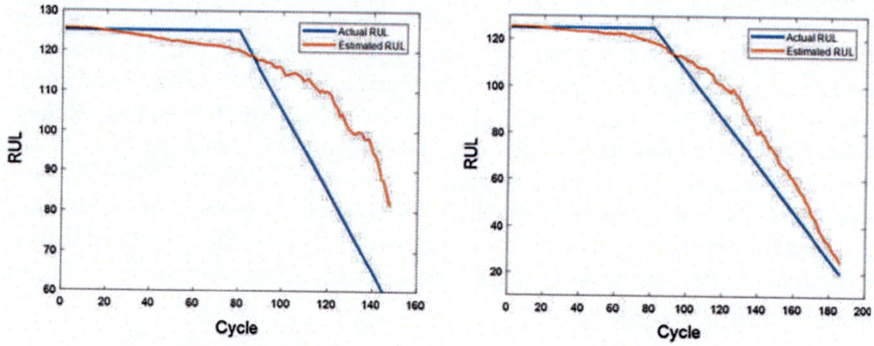

(a) Engine Unit#21 (b) Engine Unit#24

Fig. 3 RUL degradation predicted using Fuzzy Bolt© on two engines in the FD001 test data

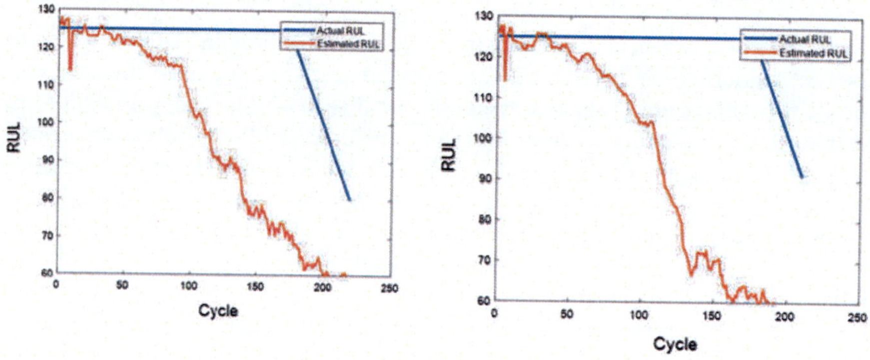

(a) Engine Unit#110 (b) Engine Unit#115

Fig. 4 RUL degradation predicted using Fuzzy Bolt© on two engines in the FD002 test data

3. Centroid defuzzification is used. This is the most commonly used defuzzification. All the rules in the FIS are taken into account when applying centroid defuzzification.

Fuzzy Bolt© tunes the membership functions and the rulebase to come up with the best system in terms of the training and validation RMSE to check for generalization. It is important that the model does not overfit to the training data. Figures 3, 4, 5 and 6 show the performance of the best models as applied a couple of engine units in each

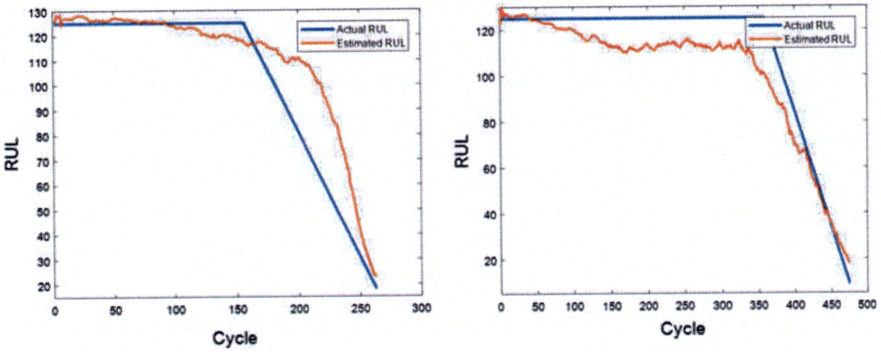

(a) Engine Unit#21 (b) Engine Unit#24

Fig. 5 RUL degradation predicted using Fuzzy Bolt© on two engines in the FD003 test data

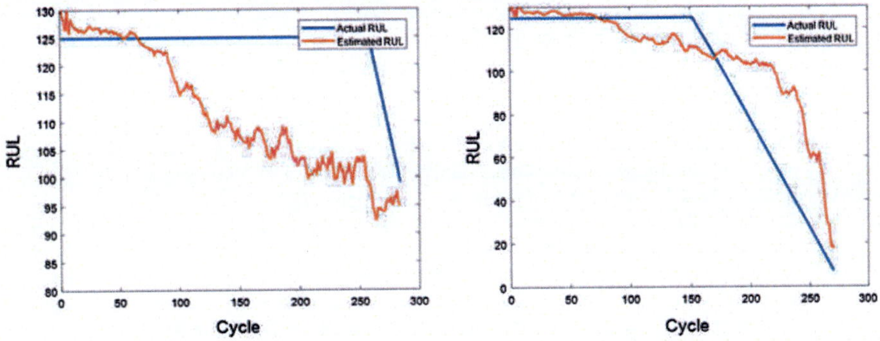

(a) Engine Unit#9 (b) Engine Unit#12

Fig. 6 RUL degradation predicted using Fuzzy Bolt© on two engines in the FD004 test data

of the four test sets. As can be seen from Figs. 3 and 5 as well as from Table 2 for FD001 and FD003, the predicted values closely match the piecewise linear RUL degradation. Similarly, the models also perform well on the other 2 sets, FD002 and FD004, although the RMSE values are slightly higher and the plots in Figs. 4 and 6 do not match as closely. It can be seen from Table 1 that the two sets of data, viz. FD001/FD003 and FD002/FD004, are different in terms of the operating conditions. This could mean that the operating conditions in FD002 and FD004 cause enough uncertainties that more variables are needed to get better predictions. It is also possible that the piecewise linear degradation is only true for the operating conditions defined for FD001 and FD003, where as another RUL degradation trend is possible for the operating conditions in FD002 and FD004. However, this is outside the scope of this paper.

5 Conclusions

In this paper, we showed the applicability of Fuzzy Bolt© to aircraft engine prognostics. It was able to successfully train four FISs for four separate engine prognostics datasets. The RUL predictions were made based on the operating conditions, 21 sensor variables and the engine cycle number. The trained FISs were able to provide an RUL prediction at each cycle with a reduced RMSE score. We also showed that RMSE score obtained using Fuzzy Bolt© were much better than that of a similarity-based approach [15] that has been proven to outperform most of the other approaches.

References

1. J.-S.R. Jang, ANFIS: adaptive-network-based fuzzy inference system. IEEE Trans. Syst. Man Cybern. **23**(3), 665–685 (1993)
2. R. Jain, N. Sivakumaran, T.K. Radhakrishnan, Design of self tuning fuzzy controllers for nonlinear systems. Expert Syst. Appl. **38**(4), 4466–4476 (2011)
3. O. Cordón, F. Gomide, F. Herrera, F. Hoffmann, L. Magdalena, Ten years of genetic fuzzy systems: current framework and new trends. Fuzzy Sets Syst. **141**(1), 5–31 (2004)
4. A. Sathyan, H.S. Harrison, A.W. Kiefer, P.L. Silva, R. MacPherson, K. Cohen, Genetic fuzzy system for anticipating athlete decision making in virtual reality, in *International Fuzzy Systems Association World Congress* (Springer, Berlin, 2019), pp. 578–588
5. A. Sathyan, O. Ma, Collaborative control of multiple robots using genetic fuzzy systems. Robotica **37**(11), 1922–1936 (2019)
6. A. Sathyan, O. Ma, Collaborative control of multiple robots using genetic fuzzy systems approach, in *Proceedings of ASME 2018 Dynamic Systems and Control Conference* (American Society of Mechanical Engineers, 2018), pp. V001T03A002–V001T03A002
7. A. Sathyan, O. Ma, K. Cohen, Intelligent approach for collaborative space robot systems, in *Proceedings of 2018 AIAA SPACE and Astronautics Forum and Exposition* (2018), p. 5119
8. N. Ernest, D. Carroll, C. Schumacher, M. Clark, K. Cohen, G. Lee, Genetic fuzzy based artificial intelligence for unmanned combat aerial vehicle control in simulated air combat missions. J. Defense Manag. **6**(144), 2167–0374 (2016)
9. A. Saxena, K. Goebel, D. Simon, N. Eklund, Damage propagation modeling for aircraft engine run-to-failure simulation, in *2008 International Conference on Prognostics and Health Management* (IEEE, 2008), pp. 1–9
10. F.O. Heimes, Recurrent neural networks for remaining useful life estimation, in *2008 International Conference on Prognostics and Health Management* (IEEE, 2008), pp. 1–6
11. G.S. Babu, P. Zhao, X.-L. Li, Deep convolutional neural network based regression approach for estimation of remaining useful life, in *International Conference on Database Systems for Advanced Applications* (Springer, 2016), pp. 214–228
12. S.K. Singh, S. Kumar, J.P. Dwivedi, A novel soft computing method for engine rul prediction. Multimedia Tools Appl. **78**(4), 4065–4087 (2019)
13. S. Zheng, K. Ristovski, A. Farahat, C. Gupta, Long short-term memory network for remaining useful life estimation, in *2017 IEEE International Conference on Prognostics and Health Management (ICPHM)* (IEEE, 2017), pp. 88–95
14. A.-B. Fahad, Predictive maintenance use of machine learning to predict machine remaining useful life (2018)
15. X. Jia, H. Cai, Y. Hsu, W. Li, J. Feng, J. Lee, A novel similarity-based method for remaining useful life prediction using kernel two sample test, in *Proceedings of the Annual Conference of the PHM Society*, vol. 11 (2019)

Optimal Number of Classes in Fuzzy Partitions

Fabian Castiblanco, Camilo Franco, J. Tinguaro Rodriguez, and Javier Montero

Abstract This paper proposes a cluster validation procedure allowing to obtain the optimal number of clusters on a set of fuzzy partitions. Such a procedure is established considering fuzzy classification systems endowed with a dissimilarity function that, in turn, generates a dissimilarity matrix. Establishing a dissimilarity matrix for the case of a crisp partition, we propose an optimization problem comparing the characteristic polynomials of the fuzzy partition and crisp partition. Based on the above, we propose a definition for the *optimal number of fuzzy classes* in a fuzzy partition. Our approach is illustrated through an example on image analysis by the fuzzy c-means algorithm.

Keywords Optimal number of fuzzy clusters · Fuzzy Classification Systems · Dissimilarity functions · Characteristic Polynomial

1 Introduction

Clustering is a popular technique for solving unsupervised classification problems. The objective of cluster analysis is to partition a given data set into a number of natural and homogeneous sets where the elements of each set are as similar as possible, and are as dissimilar as possible from those of the other sets. However, it can be hard to choose a unique mechanism for validating the obtained results.

In the search for algorithms that can identify in the best way the structure present in the data, fuzzy classification has emerged as an alternative to problems where

F. Castiblanco (✉)
Faculty of Economics and Business Sciences, Gran Colombia University, Bogotá, Colombia
e-mail: fabianalberto.castiblanco@ugc.edu.co

C. Franco
Department of Industrial Engineering, Andes University, Bogotá, Colombia

J. T. Rodriguez · J. Montero
Department of Statistics, Complutense University, Madrid, Spain

B. Bede et al. (eds.), *Fuzzy Information Processing 2020*,
Advances in Intelligent Systems and Computing 1337,
https://doi.org/10.1007/978-3-030-81561-5_12

clusters are not completely disjoint; rather, in practice, in many cases the separation of clusters is a fuzzy notion.

In any case, several interesting questions arise in the implementation of classification algorithms: How do we determine from a set of given algorithms the one which gives more valid or useful results? Or, how to determine the appropriate number of clusters on a set of data? These questions in fact refer to how may we validate that the resulting partition is a "good partition".

In particular, to answer the second question in the fuzzy case, several indices that measure fuzziness of clusters, fuzzy set structures, intracluster compactness, among others, have been established in the literature. Within the most used indexes we have, for instance, Bezdek's partition coefficient (PC) [1], partition entropy (PE) [2], modified partition coefficient [3], average silhouette width criterion [4], Fuzzy Silhouette [5].

Although the validation indices are independent of the algorithm used, several proposals have used the fuzzy c-means algorithm in their application [6–8]. Extensive reviews on the subject are presented in [9, 10] for validating fuzzy partitions. However, each index allows evaluating the performance of the clustering procedure according to different properties, making it very difficult to identify a single performance measure which adequately summarizes all the relevant information for evaluating the outcome of the clustering.

We consider that an adequate cluster validity process requires the calibration of several properties or characteristics extracted from the clusters or the obtained partition. Therefore, it is pertinent to establish a criterion that allows the partition to be evaluated both by *external* and *internal* criteria. The internal criteria consider measures that evaluate separately different properties or features inherent to the partition being considered. For instance, the indexes mentioned above correspond to internal criteria. In this case, the proposed indexes allow evaluating specific properties of the obtained structure, and there is no straightforward manner of aggregating all of them into an overall clustering index. In this way, it is up to the decision analyst to choose the property or set of properties that should be verified by the clustering procedure.

The external criteria generally consider evaluating the resulting clustering structure by comparing it to an independent partition of the data built according to our intuition about the clustering structure of the data set [11], or in the fuzzy case, by comparing it with a crisp partition. In this way, crisp partition represents the ideal case in which the edges of an image, i.e., the borders of the clusters, are clearly defined.

In this paper, we propose a cluster validation procedure considering both internal and external indexes. Additionally, our procedure evaluates more than one (internal) property of the fuzzy partition obtained.

We consider the fuzzy classification systems (FCS) proposed by [12] (see also [13]) which allows, from recursive rules, to evaluate a first set of properties of a fuzzy partition. Therefore, FCS will be the starting point for an evaluation based on internal indexes.

Subsequently, we propose to equip those systems with a dissimilarity function. From such a function, a dissimilarity matrix is obtained. We compare that matrix with a "ideal" matrix proposed from the crisp case used to represent the desired structure for a "good" partition. For such comparison we use the characteristic polynomials of the matrices obtained. Thus, we establish a comparison process with an "ideal" structure that is independent of the data. According to the above, our proposal is summarized through the definition of *the optimal number of classes* in an unsupervised fuzzy classification problem by evaluating the partitions obtained both internally and externally.

This paper is organized as follows. In the next section, some preliminary definitions are presented. In Sect. 3, we present the basic elements of the characteristic polynomials of dissimilarity matrices. We propose to evaluate the optimal number of clusters in a fuzzy partition based on the characteristic polynomials, solving an optimization problem in which it is proposed to minimize the distance between polynomials. Finally, in Sect. 4, the proposed procedure is applied to an image segmentation problem, for which fuzzy partitions are obtained by using the fuzzy c-means algorithm.

2 Preliminaries

Definition 1 ([12]) Let us assume a finite set of objects X. A *fuzzy classification system* (FCS) is a finite family C of n fuzzy classes, where each $c \in C$ has an associated membership function $\mu_c : X \to [0, 1]$, together with a recursive[1] triplet (φ, ϕ, N) such that,

1. ϕ is a standard recursive rule, where $\phi_2\,(0, 1) = \phi_2\,(1, 0) = 0$;
2. $N : [0, 1] \to [0, 1]$ is a strong negation function, i.e., a continuous strictly decreasing function such that $N(N(a)) = a, \forall\, a \in [0, 1]$;
3. φ is a standard recursive rule such that, $\forall n > 1,\ \varphi_n(a_1, \ldots, a_n) = N^{-1}[\phi_n(N(a_1), \ldots, N(a_n)], \forall (a_1, \ldots, a_n) \in [0, 1]^n$.

Fuzzy classification systems were conceived as a structure allowing the treatment of complex classification problems from recursive rules. Such systems propose indexes that allow measuring the degree of redundancy (overlapping) between classes, the degree in which all of the classes accurately cover the aspect(s) of reality under consideration, and the degree in which some classes can be disregarded.

[1] Reference [12] A recursive rule ρ is a family of aggregation operators

$$\{\rho_n : [0, 1]^n \to [0, 1]\}_{n>1}$$

such that there exist an ordering rule π and two sequences of binary operators $\{L_n : [0, 1]^2 \to [0, 1]\}_{n>1}$ and $\{R_n : [0, 1]^2 \to [0, 1]\}_{n>1}$ such that for each n and for each $(a_1, \ldots, a_n) \in [0, 1]^n$, $\rho_n(a_{\pi(1)}, \ldots, a_{\pi(n)}) = L_n(\rho_{n-1}(a_{\pi(1)}, \ldots, a_{\pi(n-1)}), a_{\pi(n)}) = R_n(a_{\pi(1)}, \rho_{n-1}(a_{\pi(2)}, \ldots, a_{\pi(n)})$.

Notice that in a fuzzy classification system each $x \in X$ has a membership degree $\mu_c(x)$ associated with each class $c \in C$. As our purpose is to analyze the fuzzy classes, from now on we consider each standard recursive rule to act on such membership degrees, that is, for any given object $x \in X$ we are interested in the sets $\phi\{\mu_c(x)/\ c \in C\}$ or $\varphi\{\mu_c(x)/\ c \in C\}$. Fuzzy classification systems were proposed taking into account that ϕ is a conjunctive recursive rule, in the sense that $\phi_n(\mu_1(x), \ldots, \mu_n(x)) = 0$, whenever $\mu_j(x) = 0$ for certain $j \in \{1, \ldots, n\}$. As a direct consequence, φ is a disjunctive recursive rule, in the sense that $\varphi_n(\mu_1(x), \ldots, \mu_n(x)) = 1$, whenever there is $\mu_j(x)$ such that $\mu_j(x) = 1$.

Example 1 We can consider a recursive triplet (φ, ϕ, N) such that,

1. $\phi_n(\mu_1(x), \ldots, \mu_n(x)) = \dfrac{3\prod_{k=1}^{n} \mu_k(x)}{1+2\prod_{k=1}^{n} \mu_k(x)}$
2. $\varphi_n(\mu_1(x), \ldots, \mu_n(x)) = \dfrac{1-\left(\prod_{k=1}^{n} (1-\mu_k(x))\right)}{1+2\prod_{k=1}^{n} (1-\mu_k(x))}$
3. $N(\mu(x)) = 1 - \mu(x)$

Following the framework proposed in [14] (see also [15]), we consider two new mappings $\sigma_n, \delta_n : [0, 1]^n \to [0, 1]$, defined for all aggregation operators ϕ_n and φ_n. In this way, $\sigma_n : [0, 1]^n \to [0, 1]$ is defined as $\sigma_n(\mu_1(x), \ldots, \mu_n(x)) = N(\varphi_n(\mu_1(x), \ldots, \mu_n(x)))$ and $\delta_n : [0, 1]^n \to [0, 1]$ is defined as $\delta_n(\mu_1(x), \ldots, \mu_n(x)) = N(\phi_n(\mu_1(x), \ldots, \mu_n(x)))$.

Notice that when the strong negation N is used on φ_n or ϕ_n, the resulting expression can be interpreted as the complement of the set of aggregated classes $\{\mu_1(x), \ldots, \mu_n(x)\}$. In particular, if $\varphi_n(\mu_1(x), \ldots, \mu_n(x))$ represents the degree of coverage of the classes, then $N(\varphi_n(\mu_1(x), \ldots, \mu_n(x)))$ represents the degree of non-coverage of the classes, understanding φ_n as a proposition and $N(\varphi_n)$ as the negation of such a proposition. In a similar way, if $\phi_n(\mu_1(x), \ldots, \mu_n(x))$ represents the degree of redundancy of the classes, then $N(\phi_n(\mu_1(x), \ldots, \mu_n(x)))$ represents the degree of non-redundancy of the classes. Let us recall from Definition 2 that if N is a strong negation operator, then $\varphi_n(\mu_1(x), \ldots, \mu_n(x)) = N[\phi_n(N(\mu_1(x)), \ldots, N(\mu_n(x)))]$ and thus, given the mapping σ_n, it holds that $\sigma_n(\mu_1(x), \ldots, \mu_n(x)) = N(N[\phi_n(N(\mu_1(x)), \ldots, N(\mu_n(x)))]$ and therefore,

$$\sigma_n(\mu_1(x), \ldots, \mu_n(x)) = \phi_n(N(\mu_1(x)), \ldots, N(\mu_n(x))). \tag{1}$$

Let us denote $G = \{\phi_n, \varphi_n, \sigma_n, \delta_n\}$.

From a fuzzy classification systems (C, φ, ϕ, N) and the negations of its operators, we obtain the degrees of grouping, overlapping, non-covering and non-overlapping of each element x in C. Now, considering the definitions proposed in [16], we address the concept of global degree,

Definition 2 ([16]) Given a universe X and a family of fuzzy classes C over this universe, we define *the global degree of covering of* X as the aggregation of the degrees of covering for all elements $x \in X$.

Such aggregation function can be of very different nature (conjunctive, disjunctive or average). In a similar way the global degrees of overlaps, non-coverage and non-overlap are defined. Let us denote by ϕ_n^T, φ_n^T, σ_n^T and δ_n^T such global degrees. In our proposal we consider the particular case of average aggregation operators to obtain such global degrees. Therefore, the following definition is presented,

Definition 3 ([17]) An aggregation operator ψ is stable for the strong negation N if

$$\psi_m\big(N(x_i), \dots, N(x_m)\big) = N\big(\psi_m(x_1, \dots, x_m)\big).$$

As a particular case, in [17] it is established that the class of generalized means which are stable for the particular negation $N(x) = (1 - x^\theta)^{\frac{1}{\theta}}$ corresponds to the family of related means, i.e., $\psi_m(x_1, \dots, x_m) = \left(\sum_i \frac{x_i^\theta}{m}\right)^{\frac{1}{\theta}}$.

Now, we consider the definition of restricted dissimilarity functions proposed in [18].

Definition 4 ([18]) A function $d : [0, 1]^2 \to [0, 1]$ is called a *restricted dissimilarity function*, if it satisfies the following conditions:

1. $d(x, y) = d(y, x), \forall x, y \in [0, 1]$
2. $d(x, y) = 1$ if and only if $x = 0$ and $y = 1$ or $x = 1$ and $y = 0$
3. $d(x, y) = 0$ if and only if $x = y$
4. For all $x, y, z \in [0, 1]$, if $x \leq y \leq z$, then $d(x, y) \leq d(x, z)$ and $d(y, z) \leq d(x, z)$

We consider Definition 4 because it generalizes the concept of distance restricted to values in the interval [0, 1].

To end this section, remember a classic definition of linear algebra,

Definition 5 Let A be an $n \times n$ matrix. *The characteristic polynomial* of A is the function $f(\lambda)$ given by $f(\lambda) = det(A - I_n\lambda)$.

3 Optimal Number of Classes in a Fuzzy Partition

In this section, we address the concept of characteristic polynomial on dissimilarity matrices. In particular, considering a fuzzy classification system, we establish a comparison process based on a dissimilarity function and its characteristic polynomial in order to establish the optimal number of classes in a fuzzy partition.

3.1 Characteristic Polynomial of a Dissimilarity Matrix

Given a finite set of objects X and a fuzzy classification system (C, φ, ϕ, N) with n classes, consider the set $G = \{\phi_n^T, \varphi_n^T, \sigma_n^T, \delta_n^T\}$ where the global degrees have been obtained through an average operator ψ_m stable for the strong negation N of the fuzzy classification system. On such a set, a dissimilarity relation d is established in such a way that the following dissimilarity matrix for the relation d is obtained,

$$A_{d_n} = \left(\begin{array}{c|cccc} d & \phi_n^T & \sigma_n^T & \delta_n^T & \varphi_n^T \\ \hline \phi_n^T & 0 & \beta & \pi & \alpha \\ \sigma_n^T & \beta & 0 & \gamma & \rho \\ \delta_n^T & \pi & \gamma & 0 & \tau \\ \varphi_n^T & \alpha & \rho & \tau & 0 \end{array} \right)$$

By Definition 4, it is immediate that such a matrix is symmetric and the elements of the main diagonal are all zeros.

Proposition 1 *Given a set X with m elements and a fuzzy classification system* (C, ϕ, φ, N) *with n classes. Consider the set* $G = \{\phi_n^T, \sigma_n^T, \delta_n^T, \varphi_n^T\}$ *and a dissimilarity function d. If* $d(x, y) = d(N(x), N(y))$ *then,* $d(\varphi_n^T, \delta_n^T) = d(\phi_n^T, \sigma_n^T)$ *and* $d(\delta_n^T, \sigma_n^T) = d(\phi_n^T, \varphi_n^T)$*. Therefore, the following dissimilarity matrix for relation d is obtained,*

$$A_{\bar{d}_n} = \left(\begin{array}{c|cccc} & \phi_n^T & \sigma_n^T & \delta_n^T & \varphi_n^T \\ \hline \phi_n^T & 0 & \beta & \pi & \alpha \\ \sigma_n^T & \beta & 0 & \alpha & \rho \\ \delta_n^T & \pi & \alpha & 0 & \beta \\ \varphi_n^T & \alpha & \rho & \beta & 0 \end{array} \right)$$

where $\bar{d}$ *denote a dissimilarity function fulfilling the property for the strong negation N.*

Proof Let's consider the first equality, $d(\varphi_{in}^T, \delta_{in}^T) = d(\phi_{in}^T, \sigma_{in}^T)$. Let ψ_m be the average operator (stable for strong negation N) such that $\varphi_n^T = \psi_m\Big(\varphi_{1n}, \ldots, \varphi_{mn}\Big)$, i.e., by means of ψ_m, the global degree of covering is obtained. From Definition 2 and Eq. (1), we have that for each i with $i = 1, \ldots, m$ it hold that,

$$\varphi_{in}\Big(\mu_1(x_i), \ldots, \mu_n(x_i)\Big) = N\Big(\phi_{in}\Big(N(\mu_1(x_i)), \ldots, N(\mu_n(x_i))\Big)$$

and

$$\phi_{in}\Big(N\big(\mu_1(x_i)\big), \ldots, N\big(\mu_n(x_i)\big)\Big) = \sigma_{in}\Big(\mu_1(x_i), \ldots, \mu_n(x_i)\Big)$$

therefore we have that,

$$\varphi_{in}\Big(\mu_1(x_i), \ldots, \mu_n(x_i)\Big) = N\Big(\sigma_{in}\Big(\mu_1(x_i), \ldots, \mu_n(x_i)\Big)\Big),$$

thus, as ψ_m is stable for the strong negation, it holds that,

$$\varphi_n^T = \psi_m\Big(N(\sigma_{1n}), \ldots, N(\sigma_{mn})\Big) = N\Big(\psi_m\Big(\sigma_{1n}, \ldots, \sigma_{mn}\Big)\Big) = N(\sigma_n^T)$$

In a similar way, remember that,

$$\delta_{in}\Big(\mu_1(x_i), \ldots, \mu_n(x_i)\Big) = N\Big(\phi_{in}\Big(\mu_1(x_i), \ldots, \mu_n(x_i)\Big)\Big),$$

therefore,

$$\delta_n^T = \psi_m\Big(\delta_{1n}, \ldots, \delta_{mn}\Big) = \psi_m\Big(N\Big(\phi_{1n}, \ldots, \phi_{mn}\Big)\Big) = N\Big(\psi_m\Big(\phi_{1n}, \ldots, \phi_{mn}\Big)\Big) = N(\phi_n^T).$$

According to the above, $d(\varphi_n^T, \delta_n^T) = d(N(\sigma_n^T), N(\phi_n^T)) = d(\sigma_n^T, \phi_n^T)$.

In a similar way, the second equality $d(\delta_n^T, \sigma_n^T) = d(\phi_n^T, \varphi_n^T)$ is fulfilled because $\delta_n^T = N(\phi_{in})$ and $\sigma_n^T = N(\varphi_n^T)$. □

In Proposition 1, a restricted dissimilarity function such that $d(x, y) = d(N(x), N(y))$ has been used. An example of such functions is shown below.

Example 2 The function $d(x, y) = |x - y|$ is a restricted dissimilarity function such that $d(x, y) = d(N(x), N(y))$.

Notice that the dissimilarity matrix $A_{\bar{d}_n}$ obtained in Proposition 1, is defined for each set of classes C with n classes, where $n = 2, \ldots, m$. For each of these matrices, we compute its characteristic polynomial in the λ variable according to Definition 5. We will denote by $p_{\bar{d}_n}(\lambda)$ such polynomial.

$$p_{\bar{d}_n}(\lambda) = \lambda^4 - \Big(2\alpha^2 + 2\beta^2 - \pi^2 - \rho^2\Big)\lambda^2 - 4\alpha\beta\Big(\pi + \rho\Big)\lambda + \Big((\alpha + \beta)^2 - \pi\rho)((\alpha - \beta)^2 - \pi\rho\Big) \tag{2}$$

Naturally, each polynomial is obtained for each set of classes C with n classes. About Eq. 2 some key aspects are immediate:

1. The coefficient of λ^3 is zero because the trace of the dissimilarity matrix is zero.
2. The determinant of the matrix is $(\alpha + \beta)^2 - \pi\rho)(\alpha - \beta)^2 - \pi\rho)$.
3. The roots of the polynomial are given by

$$\lambda_{1,2} = \frac{1}{2}p + \frac{1}{2}r \pm \frac{1}{2}\sqrt{(2\alpha + 2\beta)^2 + (p - r)^2}$$
$$\lambda_{3,4} = -\frac{1}{2}p - \frac{1}{2}r \pm \frac{1}{2}\sqrt{(2\alpha + 2\beta)^2 + (p - r)^2}$$

Therefore, the polynomial has all its roots real.

3.2 Optimization Problem Related to Characteristic Polynomials

Considering the fuzzy classification system proposed in [12], some basic quality criteria on partition are given in advance. In principle, it is desirable that global degrees of coverage be greater than global degrees of overlap. Therefore, it is desirable that the degrees of non-coverage be less than the degrees of non-overlap. In this perspective, if we compare all the global degrees of fuzzy partition (through a dissimilarity function), we propose to consider as an ideal case, which can be seen as case crisp, or threshold, the situation described by the matrix A_c.

$$A_c = \left(\begin{array}{c|cccc} & \phi_n^T & \sigma_n^T & \delta_n^T & \varphi_n^T \\ \hline \phi_n^T & 0 & 0 & 1 & 1 \\ \sigma_n^T & 0 & 0 & 1 & 1 \\ \delta_n^T & 1 & 1 & 0 & 0 \\ \varphi_n^T & 1 & 1 & 0 & 0 \end{array}\right)$$

According to the above, we are considering a partition in which the global degree of overlap is equal to zero and the global degree of grouping is equal to 1. Therefore, for such dissimilarity matrix we have the following characteristic polynomial,

$$p_c(\lambda) = \lambda^4 - 4\lambda^2. \tag{3}$$

Considering Eqs. 2 and 3, we propose the following optimization problem,

$$\arg\min_n \left\{ \int_{-a}^{a} \left| p_c(\lambda) - p_{\bar{d}_n}(\lambda) \right| d\lambda \right\} \tag{4}$$

with $n = 2, \ldots, k$ with $k \leq m$ and $a \geq 2$. For each n it is immediate that the improper integral $\int_{-\infty}^{\infty} \left| p_c(\lambda) - p_{\bar{d}_n}(\lambda) \right| d\lambda$ is divergent. Similarly, it is true that $p_c(\lambda) < p_{\bar{d}_n}(\lambda)$ for all $\lambda \in [2, \infty)$ or $(-\infty, 2]$. The choice of $\lambda_1 = 2$ and $\lambda_2 = -2$ is not arbitrary because such values are two roots of $p_c(\lambda)$.

Therefore, such a problem is interpreted as follows. Find the set of classes C_n with n classes such that their characteristic polynomial is as close as possible to the characteristic polynomial of the threshold, or ideal case.

Thus, if we consider a set of fuzzy partitions $\mathbb{C}_p = \{C_2, \ldots, C_k\}$, where $p = k - 1$ is the number of partitions obtained through an iterative, non-hierarchical classification algorithm and each partition C_n has n classes, we propose the following definition.

Definition 6 Given a set X with m elements and the set of fuzzy partitions $\mathbb{C}_p = \{C_2, \ldots, C_k\}$, with $k < m$. Let (C_n, ϕ, φ, N) be a fuzzy classification system with ϕ, φ, N fixed for each C_n and $n = 2, \ldots, k$. The *optimal number of fuzzy classes* for X, denoted by n_o, is given by,

$$n_o = \arg\min_n \left\{ \int_{-a}^{a} \left| p_c(\lambda) - p_{\bar{d}_n}(\lambda) \right| d\lambda \right\}, \text{ with } a \geq 2.$$

Notice that under Definition 6 we start with a set of p partitions where $p = k - 1$ and such set is bounded by $m - 1$ partitions. Therefore, since the improper integral is divergent, for different values of p we can obtain different optimal values.

Despite the above, in the context of cluster validation it is desirable that such an optimum correspond to the partition with the fewest number of classes. Therefore, we select such optimal as follows: each set of partitions $\mathbb{C}_p$ must start with the set of classes C_2 and increase the following partition by one class (according to the algorithm used). Thus, given two sets of partitions $\mathbb{C}_{p_1}$ and $\mathbb{C}_{p_2}$ on X, such that $\mathbb{C}_{p_1} = \{C_2, \ldots, C_k\}$, and $\mathbb{C}_{p_2} = \{C_2, \ldots, C_r\}$ with $r > k$. If n_{ok} and n_{or} denote the optimal number of fuzzy classes for $\mathbb{C}_{p_1}$ and $\mathbb{C}_{p_2}$ respectively, and $n_{ok} < n_{or}$. Then, n_{ok} is the optimal number of fuzzy classes on X.

4 Application

In order to apply Definition 8, we have selected the image presented in Fig. 1, considering the unsupervised classification problem of obtaining classes of similar pixels through the fuzzy c-means algorithm. Such an image is interesting in the framework of our proposal because some edges are not clearly defined, and the separation between background and object is not completely clear. That is, separating between sand and fish, in some areas, can be hard.

We consider the recursive triple of Example 1, the restricted dissimilarity function of Example 2, and for practical purposes, let $a = 2$ and let $n = 5$, i.e, we consider the set $\mathbb{C}_4 = \{C_2, \ldots, C_5\}$.

We start computing the corresponding characteristic polynomial of the matrix obtained under the conditions of Proposition 1 for each n. Subsequently, we compute $\int_{-2}^{2} \left| p_c(\lambda) - p_{\bar{d}_n}(\lambda) \right| d\lambda$ for each n. Table 1 summarizes the results obtained.

Similarly, Fig. 2 shows the characteristic polynomials obtained for $A_{\bar{d}_n}$ with $n = 2, 3, 4, 5$ and the characteristic polynomial of the threshold, i.e., $p(\lambda) = x^4 - 4x^2$.

Fig. 1 Image extracted from the Berkeley Segmentation Dataset (BSDS500) [19]

Table 1 Characteristic polynomial for c_2, c_3, c_4, c_5

Classes	Characteristic polynomial	$\int_{-2}^{2} p_c(\lambda) - p_{\bar{d}_n}(\lambda)$
C_2	$x^4 - 1.2x^2 - 1.05 \cdot 10^{-15}x - 2.3 \cdot 10^{-31}$	14.91
C_3	$x^4 - 2.34x^2 - 1.06x - 0.119$	8.69
C_4	$x^4 - 2.35x^2 - 1.17x - 0.145$	8.65
C_5	$x^4 - 1.34x^2 - 0.74x - 0.102$	13.82

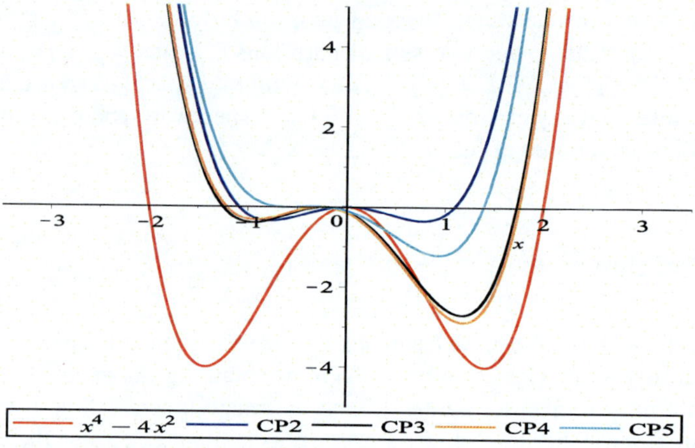

Fig. 2 Characteristic polynomials for threshold and for $A_{\bar{d}_2}$, $A_{\bar{d}_3}$, $A_{\bar{d}_4}$ and $A_{\bar{d}_5}$. We denote CP2, CP3, CP4, and CP5 respectively

Under the procedure above, we may conclude that $n = 4$, i.e., the set with four classes obtained through the fuzzy c-means algorithm, is the optimal number of classes for Fig. 1. In other words, the fuzzy partition with four classes is the one that best classifies similar pixels.

The corresponding classes of such fuzzy partition for Fig. 1 are presented in Figs. 3, 4, 5 and 6. Such classes were obtained by applying the fuzzy c-means algorithm. The gray scale represents the membership degree of each pixel to the class, where black = 0 and white = 1. We interpret the obtained classes as follows.

The classes shown in Figs. 3 and 4 partition the original images of Fig. 1 into two regions. The class obtained in Fig. 3 will be called "black background class" and the class obtained in Fig. 4 will be called "sand background class". Therefore, we conclude that in general, Fig. 1 is composed of a part completely black and another part with different intensities of color (sand).

However, "sand background class" has been segmented into two new classes (which do not contain the "black background class"): the class obtained in Fig. 5 will be called "object class" and the class obtained in Fig. 6 will be called "sand without object class.".

Fig. 3 Black background class. Black part and the remaining similar color intensities

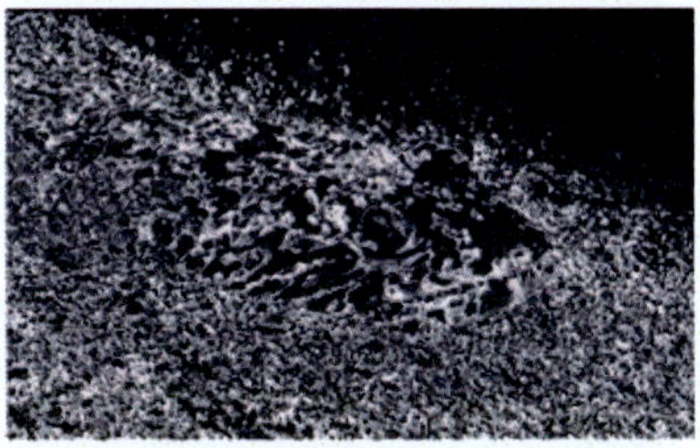

Fig. 4 Sand background class. Sand and similar color intensities

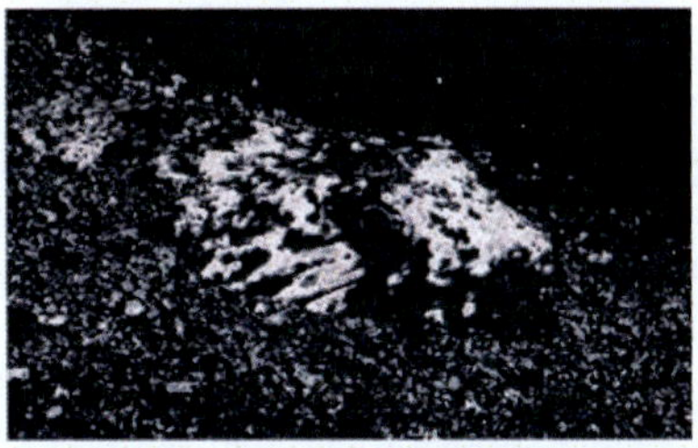

Fig. 5 Object class. Object and its own color intensities

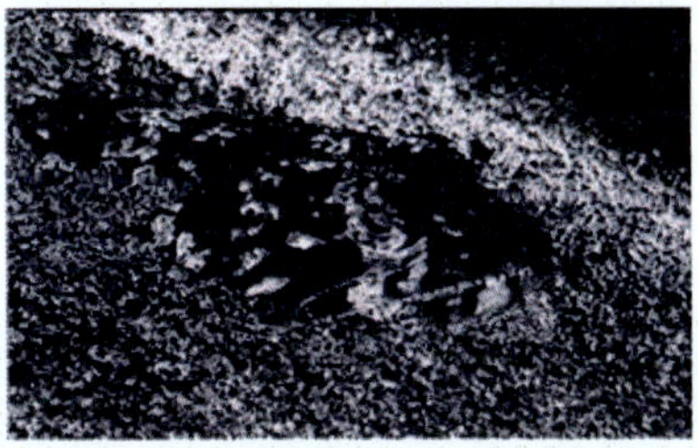

Fig. 6 Sand class without object. Sand and its own color intensities

These last two classes naturally share similar intensities of color. This fact is clear in "sand background class". However, although there are classes that share pixels of similar intensity, the segmentation of "sand background class" into two additional classes is interpreted as the presence of unique color intensities of the object and unique color intensities of the sand.

5 Final Comments

Through this paper, we propose a definition for the optimal number of classes in a fuzzy partition considering fuzzy classification systems. For this purpose, in a first first stage we provide FCS with a function of dissimilarity allowing the comparison of the global degrees of coverage, overlapping, non-covering and non-overlapping of a set of classes. Subsequently, based on the dissimilarity function, we establish the respective dissimilarity matrix for each set of classes and compute its characteristic polynomial. Finally, considering a threshold (matrix of dissimilarity of the global degrees in the crisp case) and its corresponding characteristic polynomial, we propose to find the characteristic polynomial whose distance to the threshold is minimal.

Some considerations are established under our proposal: (1) The distance between polynomials has been added from the classical methodology of area of the region bounded by two curves. In this sense, the resulting improper integrals are divergent and for this reason, intervals of the form $[-a, a]$ have been considered for comparison. Because the cut-off points between the polynomial considered and the threshold ($p_c(\lambda) = p_{\bar{d}_n}(\lambda)$) are always in the interval $[-2, 2]$, equal or greater intervals have been considered. (2) Our proposal defines the optimal number of classes of a fuzzy partition considering a given set of partitions. Therefore, given a different set of partitions, the optimal number may be different. For this reason, following the principle of parsimony, we propose to select the partition with the least optimal number.

As future work, we propose adding a criterion allowing to evaluate the optimal number of classes of a fuzzy partition, based on the relevance of the classes. This is, given an optimal value (under our proposal) and another partition with a value very close to it, to determine the degree of relevance of the classes for each partition. For instance, notice that under our application, C_3 and C_4 have very close values. It may be possible to delete a class from C_4 without affecting the quality of the partition or even select C_3 as the partition with the optimal number of classes. Similarly, we propose to compare the results obtained with both internal and external classical indexes.

Acknowledgements This research has been partially supported by the Government of Spain (grant PGC2018-096509-B-I00) Complutense University (UCM Research Group 910149) and Gran Colombia University (grant JCG2019-FCEM-01).

References

1. J.C. Bezdek, Cluster validity with fuzzy sets. J. Cybern. **3**, 58–73 (1973)
2. J.C. Bezdek, *Pattern Recognition with Fuzzy Objective Function Algorithms* (Springer, New York, 1981)
3. R.N. Dave, Validating fuzzy partitions obtained through c-shells clustering. Pattern Recogn. Lett. **17**, 613–623 (1996)
4. P.J. Rousseeuw, Silhouettes: a graphical aid to the interpretation and validation of cluster analysis. J. Comput. Appl. Math. **20**, 53–65 (1987)

5. R.J.G.B. Campello, E.R. Hruschka, A fuzzy extension of the silhouette width criterion for cluster analysis. Fuzzy Sets Syst. **157**, 2858–2875 (2006)
6. A. Suleman, Measuring the congruence of fuzzy partitions in fuzzy c-means clustering. Appl. Soft Comput. **52**, 1285–1295 (2017)
7. E. Hullermeier, M. Rifqi, S. Henzgen, R. Senge, Comparing fuzzy partitions: a generalization of the rand index and related measures. IEEE Trans. Fuzzy Syst. **20**, 546–556 (2012)
8. K.L. Wu, M.S. Yang, A cluster validity index for fuzzy clustering. Pattern Recogn. Lett. **26**, 1275–1291 (2005)
9. M. Halkidi, Y. Batistakis, M. Vazirgiannis, On clustering validation techniques. J. Intell. Inf. Syst. **17**, 107–145 (2001)
10. W. Wang, Y. Zhang, On fuzzy cluster validity indices. Fuzzy Sets Syst. **158**, 2095–2117 (2007)
11. D.T. Anderson, J.C. Bezdek, M. Popescu, J.M. Keller, Comparing fuzzy, probabilistic, and possibilistic partitions. IEEE Trans. Fuzzy Syst. **18**, 906–918 (2010)
12. A. Amo, J. Montero, G. Biging, V. Cutello, Fuzzy classification systems. Eur. J. Oper. Res. **156**, 495–507 (2004)
13. A. Amo, D. Gomez, J. Montero, G. Biging, Relevance and Redundancy in fuzzy classification systems. Mathware Soft Comput. **8**, 203–216 (2001)
14. F. Castiblanco, C. Franco, J. Montero, J.T. Rodríguez, Relevance of classes in a fuzzy partition. A study from a group of aggregation operators, in *Fuzzy Information Processing*, ed. by G.A. Barreto, R. Coelho (Springer International Publishing, Cham, 2018), pp. 96–107
15. F. Castiblanco, C. Franco, J. Montero, J. Tinguaro Rodríguez, Aggregation operators to evaluate the relevance of classes in a fuzzy partition, in *Fuzzy Techniques: Theory and Applications*. ed. by R.B. Kearfott, I. Batyrshin, M. Reformat, M. Ceberio, V. Kreinovich (Springer International Publishing, Cham, 2019), pp. 13–21
16. F. Castiblanco, D. Gómez, J. Montero, J.T. Rodríguez, Aggregation tools for the evaluation of classifications, in *IFSA-SCIS 2017 - Joint 17th World Congress of International Fuzzy Systems Association and 9th International Conference on Soft Computing and Intelligent Systems* (IEEE, Otsu, Japan, 2017), pp. 1–5
17. J. Fodor, M. Roubens, *Preference Modelling and Multicriteria Decision Support*. Theory and Decision Library. SeriesD: System Theory, Knowledge Engineering and Problem Solving (1994)
18. H. Bustince, E. Barrenechea, M. Pagola, Relationship between restricted dissimilarity functions, restricted equivalence functions and normal EN-functions: image thresholding invariant. Pattern Recogn. Lett. **29**, 525–536 (2008)
19. D. Martin, C. Fowlkes, D. Tal, J. Malik, A database of human segmented natural images and its application to evaluating segmentation algorithms and measuring ecological statistics. Proc. IEEE Int. Conf. Comput. Vis. **2**, 416–423 (2001)

Consistence of Interactive Fuzzy Initial Conditions

Vinícius F. Wasques, Nilmara J. B. Pinto, Estevão Esmi, and Laécio Carvalho de Barros

Abstract This article investigates the consistence of initial conditions present in mathematical models under uncertainty. These conditions are given in form of a linear equation and the uncertainty is described by fuzzy numbers. This work provides an algorithm that constructs triangular fuzzy numbers satisfying the fuzzy linear equation. Moreover, it shows that the initial conditions should not be chosen aleatory, instead, these conditions must be given by interactive fuzzy numbers. An example is presented in order to illustrate this method.

Keywords Fuzzy linear equation · Fuzzy initial conditions · Interactive fuzzy numbers

1 Introduction

Mathematical models can be used to describe and understand the behaviour of phenomenon of nature. For example, one can describe the growth population using differential equations. In this case, if one considers that a population with P_0 individuals has enough food and space, then one expects that this population grows indefinitely. From the mathematical point of view, this dynamic grows according to the exponential function.

V. F. Wasques (✉)
São Paulo State University, Rio Claro, São Paulo, Brazil
e-mail: vwasques@outlook.com

N. J. B. Pinto · E. Esmi · L. C. de Barros
University of Campinas, Campinas, São Paulo, Brazil
e-mail: nilmarabiscaia@gmail.com

E. Esmi
e-mail: eelaureano@gmail.com

L. C. de Barros
e-mail: laeciocb@ime.unicamp.br

B. Bede et al. (eds.), *Fuzzy Information Processing 2020*,
Advances in Intelligent Systems and Computing 1337,
https://doi.org/10.1007/978-3-030-81561-5_13

Several phenomenon are difficult to describe mathematically, since there are many subjectivity and uncertainty parameters. For instance, consider a population with $P(t)$ individuals at time t. Now let us consider a disease that spreads with physical contact such as the flu, Ebolavirus, or recently, the novel coronavirus disease [1]. Since there is a delay for the symptoms to appear after the infection, it is difficult to determinate how many individuals are contaminated at each instant time t_0.

The Fuzzy Set Theory can be used to describe these aforementioned imprecisions, in contrast to deterministic models. For example, let us consider a population P with ten individuals. Suppose that there is a certain disease with three confirmed cases of infected individuals. So we expected that there is in fact *around* three infected individuals and *around* seven susceptible individuals. These linguistic variables, which incorporate the uncertain of the phenomena, can be modeled by fuzzy numbers. One of the most used fuzzy numbers is the triangular fuzzy number [2].

The uncertain of the infected (I) individuals number implies in the uncertain of susceptible (S) individuals number. However, the total population is exact. Thus, this problem boils down on a simply equation

$$I + S = N, \tag{1}$$

where N is the (real) number of the total population and I and S are the (fuzzy) numbers of the infected and susceptible individuals.

Since the left side of Eq. (1) has fuzzy numbers, the sum presented in Eq. (1) must be adapted for fuzzy numbers. One could try to solve this problem by taking $S = N - I$, however $S = N - I$ must solve the original Eq. (1), which does not always occur, because the standard sum between two fuzzy numbers does not produce real numbers as a result [3]. In fact, $S + I = N$ is equivalent to $S = N - I$, where $N \in \mathbb{N}$ if, and only if, the arithmetic operations "+" and "−" are interactive [4].

This paper investigates the consistence of the choices of the values S_0 and I_0, in order to guarantee that the final result produces a real number. Moreover, an algorithm to build I_0 and S_0 such that N_0 in (1) is a real number will be provided.

This work is divided as follows. Section 2 contains fuzzy set theory and interactivity concepts. Section 3 discusses about the choice on type of interactivity between fuzzy numbers. Section 4 presents solutions to fuzzy linear equations with crisp right side, providing some examples and a discussion about symmetry of fuzzy numbers.

2 Preliminaries

This section presents a brief summary of the basic concepts of the Fuzzy Set Theory and the arithmetic for interactive fuzzy numbers.

A fuzzy set A of a universe X is characterized by a function $\mu_A : X \to [0, 1]$ called membership function, where $\mu_A(x)$ represents the membership degree of x in A for all $x \in X$ [5]. For notational convenience, the symbol $A(x)$ will be used to denote $\mu_A(x)$. The class of fuzzy subsets of X is denoted by $\mathcal{F}(X)$. Note that each classical subset of X can be uniquely identified with the fuzzy set whose membership function is given by its characteristic function.

The α-cuts of a fuzzy set $A \subseteq X$, denoted by $[A]^\alpha$, are defined as follows [2]

$$[A]^\alpha = \{x \in X : A(x) \geq \alpha\}, \ \forall \alpha \in (0, 1].$$

In addition, if X is also a topological space, then the 0-cut of A is defined by

$$[A]^0 = cl\{x \in X : A(x) > 0\},$$

where cl Y, $Y \subseteq X$, denotes the closure of Y.

An important subclass of $\mathcal{F}(\mathbb{R})$, denoted by $\mathbb{R}_\mathcal{F}$, is the class of fuzzy numbers which includes the sets of the real numbers as well as the set of the bounded closed intervals of $\mathbb{R}$.

A fuzzy set A of $\mathbb{R}$ is said to be a fuzzy number if all α-cuts are bounded, closed and non-empty nested intervals for all $\alpha \in [0, 1]$ [2]. This definition allows to write the α-cuts of the fuzzy number A by $[A]^\alpha = [a_\alpha^-, a_\alpha^+]$.

The class of fuzzy numbers such that a_α^- and a_α^+ are continuous function with respect to α, is denoted by $\mathbb{R}_{\mathcal{F}C}$. Note that every triangular fuzzy number is an example contained in $\mathbb{R}_{\mathcal{F}C}$. Recall that a triangular fuzzy number A is denoted by the triple $(a; b; c)$ for some $a \leq b \leq c$. By means of α-cuts, one obtains that $[A]^\alpha = [a + \alpha(b - a), c - \alpha(c - b)], \ \forall \alpha \in [0, 1]$.

The *width* of a fuzzy number A is defined by $width(A) = a_0^+ - a_0^-$. The fuzzy number A is symmetric with respect to $x \in \mathbb{R}$ if $A(x - y) = A(x + y)$, $\forall y \in \mathbb{R}$. If there is no $x \in \mathbb{R}$ such that this property is satisfied, we say that A is non-symmetric [6].

A fuzzy relation R between two universes X and Y is given by the mapping $R : X \times Y \to [0, 1]$, where $R(x, y) \in [0, 1]$ is the degree of relationship between $x \in X$ and $y \in Y$.

A fuzzy relation $J \in \mathcal{F}(\mathbb{R}^n)$ is said to be a joint possibility distribution (JPD) among the fuzzy numbers $A_1, \ldots, A_n \in \mathbb{R}_\mathcal{F}$ if

$$A_i(y) = \bigvee_{x : x_i = y} J(x), \ \forall y \in \mathbb{R}, \tag{2}$$

for all $i = 1, \ldots, n$.

One example of JPD is the t-norm-based joint possibility distribution whose definition is given as follows. Let t be a t-norm, that is, an associative, commutative and increasing operator $t : [0, 1]^2 \to [0, 1]$ that satisfies $t(x, 1) = x$ for all $x \in [0, 1]$. For example, the minimum operator ($t = \wedge$) is a t-norm.

The fuzzy relation J_t given by

$$J_t(x_1, \ldots, x_n) = A_1(x_1) \ t \ \ldots \ t \ A_n(x_n) \tag{3}$$

is called the t-norm-based joint possibility distribution of $A_1, \ldots, A_n \in \mathbb{R}_\mathcal{F}$.

Definition 1 The fuzzy numbers $A_1, \ldots, A_n$ are said to be non-interactive, if Eq. (2) is satisfied for $J = J_\wedge$. Otherwise, that is, if J satisfies Eq. (2) and $J \neq J_\wedge$, then $A_1, \ldots, A_n$ are called interactive.

Definition 1 reveals that the interactivity among the fuzzy numbers $A_1, \ldots, A_n$ arises from a given joint possibility distribution. The concept of interactivity resembles the relation of dependence in the case of random variables.

The sup-J extension principle is a mathematical tool that extends classical functions to fuzzy functions. Moreover, it takes the interactivity between fuzzy numbers into account. The definition is presented as follows [7, 8].

Definition 2 Let $J \in \mathcal{F}(\mathbb{R}^n)$ be a joint possibility distribution of $(A_1, \ldots, A_n) \in \mathbb{R}^n_{\mathcal{F}}$ and $f : \mathbb{R}^n \to \mathbb{R}$. The sup-J extension of f at $(A_1, \ldots, A_n) \in \mathbb{R}^n_{\mathcal{F}}$, denoted $f_J(A_1, \ldots, A_n)$, is the fuzzy set defined by:

$$f_J(A_1, \ldots, A_n)(y) = \bigvee_{(x_1,\ldots,x_n)\in f^{-1}(y)} J(x_1, \ldots, x_n), \tag{4}$$

where $f^{-1}(y) = \{(x_1, \ldots, x_n) \in \mathbb{R}^n : f(x_1, \ldots, x_n) = y\}$.

Note that, if the fuzzy numbers are non-interactive, then the sup-J extension principle boils down to the Zadeh's extension principle [9]. This means that the sup-J extension is a generalization of the Zadeh's extension. The sup-J extension gives rise to the arithmetic on interactive fuzzy numbers, considering f as an arithmetic operator.

There are other types of interactivity besides the ones obtained from the t-norm based joint possibility distributions [4, 7, 10]. For instance, there exists a type of interactivity raised from the concept of completely correlation. This concept was introduced by Fullér and Majlender [7] but only for two fuzzy numbers. Subsequently, the authors of [11] proposed a generalization of this notion for n fuzzy numbers, $n > 2$.

The fuzzy numbers $A_1, \ldots, A_n$ are said to be linearly interactive (completely correlated) if there exist $q = (q_1, \ldots, q_n), r = (r_1, \ldots, r_n) \in \mathbb{R}^n$ with $q_1 q_2 \ldots q_n \neq 0$ such that the corresponding joint possibility distribution $J = J_{\{q,r\}}$ is given by

$$J_{\{q,r\}}(x_1, \ldots, x_n) = A_1(x_1)\chi_U(x_1, \ldots, x_n) = \ldots = A_n(x_n)\chi_U(x_1, \ldots, x_n), \tag{5}$$

for all $(x_1, \ldots, x_n) \in \mathbb{R}^n$, where χ_U stands for the characteristic function of the set $U = \{(u, q_1 u + r_1, \ldots, q_n u + r_n) : \forall u \in \mathbb{R}\}$.

The JPD given by (5) can be used to describe fuzzy dynamic systems that consider interactivity [4, 11]. However, $J_{\{q,r\}}$ can only be applied to fuzzy numbers that have a co-linear relationship among their membership functions, which means that it can not be used to fuzzy numbers that do not have the same shape. For example, the fuzzy numbers (0; 1; 2) and (3; 4; 6) are not linearly interactive [12].

Alternatively, Esmi et al. [10] employed a parametrized family of joint possibility distributions $\mathcal{J} = \{J_\gamma : \gamma \in [0, 1]\}$ in order to define interactive additions for fuzzy numbers. The authors of [12] used this family of JPDs to produce a numerical solution of a fuzzy initial value problem and verified that $\mathcal{J}$ is more embracing than $J_{\{q,r\}}$.

This paper focuses on the distribution $J_0 \in \mathcal{J}$, whose definition is given as follows. Let A_1 and A_2 be two fuzzy numbers in $\mathbb{R}_{\mathcal{F}_C}$ and the function g^i defined in [10]

$$g^i(z, \alpha) = \bigwedge_{w \in [A_{3-i}]^\alpha} |w + z|,$$

for all $z \in \mathbb{R}$, $\alpha \in [0, 1]$ and $i \in \{1, 2\}$.

Also, consider the sets R^i_α and $L^i(z, \alpha)$,

$$R^i_\alpha = \begin{cases} \{a^-_{i_\alpha}, a^+_{i_\alpha}\} & \text{if } 0 \leq \alpha < 1 \\ [A_i]^1 & \text{if } \alpha = 1 \end{cases} \quad \text{and} \quad L^i(z, \alpha) = [A_{3-i}]^\alpha \cap [-g^i(z, \alpha) - z, g^i(z, \alpha) - z].$$

Finally, J_0 is defined by

$$J_0(x_1, x_2) = \begin{cases} A_1(x_1) \wedge A_2(x_2), & \text{if } (x_1, x_2) \in P \\ 0, & \text{otherwise}, \end{cases} \tag{6}$$

with $P = \cup_{i=1}^2 \cup_{\alpha \in [0,1]} \{(x_1, x_2) : x_i \in R^i_\alpha \text{ and } x_{3-i} \in L^i(x_i, \alpha)\}$.

Note that the set P establishes which pairs $(x_1, x_2) \in \mathbb{R}^2$ satisfy $J_0(x_1, x_2) > 0$. Since P is a proper subset of $\mathbb{R}^2$, one obtains that $J_0 \neq J_\wedge$ [13].

Esmi et al. [10] proved that the fuzzy relation J_0, given by (6), is a joint possibility distribution of A_1 and A_2. Sussner et al. [14] employed shifts for the fuzzy numbers, in order to define a new family of parametrized joint possibility distributions that can be used to control the width of the corresponding interactive addition.

Definition 3 Let $A \in \mathbb{R}_\mathcal{F}$. The translation of A by $k \in \mathbb{R}$ is defined as the fuzzy number $\tilde{A}(x) = A(x + k)$, $\forall x \in \mathbb{R}$.

Definition 3 gives raise to a new JPD, which incorporates the translation of fuzzy numbers [14].

Theorem 1 *Given $A_1, A_2 \in \mathbb{R}_\mathcal{F}$ and $c = (c_1, c_2) \in \mathbb{R}^2$. Let $\tilde{A}_i \in \mathbb{R}_\mathcal{F}$ be such that $\tilde{A}_i(x) = A_i(x + c_i)$, $\forall x \in \mathbb{R}$ and $i = 1, 2$. Let $\tilde{J}_0$ be the joint possibility distribution of fuzzy numbers $\tilde{A}_1, \tilde{A}_2 \in \mathbb{R}_\mathcal{F}$ defined as Eq. (6). The fuzzy relation J_0^c given by*

$$J_0^c(x_1, x_2) = \tilde{J}_0(x_1 - c_1, x_2 - c_2), \ \forall (x_1, x_2) \in \mathbb{R}^2, \tag{7}$$

is a joint possibility distribution of A_1 and A_2.

From now on, the joint possibility distribution used in this paper is the one provided by Theorem 1. For simplicity of notation, the JPD J_0^c will be denoted by J_0.

From the definition of the sup-J extension principle, it is possible to establish an arithmetic for interactive fuzzy numbers. For example, the interactive sum and difference are respectively given by

$$(A_1 +_J A_2)(y) = \bigvee_{x_1+x_2=y} J(x_1, x_2) \tag{8}$$

and

$$(A_1 -_J A_2)(y) = \bigvee_{x_1-x_2=y} J(x_1, x_2) \tag{9}$$

where J is an arbitrary JPD of A_1 and A_2.

In the case where $J = J_0$, the operations defined as in (8) and (9) are denoted by $A_1 +_0 A_2$ and $A_1 -_0 A_2$, respectively. The next examples illustrate that these arithmetic operations have special properties.

Example 1 Let be the triangular fuzzy numbers $A_1 = (1; 2; 3)$ and $A_2 = (2; 3; 4)$. Thus,

$$A_1 +_0 A_2 = 5,$$

where 5 stands for the fuzzy number 5, whose membership function is given by the characteristic function $\chi_{\{5\}}$.

Example 2 Let be the triangular fuzzy numbers $A_1 = (1; 2; 3)$ and $A_2 = (4; 6; 8)$. Thus,

$$A_1 +_0 A_2 = (7; 8; 9).$$

Example 1 illustrates that from the interactive sum, via J_0, it is possible to obtain a real number as a sum of two fuzzy numbers, in contrast to the standard arithmetic sum. This real result, however, not always occurs as can be seen in Example 2.

Example 3 Let be the triangular fuzzy numbers $A_1 = (4; 5; 6)$ and $A_2 = (1; 2; 3)$. Thus,

$$A_1 -_0 A_2 = 3,$$

where 3 stands for the fuzzy number 3, whose membership function is given by the characteristic function $\chi_{\{3\}}$.

Example 4 Let be the triangular fuzzy numbers $A_1 = (4; 6; 8)$ and $A_2 = (1; 2; 3)$. Thus,

$$A_1 -_0 A_2 = (3; 4; 5).$$

Example 3 illustrates that the interactive difference between two fuzzy numbers may also result in a real number.

It is interesting to observe that in Example 4 the following equalities hold true

$$A_1 -_0 A_2 = A_1 -_g A_2 = A_1 -_{gH} A_2 = A_1 -_H A_2, \tag{10}$$

where the differences $-_H$, $-_{gH}$ and $-_g$, represent the Hukuhara, generalized Hukuhara and generalize differences, respectively [15]. In fact, if the Hukuhara difference between two fuzzy numbers in $\mathbb{R}_{\mathcal{F}_C}$ exists, then the equalities given by Eq. (10) are always satisfied [13]. This means that the Hukuhara difference and its generalizations are particular types of interactive arithmetic operations.

The next section discusses the previews works that contributed for the consistence of initial fuzzy conditions and the importance of JPD choice.

3 Choice of Joint Possibility Distribution

There are several works in the literature that study fuzzy linear equations. In particular, Esmi et al. [16] investigated this topic considering that the arithmetic operation is given by an interactive sum. This equation is given by

$$X +_J B = C, \tag{11}$$

where $B, C \in \mathbb{R}_{\mathcal{F}}$, X is the free variable and J is some joint possibility distribution between X and B.

The existence and uniqueness of the solution for Eq. (11) depend on the choice of the JPD J, since the variable X is constructed in terms of J. This means that the independent variable is in fact the joint possibility distribution J.

They have shown that an interactive sum between two fuzzy numbers with equal shapes may result in a fuzzy number with a different shape. For example, an interactive sum between two Gaussian fuzzy numbers may result in a triangular fuzzy number (for more details of a Gaussian fuzzy number, the reader can refer to [2]). This fact induces that it is possible to obtain a real number as result of an interactive sum between two fuzzy numbers and this consequence is associated with the choice of the JPD [16].

Carlsson et al. [8] studied completely correlated fuzzy numbers, which is a particular class of interactive fuzzy numbers. From this type of interactivity, one may reach in conclusions about the linear equation

$$A +_{J_{\{q,r\}}} B = C, \tag{12}$$

where $J_{\{q,r\}}$ is the JPD given by (5).

In this case, the fuzzy numbers A and B must satisfy $[B]_\alpha = q[A]_\alpha + r$, for all $\alpha \in [0, 1]$. Thus, the interactive sum between A and B can be written in terms of α-cuts [4] by

$$[A +_{J_{\{q,r\}}} B]^\alpha = [(q+1)a_\alpha^- + r, (q+1)a_\alpha^+ + r]. \tag{13}$$

Therefore, $C = r$ if, and only if, $q = -1$. This means that Eq. (12) may result in a real number and, in this case, the fuzzy number B is given by

$$B = -A + r. \tag{14}$$

The joint possibility distribution $J_{\{q,r\}}$ ensures that the problem investigated in this work can be solved and it exhibits the solution. However, as it was pointed out, the JPD $J_{\{q,r\}}$ is restricted by the shapes of the fuzzy numbers A and B. This paper focuses on J_0 in order to find broader solutions, which will be done in the next section.

4 Consistence of Fuzzy Initial Conditions

This section presents solutions for fuzzy linear equations with fuzzy variables and real coefficients

$$\alpha_1 X_1 +_0 \alpha_2 X_2 = r, \tag{15}$$

where $\alpha_1, \alpha_2, r \in \mathbb{R}$, with $\alpha_1, \alpha_2 \neq 0$, and the triangular fuzzy variables are $X_1 = (a; b; c)$ and $X_2 = (d; e; f)$.

The next theorem provides a characterization for the interactive sum $+_0$ between two triangular fuzzy numbers.

Theorem 2 *Let $A = (a; b; c)$ and $B = (d; e; f)$ be triangular fuzzy numbers. Let J_0 be the joint possibility distribution between A and B, given by (7). Thus*

$$A +_0 B = \begin{cases} ((a+f) \wedge (b+e); b+e; (b+e) \vee (c+d)), & \textit{if } width(A) \geq width(B) \\ ((c+d) \wedge (b+e); b+e; (b+e) \vee (a+f)), & \textit{if } width(A) \leq width(B) \end{cases}. \tag{16}$$

Proof Since A and B are triangular fuzzy numbers, we have $A, B \in \mathbb{R}_{\mathcal{F}_C}$. Thus, Theorem 3 of [13] ensures that

$$A^{k_A} +_0 B^{k_B} = A -_0 (-B) - k_A - k_B, \tag{17}$$

where k_A and k_B are the midpoints of the 1-cut of A and B, respectively.

Thus, from the definition of the JPD J_0 we have

$$\begin{aligned} A^{k_A} +_0 B^{k_B} &= (\min(a+f, c+d, b+e); b+e; \max(a+f, c+d, b+e)) \\ &\quad -k_A - k_B, \end{aligned}$$

since for $A = (a; b; c)$ and $B = (d; e; f)$, it follows that

$$A -_0 (-B) = (\min(a - (-f), c - (-d), b - (-e)); b - (-e);$$
$$\max(a - (-f), c - (-d), b - (-e)).$$

A combination of Proposition 3 and Theorem 3 of [13] leads to the following

$$\begin{aligned} A +_0 B &= A^{k_A} +_0 B^{k_B} + k_A + k_B \\ &= (\min(a + f, c + d, b + e); b + e; \max(a + f, c + d, b + e)). \end{aligned}$$

On the one hand, if $width(A) \geq width(B)$, then $c + d \geq a + f$ and we obtain

$$A +_0 B = (\min(a + f, b + e); b + e; \max(c + d, b + e))$$

On the other hand, if $width(A) \leq width(B)$, then $c + d \leq a + f$ and we obtain

$$A +_0 B = (\min(c + d, b + e); b + e; \max(a + f, b + e)),$$

which concludes the proof.

In Eq. (15) there are four possibilities to consider, which boil down in two cases.

1. If $\alpha_1, \alpha_2 > 0$ (or if $\alpha_1, \alpha_2 < 0$), then the sum is

$$(\alpha_1 a; \alpha_1 b; \alpha_1 c) +_0 (\alpha_2 d; \alpha_2 e; \alpha_2 f) = r$$
$$(\text{or } (\alpha_1 c; \alpha_1 b; \alpha_1 a) +_0 (\alpha_2 f; \alpha_2 e; \alpha_2 d) = r).$$

Since $r \in \mathbb{R}$, from Theorem 2 the fuzzy variables $(a; b; c)$ and $(d; e; f)$ must satisfy

$$\begin{cases} \alpha_1 a + \alpha_2 f = r \\ \alpha_1 b + \alpha_2 e = r \\ \alpha_1 c + \alpha_2 d = r \end{cases}.$$

Therefore, given b and e such that $\alpha_1 b + \alpha_2 e = r$, from Eqs. (18) and (19) a, f, c and d can be derived:

$$\alpha_1(b - a) = \alpha_2(f - e) \quad (18)$$
$$\alpha_1(c - b) = \alpha_2(e - d). \quad (19)$$

The choices of a, b, c, d, e, f are not arbitrarily. Instead, they are constructed according to the following algorithm.

1a. From a given value b (or e) in Eq. (15), the other value e (or b) is determined by

$$e = \frac{r}{\alpha_2} - \frac{\alpha_1}{\alpha_2} b. \quad (20)$$

1b. The choice of $a < b$, implies in the determination of f by

$$f = \frac{\alpha_1}{\alpha_2}(b - a) + e.$$

1c. The choice of $c > b$, implies in the determination of d by

$$d = \frac{\alpha_1}{\alpha_2}(b - c) + e.$$

Note that the choices are attached in pairs a and f, b and e, c and d.

2. If $\alpha_1 > 0$ and $\alpha_2 < 0$ (or if $\alpha_1 > 0$ and $\alpha_2 < 0$), then the sum in (15) is

$$(\alpha_1 a; \alpha_1 b; \alpha_1 c) +_0 (\alpha_2 f; \alpha_2 e; \alpha_2 d) = r$$
$$(\text{or } (\alpha_1 c; \alpha_1 b; \alpha_1 a) +_0 (\alpha_2 d; \alpha_2 e; \alpha_2 f) = r).$$

Since $r \in \mathbb{R}$, from Theorem 2 it follows that

$$\begin{cases} \alpha_1 a + \alpha_2 d = r \\ \alpha_1 b + \alpha_2 e = r \\ \alpha_1 c + \alpha_2 f = r \end{cases}.$$

In other words,

$$\alpha_1(b - a) = -\alpha_2(e - d) \tag{21}$$
$$\alpha_1(c - b) = -\alpha_2(f - e). \tag{22}$$

The algorithm to find a solution is the following.

2a. To choose $b \in \mathbb{R}$, so the value of e is calculated by (20).

2b. To choose a, which implies in the determination of d by

$$d = \frac{\alpha_1}{\alpha_2}(b - a) + e.$$

2c. The choice of $c > b$, implies in the determination of f by

$$f = \frac{\alpha_1}{\alpha_2}(b - c) + e.$$

In this second case, the pairs attached changed: a is now connected with d, and c is connected with f.

Next, two examples are provided in order to illustrate the construction of this method.

Example 5 Consider the fuzzy linear equation given by $X_1 +_0 (-X_2) = 2$. This equation enters in Case 2, since $\alpha_1 = 1$ and $\alpha_2 = -1$. The solution is $X_1 = (a; b; c)$

and $X_2 = (a - 2; b - 2; c - 2)$, therefore $X_2 = X_1 - 2$, where 2 stands for the real number 2.

Example 6 Consider the fuzzy linear equation given by $3X_1 +_0 2X_2 = 10$. This equation enters in Case 1, since $\alpha_1 = 3$ and $\alpha_2 = 2$. The solution is $X_1 = (a; b; c)$ and $X_2 = \left(5 - \frac{3}{2}c; 5 - \frac{3}{2}b; 5 - \frac{3}{2}a\right)$.

Note that in Example 5, the fuzzy number X_2 is given by a translation of the given fuzzy number X_1. On the other hand, in Example 6, the fuzzy number X_2 is given by performing a reflection and a translation on X_1. Also note that once the fuzzy number X_1 is fixed, the fuzzy number X_2 is unique determined.

It is interesting to observe the relation of symmetry of the fuzzy numbers in this algorithm. Examples 5 and 6 reveal that, if X_1 is (non) symmetric, then X_2 must be (non) symmetric as well. This fact can be verified directly from the proposed algorithm.

The next example illustrates the restrictions imposed by this algorithm in modeling.

Example 7 Consider a population with ten individuals and a contagious disease such as the HIV. Suppose that there are "around" three infected individuals, which implies that there are "around" seven susceptible individuals. If one describes the linguistic variable "around" three by the fuzzy number (2; 3; 4), then the algorithm establishes that "around" seven must be modeled by the fuzzy number (6; 7; 8), in order to guarantee that the initial condition $(2; 3; 4) +_0 (6; 7; 8) = 10$ be satisfied.

On the other hand, if one chooses to describe "around" three by the fuzzy number (2; 3; 5), then the algorithm ensures that "around" seven must be (5; 7; 8).

From the consistence point of view, it was shown here that there are criteria to consider triangular fuzzy numbers as fuzzy initial conditions in biological models. The same applies in chemical and physical problems.

5 Final Remarks

This manuscript presented a discussion about the consistence of the choice of fuzzy numbers in modeling. This discussion emerged from biological/epidemiological problems where there is uncertainty at the input, but the output is precisely determined. One example of this problem is to estimate the number of infected individuals, in a population that the total number of individuals is known. The imprecision in the number of infected individuals leads to the uncertainty of the number of susceptible individuals. This type of problem motivates the study of fuzzy linear equations.

This paper investigated equations with a sum between two fuzzy numbers on the one side and a risp number on the other side. It is well known in the literature that the interactive sum produces this type of result, in contrast to the standard sum [3]. There are at least two interactive sums that may solve this problem, the sums $+_L$ [4] and $+_0$ [13]. This work focuses in the interactive sum $+_0$, since it is more embrace than $+_L$ [12]. Moreover, this paper studies fuzzy numbers with triangular shapes, since the triangular fuzzy numbers are one of the most types of fuzzy numbers used in modeling.

It was provided an algorithm in order to construct the fuzzy numbers, where the sum between them results in a real number. Moreover, this construction implies in the study of the symmetry of the fuzzy numbers. This paper shows that in mathematical models, where the fuzzy numbers are used to describe the uncertainty in the initial condition, the choice must not be aleatory. There exists a dependence between the initial conditions, where this relation of dependence is known in the context of the Fuzzy Set Theory as interactivity.

Acknowledgements Pinto, N. J. B thanks to CAPES for Financial support under grant no 1691227, Esmi, E. thanks FAPESP under grant no 2016/26040-7 and Barros, L. C. thanks CNPq under grant no 306546/2017-5.

References

1. L. Edelstein-Keshet, *Mathematical Models in Biology* (Society for Industrial and Applied Mathematics, Philadelphia, 2005)
2. L.C. Barros, R.C. Bassanezi, W.A. Lodwick, A first course in fuzzy logic, fuzzy dynamical systems, and biomathematics, in *Studies in Fuzziness and Soft Computing* (Springer, Berlin, 2017)
3. V.F. Wasques, E. Esmi, L.C. Barros, B. Bede, Comparison between numerical solutions of fuzzy initial-value problems via interactive and standard arithmetics, in *Fuzzy Techniques: Theory and Applications*, vol. 1000 (Springer International Publishing, Cham, 2019), pp. 704–715
4. L.C. Barros, F.S. Pedro, Fuzzy differential equations with interactive derivative. Fuzzy Sets Syst. **309**, 64–80 (2017)
5. L.A. Zadeh, Fuzzy sets. Inf. Control **8**, 338–353 (1965)
6. E. Esmi, F. Santo Pedro, L.C. Barros, W. Lodwick, Fréchet derivative for linearly correlated fuzzy function. Inf. Sci. **435**, 150–160 (2018)
7. R. Fullér, P. Majlender, On interactive fuzzy numbers. Fuzzy Sets Syst. **143**, 355-369 (2004)
8. C. Carlsson, R. Fuller, P. Majlender, Additions of completely correlated fuzzy numbers. IEEE Int. Conf. Fuzzy Syst. **1**, 535–539 (2004). https://doi.org/10.1109/FUZZY.2004.1375791
9. L.A. Zadeh, Concept of a linguistic variable and its application to approximate reasoning - I. Inf. Sci. **8**, 199–249 (1975)
10. E. Esmi, P. Sussner, G.B.D. Ignácio, L.C. Barros, A parametrized sum of fuzzy numbers with applications to fuzzy initial value problems. Fuzzy Sets Syst. **331**, 85–104 (2018)
11. V.F. Wasques, E. Esmi, L.C. Barros, F.S. Pedro, P. Sussner, Higher order initial value problem with interactive fuzzy conditions. IEEE Int. Conf. Fuzzy Syst. (FUZZ-IEEE) (2018) 1–8, https://doi.org/10.1109/FUZZ-IEEE.2018.8491465
12. V.F. Wasques, E. Esmi, L.C. Barros, P. Sussner, Numerical solutions for bidimensional initial value problem with interactive fuzzy numbers, in *Fuzzy Information Processing* (Springer International Publishing, Cham, 2018), pp. 84–95.

13. V.F. Wasques, E. Esmi, L.C. Barros, P. Sussner, The generalized fuzzy derivative is interactive. Inf. Sci. **519**, 93–109 (2020)
14. P. Sussner, E. Esmi, L.C. Barros, Controling the width of the sum of interactive fuzzy numbers with applications to fuzzy initial value problems. IEEE Int. Conf. Fuzzy Syst (FUZZ-IEEE). 1453–1460 (2016). https://doi.org/10.1109/FUZZ-IEEE.2016.7737860
15. B. Bede, *Mathematics of Fuzzy Sets and Fuzzy Logic* (Springer, Berlin, 2013)
16. E. Esmi, L.C. de Barros, V.F. Wasques, Some notes on the addition of interactive fuzzy numbers, in *Fuzzy Techniques: Theory and Applications. Advances in Intelligent Systems and Computing*, vol. 1000 (International Publishing, Cham, 2019), pp. 246–257

An Approximate Perspective on Word Prediction in Context: Ontological Semantics Meets BERT

Kanishka Misra and Julia Taylor Rayz

Abstract This paper presents an analysis of a large neural network model—BERT, by placing its word prediction in context capability under the framework of Ontological Semantics. BERT has reportedly performed well in tasks that require semantic competence without any explicit semantic inductive bias. We posit that word prediction in context can be interpreted as the task of inferring the meaning of an unknown word. This practice has been employed by several papers following the Ontological Semantic Technology (OST) approach to Natural Language Understanding. Using this approach, we deconstruct BERT's output for an example sentence and interpret it using OST's fuzziness handling mechanisms, revealing the degree to which each output satisfies the sentence's constraints.

1 Introduction

Recent progress made by deep learning approaches in natural language processing (NLP) have led to the emergence of highly parameterized neural network models that represent a word in its context, collectively known as contextualized word embeddings (CWE). The goal of these embeddings is to adapt to context (described by sentences) for the same word. This means that a word *table* should be represented differently depending whether it is furniture or chart. One such CWE, BERT [1] learns word representations by using a training procedure known as Masked Language Modelling, which is similar to Cloze Tasks [20]. In this task, a word in a sentence (typically called a "cloze sentence") is hidden or "masked" and the task is to identify the masked word given the context it occurs in, an example is shown in (1). For this example BERT predicts the word bank in place of the mask with greater than 0.96 probability.

K. Misra (✉) · J. T. Rayz
Purdue University, West Lafayette, IN 47906, USA
e-mail: kmisra@purdue.edu

J. T. Rayz
e-mail: jtaylor1@purdue.edu

B. Bede et al. (eds.), *Fuzzy Information Processing 2020*,
Advances in Intelligent Systems and Computing 1337,
https://doi.org/10.1007/978-3-030-81561-5_14

(1) I went to the ______ to withdraw some money.

Unfortunately, BERT representations and mapping to the word, while robust and impressive in scale, can be somewhat questionable in quality. In this paper we deconstruct the task of predicting a word in context by borrowing from the school of Ontological Semantics [10], and its latest product, the Ontological Semantic Technology (OST) [5, 11, 15], which is inherently fuzzy in nature [16]. We analyze simple cloze sentences by making fuzzy inferences with the help of the OST system and represent the outputs of BERT by their corresponding concepts that form various solutions to the cloze task depending on their fuzzy membership which is calculated based on the concepts that occur in their context.

2 Bidirectional Encoder Representations from Transformers (BERT)

BERT is a language representation neural network model that learns to represent words in sentences by jointly conditioning on words to the left of a target as well as to the right. The representation of a word (a vector) is computed by estimating word probabilities in context, and thus the model produces context-sensitive or contextualized representations of words. Its underlying architecture is based on Transformers [21], which enables it to represent each word as a function of words occurring in its context. The model comes in two variants—BERT-base and BERT-large, differing in the total number of parameters—110 M and 340 M respectively. Although the exact nature of these outputs is largely unknown, [13] found BERT's representations of words with similar senses to cluster together in vector space, signaling to some extent that BERT captures sense-specific properties of words through its training mechanism. As a result, BERT advanced the state-of-the-art in NLP (at the time of its publication) by facilitating fine-tuning on a wide variety of language tasks such as Question Answering, Natural Language Inference, etc. The model accepts two sentences as input during each step and is jointly optimized using the following objectives: (1) Masked Language Modelling (MLM), inspired by the cloze task, in which the model uses its context to predict hidden tokens in the input, and (2) Next Sentence Prediction, in which the model predicts whether the second sentence follows the first sentence. Due to its MLM objective, BERT is not considered to be an incremental language model (such as models using Recurrent Neural Networks or its variants) that form sentences by predicting words one by one in a single and fixed direction (left to right) and contain a sequential inductive bias. In our analysis of BERT in this paper, we will analyze the BERT-base model, but this can be extended to any similar language model.

2.1 Semantic Capabilities of BERT

Fine tuning BERT has resulted in incremental performance on NLP tasks that require a high linguistic competence. As a result, a myriad of methods have been used to probe BERT for the various linguistic properties that it captures. A majority of such methods have focused on BERT's knowledge of syntactic phenomena such as number agreement [8] and garden-path [14]. While BERT shows syntactic competence on a variety of tasks, it has been found to be less sensitive to an analysis of semantic properties.

Such tasks can be adapted from adjacent disciplines that test human competence. Adapting from the psycholinguistic apparatus of human sentence processing—the N400 experiment—Ettinger [2] developed a suite of tests to analyze BERT's sensitivity for semantic phenomenon observed in humans. BERT was found to show a level of insensitivity in several of these tasks. Specifically, for tests of negation, BERT is unable to assign a lower probability for *bird* as compared to a nonsensical—in the given context—word *tree* in the stimulus: *A robin is not a* ___. Further, it generated non-sequitur situations in tests for commonsense inference, such as predicting words such as *gun* in the stimuli: *The snow had piled up on the drive so high that they couldn't get the car out. When Albert woke up, his father handed him a* ___. However, it showed positive results in attributing nouns to their hypernyms[1] and was sensitive to role-reversal stimuli, such as assigning higher probability to the word *served* in the stimuli - *the restaurant owner forgot which customer the waitress had* ___ as opposed to its role-reversed counterpart, *the restaurant owner forgot which waitress the customer had* ___.

Misra, Ettinger, and Rayz [9] investigated the degree to which BERT borrows from lexical cues in the context of a missing word position. In example (2), when the sentence is preceded by a minimal lexical cue of *delicate*, BERT is able to predict *fragile* in place of the ___ with higher probability as compared to when the sentence is preceded by an unrelated word, *salad*.

(2) a. **delicate.** It was a very ___ tea set.
    b. **salad.** It was a very ___ tea set.

3 Ontological Semantic Technology

Unlike BERT whose knowledge is based on a corpus, albeit very large, Ontological Semantics is based on human knowledge and the ontology is hand-crafted. Raskin et al. [11] argued that the relative time of acquisition is acceptable for a semantic system and recent views on deep learning [7] agree that a knowledge-based approach could improve deep learning systems. With this in mind, we outline the differences in results between a very large scale DL architecture and a very small knowledge-based one.

---

[1] Is-a relationship, for example: *a dog* **is a** *mammal*.

Ontological Semantic Technology [5, 11, 15] is a meaning-based Natural Language Understanding system that consists of several repositories of world and linguistic knowledge. The main static resources consist of: a language independent ontology—a graph whose nodes are concepts and edges are the various relations between them; a lexicon per supported language, that defines word senses of a language by anchoring them with an appropriate concept or property in the ontology. OST processing is event-driven, usually selected from the main verb in the sentence. Once the event's sense is disambiguated, a Text Meaning Representation (TMR) is produced, and stored in the Information Repository. This repository is used in processing of further text, depending on the application.

3.1 On OST's Fuzzy Nature

OST is fuzzy in nature [12, 16] as most of the processing is driven by so-called facets that represent various membership degrees of a particular event, as described by information in the sentence. The memberships themselves are derived from a location of a concept, recovered from a sentence, in an ontological hierarchy, based on the defined facet and filler combination. While the explicit hierarchical nature is easy to navigate for fillers of OST facets [16] it is worth mentioning that the same procedure can be applied to concepts that can be virtually formed with the help of ontological properties. Such construction of virtual nodes for a crisp ontology was explained in [15]. A crisp ontology, however, always has a membership degree of 1 for every acceptable concept, thus it is worth to address the fuzzy virtual ontology here.

When an ontological event is defined, its semantic roles are filled with concepts, defined in the ontology. For each property, each (facet, filler) pair is a pointer to a concept and its descendants with a membership degree of a pointed concept defined by a facet. OST has four facets: `default`, `sem`, `relaxable-to`, and `not`. `default` has the largest membership degree, 1; `sem` has a smaller membership degree, `relaxable-to` approaches 0, while not membership is exactly 0. The ontological hierarchy, shown in Fig. 1 shows a hierarchy of concepts that can be used in a given event E. The grey concepts will be used explicitly in a definition of a property in the ontology, such as $P(\text{default}(x))(\text{sem}(a))(\text{relaxable-to}(p))$. The membership degrees of other concepts, indicated as single circles, will be calculated according to the formula showed in [18].

Concepts indicated with double circles are virtual—they are not defined by a knowledge engineer but rather taken from a lexical knowledge of language. These are calculated per language and do not have to be stored. Their membership degree is calculated as if they were explicitly defined. In other words, if a knowledge engineer were to place node z into the ontology, the calculation of its membership in an event E should not change. This gives us flexibility when working with several languages at a time.

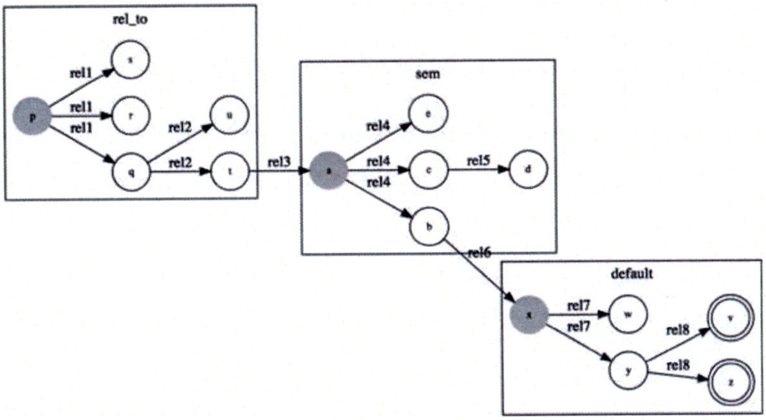

Fig. 1 Hierarchy of concepts (nodes) with properties (edges) and facets (boxes) and virtual nodes (double circles)

For example, consider a concept WASH. Since any physical object of an appropriate size can be washed, its `sem` facet for a property THEME is likely to be PHYSICAL- OBJECT. However, we may see a sense in some lexicon that restricts a verb anchored in WASH by adding a property to WASH, such as INSTRUMENT with a filler LAUNDRY- DETERGENT. Now suppose an INSTRUMENT of WASH is SOAP, defined as `sem`. SOAP, however, can have different children, such as HAND- SOAP, SHAMPOO, etc. When a lexicon sense of WASHING- WITH- LAUNDRY- DETERGENT is found, its filler, LAUNDRY- DETERGENT, would be placed as a virtual node of a hierarchy, as demonstrated in Fig. 1. This node, defined as INSTRUMENT- OF(WASHING- WITH- LAUNDRY- DETERGENT) will be used whenever appropriate in a calculation of sentence acceptability.

4 Masked Word Prediction as Guessing of an Unknown Word's Meaning

Masked word prediction forms the basis of how BERT learns word representations, which are further used in high-level NLP tasks to produce substantial improvements in terms of performance (as reported). Our goal in this work is to analyze BERT's word prediction in context by viewing it from the lens of OST's fuzzy inference capabilities. There can be several ways to infer what will appear in the place of the masked token. From the distributional semantics point of view, words that appear in the same context tend to have similar meanings [3, 4]. An example borrowed from Jurafsky and Martin [6] is presented in (3).

(3) a. *Ongchoi* is delicious sauteed with garlic.
b. *Ongchoi* is superb over rice.
c. *spinach* sauteed with garlic over rice.
d. *chard* stems and leaves are delicious.
e. *collard greens* and other salty leafy greens.

Since the unknown word *ongchoi* occurs in similar contexts as *spinach, chard,* and *collard greens,* it can be inferred that it is a green leafy vegetable, similar to those mentioned before. While distributional semantics presents a case for statistical approaches, Taylor, Raskin and Hempelmann [17, 19] present a computational semantic approach using OST. Like [1], they formulate the process of acquiring the meaning of an unknown word as a cloze task and produce TMRs by analyzing exemplar contexts consisting of the unknown word. Here, the functional details in the unknown word's context (usually a sentence) determine the basis of understanding the meaning of the unknown word. The example they analyze is a sentence with the verb *rethink* (4a), and the task is to understand the meaning of its direct-object, the new curtains which is replaced with a *zzz* in (4b) to indicate that it is unknown.

(4) a. She decided she would rethink *the new curtains* before buying them for the whole house.
b. She decided she would rethink *zzz* before buying them for the whole house.
c. She decided she would rethink the new ___ before buying them for the whole house.

Based on the TMR representation presented in their paper (shown below), the word *zzz* references the concept that must satisfy certain constraints: (i) it is something that can be rethought, (ii) it carries the semantic role - theme of BUY (iii) it is located in a HOUSE. Note that the paper determines the concept the unknown word evokes as opposed to the word itself. The word could be anything that satisfies those constraints: any kind of furniture - *chair, table, desk, sofa, etc.* or a decorative item such as a *painting*.

```
(DECIDE
    (AGENT(HUMAN(GENDER(FEMALE)))
        (THEME(CONSIDER- INFO(ITERATION(MULTIPLE))
            (AGENT(HUMAN(GENDER(FEMALE)
            (THEME(???))
            (BEFORE(BUY
                (THEME(???(HAS- LOCALE(HOUSE))))))
)))
```

In BERT's case, this instance would be formulated as (4c), where only the word *curtains* has been masked due to BERT's limited capability to only decode one token. Nevertheless, the example still holds as the only change in the input is an addition of the adjective *new* to describe the object. A selective list of BERT's predictions for (4c) is shown in Table 1. We see BERT assigns high probability to items that can

Table 1 Selective list of word probabilities for (4c) as estimated by BERT-base

Rank	Token	Probability	Rank	Token	Probability
1	Clothes	0.1630	21	design	0.0067
2	Designs	0.1320	22	curtains	0.0063
13	Paintings	0.0131	23	gifts	0.0060
16	Furniture	0.0111	24	wardrobe	0.0057
17	Pictures	0.0101	25	products	0.0049
18	Books	0.0096	26	toys	0.0047
19	Decorations	0.0078	28	photos	0.0041
20	Arrangements	0.0070	30	decor	0.0040

be bought for a house: *paintings, furniture, decorations, etc.*, and even the original masked word, *curtains*. Interestingly, the highest probability is assigned to *clothes*, which is anomalous but could be considered valid if *house* is metonymically referring to the people living in the house, i.e., she is buying clothes for all of them. Whether these predictions are due to purely statistical patterns or something close to true language understanding remains an open research endeavor. We posit that a system that truly understands natural language should assign approximately equal scores to objects that are semantically and syntactically plausible in the sentence. Such a phenomenon is manifested in OST's interpretation of sentences, where the meaning resolution is performed in a structured manner, using TMRs. At the same time, acquiring concepts for the ontology in an accurate manner presents a few challenges, such as an extensive training by a master ontologist.

5 Deconstructing BERT's Output Using OST and Fuzzy Inference

In this section, we interpret BERT's output for an example cloze sentence using OST's fuzzy inference mechanism. We first describe our procedure, and then present the interpretation of our example sentence.

Procedure Owing to the fact that OST is event-driven, we represent the sentence along the event that affects the missing word, E. The event is represented as a minimal-script where its various case-roles are listed based on the given sentence, as follows:

E
    AGENT:
    THEME:
    INSTRUMENT:
    ...

Assuming we do not possess a priori knowledge regarding the sense of the event, and so we decompose E into its possible senses {E-v_1, E-v_2, ..., E-v_n}. For each sense of the concept of E, we compute the fuzzy membership values of the concepts that can occupy the missing position, provided by BERT. This is denoted by $\mu_R(\text{E-v}_i, c)$ where μ_R is the membership of concept c that participates in the relation R for the ith sense of the event, E-v_i. For the sake of simplicity, we only consider the top-5 words predicted by BERT. Finally, in the same vein as [17], we compute the syntactic and semantic acceptability of the sentence (sent) formed by choosing each of the concept denoted by BERT's prediction as follows:

$$\mu_{\text{syntax}} = \min_{\text{phr}\in\text{sent}} \max_{x,y\in\text{phr}} [\mu_{\text{phr}}(x, y)],$$

$$\mu_{\text{semantics}} = \min_{R\in\text{sent}} \max_{x,y\in R} [\mu_R(x, y)],$$

$$\mu_{\text{acceptability}} = \min[\mu_{\text{syntax}}, \mu_{\text{semantics}}],$$

where $\mu_{\text{acceptability}}$ denotes the overall acceptability membership value, and μ_{syntax} and $\mu_{\text{semantics}}$ denote individual membership values for the sentence for its syntax and semantics, respectively. For a detailed analysis of how these values are obtained, please refer to Taylor et al. [17]. We do not choose any final sentence using the acceptability scores, instead, the list of acceptability memberships provide us with relative scores for the concepts evoked by BERT's predictions and help us decipher the extent to which each concept fits into the contextual constraints of the sentence.

Interpretation Example Let's consider the following example:

(5) She quickly got dressed and brushed her ___.

In its predictions, BERT attributes words that denote concepts that can be the theme of BRUSH (assuming the act-of-cleaning sense of the concept). It predicts *teeth* with the highest probability, alluding to the possibility that a similar sentence describing a person's morning routine has been observed during BERT's training procedure. Following *teeth* are *hair* (the word originally present in the sentence), *face, ponytail,* and *dress*. Assigning a considerably higher probability to *teeth* as opposed to *hair* can be attributed to BERT's statistical bias which is determined by the corpus it was trained on. While the sentence has two events (DRESS and BRUSH), we will only work with the one that is most concerned with the missing word—BRUSH. This event can be represented as the following minimal-script:

BRUSH
  AGENT: HUMAN
    GENDER: FEMALE
  THEME: ___
  INSTRUMENT: NONE

Table 2 Top-5 predicted words for (5) as estimated by BERT-base

Rank	Token	Probability
1	Teeth	0.8915
2	Hair	0.1073
3	Face	0.0002
4	Ponytail	0.0002
5	Dress	0.0001

Further, consider the following senses of BRUSH (as a verb):

1. Act of cleaning [*brush your teeth*]
2. Rub with brush [*I brushed my clothes*]
3. Remove with brush [*brush dirt off the jacket*]
4. Touch something lightly [*her cheeks brushed against the wind*]
5. ...

Only the first two senses are applicable for the words shown in Table 2. In this analysis, we will interpret the first sense of the event, BRUSH-v1, since the same procedure can be applied to interpret any other sense of the event.

Considering BRUSH-v1, we have four concepts that can have the property, THEME-OF BRUSH: TEETH, HAIR, FACE, PONYTAIL. Notice that PONYTAIL is a descendent of HAIR and its membership for THEME- OF BRUSH-v1 would be slightly lower than that of HAIR. Since the instrument of BRUSH-v1 is missing here, TEETH, HAIR, and FACE have the same membership value as shown in Fig. 2. While all these concepts have high-membership, none of them can be a `default`. We also consider the `relaxable-to` facet here as we want to restrict non-physical objects from being counted as THEME of BRUSH-v1. The memberships of the concepts denoted by these words would be ordered as follows:

$$\mu_{\text{theme}}(\text{TEETH}) = \mu_{\text{theme}}(\text{FACE}) = \mu_{\text{theme}}(\text{HAIR}) > \mu_{\text{theme}}(\text{PONYTAIL})$$

However, consider the following sentences:

(6) a. She quickly got dressed and brushed her ___ with a comb.
b. She quickly got dressed and brushed her ___ with a toothbrush.

These examples further constrain the membership values for the concept that satisfies the THEME- OF BRUSH-v1 relation by adding an INSTRUMENT- OF relation. To account for the instrument, we traverse down the virtual hierarchy of BRUSH-v1, a subset of which is shown in Fig. 3. As mentioned in Sect. 3.1, two new virtual nodes are created when BRUSH-v1 is endowed with an instrument (either COMB or TOOTHBRUSH). With this new knowledge, the membership value for certain concepts is elevated to the `default` facet. At the same time, the membership of all other concepts that can no longer be the theme of BRUSH- WITH- [INSTRUMENT] is lowered. For descendants of the `default`, the membership for THEME- OF BRUSH- WITH- [INSTRUMENT] would

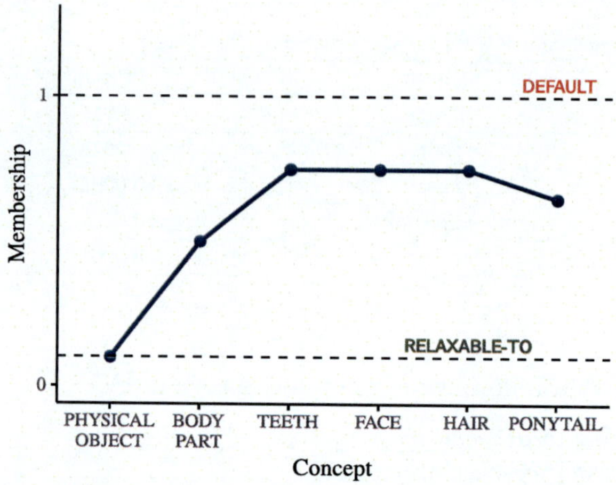

Fig. 2 Membership values of the various concepts that could be theme of BRUSH-V1. Note that none of the concepts are a `default` but have high membership when the instrument of BRUSH-v1 is not present. Descendants of all such concepts (such as PONYTAIL, which is a child of HAIR) have slightly lower membership. The concept BODY-PART is added to indicate relative position, close to TEETH, etc. and distant from PHYSICAL-OBJECT

Fig. 3 Virtual nodes created in the hierarchy of BRUSH-v1 when it is endowed with an INSTRUMENT-OF relation. These nodes alter the membership values for concepts that can satisfy the relation, THEME-OF for BRUSH-v1, and assign new scores to them depending on the value of the INSTRUMENT-OF BRUSH-v1

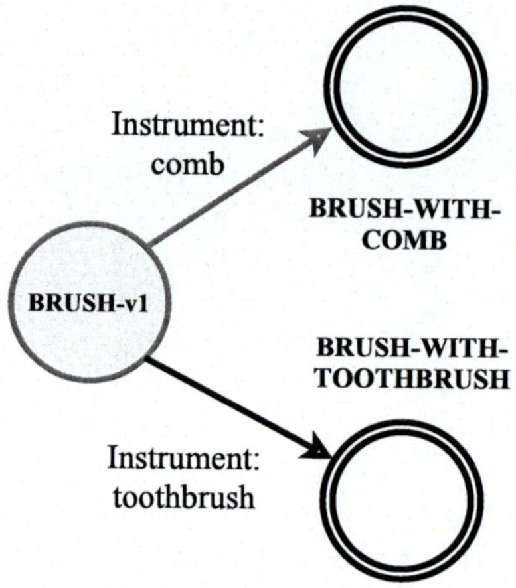

increase relative to their membership for THEME-OF BRUSH. This can be summarized by the following for concepts HAIR, TEETH, and PONYTAIL:

$$\mu_{\text{theme}}(\text{BRUSH-WITH-TOOTHBRUSH, TEETH}) = 1$$
$$\mu_{\text{theme}}(\text{BRUSH-WITH-COMB, HAIR}) = 1$$
$$\mu_{\text{theme}}(\text{BRUSH-WITH-COMB, PONYTAIL}) > \mu_{\text{theme}}(\text{BRUSH, PONYTAIL})$$

BERT's outputs for the sentences in example (6) is shown in Table 3.

Table 3 BERT-base probabilities for words predicted in (5) but with (6a) and (6b) as inputs

BRUSH- WITH- COMB (6a)			BRUSH- WITH- TOOTHBRUSH (6b)		
Rank	Token	Probability	Rank	Token	Probability
1	Hair	0.8704	1	teeth	0.9922
2	Teeth	0.1059	2	hair	0.0052
3	Face	0.0210	3	face	0.0019
12	Ponytail	<0.0001	31	ponytail	<0.0001
27	Dress	<0.0001	98	dress	<0.0001

We see the probabilities estimated by BERT for the event of BRUSH with and without an INSTRUMENT- OF relation are considerably different. BERT assigns the highest probability to each event's `defaults`, denoted by words *hair* and *teeth* respectively. Judging from the top-predicted word in both events, BERT follows the same pattern as our interpretation of assigning higher score to the `default`. We compare the rest of the top-5 predicted words by following the discussion about THEME- OF BRUSH- WITH- [INSTRUMENT]. For the event of BRUSH- WITH- COMB, the following acceptability measures emerge:

$$\mu_{\text{syntax}}(\text{THEME- OF BRUSH- WITH- COMB}, \{\text{top-5-predictions}\}) = 1 \quad (1)$$

$$\mu_{\text{semantics}}(\text{THEME- OF BRUSH- WITH- COMB}, \text{HAIR}) = 1 \quad (2)$$

$$\mu_{\text{semantics}}(\text{THEME- OF BRUSH- WITH- COMB}, \{\text{TEETH, FACE, DRESS}\}) < \mu_{\text{semantics}}(\text{BRUSH- WITH- COMB}, \textit{descendent-of}(\text{HAIR})) \quad (3)$$

Every prediction in Table 3 is considered syntactically acceptable for both examples since all the predicted words are nouns in the example sentences. The difference lies in the overall acceptability scores with respect to semantics. The `default`, HAIR will have the highest score, 1. Following HAIR should be its descendants, in this case, PONYTAIL. Every other word and the concepts evoked by them should be scored considerably lower since they would be considered semantically anomalous as THEME- OF BRUSH-v1. Hence,

$$\mu_{\text{acceptability}}(\text{HAIR}) > \mu_{\text{acceptability}}(\text{PONYTAIL}) = \mu_{\text{acceptability}}(\textit{descendent-of}(\text{HAIR})) > \mu_{\text{acceptability}}(\{\text{all-other-predictions}\})$$

From the predictions for BRUSH- WITH- COMB, we find BERT's outputs to deviate from our acceptability measures. PONYTAIL is scored considerably lower than words denoting concepts with much lower membership values for THEME- OF BRUSH- WITH- [INSTRUMENT]. Why such a situation arises is subject to further quantitative evaluation, including the investigation of BERT's internal mechanisms. Our takeaway here is that the statistical patterns learnt by BERT could prevent it from demarcating concepts that satisfy a semantic relation, as exemplified in the above evaluation.

6 Conclusion

In this paper, we have adopted an Ontological Semantics [10] approach to analyze and interpret the output of a black-box neural network model, BERT [1]. We evaluate BERT's output for the task of predicting the word occupying a missing position (indicated by a ___) in a sentence, which is in the same vein as "guessing the meaning of an unknown word in context," a task proposed earlier [19]. To this end, we utilize a realization of Ontological Semantics, the OST system and its inherent ability to make fuzzy inferences about the concept occupying the missing position in the sentence. This provides a mechanism for us to quantify the degree to which a concept is evoked by the unknown word in its context based on the functional relationships of its surrounding items [17, 19]. Using an exemplar sentence, we discussed the event of BRUSH [act of cleaning] and modified it by adding an INSTRUMENT- OF relation. While we found BERT to change its top-predicted word when the instrument of the event changed, it was unable to show structural (semantics-wise) phenomena. This was evidenced by BERT scoring a descendent of HAIR: PONYTAIL lower than a nonsensical concept (in the given instance)—TEETH, indicating a hindrance displayed by its training process with respect to learning true semantic understanding.

References

1. J. Devlin, M.W. Chang, K. Lee, Toutanova, K.: BERT, Pre-training of deep bidirectional transformers for language understanding (2019), pp. 4171–4186
2. A. Ettinger, What BERT is not: lessons from a new suite of psycholinguistic diagnostics for language models. Trans. Assoc. Comput. Ling. **8**, 34–48 (2020)
3. J.R. Firth, A synopsis of linguistic theory 1930-1955. Studies in linguistic analysis (1957)
4. Z.S. Harris, Distributional structure. Word **10**(2–3), 146–162 (1954)
5. C.F. Hempelmann, J.M. Taylor, V. Raskin, Application-guided ontological engineering, in *ICAI 2010: Proceedings of the 2010 International Conference on Artificial Intelligence* (Las Vegas NV, July 12–15, 2010), pp. 843–849
6. D. Jurafsky, *Speech & Language Processing*, 3rd edn. (2020)
7. J. Launchbury, A DARPA perpective on artificial intelligence (2019). https://www.darpa.mil/attachments/AIFull.pdf
8. T. Linzen, E. Dupoux, Y. Goldberg, Assessing the ability of LSTMs to learn syntax-sensitive dependencies. Trans. Assoc. Comput. Ling. **4**, 521–535 (2016)
9. K. Misra, A. Ettinger, J.T. Rayz, Exploring BERT's sensitivity to lexical cues using tests from semantic priming, in *Proceedings of the 2020 Conference on Empirical Methods in Natural Language Processing: Findings* (2020), pp. 4625–4635
10. S. Nirenburg, V. Raskin, *Ontological Semantics* (MIT Press, New York, 2004)
11. V. Raskin, C.F. Hempelmann, J.M. Taylor, Guessing vs. knowing: The two approaches to semantics in natural language processing (2010), pp. 642–650

12. V. Raskin, J.M. Taylor, in *The (not so) Unbearable Fuzziness of Natural Language: The Ontological Semantic Way of Computing with Words* (IEEE, 2009), pp. 1–6
13. E. Reif, A. Yuan, M. Wattenberg, F.B. Viegas, A. Coenen, A. Pearce, B. Kim, Visualizing and measuring the geometry of BERT, in *Advances in Neural Information Processing Systems* (2019), pp. 8592–8600
14. M. van Schijndel, T. Linzen, Modeling garden path effects without explicit hierarchical syntax (2018)
15. J.M. Taylor, C.F. Hempelmann, V. Raskin, On an automatic acquisition toolbox for ontologies and lexicons in ontological semantics (2010), pp. 863–869
16. J.M. Taylor, V. Raskin, Fuzzy ontology for natural language, in *2010 Annual Meeting of the North American Fuzzy Information Processing Society* (IEEE, 2010), pp. 1–6
17. J.M. Taylor, V. Raskin, Understanding the unknown: unattested input processing in natural language, in *2011 IEEE International Conference on Fuzzy Systems (FUZZ-IEEE 2011)* (IEEE, 2011), pp. 94–101
18. J.M. Taylor, V. Raskin, in *Conceptual Defaults in Fuzzy Ontology* (IEEE, 2016), pp. 1–6
19. J.M. Taylor, V. Raskin, C.F. Hempelmann, Towards computational guessing of unknown word meanings: the Ontological Semantic approach, in *Proceedings of the Annual Meeting of the Cognitive Science Society*, vol. 33 (2011)
20. W.L. Taylor, "Cloze procedure": a new tool for measuring readability. J. Q. **30**(4), 415–433 (1953)
21. A. Vaswani, N. Shazeer, N. Parmar, J. Uszkoreit, L. Jones, A.N. Gomez, Ł. Kaiser, I. Polosukhin, Attention is all you need, in *Advances in Neural Information Processing Systems* (2017), pp. 5998–6008

Using Fuzzy Sets to Assess Differences in Online Grooming Conversations with Victims, Decoys, and Law Enforcement

Tatiana Ringenberg, Julia Taylor Rayz, and Kathryn Seigfried-Spellar

Abstract The sexual solicitation of minors online is a known and growing problem. The process by which adults attempt to entice minors into sexual situations both online and offline is called grooming. Online grooming is a complex, non-linear process consisting of six interweaving stages. While the stages of grooming are well-understood in the literature, little work has focused on the interplay and overlap of the grooming stages throughout grooming conversations. Previous researchers have identified key aspects of the grooming process by annotating and analyzing grooming conversations. However, traditional annotation methods are unable to express the fuzzy nature of the grooming process, as annotation per line is generally limited to a single crisp grooming stage. This paper addresses the gap in literature by describing an annotation method by which the complexities and differences in grooming may be coded and examined. Three conversation types from the domain are annotated to demonstrate the applicability of the annotation protocol. The method presented in this paper results in fuzzy sets for each of the grooming stages within each of the three conversation types. In this paper, the fuzzy annotation protocol and protocol considerations are discussed. Following this discussion, six chats (two offender-victim, two offender-decoy, and two offender-Law Enforcement) chats are analyzed using the fuzzy annotation protocol in order to demonstrate the applicability of fuzzy sets for this task.

Keywords Fuzzy annotation · Natural language · Online grooming

T. Ringenberg (✉) · J. T. Rayz · K. Seigfried-Spellar
Purdue University, West Lafayette, IN, USA
e-mail: tringenb@purdue.edu

J. T. Rayz
e-mail: jtaylor1@purdue.edu

K. Seigfried-Spellar
e-mail: kspellar@purdue.edu

B. Bede et al. (eds.), *Fuzzy Information Processing 2020*,
Advances in Intelligent Systems and Computing 1337,
https://doi.org/10.1007/978-3-030-81561-5_15

1 Introduction

Tagged corpora are the foundation on which many Natural Language Processing (NLP) tasks are built including part of speech tagging [1], semantic role labeling [2], sentiment analysis [3], and text classification problems [4]. Additionally, researchers often use annotation to identify aspects of the grooming process [5–7]. In general annotation can be completed manually [5], semi-automatically [8], or automatically [9]. Semi-automatic tagging and automatic tagging are often preferable due to the expensive nature of manual tagging [10]. However, for some complex tasks semi-automatic and automatic tagging may not be possible; though time consuming and costly, manual annotation continues to provide more accurate labeling, especially for complex tasks [11]. Within domains outside of NLP, annotation is often referred to as "coding." "Codes" will be used in this paper to describe categories of information which are being annotated.

While manual annotation is preferable for complex tasks, it is not without its shortcomings. Manual annotators face many challenges related to complexity including code dependencies [11], number of codes to annotate [12], and the amount of text corresponding to a code (which is known as code span) [13, 14]. Code independence refers to defining codes in such a way that there is no overlap between codes [11]. This does not mean a situation may not call for multiple codes, but rather the definitions of each code should not overlap. Code independence is crucial for ensuring repeatability of a coding task both within a single annotator and when other annotators are introduced [11]. The choice of the number of codes is difficult for a coding task because annotators can only remember so many codes at once [12]. Code span refers to the determination of where the coding of data starts and stops and has been found to be difficult—due to its inherent fuzziness in communication—but crucial for increasing reliability of annotation [13, 14].

One possible solution for the shortcomings of manual annotation of code spans is the incorporation of fuzzy sets. Fuzzy sets are useful in cases in which events are not sharply definable [15]. Fuzzy sets allow for partial membership of an object or utterance in multiple classes [16], which is useful in our task for identifying multiple stages of grooming. Additionally, creating fuzzy membership functions allows one to examine meaningful properties of various classes such as intersection and containment [16]. Fuzzy sets have historically been used for multiple purposes in NLP including the assignment of textual or visual content to multiple concepts within an ontology [10] and to sentiment analysis [17].

In this study, we explore a methodology for the use of fuzzy set annotations for a complex task—the identification of the child sexual grooming stages in online sexual solicitation conversations within three types of scenarios: predators and victims, predators and decoys, and predators and police officers. The term "victims" refers to minors who are exploited by adult online sexual solicitors; "predators" refers to individuals who sexual solicit a minor; "decoys" refers to adult non-law enforcement individuals who are part of an organization known as Perverted Justice [5]. These individuals pretend to be a minor online in order to identify possible predators. Decoys go through training to identify offender behavior but the training is neither public nor the same as the training received by law enforcement. Finally, decoys often provide the details of a case to law enforcement. Law enforcement refers to policing agencies which pretend to be minors in order to find potential predators online. While both decoys and law enforcement are adults, the training of the individuals is not the same.

Child sexual grooming, hereafter referred to grooming, is the process by which an adult uses deceptive trust and friendship formation in order to lure a minor into engaging in sexual episodes [18]. The online grooming process consists of five overlapping steps including friendship forming, relationship forming, exclusivity, risk assessment, and the sexual stage [19]. A sixth stage known as the meeting or conclusion stage is often added as well [5, 6]. The grooming process is both transient [18] and iterative [19] with stages often overlapping and varying in term usage [7]. Each line in a grooming conversation may have full, partial, or no membership in multiple grooming stages. As a result, the membership of a given line is not sharply definable. With traditional annotation methods [11], identification of the most appropriate stage would be difficult due to this complexity and overlap of stages. Additionally, fine-grained detail related to the grooming process would be lost. With an appropriate annotation protocol, fuzzy sets may provide a means for handling the complexity of grooming stages by allowing annotators to indicate the degree to which a line in a conversation is a member of each of the grooming stages. Understanding the overlap and the content of the overlap in these stages could improve automatic detection methods as well as provide a better understanding of how offenders adjust strategies when engaging in manipulation.

The aim of this study is to fill the gap in literature by describing a method for annotating grooming stages using fuzzy sets for the three types of conversations in this domain: offender-victim, offender-decoy, and offender-law enforcement.

Ringenberg et al. [20] described a method for identifying fuzzy sets for assessing risk in offender conversations with decoys. This study extends the study by examining not risk but rather the stages of the grooming process as defined in [5, 19]. Further, in this protocol fuzzy sets are annotated not just for decoys but also for victims and police officers. The authors believe using fuzzy sets to annotate the stages of the grooming process may highlight some of the differences in the progression of the chats posited in previous research including the openness of the officer to quickly meet [21], differences in responses to sexual solicitations [21], and differences in the progression of a relationship [21]. Additionally, the creation of fuzzy sets could help researchers determine how investigative decisions made by officers and decoys mold and change the progression of a conversation [22].

In this paper, we first describe our codebook for the grooming stages and how fuzzy codes are to be applied to the lines of a chat conversation. Then, we discuss the results of annotating a small corpus of two offender-victim conversations, two offender-decoy conversation, and two offender-officer conversations using the protocol. Three of the conversations in this study were annotated by two annotators to ensure reliability of the coding scheme. Using the protocol, we derive the shapes of the fuzzy membership functions for each stage in each conversation type and discuss the progression and intermingling between the stages in each of the chats. The goal

Table 1 Names and descriptions of grooming stages

Stage	Content
Friendship forming	Basic exchange of information including name, profile information, general location, age, gender, and family details [5, 19]
Relationship forming	Content related to interests including hobbies [5, 19], family interactions [19], friends [19] and school [5, 19]
	Understanding of difficulties the child might have [25]
	Compliments, image exchanges, and activities of victim [5]
Exclusivity	References to love, exclusiveness, strong emotions, or a future together [5]
	Trust building content [5, 19] such as secrets and self disclosures [19]
Risk Assessment	Questions around location and accessibility [19, 25]
	Attempts to hide evidence (delete things), confirming the age difference, and questions related to location of family [5]
Sexual	Sexually-oriented compliments, images, and bodily descriptions [5]
	Progressively graphic messages and hypothetical questions and descriptions, questions about sexual past, and fantasy enactment [19]
Meeting	Suggestions of meeting, hypothetical of meeting, plans to meet [5, 26]
	Ensuring child will come alone [5]

of this work is to provide a use case for the coding protocol as well as discuss some of the challenges faced when using fuzzy sets to annotate grooming. As such, emphasis is placed on the chosen methodology and the challenges experienced there within.

2 Relevant Literature

Grooming refers to the manipulative communication process used by adults to entice minors into engaging in sexual activity through the use of trust and friendship [18]. While there are multiple models of grooming [19, 23, 24], the online grooming process was originally identified by O'Connell [19]. A summary of the content present in each of the stages of the online grooming process is provided in Table 1.

Part of the difficulty in identifying stages of the grooming process stems from the non-linearity of the stages within a chat conversation [5–7, 19]. For instance, researchers have noted sexual conversation or inquiry will sometimes begin early on in the chat [6, 7] along with risk assessment [7]. Finally, relationship forming and references to meetings and activities to be performed during a meeting occur throughout chat conversations [5].

Several factors may affect sequencing including the goal of the offender [21, 27, 28] and the category of the victim (victim, decoy, or officer) [5, 21, 22, 29–31]. The goal of the offender is frequently divided into contact-driven and fantasy-driven [21, 27, 28]. Contact driven offenders aim to meet the minor in the physical world while fantasy-driven offenders aim to engage in either cybersex or merely the grooming process itself [21].

Law enforcement, decoys, and victims are partially thought to be different due to differences in motivation [21]. The goal of law enforcement and decoys is to gather evidence against the potential offender to which they are speaking, often in a timely manner [21]. As a result, researchers posit law enforcement and decoys are more willing to respond positively to uncomfortable messages than an at-risk teen would be [21, 29]. Researchers have noted some differences in conversations may stem from the fact that offenders speaking to victims would have to deal with actual rejections and distrust from the victims and thus would engage in more drastic measures to gain compliance [31]. Finally, law enforcement officers and the decoys who chatted in this study had prior training related to the grooming process, which gave them an understanding and context for these conversations which the victim would not have [21]. Such context could influence the offender's communication as well [22].

3 Methodology

The authors annotated stages of the grooming process as shown in Table 1. The descriptions in Table 1 were chosen for their consistency with the descriptions as provided by O'Connell's model of online grooming [19]. The meeting code was added in addition to the five stages due to its support in literature [5, 6, 19] and its possible relevance in differentiating between victims, law enforcement, and decoys.

In addition to the main stages of the online grooming process and meeting, a code for Sexual Stage Non-Consent was added to the study to account for attempts to depart from the sexual stage. For instance, in one of the conversations the solicitor asked for sexual images of the officer and the officer stated they did not feel comfortable providing the image.

When annotating a corpus, it is necessary to ensure the codes themselves are independent [11]. While the codes for each grooming stage are independent, the chat lines to which they are applied are not. The easiest case for annotators occurs when lines only pertain to a single theme. For instance, chat lines in which the participants are discussing non-sexual interests would be labeled as having full membership in the relationship forming stage.

Full membership of a single stage set is the most straightforward of situations. However, many lines are not as easily discernible. As a result, applying a single code to each line is not representative of the process; a line may be part of the set for more than one stage.

We believe a chat may be a member of multiple stages in two situations. First, some lines may contain a response to the previous line while also including a change of topic. We found this scenario often occurred in situations in which the participants were moving between the sexual and risk assessment stages. Second, some instances of the stages may subsume others. The authors found one of the instances of this to occur when participants were discussing meeting and would incorporate the risks and mitigation of the risks into the discussion of meeting. Thus, the entire discussion would be coded with full membership in the meeting stage and the lines which also pertained to risk and mitigation would additionally be coded with full membership in risk assessment.

In addition to multiple codes occurring within the same line, we have found the membership of a line to a particular stage can be less than full. Again, this appears to occur in two situations. First, when an individual provides a non-committal response to the other participant in the chat, such as a smiley or utterance such as "hmm". In this case, the response does pertain to the stage but does not provide substance. Non-committal lines were labeled by annotators as partial membership. The exact degree of membership was later calculated per line using trapezoidal membership functions as described below. Second, lines often serve to prompt a theme, even though those lines are not directly on-theme. For instance, participants may prompt a discussion of sexual history with an inquiry of whether or not they could ask a personal question. The act of asking about a personal question is a gentle lead-in to a sensitive topic and thus is captured by annotating the line as having partial membership to the sexual stage. The exact degree of membership is later calculated using a trapezoidal membership function.

In addition to rules surrounding the labeling of membership of each chat line, the authors established some general guidelines for annotation. To minimize drift in annotation, a single chat was coded for the grooming stages at one time. Coding large amounts of data has been associated with decreased reliability in coding practices and thus the annotator took steps to limit exhaustion due to coding [12].

The annotators also adhered to the following:

- Only annotate lines pertaining to a grooming stage.
- When labeling a stage, do not label a line as partial membership to that stage if the line occurs between two lines of full membership to that stage.
- Review the codebook before each session.
- Note any difficulties during the process.

Once all of the chat conversations were labeled for the grooming stage sets, fuzzy membership functions were created for each stage. Figure 2a illustrates the components of the trapezoidal fuzzy membership function, discussed below.

As in Ringenberg et al. [20], a trapezoidal membership function similar to that shown in Fig. 1 was chosen to represent the shape of each stage within the chat. The trapezoidal shape was chosen due to the tendency of stages to intermingle and due to the partial membership of lines before and after groups of lines with full membership in a stage.

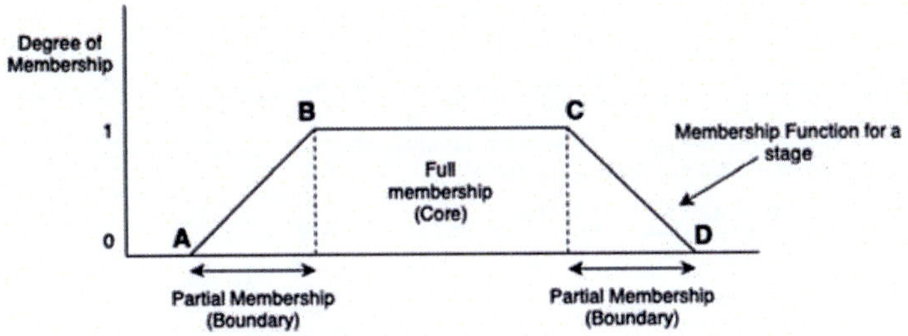

Fig. 1 Partial fuzzy membership function created for the online grooming stages

Table 2 Breakdown of membership for excerpt

Message	Value in Fig. 1
Solicitor: i washed my car all nice and pretty today too	A
Decoy: ohhh i would have washed ur car too :)	
Solicitor: well thats hard for u to do when u cant get away	B
Solicitor: hehe	
Decoy: i kno	
Decoy: :(	C
Solicitor: :-/	
Decoy: hehe	
Decoy: how do u get those cute froggie smilies?	D

Membership functions were created by first pulling all "chunks" for a stage within a chat conversation. The starting and tail boundaries of the membership functions were calculated using the partial membership codes at the start and end of each chunk This is exemplified in Table 2.

4 Preliminary Analysis

The two annotators initially coded three conversations. Interrator reliability for each of the stages was calculated with simple agreement between the annotators on all lines labeled "none," "partial," or "full" membership. In the future, a more complex measure of agreement such as Cohen's Kappa will be utilized [32]. Agreement for each stage was greater than 80%, though we acknowledge such agreement could decrease when chance agreement is taken into consideration. Lowest agreement occurred in the sexual stage (82.9%) and the risk assessment stage (83.9%). Disagreements were discussed and adjustments to the codebook were made before the first annotator annotated the three remaining chats.

Table 3 Descriptive information for solicitor conversations

Chat number	Chat type	Number of lines	Total stage chunks	Multi-membership (%)
Chat 1	Solicitor-Officer	308	41	18.6
Chat 2	Solicitor-Officer	102	23	42
Chat 3	Solicitor-Decoy	481	52	15.7
Chat 4	Solicitor-Decoy	1358	112	4.2
Chat 5	Solicitor-Victim	596	53	14.9
Chat 6	Solicitor-Victim	333	41	15.4

The six chats in this study came from the three conversation types discussed earlier in this paper. Each of the six conversations involved a unique solicitor and a unique victim, decoy, or law enforcement. All law enforcement members are from the US as are all of the chats.

Following annotation and before the creation of fuzzy membership functions, the authors combined chunks which were within three lines of one another to reflect the continuous nature of the grooming process. The membership functions were then generated based on the resulting annotations. Table 3 summarizes the number of lines, number of chunks, and percentage of the chat with lines having multiple membership. The membership functions produced by this protocol allow visualization of the progression of conversations interaction of stages. Some of the individual differences in each chat are discussed below.

Figure 2a shows the interaction of all of the stages for Chat 1 between an offender and law enforcement. The chat begins with a combination of friendship and relationship forming behaviors with the incorporation of some sexual discussion early on. This is consistent with Chat 2 in which the conversation begins with sexual comments and transitions into friendship forming. The use of sexual discourse at the beginning of the chat appears to be present in both the offender-law enforcement and offender-victim chats, but not as prevalent in the two offender-decoy chats. Due to the sample size of this study, it is unclear whether this is a pattern or a series of unique circumstances resulting from individual differences. However, future research could use the protocol presented in this paper to further examine the use of, and responses to, the sexual stage at the start of a chat (Fig. 3).

Chat 2 (Fig. 2b), begins with the sexual stage and transitions into friendship forming. It does not appear that this conversation focuses on forming a relationship or trust, as the chat is dominated by sexual content, both consensual and non-consensual, with few relationship and friendship forming clusters. Chat 3 (Fig. 2c), on the other hand, contains a mixture of sexual and non-sexual content. The chat begins with friendship forming and transitions into relationship forming, which continues to occur in short bursts throughout the chat. While evidence of the sexual stage exists early in the chat, sexual discussions occur with greater frequency as the chat progresses.

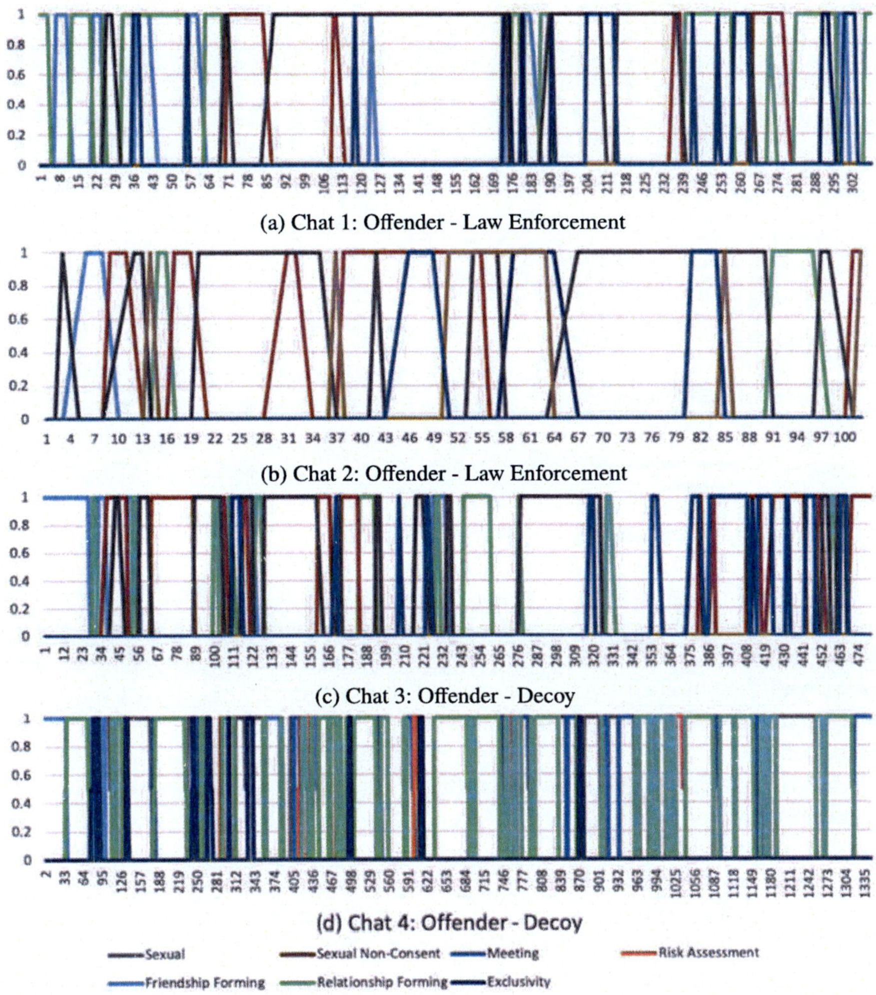

Fig. 2 Fuzzy sets visualization for offender-law enforcement and decoy chats

While friendship and relationship forming are prevalent, it does not appear exclusivity occurs frequently within the chat. In this chat, risk assessment and the sexual stage have noticeable overlap along with risk assessment and meeting.

Chat 4 (Fig. 2d) appears to have consistent risk assessment, meeting, and relationship stages throughout. This conversation appears to be a mixture between the stages with no specific emphasis on any given stage. While Chat 4 contained the stages mixed through the entire chat conversation, Chat 5 is predominantly sexual. The sexual stage chunks of the chat appear to be long and thus prolonged sexual conversation appears to occur.

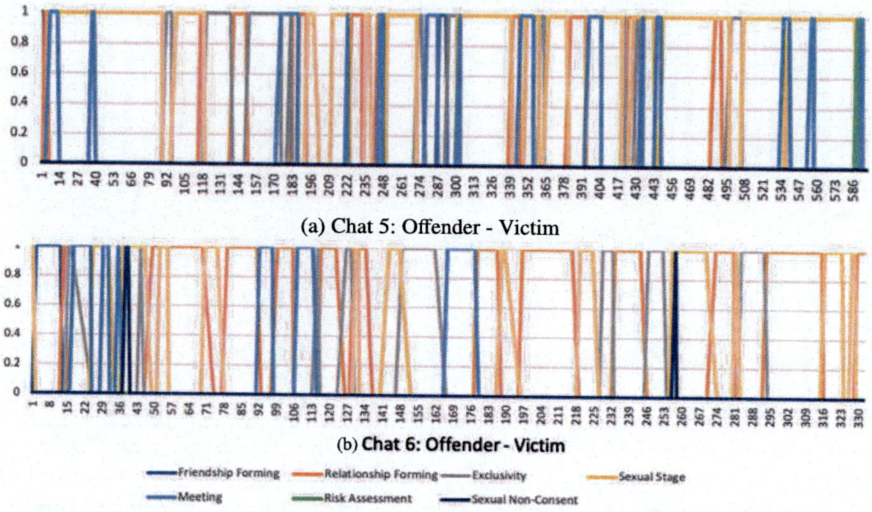

(a) Chat 5: Offender - Victim

(b) Chat 6: Offender - Victim

Fig. 3 Fuzzy sets visualization for offender-victim chats

The sexual stage was often introduced through the discussion of meeting. Either the offender or the victim frequently begins a line of sexual questioning or scenarios with a reference to visiting one another. Additionally, the sexual stage was reinforced with relationship forming and exclusivity often occurring between sexual discussions.

Chat 6 (Fig. 2) is one of the few chats which contained sexual non-consent content. It appears non-consensual discussions occurred twice in the chat. Following the first non-consensual sexual discussion, it appears an attempt at establishing exclusivity between the pair was made, followed by a reinforcement of the relationship.

However, it appears the second non-consensual sexual discussion was not followed by an attempt to reinforce the relationship but additional sexual discussion. The sexual stage and relationship forming, as in Chat 5, appear closely tied together. In Chat 5, larger chunks of sexual discussion appeared to occur while Chat 6 contains short bursts of sexual discussion.

5 Conclusions and Future Work

In this paper, the authors outlined a manual annotation protocol for using fuzzy sets to identify the membership of chat lines in the grooming process of the three chat types (offender-victim, offender-decoy, offender-solicitor). Two annotators labeled three chats and compared agreement to ensure the repeatability of the protocol.

The authors believe the creation of fuzzy sets make the sequence of, and relationships between, stages in the grooming process more apparent than traditional manual annotation methods. In a process as complicated as grooming, detailed information on the interaction of stages within the process is crucial.

While the protocol has many benefits for understanding fine-grained details about the grooming process, there are also some challenges in using the protocol. As in traditional annotation methods [13, 14], identification of code span is difficult. The annotators found difficulties in code span manifested in the identification of partial membership. Often, the authors were able to identify the same span of text contained a given stage but the agreement on what portion of the span constituted partial membership was a source of disagreement early-on in annotation. This was handled by adjusting the codebook with more examples and with explicit guidelines on partial membership.

In future work, the authors will apply the fuzzy set annotation protocol to additional themes which are found within the grooming process. The authors will also use the protocol to examine similarities and differences in the three chat types. In doing so, the authors hope to increase the understanding of the flow of the grooming process as a whole but also an understanding of the differences between victims, decoys, and law enforcement.

Acknowledgements The authors would like to acknowledge Lin Sun Fa for coding three of the chats as the second annotator. This research was partially supported by the DOJ grant "Strengthening Investigative Tools for Technology For Combating Child Sexual Exploitation."

References

1. T. Charoenporn, V. Sornlertlamvanich, H. Isahara, Building a large thai text corpus-part of speech tagged corpus: Orchid, in *Natural Language Processing Pacific Rim Symposium*. Citeseer (1997), pp. 509–512
2. E. Kim, R. Klinger, Who feels what and why? Annotation of a literature corpus with semantic roles of emotions (2018), pp. 1345–1359
3. M. Abdul-Mageed, M.T. Diab, Awatif: a multi-genre corpus for modern standard arabic subjectivity and sentiment analysis. in *Language Resources and Evaluation Conference*, vol. 515 (2012), pp. 3907–3914
4. J. Golbeck, Z. Ashktorab, R.O. Banjo, A. Berlinger, S. Bhagwan, C. Buntain, P. Cheakalos, A.A. Geller, R.K. Gnanasekaran, R. R. , Gunasekaran et al., "A large labeled corpus for online harassment research," in (2017), pp. 229–233
5. A. Gupta, P. Kumaraguru, A. Sureka, Characterizing pedophile conversations on the internet using online grooming (2012)
6. J.A. Kloess, S. Seymour-Smith, C.E. Hamilton-Giachritsis, M.L. Long, D. Shipley, A.R. Beech, A qualitative analysis of offenders' modus operandi in sexually exploitative interactions with children online. Sexual Abuse **29**(6), 563–591 (2017)
7. P.J. Black, M. Wollis, M. Woodworth, J.T. Hancock, A linguistic analysis of grooming strategies of online child sex offenders: implications for our understanding of predatory sexual behavior in an increasingly computer-mediated world. Child Abuse & Neglect **44**, 140–149 (2015)
8. P. Viszlay, J. Staš, T. Koctúr, M. Lojka, J. Juhár, An extension of the slovak broadcast news corpus based on semi-automatic annotation (2016), pp. 4684–4687
9. C. Fabian Benitez-Quiroz, R. Srinivasan, A.M. Martinez, Emotionet: an accurate, real-time algorithm for the automatic annotation of a million facial expressions in the wild (2016), pp. 5562–5570

10. C. De Maio, G. Fenza, M. Gallo, V. Loia, S. Senatore, Formal and relational concept analysis for fuzzy-based automatic semantic annotation. Appl. Intell. **40**(1), 154–177 (2014)
11. P.S. Bayerl, H. Lüngen, U. Gut, K.I. Paul, Methodology for reliable schema development and evaluation of manual annotations (2003)
12. P.S. Bayerl, K.I. Paul, What determines inter-coder agreement in manual annotations? A meta-analytic investigation. Comput. Ling. **37**(4), 699–725 (2011)
13. V. Stoyanov, C. Cardie, Annotating topics of opinions (2008), pp. 3213–3217
14. M. Bada, M. Eckert, D. Evans, K. Garcia, K. Shipley, D. Sitnikov, W.A. Baumgartner, K.B. Cohen, K. Verspoor, J.A. Blake et al., Concept annotation in the craft corpus. BMC Bioinf. **13**(1), 161 (2012)
15. L.A. Zadeh, Probability measures of fuzzy events. J. Math. Anal. Appl. **23**(2), 421–427 (1968)
16. L.A. Zadeh, Fuzzy sets. Inf. Control **8**(3), 338–353 (1965)
17. M. Haque, et al., Sentiment analysis by using fuzzy logic (2014), arXiv:1403.3185
18. A. Gillespie, Grooming definitions and the law. New Law J. **154**(7124), 586–587 (2004)
19. R. O'Connell, *A Typology of Child Cybersexploitation and Online Grooming Practices* (University of Central Lancashire, Cyberspace Research Unit, 2003)
20. T.R. Ringenberg, K. Misra, J.T. Rayz, Not so cute but fuzzy: estimating risk of sexual predation in online conversations, in *2019 IEEE International Conference on Systems, Man and Cybernetics (SMC)* (2019), pp. 2946–2951
21. P. Briggs, W.T. Simon, S. Simonsen, An exploratory study of internet-initiated sexual offenses and the chat room sex offender: has the internet enabled a new typology of sex offender? Sexual Abuse **23**(1), 72–91 (2011)
22. D. DeHart, G. Dwyer, M.C. Seto, R. Moran, E. Letourneau, D. Schwarz-Watts, Internet sexual solicitation of children: a proposed typology of offenders based on their chats, e-mails, and social network posts. J. Sexual Aggress. **23**(1), 77–89 (2017)
23. L.N. Olson, J.L. Daggs, B.L. Ellevold, T.K. Rogers, Entrapping the innocent: toward a theory of child sexual predators' luring communication. Commun. Theory **17**(3), 231–251 (2007)
24. T. Ward, K. Louden, S.M. Hudson, W.L. Marshall, A descriptive model of the offense chain for child molesters. J. Interpers. Viol. **10**(4), 452–472 (1995)
25. S. Aitken, D. Gaskell, A. Hodkinson, Online sexual grooming: exploratory comparison of themes arising from male offenders' communications with male victims compared to female victims. Deviant Behav. **39**(9), 1170–1190 (2018)
26. G.M. Winters, L.E. Kaylor, E.L. Jeglic, Sexual offenders contacting children online: an examination of transcripts of sexual grooming. J. Sex. Aggress. **23**(1), 62–76 (2017)
27. M.M. Chiu, K.C. Seigfried-Spellar, T.R. Ringenberg, Exploring detection of contact vs. fantasy online sexual offenders in chats with minors: statistical discourse analysis of self-disclosure and emotion words. Child Abuse & Neglect **81**, 128–138 (2018)
28. H.L. Merdian, N. Moghaddam, D.P. Boer, N. Wilson, J. Thakker, C. Curtis, D. Dawson, Fantasy-driven versus contact-driven users of child sexual exploitation material: offender classification and implications for their risk assessment. Sex. Abuse **30**(3), 230–253 (2018)
29. D. Schneevogt, E. Chiang, T. Grant, Do perverted justice chat logs contain examples of overt persuasion and sexual extortion? a research note responding to chiang and grant (2017, 2018). *Language and Law= Linguagem e Direito* **5**(1), 97–102 (2018)
30. F.-C. Chang, C.-H. Chiu, N.-F. Miao, P.-H. Chen, C.-M. Lee, J.-T. Chiang, Predictors of unwanted exposure to online pornography and online sexual solicitation of youth. J. Health Psychol. **21**(6), 1107–1118 (2016)
31. E. Chiang, T. Grant, Deceptive identity performance: Offender moves and multiple identities in online child abuse conversations. Appl. Ling. **40**(4), 675–698 (2019)
32. J. Cohen, A coefficient of agreement for nominal scales. Educ. Psychol. Meas. **20**(1), 37–46 (1960)

An Optimized Intelligent Fuzzy Fractional Order TID Controller for Uncertain Level Control Process with Actuator and System Component Uncertainty

Himanshukumar Patel and Vipul Shah

Abstract This article is about the design of controllers for Conical Tow-tank Non-interacting Level (CTTNL) system in simulation. Local linearization around the equilibrium point has been done for the nonlinear CTTNL system to obtain a linearized model transfer function. In this study, the design of fuzzy tilted-integral-derivative (FTID) controller has been proposed for the CTTNL system and its performance is compared with conventional proportional-integral-derivative (PID) controller. In this study, FTID controller parameters have been optimized through genetic algorithm (GA) and those set of values have been employed for the design of PID controller. A performance comparison between FTID and PID controller is then investigated. The analysis shows the superiority of FTID controller over PID controller in terms of IAE, ISE, ITAE and ITSE integral errors. The transient and steady state performance of the FTID controller are superior as compared to conventional PID controller. In future, the FTID controller fault tolerance capability tested on CTTNL system subject to actuator and system component (leak) faults. The detailed study of robustness in presence of model uncertainties will be incorporated as a scope of further research.

1 Introduction

Fractional-order PID (FOPID) was introduced in 1999 by Podlubny as a generalization of trivial PID controllers [1]. The transfer function of the ideal FOPID, also known as $PI^{\lambda}D^{\mu}$, is defined as [1]:

$$C_{FOPID}(s) = K_p + \frac{K_i}{s^{\lambda}} + K_d s^{\mu} \tag{1}$$

H. Patel (✉) · V. Shah
Department of Instrumentation and Control Engineering, Faculty of Technology, Dharmsinh Desai University, Nadiad 387001, Gujarat, India
e-mail: himanshupatelp32@gmail.com; himanshupatel.ic@ddu.ac.in

B. Bede et al. (eds.), *Fuzzy Information Processing 2020*,
Advances in Intelligent Systems and Computing 1337,
https://doi.org/10.1007/978-3-030-81561-5_16

where K_p, K_i, $K_d \in \mathbb{R}$ and λ, $\mu \in \mathbb{R}^+$ are unknown parameters of controller to be tuned.Where $s \in \mathbb{N}$. See, for example, [2] for the possible time-domain elucidation of the fractional powers of s in (1) and their properties. Taking into consideration the reality that FOPID has five parameters to tune, two additionally then the classical PID, and hence it is expected that it leads to a higher performance in contrast to PID particularly in dealing with problems with complex control objectives [3]. Numerous successful applications of FOPID controllers have been described in the literature so far. A few examples are motion control [4], unmanned aerial vehicle [5] and path tracking control of tractors [6].

Another fractional-order controller which is came out from the fractional calculus and nearly related to FOPID and explained in this article is the Tilt-Integral-Derivative (TID). The transfer function of TID is defined as [7]:

$$C_{TID}(s) = \frac{K_p}{s^{1/n}} + \frac{K_i}{s} + K_d s \tag{2}$$

where again K_p, K_i, and K_d are unknown real parameters to be calculated $s \in \mathbb{N}$, and n is an unknown positive integer often considered equal to 2 or 3 before tuning other parameters [7]. The s is the Laplace parameter which is defined $s \in \mathbb{N}$. The optimal output of TID controller are completely depends on tuned parameters value.A few properties of TID have been investigated in [8]. A technique for tuning its parameters is also conferred therein.

In the past, an ample variety of methods have been suggested for tuning the parameters of FOPID. Such as,tuning the parameters based on open-loop shaping [3, 9], Bode's ideal transfer function [10, 11], minimization of performance indices like Integral Square Error (ISE) and Integral Absolute Error (IAE) [12, 13] Concurrent improvement of phase and gain margin to the desired values [14] and fractional-order root-locus method [15] can be found in the literature. It is worth observing that most of all the methods developed so far for tuning FOPID controllers are based on solving a kind of optimization problem. Some investigators have used meta-heuristic optimization algorithms for solving such problems; examine for example [16–18]. This article, we used a Genetic Algorithm (GA) to determine optimum controller parameters value. In resent years artificial intelligent techniques are very popular and effectively applied on many nonlinear level control systems [19–23].

The objective of this paper is to compare FTID and PID controller performance for nonlinear level control process, and controller parameters are tuned using GA. For this reason first a novel fuzzy tilt-integration-derivative controller, which is defined by adding a fractional differentiator of fixed order to the classical TID, is introduced. The suggested approach makes it possible to use the improvement of convex optimization and fractional-order operators to solve a control problem.The main contributions of the article are printed as follows:

- Fault tolerant controller is developed using FTID and PID controller
- Fractional differentiator is added with two additional scaling parameters β_1 and β_2
- Optimal parameter values have been tuned using Genetic Algorithm

- By adding fractional order in conventional controller, degree of freedom is increased and hence it provides more flexibility and opportunity to better adjust the dynamical properties of the nonlinear control system.
- Computational burden is reduced by taking two input one output Fuzzy Inference (FIS) System
- The proposed controller is able to accommodate actuator and system component (leak) faults individually
- To established the effectiveness of the proposed FTID controller no-interacting nonlinear level control process (CTTNL) is taken
- The overall open loop response of the CTTNL system is split into 5 linear operating regions

The rest of this paper is organized as the following. The process description and mathematical-model of the CTTNL system for the particular region present in Sect. 2. Fuzzy TID controller structure and tuning parameters using GA presented in Sect. 3. Simulation results of FTID are compared with a conventional PID controller which are presented in Sects. 4, and 5 concludes the paper.

2 Process Description

The CTTNL system as shown in Fig. 1 is based on the two-tank benchmark problem that is used by a number of researchers.

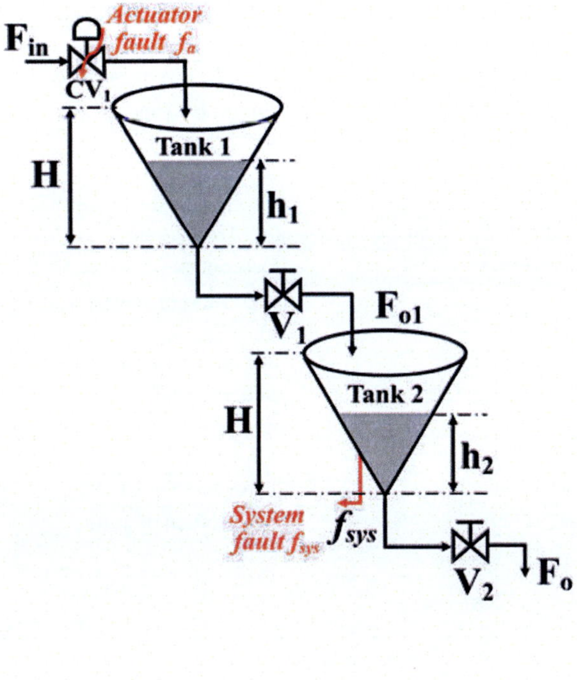

Fig. 1 Prototype structure of the CTTNL system with two faults provision [24–26]

The CTTNL system is regarded as a setup for investigating nonlinear single-input single-output (SISO) feedback control schemes hypothetically. CTTNL system subsist of two exact conical tanks (TANK 1 and TANK 2), one water pump (PUMP 1) that delivers the liquid flows F_{in} to TANK 1 through control valves CV_1. The two conical shape two tanks construct as non-interacting processes, in that the outlet flow rate of tank 1 F_{o1} is the inlet flow rate of tank 2. F_{o1} and F_o are the two output flows from TANK 1 and TANK 2 through manual operated control valves V_1 and V_2 with valve coefficients γ_1 and γ_2 respectively. The γ_1 and γ_2 are adjustable coefficients characterizing the resistance of corresponding valves opening orifice. The valve coefficient γ_i may be evaluated using Eq. (3),

$$\gamma_i = S_i a_i \sqrt{2g} \quad (i = 1, 2) \tag{3}$$

where a_i are the cross-sectional areas of valves and S_i is the discharge coefficient, which is assumed unity. The operating parameters of TCTILS are presented in Table 1. In CTTNL process, level h_2 in TANK 2 is considered as an output variable and F_{in} is considered as manipulated variable. The mathematical model of CTTNL process is given by Eqs. (4) and (5).

$$\frac{dh_1}{dt} = \left[\frac{F_{in} - \frac{1}{3} h_1 \frac{dA(h_1)}{dt} - \gamma_1 \sqrt{h_1}}{\frac{1}{3}\pi R^2 \frac{h_1^2}{H^2}} \right] \tag{4}$$

$$\frac{dh_2}{dt} = \left[\frac{F_{o1} - \frac{1}{3} h_2 \frac{dA(h_2)}{dt} - \gamma_2 \sqrt{h_2}}{\frac{1}{3}\pi R^2 \frac{h_2^2}{H^2}} \right] \tag{5}$$

where, $A(h_1)$ cross-sectional area of TANK 1 at h_1 (cm^2), $A(h_2)$ cross-sectional area of TANK 2 at h_2 (cm^2).

The mathematical model of the CTTNL system with actuator f_a and system component f_{sys} (leak) faults given by Eqs. (6) and (7):

Table 1 Operating parameters of CTTNL system

Parameter	Description	Value
R	Top radius of conical tank	19.25 cm
H	Maximum height of TANK1 and TANK2	72 cm
F_{in}	Maximum Inflow to TANK 1	132.89 cm^3/s
γ_1	Valve co-efficient of V_1	35 cm^2/s
γ_2	Valve co-efficient of V_2	18.36 cm^2/s
g	Acceleration of gravity	981 cm/s^2

Table 2 Operating range for each linearized regime of CTTNL

Regimes	F_{in} (LPH)	h_2 (cm)
Region 1	0-104.0	0–5.78
Region 2	104.0–205.0	5.78-12.19
Region 3	205.0–306.5	12.19–21.95
Region 4	306.5–406.5	21.95–34.85
Region 5	406.5–500.0	34.85–50.00

$$\frac{dh_1}{dt} = \left[\frac{\alpha F_{in} - \frac{1}{3}h_1\frac{dA(h_1)}{dt} - \gamma_1\sqrt{h_1}}{\frac{1}{3}\pi R^2\frac{{h_1}^2}{H^2}}\right] \tag{6}$$

$$\frac{dh_2}{dt} = \left[\frac{F_{o1} - \frac{1}{3}h_2\frac{dA(h_2)}{dt} - \gamma_2\sqrt{h_2} - f_{sys}}{\frac{1}{3}\pi R^2\frac{{h_2}^2}{H^2}}\right] \tag{7}$$

where $\alpha \in [0, 100]\%$ it is a adjustable factor for actuator fault f_a in (CV_1) and $f_{sys} \in [0, 100]\ \text{cm}^3/\text{s}$ is a leak flow-rate from Tank 2.

Since, the cross-sectional area of the conical tank has varied with respect to change in operating point, the process parameters will also change. Hence, the overall CTTNL nonlinear process ware divided into 5 operating regions, in that particular operating region the model gives linear response. However, the controller parameters are different for different operating region. Therefor, we consider operating region 2 for CTTNL process and destined FTID and PID controllers. The operating range of the CTTNL system is given in Table 2 which is based on open loop response of the CTTNL.

From the different operating range of CTTNL system, we consider operating **region 2** of CTTNL system. Now the various steps involved in designing FTID and conventional PID controller are listed below.

step 1: Select the Design Level of Operation (DLO)
step 2: Obtain the T.F model for DLO (2)
step 3: Obtain optimal FTID & PID controller parameters using GA

The center point of second linearized regime for h_2 (refer Eq. 4—$F_{in} = 154.5$ LPH, $h_2 = 5.985\,\text{cm}$ is considered as Design Level of Operation (DLO (2)). The transfer function model for specified DLO range 2 is obtained and is reported Eq. 8.

$$G_{DLO(2)} = \frac{0.05279}{s^2 + 42.18s + 0.7589} \tag{8}$$

It is reasonable to design the FTID and PID controller based on the middle level of operation (the center point of second linearized regime) because this will yield a compromise in performance over the range of dynamic behavior. Once the model of the CTTNL system is found then, parameter of the controllers is tuned using genetic algorithm (GA). The next section of the article presents proposed FTID structure and GA steps for tuning the optimum parameters.

3 Fuzzy TID Controller

In a tilt integral derivative (TID) controller, the proportional component (K_P) of PID controller is multiplied with a tilted component having a transfer function $s^{-1/n}$. The derivative mode of PID controller, however advances the stability of the system and raises the pace of the controller response, but also produces undesirable sized control inputs to the plant.

The transfer function of TID is defined as [7]:

$$C_{TID}(s) = \frac{K_p}{s^{1/n}} + \frac{K_i}{s} + K_d s \tag{9}$$

where again K_p, K_i, and K_d are unknown real parameters to be calculated, s is the Laplace operator $s \in \mathbb{N}$ and n is an unknown positive integer often considered equal to 2 or 3 before tuning other parameters [7].

In this article, we proposed novel fuzzy TID controller and the schematic is depicted in Fig. 2. The β_1 and β_2 is adjustable or tuning parameters of the fuzzy fractional differentiator which are reduce computational burden of FIS as well as adding more degree of freedom for controller. Also the asymmetrical membership function for input and output of the FLC is depicted in Fig. 3. In the FLC the Membership function described by linguistic variable name like 'NB-Negative Big', 'NS-Negative Small', 'NL-Negative Low', 'Z-Zero', 'PL-Positive Low', 'PS-Positive Small' and 'Positive Big'.

The "IF-THEN" rules for the designed FLC is presented in Table 3.

3.1 Genetic Algorithm

GA pursue the natural progression for Drawin theory "The survival of the fittest". In 1975 John Holland proposed the GA for the first time In GA, the smallest unit of data is the gene, the data which carried by a set of genes is called individual, and

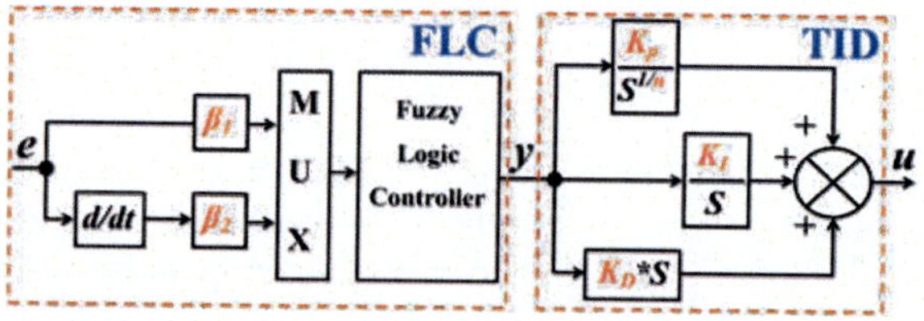

Fig. 2 Structure of the proposed Fuzzy TID controller

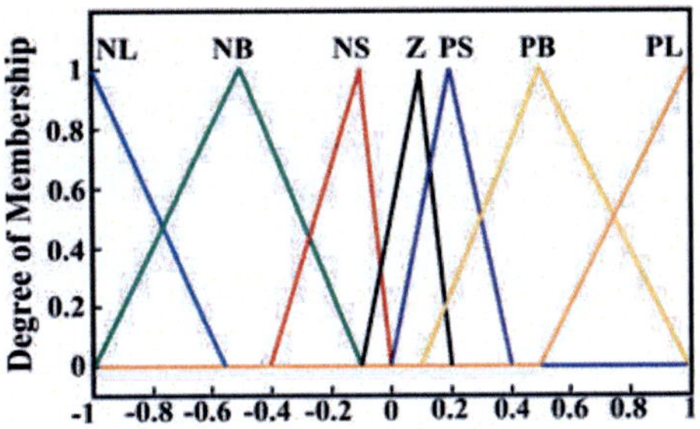

Fig. 3 Asymmetrical membership functions for e, $\dot{e}$ and FLC output

Table 3 Rule base for e, $\dot{e}$ and FLC output

e	*Derivative of e*						
	NB	NS	NL	Z	PL	PS	PB
NB	NL	NL	NL	NB	NB	NS	Z
NS	NL	NL	NB	NB	NS	Z	PS
NL	NL	NB	NB	NS	Z	PS	PB
Z	NB	NB	NS	Z	PS	PB	PB
PL	NB	NS	Z	PS	PB	PB	PL
PS	NS	Z	PS	PB	PB	PL	PL
PB	Z	PS	PB	PB	PL	PL	PL

population is a set of individuals, synonyms are chromosome and individual. GA is used to determine the optimum parameters of the fuzzy TID and PID controller to achieve the desired response of the system. The process of optimization is calculated by minimizing of objective function (J_s). In the design of an optimization technique based controller, a performance index or criterion (J_s) is required as per the desired specifications and limits. The performance criteria frequently used in the controller design are integral of squared error (ISE), integral of time multiplied squared error (ITSE), integral of absolute error (IAE) and integral of time multiplied absolute error (ITAE).

$$J_s = ISE + IAE + ITSE + ITAE \rightarrow min \tag{10}$$

The mathematical expression for the integral errors are depicted by Eq. (11) to Eq. (14)

$$IAE = \int |e|\, dt \tag{11}$$

$$ISE = \int e^2 dt \tag{12}$$

$$ITAE = \int t\,|e|\,dt \tag{13}$$

$$ITSE = \int te^2 dt \tag{14}$$

Fitness function is designed as:

$$f_g = J_{max} + J_{min} - J_g \tag{15}$$

where is J_{max} , J_{min} the largest and the smallest value of J_s respectively, and for the current population value of J denoted by J_g. f_g is the fitness function which is calculate for each population.

The optimum FTID and PID controller parameters are tuned using four integral Error as a fitness function. The steps involved in tuning FTID and PID controller parameters using GA for CTTNL are listed below.

step 1: Initialize GA parameters and FTID controller parameters
step 2: Initialize population
step 3: Evaluate fitness values of CTTNL
step 4: If termination conditions are not met, go to Step3
step 5: Stop

Here the goal is to find a set of FTID and PID parameters that will give a minimum fitness value over the period [0, t]. When this cycle is completed, produce new sets of FTID and PID values which ideally will be at the fitness level higher than the initial population of FTID and PID values. These new competent sets of FTID and PID values are then passed to the fitness evaluation function again where the above mentioned process is repeated. This way the process is cycled unceasingly until the best fitness is achieved. If the predefined termination criterion is not met, again a new population is obtained using various operators that would have better gene. The termination criterion may be formulated as the magnitude of difference between the index value of previous generation and present generation becoming less than a prespecified value.

The objective function J_s, containing six unknown variables K_p, K_i, K_d, n, β_1 and β_2 has been optimized through GA with number of iterations, population size, bit size, crossover probability and mutation probability taken as $25, 40, 10, 0.8$ and 0.125 respectively. The decision regarding the range $(0 \leq K_p \leq 50, 0 \leq K_i \leq 50, 0 \leq K_d \leq 10, 0 \leq n \leq 5, 0 \leq \beta_1 \leq 5,\ and\ 0 \leq \beta_2 \leq 5)$ of these unknown parameters has been made after performing a number of trial runs. After several experiments within the range of these parameters, the final parameters of the fine-tuned GA are run for 10 independent times to get final optimum values of the unknown variables. After optimization, the values of $K_p, K_i, K_d, n, \beta_1$ and β_2 are found to be as **3.5637**, **4.8254**, **0.3732**, **3.3563**, **1.9187** and **0.6721** respectively. These values are then utilized for designing both the FTID and PID controller.

4 Simulation Studies

In all the simulation runs, the process is simulated using the nonlinear first principle model [Eqs. (4) and (5)]. The output variables are computed by solving the nonlinear differential equations using differential equation solver in Matlab R2015a.

4.1 Regulatory Performance

Simulation studies are carried out to demonstrate the control performance of the FTID and PID Control schemes at nominal operated points. From the regulatory response, clearly shows that FTID outperforms the conventional PID controller. However the rise time of the PID controller is better as a compared to FTID. Figure 4 shows comparative regulatory response of FTID and PID controllers. Also, simulation studies are carried out to demonstrate the fault tolerance behaviour of the FTID and PID. Figures 5 and 6 depicted the control performance of the FTID and PID against the abrupt actuator and leak faults respectively.

The Performance comparison of both the controller FTID and PID are carried out based on four integral error indices for regulatory response. Figures 7 and 8 depicted graphical chart based comparison for FTID and PID controller with two faults.

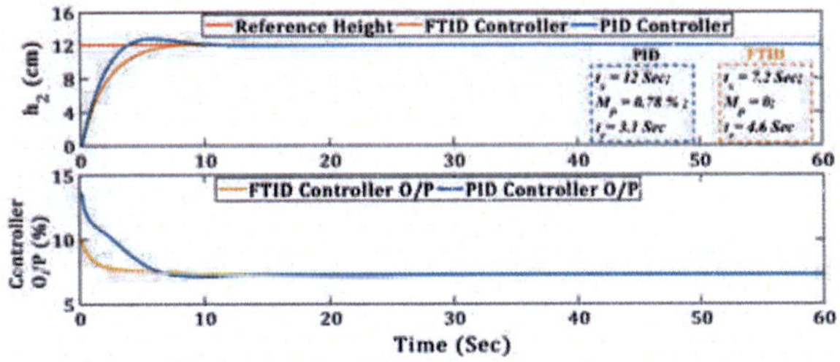

Fig. 4 Regulatory response of CTTNL system with FTID and PID without fault

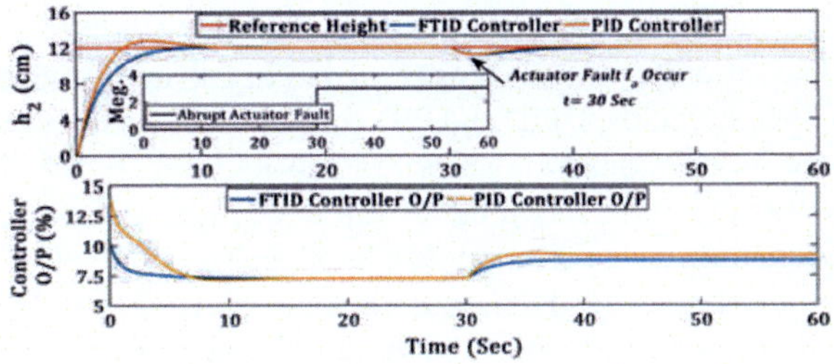

Fig. 5 Regulatory response of CTTNL system with FTID and PID with abrupt actuator f_a fault

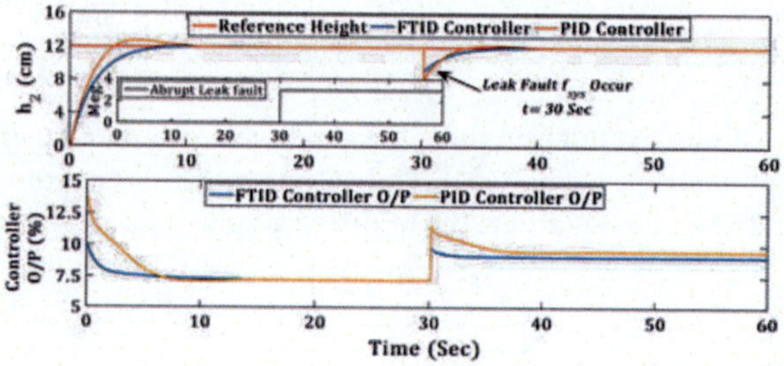

Fig. 6 Regulatory response of CTTNL system with FTID and PID with abrupt leak f_{sys} fault

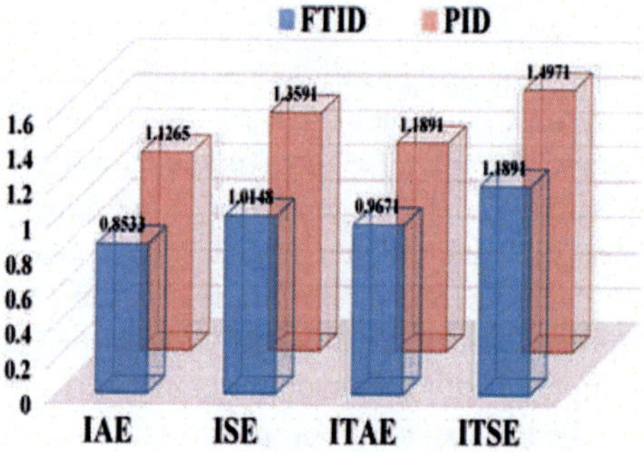

Fig. 7 Performance indices of FTID and PID for regulatory response subject to actuator faults

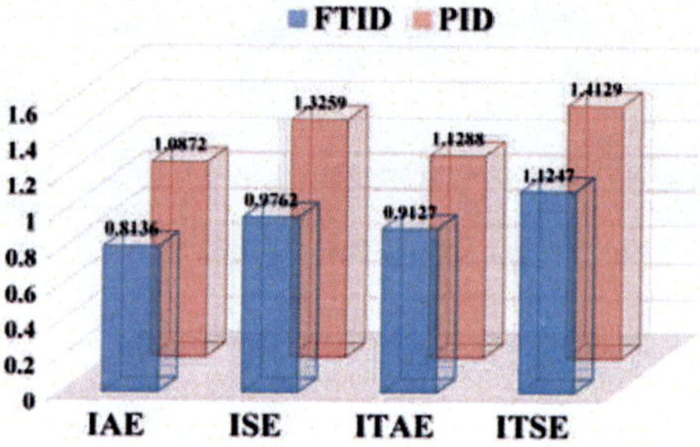

Fig. 8 Performance indices of FTID and PID for regulatory response subject to system component faults

4.2 Servo Performance

The set-point variations as shown in Fig. 9 are introduced for assessing the tracking capability of the FTID and PID control schemes. The variation in process outputs and controller outputs are shown in sub-figure of Fig. 9. The fault-tolerance capability of the controllers, two faults introduced into CTTNL system one is actuator f_a and second is system component fault f_{sys} respectively. From the response, it can be inferred that, the FTID control scheme is able to maintain the tank levels h_2 at the respective set-points better than PID control scheme. The FTID control scheme produces response with faster settling time, zero peak-overshoot compared to PID control scheme. From Fig. 9, it is inferred that the values of performance indices for FTID control scheme are found to be considerably less than PID control scheme.

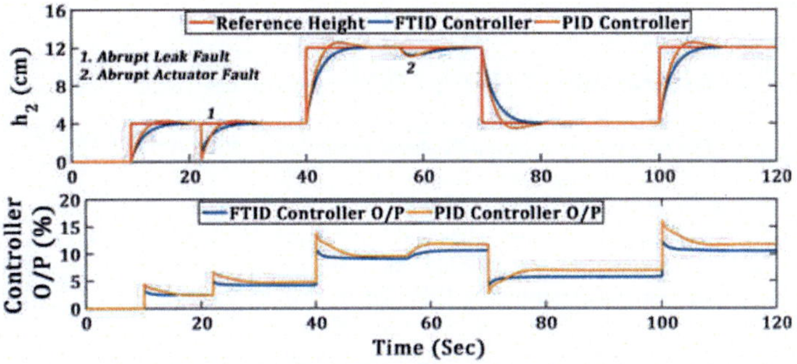

Fig. 9 Servo response of CTTNL system with FTID and PID subject to actuator and leak faults

5 Conclusions

In this paper, the authors have done caparison between proposed FTID and PID controller for nonlinear level control process (for this case study CTTNL system is taken). The simple and straightforward procedure demonstrates for optimum parameters tuning using GA. Also for designing a fuzzy tilt-integral-derivative (FTID) control scheme and proportional-integral-derivative (PID) Control scheme using regime-based multi-model control strategy for the CTTNL system, which exhibits significant nonlinear dynamics and also consider actuator and system component (leak) faults. The optimum FTID and PID controllers are designed for local region 2 linear model using the stability boundary equations method and GA. The effectiveness and practical applicability of FTID and PID schemes are demonstrated through non-stationary level trajectory tracking control of a CTTNL in the presence of two abrupt faults. The extensive simulation studies show that FTID scheme significantly outperforms PID scheme, especially when facing with set-point variations and faulty condition. From the performance indices values, it is observed that the regulatory response of FTID control scheme at region 2 operating points is found to be better than PID scheme. Further, it can be concluded that the FTID scheme helps to produce a response with zero overshoot and settles to the set-point faster in the entire operating region 2 compared to PID scheme.

The merits of FTID scheme using GA control strategy are summarized as follows: (1) it can make use of the traditional SISO linear controller design; (2) it helps to produce response with quicker rise time and settles to the set-point faster in the entire operating region 2 compared to PID scheme.

Acknowledgements I would like to express my special thanks of gratitude to my guide (Vipul Shah) who gave me the golden opportunity to make this wonderful project on the topic, which also helped me in doing a lot of Research and i came to know about so many new things I am really thankful to them.Secondly i would also like to thank my parents and friends who helped me a lot in finalizing this project within the limited time frame.

References

1. I. Podlubny, Fractional-order systems and $PI^{\lambda}D^{\mu}$-controllers. IEEE Trans. Autom. Control **44**, 208–214 (1999)
2. D. Valerio, Introduction to single-input, single-output fractional control. IET Control Theory A **5**, 1033–1057 (2011)
3. C.A. Monje, B.M. Vinagre, V. Feliu, Y.Q. Chen, Tuning and auto-tuning of fractional order controllers for industry applications. Control. Eng. Pract. **16**, 798–812 (2008)
4. H.S. Li, Y. Luo, Y. Chen, Q.A fractional order proportional and derivative (FOPD) motion controller: tuning rule and experiments. IEEE Trans. Control Syst. Technol. **18**, 516–520 (2010)
5. H. Chao, Y. Luo, L. Di, Y.Q. Chen, Roll-channel fractional order controller design for a small fixed-wing unmanned aerial vehicle. Control Eng. Pract. 18, 761–772 (2010)
6. M. Zhang, X. Lin, W. Yin, An improved tuning method of fractional order proportional differentiation (FOPD) controller for the path tracking control of tractors. Bio Syst. Eng. **116**, 478–486 (2013)
7. B.J. Lurie, Three-parameter tunable tilt-integral-derivative (TID) Controller. United States Patent (1994)
8. D. Xue, Y.Q. Chen, A comparative introduction of four fractional order controllers, in *Proceedings of the 4th World Congress on Intelligent Control and Automation*, Shanghai, China, June 10–14 (2002), pp. 3228–3235
9. Y. Luo, Y.Q. Chen, C.Y. Wang, Y.G. Pi, Tuning fractional order proportional integral controllers for fractional order systems. J. Process Control **20**, 823–31 (2010)
10. R.S. Barbosa, J.A. Tenreiro Machado, I.M. Ferreira, Tuning of PID controllers based on Bode's ideal transfer function. Nonlinear Dyn. **38**, 305–321 (2004)
11. T.N.L. Vu, M. Lee, Analytical design of fractional-order proportional-integral controllers for time-delay processes. ISA Trans **52**, 583–591 (2013)
12. F. Merrikh-Bayat, General rules for optimal tuning the $PI^{l}ambdaD^{m}u$ controllers with application to first-order plus time delay processes. Can. J. Chem. Eng. **90**, 1400–1410 (2012)
13. F. Padula, A. Visioli, Optimal tuning rules for proportional-integral-derivative and fractional-order proportional-integral-derivative controllers for integral and unstable processes. IET Control Theory A **6**, 776–786 (2012)
14. R. El-Khazali, P.I. Fractional-order, Fractional-order $PI^{\lambda}D^{\mu}$ controller design. Comput. & Math. Appl. **66**, 639–646 (2013)
15. M. Karimi-Ghartemani, F. Merrikh-Bayat, Method for designing $PI^{lambda}D^{mu}$ stabilisers for minimum-phase fractional-order systems. IET Control Theory A **4**, 61–70 (2010)
16. A. Biswas, S. Das, A. Abraham, S. Dasgupta, Design of fractional-order $PI^{\lambda}D^{\mu}$ controllers with an improved differential evolution. Eng. Appl. Artif. Intell. **22**, 343–350 (2009)
17. Z. Chen, X. Yuan, B. Ji, P. Wang, H. Tian, Design of a fractional order PID controller for hydraulic turbine regulating system using chaotic non-dominated sorting genetic algorithm II. Energy Convers. Manage. **84**, 390–404 (2014)
18. C.H. Lee, F.K. Chang, Fractional-order PID controller optimization via improved electromagnetism-like algorithm. Expert Syst. Appl. **37**, 8871–8878 (2010)
19. R.P. Himanshukumar, A.S. Vipul, A fault-tolerant control strategy for non-linear system: an application to the two tank canonical noninteracting level control system, in *IEEE Distributed Computing, VLSI, Electrical Circuits and Robotics (DISCOVER-2018)*, Mangalore (Mangaluru). India 13–14, 64–70 (2018)
20. R.P. Himanshukumar, Fault, tolerant controller using. interval, type-2 TSK logic control systems: application to three interconnected conical tank system, in *Fuzzy Techniques: Theory and Applications*, ed. by R. Kearfott, I. Batyrshin, M. Reformat, M. Ceberio, V. Kreinovich. IFSA, NAFIPS 2019. Advances in Intelligent Systems and Computing, vol. 1000 (Springer, Cham, 2019)
21. R.P. Himanshukumar, Fault, tolerant control using. Interval, type-2 takagi-sugeno fuzzy controller for nonlinear system, in *Intelligent Systems Design and Applications. ISDA 2018*, ed.

by A. Abraham, A. Cherukuri , P. Melin, N. Gandhi. Advances in Intelligent Systems and Computing, vol. 941 (Springer, Cham, 2018)

22. R.P. Himanshukumar, Stable fault tolerant controller design for takagi-sugeno fuzzy model-based control systems via linear matrix inequalities: three conical tank case study. Energies **12**(11), 2221 (2019). https://doi.org/10.3390/en12112221
23. R.P. Himanshukumar, Fault detection and diagnosis methods in power generation plants - the Indian power generation sector perspective: an introductory review. J. Energy Manag. **2**, 31–39 (2018)
24. R.P. Himanshukumar, A.S. Vipul, Actuator and system component fault tolerant control using interval type-2 Takagi-Sugeno fuzzy controller for hybrid nonlinear process. Int. J. Hybrid Intell. Syst. **15**(03), 143–153 (2019)
25. R.P. Himanshukumar, A.S. Vipul, A passive fault-tolerant control strategy for a non-linear system: an application to the two tank conical non-interacting level control system. MASKAY **9**(1), 1–8 (2019)
26. R.P. Himanshukumar, A.S. Vipul, A fractional and integer order PID controller for nonlinear system: two non-interacting conical tank process case study, in *Advances in Control Systems and its Infrastructure*, ed. by A. Mehta, A. Rawat, P. Chauhan. Lecture Notes in Electrical Engineering, vol. 604 (Springer, Singapore)

Fuzzy Redundancy Mechanism to Enhance the Resilience of IoT-Based HPC Systems

Amira Abo Hozaifa and Ahmed Shawky Moussa

Abstract The Internet of Things has become one of the buzzwords in the fields of computer science and technology. The number of connected devices has been growing rapidly which introduces both, the need and the opportunity for solutions of large-scale complex problems especially in the era of Big Data. The rapid increase of the number of connected IoT devices constitutes a tremendous computational power. The authors contend that this IoT-based computational power can be used for high-performance parallel computing. However, the dynamic behaviour of IoT systems, where nodes frequently drop off, may be an obstacle against achieving parallel processing on IoT devices. In this paper, the authors discuss this problem and propose a fuzzy redundancy-based solution to increase the resilience and enable the use of IoT systems in parallel processing.

Keywords Fuzzy classes · Resilience · IoT · HPC · Parallel computing

1 Introduction

The Internet of Things (IoT) became a trending research direction and has been gaining attraction in industry [1–5]. IoT devices can be found everywhere in our life, they are used in hospitals, homes, transportation, tourist places, and can be deployed anywhere to detect fires and report environmental changes. The number of connected IoT devices has been growing rapidly. It is estimated that the number of connected devices will grow from 23 billions in 2018 to 75 billions in 2025 [6].

Nowadays, the computational problems became more complicated. High Performance Computing (HPC) systems are needed to solve these problems with better accuracy in less time. Due to the limitations that made it impossible to cope with

A. Abo Hozaifa (✉) · A. S. Moussa
FCAI-CU Department of Computer Science, Cairo University, Cairo, Egypt
e-mail: a.hozaifa@fci-cu.edu.eg

A. S. Moussa
e-mail: a.moussa@fci-cu.edu.eg

B. Bede et al. (eds.), *Fuzzy Information Processing 2020*,
Advances in Intelligent Systems and Computing 1337,
https://doi.org/10.1007/978-3-030-81561-5_17

Moore's law anymore, parallel HPC systems became a must in order to solve the large-scale problems such as Big Data, Weather Forecast, Bioinformatics and Astronomy, to name a few. According to top 500 [7], Petascale supercomputers are existing but they seem to be not enough. In 2012, IBM reported that it is a must to build an Exascale computer before 2024 [8]. Finally, Intel and the U.S. Department of Energy are planning to deliver the first supercomputer with a performance of one exaflop in the United States called "Aurora" in 2021 [9]. Hence, it is concluded that building Exascale supercomputers has become a must not a luxury.

Building an Exascale supercomputer is being challenged by the cost of power consumption. In the U.S., even if the supercomputer is powered by the least expensive power which costs at least 5 cents/KWH, operating the system will cost $1M per Megawatt per year, a substantial part of the total cost of the supercomputer operation. Therefore, The U.S. Department of Energy (DOE) set 20 MW as an upper limit for the Exascale system power consumption in order to achieve cost feasibility [10].

Based on the above exposition, the authors propose the idea of using the wide spreading of IoT devices for achieving high performance computing in order to make use of their computational power that may not be fully utilized. Through this research, the authors propose an IoT-based HPC system, where the parallel processing is performed over a cluster that can be constructed by IoT devices, and investigate the potential system resilience hazard problem and its solution.

Through this paper, Sect. 2 defines and discusses the research problem, namely the system resilience, associated with using IoT for parallel computing. Section 3 reviews the related literature in the area of IoT and Parallel Computing. Section 4 provides the proposed solution, Sect. 5 covers the experimental study and results. Finally, Sect. 6 concludes the research and its possible future extensions.

2 Problem Definition

Using the IoT connected devices in performing parallel computing can be promising for building a large-scale parallel computing system without the need for extra equipment and energy consumption. But as it is well known, the IoT systems have dynamic behaviour. The dynamic changes that occur in the IoT systems include the devices' connection state, where one or more devices may be disconnected at some instance of time [11]. This disconnection can occur due to many reasons such as network disconnection, power off or going out of coverage. In that case, the system may be blocked on an operation being performed at one of the disconnected devices.

Therefore, constructing a parallel cluster using the commercially used IoT devices will be obstructed by the dynamic nature of the IoT system components which in turn results in resilience problems which cannot be compromised in an HPC system. Thus, enhancing the system resilience will be the main problem for building the dynamic cluster of IoT devices. More precisely, the system needs to be able to recover lost calculations and data due to nodes' drop off at run time to guarantee completed correct solution. Yet, we must not forget that the cluster is intended to parallelize and speed up the computation. Hence, the proposed solution must not introduce overhead which may degrade the performance.

3 Related Work

Having done a thorough search in the literature, there are many projects and research articles on the use of parallel computing to solve IoT systems' problems [12–17]. However, the use of dynamic IoT infrastructure to solve HPC problems does not exist to the best of our knowledge.

There are some projects and researches that built static HPC clusters using IoT devices [18–21]. In 2018 Jeremy Singer et. al., built a cluster of Raspberry Pi single boards. The main target was to overcome the power consumption problem and provide a cheap and portable cluster that can be used for educational purposes, and can be beneficial for edge and portable computing [18].

Using a wide spread highly connected devices such as mobile phones in building an HPC system can be related to the proposed work in the current paper. In 2016, Ehab, Moussa, and Korany built a parallel computing cluster of Android mobile devices. Given that the mobile devices are often idle specially for the corporate phones, the device battery may be consumed and needs recharging only for keeping the device on. The authors built parallel computing cluster to perform parallel computing making use of the mobile devices which may be idle for most of the time [22].

The previous studies built parallel computing systems that are static without handling the probability of any node disconnection. This is mainly because they used IoT devices and component but not IoT systems. But The research in this paper targets building a dynamic cluster made of up and running IoT systems. Therefore, the node drop off from the HPC system is considered and a solution to enhance the resilience of the system is proposed.

4 The Proposed Solution

Based on the given problem definition, the authors expect that applying redundancy will be a promising solution for achieving dependability and enhancing system resilience. In [23], David A Patterson and John L Hennessy introduced redundancy as one of the eight great ideas in computer architecture. Also various replication techniques have been widely used in detecting and correcting errors in HPC systems [24–26].

In a nutshell, the proposed solution states that, every node in the system has a number of duplicates. The duplicates for each node are selected using a simple procedure, where the nodes are divided into groups of nodes such that the group size is equal to the number of duplicates +1. In MPI, each process in the system has a numeric ID, called “rank”. The rank is an integer in the range of [0, N−1], where N is the number of nodes in the system. The duplicates of a certain node are determined such that, in a system where the number of duplicates = 3, the duplicates of a node having rank 2 are 1, 3 and 4 where the nodes of this group are 1, 2, 3 and 4.

Each node exchanges its calculations state with its duplicates when a certain portion of the problem is done. In this case the loss of the whole calculations of a certain node in case of failure is avoided. Every node has four checkpoints on which the node should notify its duplicates by its state. For example, assuming that the problem size per process is N, each node sends its calculations state every N/4. Therefore, the maximum size of lost calculations will be N/4. Finally, when a node fails, the duplicate having least load is chosen to resume the failed node’s task.

The highest dependability can be achieved by 100% duplication where all the nodes in the system are duplicates for each other. This will cause a dramatic degrading in the system performance due to the communication overhead. In addition, The whole data may not be able to be duplicated at all the nodes because of storage limitations.

On the other extreme, the highest performance where no communication overhead exists can be obtained by 0% duplication. In such case nodes have no duplicates at all, and consequently no status and load exchange required which results in zero communication overhead. However, this may lead to easy failure of the whole system in case of any node disconnection.

Therefore, selecting the number of duplicates such that system resilience is achieved with acceptable communication overhead requires an intelligent dynamic algorithm that tunes the number of duplicates during the program run according to the system status. The number of duplicates can be tuned according to the overall system load.

The authors of the current paper suggest tuning the number of duplicates according to the average system load, where the number of duplicates increases as the system load decreases and vice versa. However, this proposed solution posed another trade-off problem. Changing the number of duplicates with every slight change in the average system load is expected to overload the system by too frequent update messages and system adaptation to the new duplication level.

Based on the above trade-off between system resilience and performance represented in the trade-off and dynamic changes in system load, the authors propose a class based categorization for tuning the number of duplicates during the system run, where the system load is grouped into ranges, or classes, and the number of duplicates changes when the average system load changes from one class to another. Figure 1 illustrates the division of the system average load into five classes (Very light load, Light load, Medium load, Heavy load and Very heavy load), represented as (Class A, Class B, Class C, Class D and Class E) respectively.

Despite the fact that the proposed class based load categorization is expected to reduce frequency of the duplication level update, the frequent changes in load around the boundary between two neighbouring classes will lead to floods of update messages. Figure 1 illustrates the wiggling problem around class boundary. A frequent

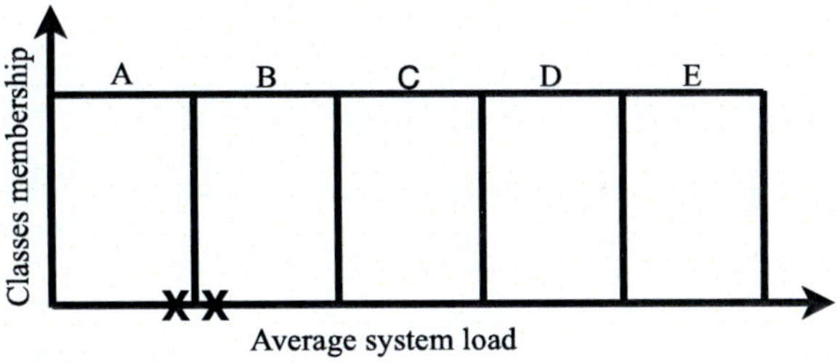

Fig. 1 Five load classes

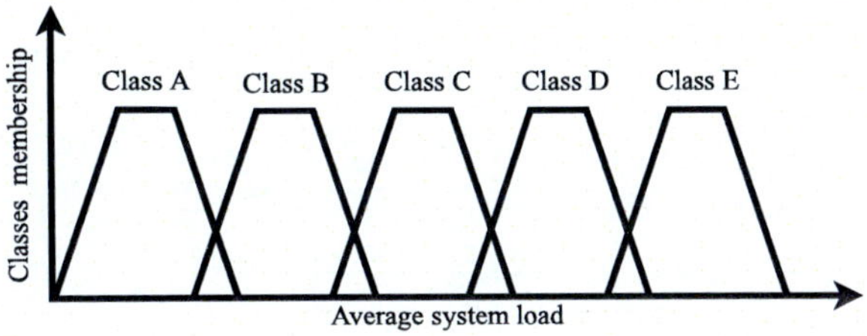

Fig. 2 Load fuzzy classes

slight change around the class boundaries will cause multiple changes in the number of duplicates for a non significant change.

To solve this problem, the authors propose using overlapping fuzzy classes. If the value of the average system load changes between two neighbouring classes but within the overlapping region, the number of duplicates will not be changed. The advantage of this approach is that the overlapping areas will absorb the frequent changes in classes in case the load value wiggles between two neighbouring class boundaries. Figure 2 illustrates the new categorization of the loads using overlapping fuzzy classes.

During the program run, the master-slave paradigm is followed, where the node with rank 0 acts as the master node which is responsible for tuning the duplicates and problem distribution. The load of each node is being monitored by itself. When the load changes by certain threshold, a message is sent to the master node notifying by the new load. At this point, the average system load is calculated according to the update message, then the new fuzzy class is determined according to the new calculated load. If the class changed, number of duplicates is changed and the master node notifies the slaves by the new number of duplicates. Figure 3 illustrates how duplicates tuning is performed.

Monitoring the CPU utilization can be done using many Linux commands that extract the values from the /proc/stat file that has the following format [27]:

cpu 293245 1819 63033 1589779 1461 0 13678 0 0 0

cpu0 73865 450 15056 398726 457 0 6063 0 0 0

cpu1 73658 458 15610 397006 291 0 3453 0 0 0

The cpu line represents the sums of the cpuN lines below cpu. The numbers on each row represent the time in 1/100 second unit for: processes executing in user

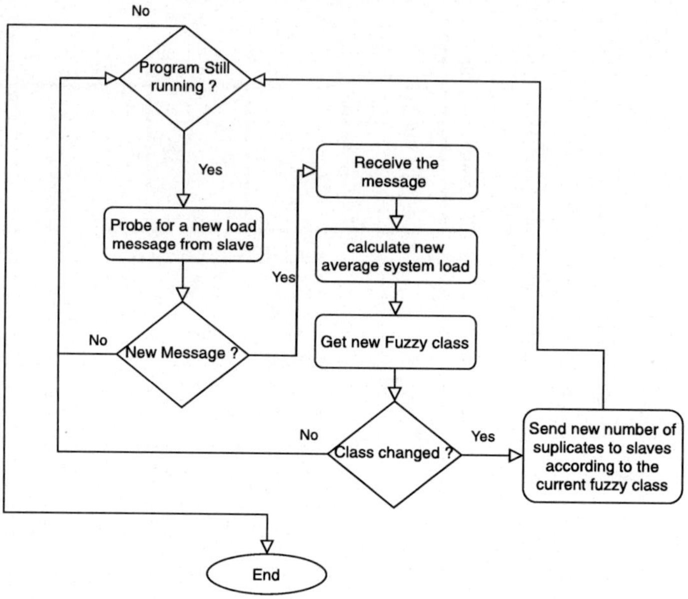

Fig. 3 Duplicates tuning algorithm

mode, niced processes (a process feature in linux), idle time, I/O waiting time, service interrupt time, and software interrupt time, respectively.

Calculating the total time spent by the CPU since first boot can be done by adding all the columns of the given core. Also calculating the time spent by the CPU in any mode other than the idle one can be done by adding all the columns except for the forth one. Calculating the average load of a single core of the CPU can be performed dividing the CPU time in non idle operations by the total time.

At run time, each node monitors its own CPU load. When the load changes, the node sends updates of its new state to its duplicates and the master node. It is expected in this case that continuous slight changes in the CPU load will lead to too frequent update messages exchanged between the nodes in the system. To avoid this problem, the authors decided that the updates are to be sent only when the change in the load exceeds certain effective threshold. For the current experimentation, the threshold was set to 20% change in the CPU load. For example, if the duplicates are notified by 40% CPU load, the next notification should be sent when the load becomes either (20% or less) or (60% or more).

5 Testing and Evaluation

Testing of the proposed solution was performed by simulation running on a computer with 8 Cores and 12 GB RAM. The program simulates running processes on IoT nodes, where nodes may be dropping out of the system at run time. The proposed solution is supposed to use the fuzzy mechanism to enhance the resilience of the system at the minimum computational cost possible to maintain the best possible performance.

The simulation algorithm is designed and run with parallel Pi calculation program. The MPI parallel program pseudocode for calculating Pi is as follows:

```
input number of intervals (n)
broadcast n
init step = 1.0 / n , sum = 0, piAtCurrentProc = 0, finalPi = 0, i = 0
calc. process portion (portion = n / number of processes)
for i = 0 to process portion
  calc. temp value (x = step * (i - 0.5))
  sum += 4.0 / (1.0 + x*x)
piAtCurrentProc = step * sum
reduce piAtCurrentProc to finalPi with SUM operator
```

The number of the nodes to be dropped is sent as a parameter to the simulator. During the simulation run, the nodes that will be dropped will be selected randomly, and the iterations in which the nodes will be dropped are randomly generated. Figure 4 is a flow chart of the simulation algorithm.

For the first round of experiments, the number of duplicates is given as an input to the simulator in order to determine the suitable number of duplicates for each fuzzy class. The simulator was run multiple times with different duplication levels.

Figures 5 and 6 show the probability (out of 50 runs) of success in case of 5 to 10 and 10 to 50 duplicates respectively when dropping 25%, 50% and 75% of the system nodes. These runs are performed for 121 processes (120 slaves).

The results proved that duplication can enhance the system resilience such that 100% system resilience can be achieved by 5 duplicates (4–5% duplication) when 25% of the system nodes fail. 10 duplicates (form 8% to 10% duplication) can achieve 100% resilience when dropping 50% of the system nodes. But in case of dropping 75% of the system nodes, it is noticed that 10 duplicates will not be sufficient for achieving more than 80% system resilience. Also it is noticed that 36 duplicates (30% duplication) and more can achieve 100% system resilience for 25%, 50% and 75% system drops.

For the second round of runs, time is measured for different duplication levels in order to evaluate the effect of duplication on performance. Figures 7 and 8 illustrate the change in performance that occurs when changing the duplication level.

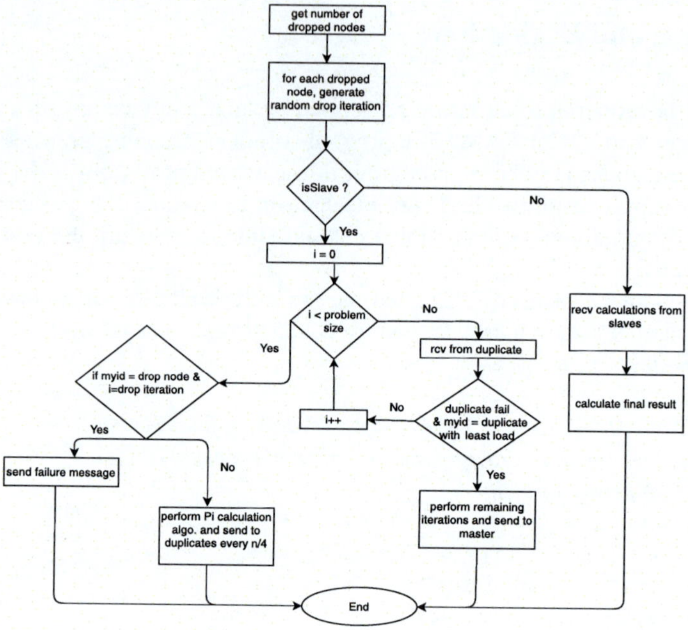

Fig. 4 Simulation algorithm

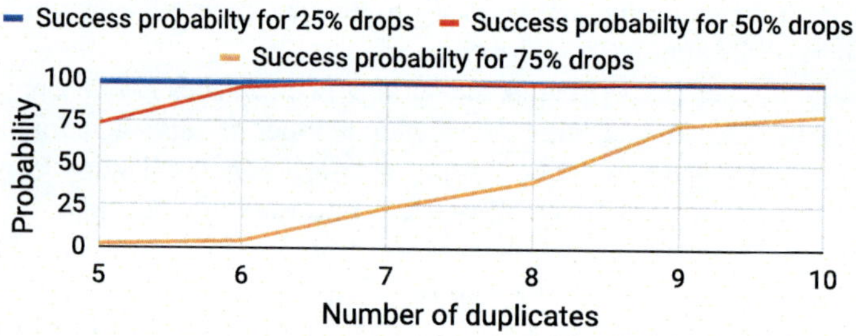

Fig. 5 Probability of system success when tuning number of duplicates between 5 and 10 in 120 slaves system

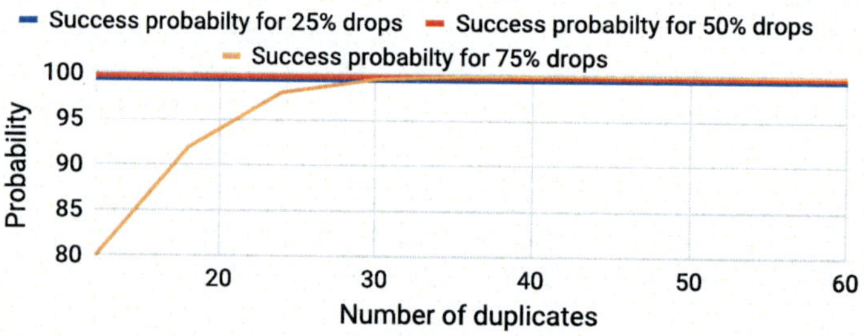

Fig. 6 Probability of system success when tuning number of duplicates between 12 (10%) and 60 (50%) in 120 slaves system

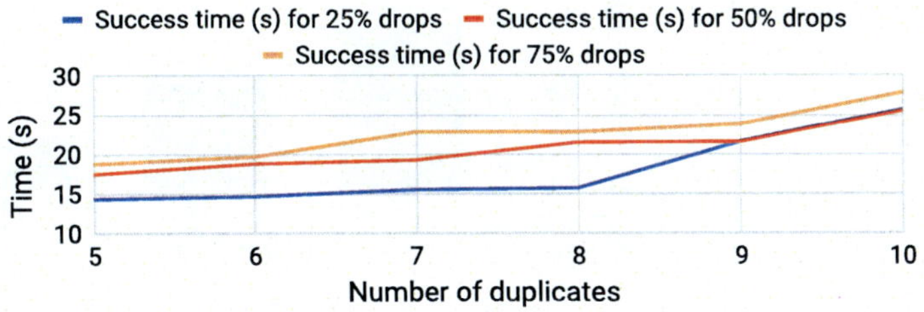

Fig. 7 Performance of successful runs when tuning number of duplicates between 5 and 10 in 120 slaves system

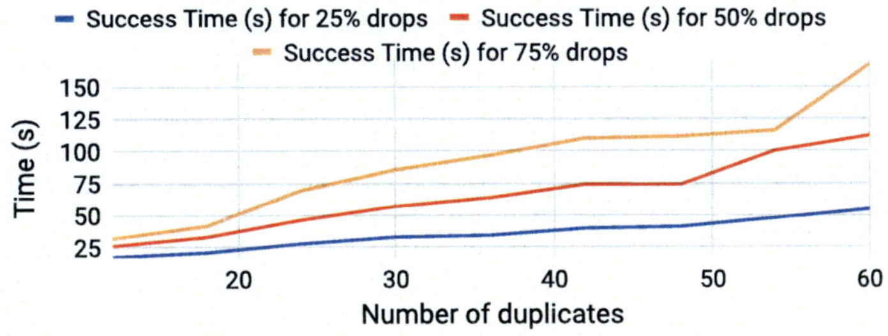

Fig. 8 Performance of successful runs when tuning number of duplicates between 10 and 50 in 120 slaves system

Fig. 9 Fuzzy categorization of system load

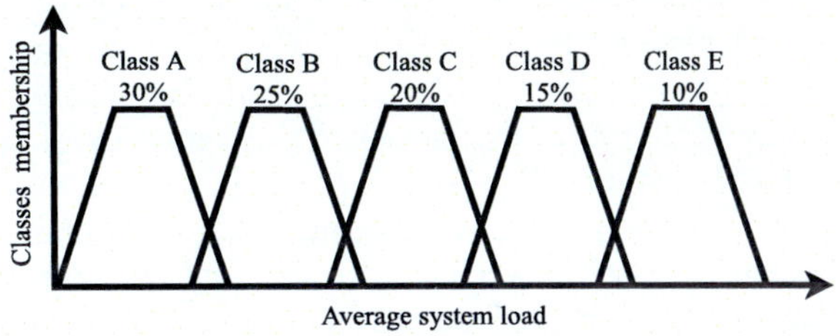

According to the results of the previous rounds, the third round of experiments is done after determining the duplication level for each class. Since the system resilience saturates at 100% when setting the duplication level at 30% or more, the maximum duplication level is set to 30%. Figure 9 illustrates the duplication level for each fuzzy class.

The third round of runs is done using fuzzy classes which achieved resilience (90%) with 75% system drops and 96% resilience for 50% and 100% resilience for 25% drops. Figure 10 illustrates the success probability of the proposed system for 25% (30 drops), 50% (60 drops), and 75% (90 drops) system drops.

More drops in the system affects the average system load, especially when dropping nodes with light loads, this results in increase the number of duplication update notifications. Figure 11 illustrates the effect of the number of drops on system duplication level updates for 25% (30 drops), 50% (60 drops), and 75% (90 drops) as above.

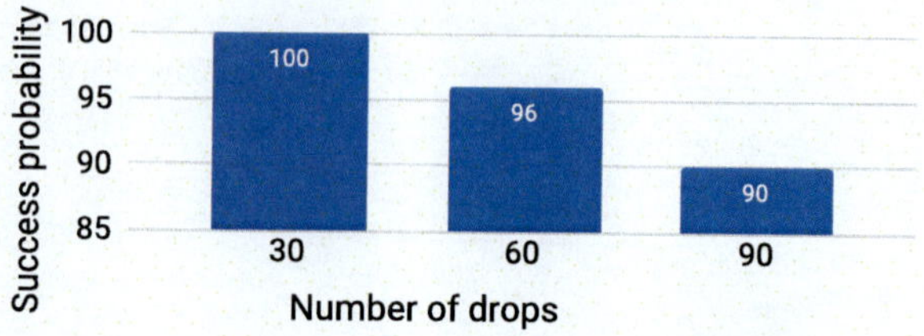

Fig. 10 Success probability versus Number of drops

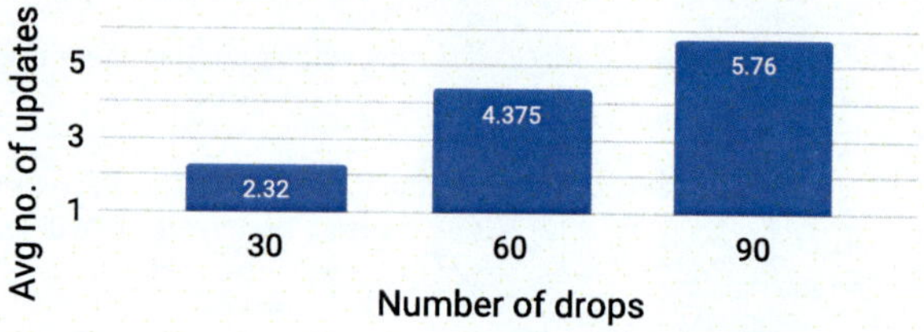

Fig. 11 Average number of updates versus Number of drops

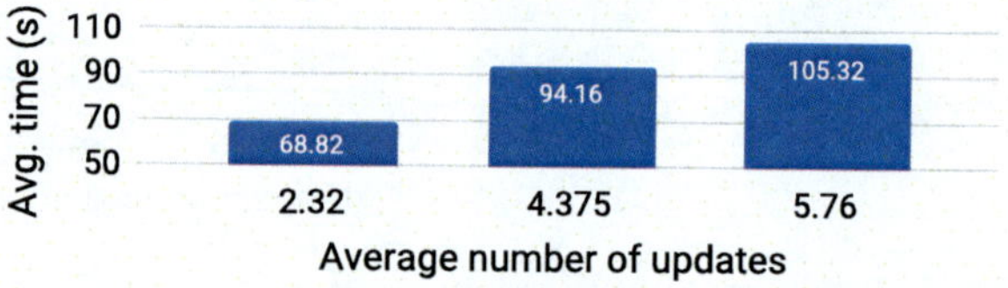

Fig. 12 Average time versus Average number of updates

The duplication level updates during the system run affected the performance. The more duplication level update messages, the more communication between nodes for exchanging the status and system adaptation leading to performance degradation. Figure 12 illustrates the impact of the frequency of updating the duplication level and performance.

Final round of runs is performed using crisp classes instead of fuzzy classes in order to compare between performance and update frequency for the two techniques. Fuzzy classes provided less number of update messages as well as better performance and slightly higher resilience. Figures 13, 14 and 15 illustrate the comparison between the effect of crisp and fuzzy techniques system on the update rate, system performance and system resilience, respectively.

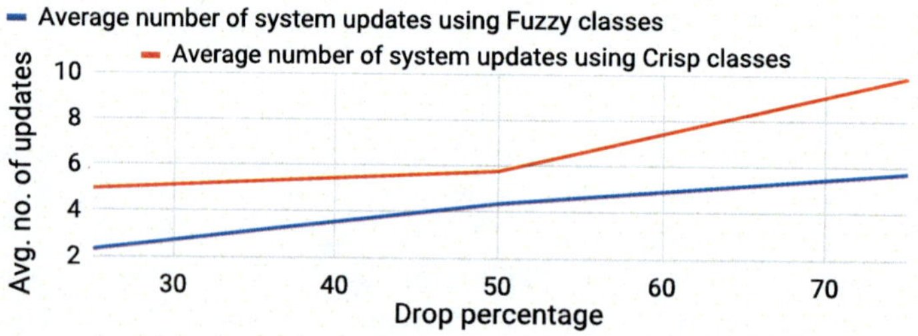

Fig. 13 Number of duplication level updates in the system using fuzzy classes and crisp classes

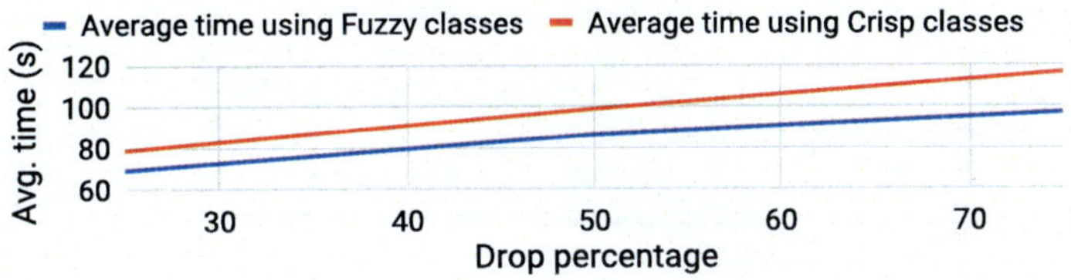

Fig. 14 System performance using fuzzy classes and crisp classes

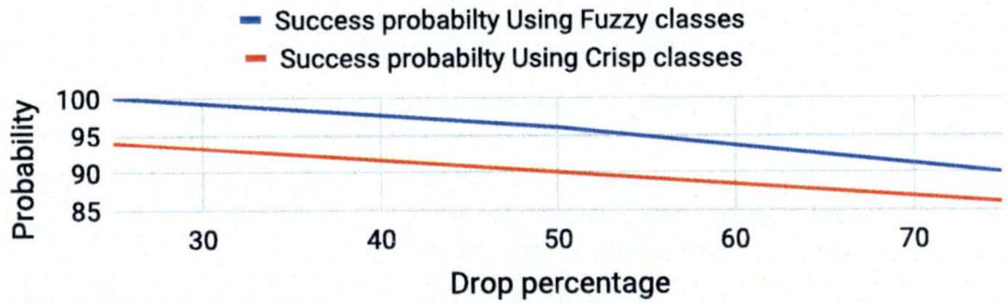

Fig. 15 System resilience using fuzzy classes and crisp classes

6 Conclusion and Future Work

Enhancing the IoT systems computational resilience by applying redundancy is proved to be a promising solution to achieve parallel computing over unstable IoT systems. But achieving system dependability and performance requires a compromise between redundancy level and the resultant overhead.

The authors in this paper proposed a dynamic mechanism to cope with the frequent changes in system load and node stability. The experiments proved that the system resilience saturates at 100% when the duplication level reaches 30%; no need for higher redundancy. Accordingly, the current research proposed a fuzzy based dynamic duplication mechanism where the maximum duplication level is set to 30% of the system size. The proposed fuzzy mechanism achieved 90% system reliance in extreme case of 75% failure of the system nodes. Also this fuzzy mechanism was compared to a crisp classes based one, and the fuzzy mechanism provided 16% higher performance with 42% less communication.

Future extension will add the performance consideration when nodes may be added at run time, in addition to testing real IoT clusters.

References

1. S. Chen et al., A vision of IoT: applications, challenges, and opportunities with china perspective. IEEE Int. Things J. **1**, 349–359 (2014)
2. S. Kumar Panda et al., IoT Retrofitting approach for the food industry, in *24th IEEE International Conference on Emerging Technologies and Factory Automation (ETFA), Zaragoza, Spain* (2019), pp. 1639–1642. https://doi.org/10.1109/ETFA.2019.8869093

3. L. Yaqiong et al., IoT based Omni-Channel Logistics Service in Industry 4.0, in *IEEE International Conference on Service Operations and Logistics, and Informatics (SOLI)*. Singapore (2018), pp. 240–243. https://doi.org/10.1109/SOLI.2018.8476708
4. H. Khaleel et al., Heterogeneous applications, tools, and methodologies in the car manufacturing industry through an IoT approach. IEEE Syst. J. **11**, 1412–1423 (2017)
5. T. Alladi et al., Blockchain applications for industry 4.0 and industrial IoT: a review. IEEE Access **7**, 176935–176951 (2019)
6. Internet of Things (IoT) connected devices installed base worldwide from 2015 to 2025, https://www.statista.com/statistics/471264/IoT-number-of-connected-devices-worldwide/. Accessed 13 Apr 2020
7. TOP500 LIST, https://www.top500.org/lists/2019/11/. Accessed 13 Apr 2020
8. D. Cooper, IBM: 'We must build an Exascale computer before 2024' (2012), https://www.engadget.com/2012/04/02/ibm-we-must-build-an-exascale-computer-before-2024-video/. Accessed 13 Apr 2020
9. Intel, https://newsroom.intel.com/news-releases/u-s-department-energy-intel-deliver-first-exascale-supercomputer/#gs.o8clf6. Accessed 13 Apr 2020
10. S. Horst, Barriers to exascale computing, in *High Performance Computing for Computational Science - VECPAR, VECPAR 2012*, ed. by M. Daydé, O. Marques, K. Nakajima. Lecture Notes in Computer Science, vol. 7851. (Springer, Berlin, 2012)
11. K. Patel, S.M. Patel, Internet of things-IoT: definition, characteristics, architecture, enabling technologies, application & future challenges. Int. J. Eng. Sci. Comput. **6**, 6122–6131 (2016)
12. M. Riduan Abid, R. Lghoul, D. Benhaddou, ICT for renewable energy integration into smart buildings: IoT and big data approach, in *Proceedings of IEEE Africon* (2017), pp. 856–861
13. Z. Yang, Hierarchical computing: a high performance computing architecture for data-processing, in *IoT Era. 4th International Conference on Systems and Informatics* (2017), pp. 1698–1702
14. L. Cimino et al., IoT and HPC integration: revision and perspectives, in *7th Brazilian Symposium on Computing Systems Engineering* (2017)
15. D. El Baz, IoT and the need for high performance computing, in *International Conference on Identification, Information and Knowledge in the Internet of Things, China* (2014), pp. 1–6
16. I. Saenko et al., Parallel processing for big heterogenous data for security monitoring of IoT networks, in *25th Euromicro International Conference on Parallel, Distributed and Network-based Processing (PDP)*, At St.-Petersburg, Russia (2017)
17. K. Niranjana Devi, R. Muthuselvi, Parallel processing of IoT healthcare applications, in *10th International Conference on Intelligent Systems and Control, Coimbatore, India* (2016), pp. 1–6
18. J. Singer et al., Next generation single board clusters, in *NOMS IEEE/IFIP Network Operations and Management Symposium* (2018)
19. D. Kumar, S. Memon, L. Thebo, Design, implementation & performance analysis of low cost high performance computing (HPC) clusters, in *2018 12th International Conference on Signal Processing and Communication Systems (ICSPCS), Cairns, Australia* (2018), pp. 1–6. https://doi.org/10.1109/ICSPCS.2018.8631769
20. C. Baun, Mobile clusters of single board computers: an option for providing resources to student projects and researchers. SpringerPlus **5**, 360 (2016)
21. A. Pfalzgraf, J. Driscoll, A low-cost computer cluster for high-performance computing education (2014), pp. 362–366. https://doi.org/10.1109/EIT.2014.6871791
22. D. Ehab Attia, A. Mohamed ElKorany, A. Shawky Moussa, High performance computing over parallel mobile systems. Int. J. Adv. Comput. Sci. Appl. **7**, 99–103 (2016)
23. D. Andrew Patterson, J. Leroy Hennessy, Computer abstractions and technology, in *Computer Organization and Design*, 5th edn. USA: MK, ch.1, sec. 1.2 (2014), p. 12
24. O. Subasi et al., Designing and modelling selective replication for fault-tolerant hpc applications (2017), pp. 452–457
25. C. George, S. Vadhiyar, Fault tolerance on large scale systems using adaptive process replication. IEEE Trans. Comput. **64**, 2213–2225 (2015)

26. B. Mills et al., Energy consumption of resilience mechanisms in large scale systems (2014), pp. 528–535
27. /proc/stat explained, http://www.linuxhowtos.org/System/procstat.htm. Accessed 13 Apr 2020

Carbon Emissions Trading as a Constraint in a Fuzzy Optimization Problem

Nilmara de Jesus Biscaia Pinto, Estevão Esmi, and Laécio Carvalho de Barros

Abstract This work presents some climate changing policies and intends to model carbon emissions trading as a constraint in a fuzzy optimization problem. Bellman-Zadeh decision scheme is used within Zimmermann approach in order to estimate the gain in carbon markets notwithstanding to lost in entities welfare function, seen as the objective function.

Keywords Fuzzy optimization problem · Climate changing · Carbon emissions trading

1 Introduction

Climate change is nowadays one of the most studies subject and the cause of amount of international negotiations [13]. Kyoto protocol (1995), signed by 192 parties, is one result from these arrangements. This pact created the Emissions Trading Systems (ETS), which converts greenhouse gas emissions in a new commodity: a mean to facility countries to reach the target of reducing greenhouse gas emissions [15].

Despite the efforts, CO_2 emissions continuous to grow [16], so countries on United Nations Framework Convention on Climate Change (UNFCCC) still have to work on solutions to reduce the gap between the reduction target and the actual emissions. Furthermore pricing carbon still being a strategy to be used by policy makers and companies [8].

N. de Jesus Biscaia Pinto (✉) · E. Esmi · L. C. de Barros
University of Campinas, Institute of Mathematics, Statistics and Scientific Computing, Campinas, São Paulo, Brazil
e-mail: nilmarabiscaia@gmail.com

E. Esmi
e-mail: eelaureano@gmail.com

L. C. de Barros
e-mail: laeciocb@ime.unicamp.br

B. Bede et al. (eds.), *Fuzzy Information Processing 2020*,
Advances in Intelligent Systems and Computing 1337,
https://doi.org/10.1007/978-3-030-81561-5_18

Reducing carbon emissions does not depend only on aiming from policy makers. There are other obstacles, such as energy transition to post-fossil fuel and cultural changes [10, 11]. Moreover it is necessary to find an equilibrium between gains in climate policies and losses in financial stability [5]. For this reason fuzzy approach may be used in order to model subjectivity and duality in decision process [12].

The purpose of this article is to represent the social cost of carbon [14] in terms of fuzzy variables; in other words, the intention is modelling decision process for policy makers by a fuzzy optimization problem, according to Bellman-Zadeh decision scheme [4], assigning welfare (utility) as the objective function and carbon trading as a constraint.

From extending the classical notion of maximum by Zadeh's extension principle, which leads to Zimmermann's method [19, 20], it is possible to estimate which decision is better. For instance: (1) to maximize the welfare, but increasing CO_2 emissions and with zero profit in carbon trading; (2) to lost in welfare, but earning a lot of in carbon market; (3) or an intermediate choice between (1) and (2).

The article is divided in four sections. Section 2 contains more detailed information about how carbon market works; Sect. 3 presents fuzzy optimization problem and decision. Section 4 presents the application of the decision method in an Emission Trading problem.

2 Carbon Market

Kyoto's protocol and Paris Agreement put nations on movement towards preserving the environment. Many research groups are forecasting and arguing about solutions for carbon emission—one of the greenhouse gas emissions. Carbon market or carbon trading arouse in the context of pricing carbon [9].

Emission Trading System (ETS) is based on emission targets, say T (in million tons of carbon dioxide $MtCO_2$). If an agent pollutes less than T, say E, then the agent may sell the difference $T - E$ in the Carbon Market in form of carbon credit. On the other hand, if an agent pollutes a quantity F more than T, then it is possible to buy the difference $F - T$ in Carbon Market, in order to avoid greater sanctions. The scheme is depicted in Fig. 1.

The carbon market became a lucrative market, since the objective is to decrease the target T, increasing the price of carbon credit. Nevertheless there are costs in

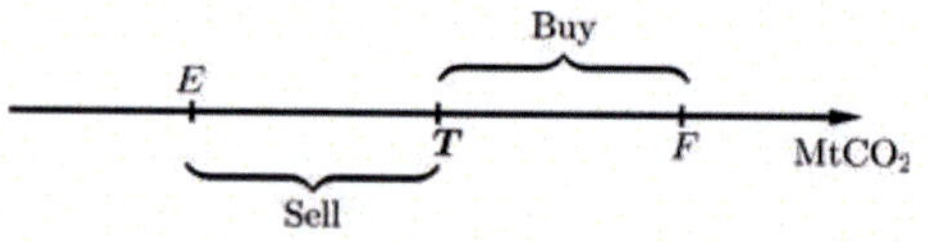

Fig. 1 Scheme of ETS showing $F - T$ as a difference that must be bought and the difference $T - E$ that may be sold in Carbon Market

embedding less carbon emissions and a fair deliberation is been made during the last decades about how much this ETS will decrease the social welfare, or in other words, what is the benefit-cost analysis on the social cost of carbon [14].

The problem consists in finding an equilibrium between losing welfare and reducing emissions. There are lots of works dealing with this subject from the mathematical point of view [14], however here we include the subjectivity in the decision of reducing emissions [12]. Furthermore, we describe the economical gain and losses in terms of fuzzy optimization objects.

3 Fuzzy Optimization Problems

Fuzzy Set Theory [17] was introduced in order to describe subjectivity and uncertainty presented in phenomena of nature, in mathematical terms. A fuzzy subset A of an universe U is associated a membership function $A : U \to [0, 1]$. Each $A(x) \in [0, 1]$ represents the degree to which the element x concerns the set A. The class of fuzzy sets of U is represented by $\mathcal{F}(U)$.

For example, we can consider A as the set of countries that emit many million tons of carbon dioxide ($MtCO_2$) in 2018. According to Global Carbon Atlas [6], China definitely is in the set A, because it is the country that most emitted $MtCO_2$–10065. On the other hand, other countries (see Table 1), such as Germany and Iran, may or may not be in the set A. They emitted much less CO_2 in 2018, in comparison with China, but still are being the sixth and seventh in the list, so it is possible to assign degrees to which they are in the set A. Germany has degree bigger than Iran, but smaller than the top five countries.

The α-cuts of a fuzzy set $A \in \mathcal{F}(U)$, denoted by $[A]_\alpha$, are defined as $[A]_\alpha = \{x \in X : A(x) \geq \alpha\}$, $\forall \alpha \in (0, 1]$. In addition, if U is also a topological space, then we can define the 0-cut of A by $[A]_0 = cl\{x \in U : A(x) > 0\}$, where cl X, $X \subseteq U$, denotes the closure of X. A fuzzy set $A \in \mathcal{F}(\mathbb{R})$ is called a fuzzy number if its $\alpha-$cuts are closed, bounded and non-empty nested intervals for all $\alpha \in [0, 1]$ [2]. Since each $\alpha-$cut of a fuzzy number A is an interval, we can write $[A]_\alpha = [a_\alpha^-, a_\alpha^+]$. The class of fuzzy numbers is denoted by $\mathbb{R}_\mathcal{F}$.

Table 1 CO_2 Territorial emissions in 2018 ($MtCO_2$)[a]

Rank	Country	$MtCO_2$
1	China	10065
2	United States of America	5416
3	India	2654
4	Russian	1711
5	Japan	1162
6	Germany	759
7	Iran	720

[a]Top countries in emissions of CO_2 in 2018. *Source* Global Carbon Atlas [6]

Bellman and Zadeh [4] studied optimization problems in the form

$$\begin{aligned} &\max\ f(x) \\ &\text{s. t. } \ x \text{ is } C, \end{aligned} \tag{1}$$

where C is a constraint, and here we consider $f : U \to \mathbb{R}$. They considered constraints (C) and goal (O) with the same level of importance. Solving (1) is equivalent to solve

$$D = O \cap C, \tag{2}$$

with D representing a possible decision to (1).

The best decision would be found by taking the maximum degree of decision in (2). That is, the optimal decision is obtained by looking for the points x that attain the maximum in D, i.e., the maximizing decision is given by argument from

$$\max_{x \in X} D(x) = \max_{x \in X} \min\{O(x), C(x)\}. \tag{3}$$

Expression (3) means that $\overline{x}$ is the maximizing decision when satisfies

$$\overline{x} = \arg\max_{x \in X} D(x). \tag{4}$$

The minimum in decision (2) is the standard choice, but could be made by other t-norms besides minimum one, which would leads to different decisions sets and numerical solutions.

It is necessary to assign membership functions to O and C sets. It can be done by Zadeh's extension principle [18], which can be viewed as a mathematical method to extend a function $f : X \to Y$ to a function that have fuzzy sets as arguments.

Definition 1 Let $f : X_1 \times \ldots \times X_n \to Y$. The Zadeh's extension of f at $(A_1, \ldots, A_n) \in \mathcal{F}(X_1 \times \ldots \times X_n)$ is the fuzzy set $\widehat{f}(A_1, \ldots, A_n) \in \mathcal{F}(Y)$ whose membership function is given by

$$\hat{f}(A_1, \ldots, A_n)(y) = \sup_{(x_1, \ldots, x_n) \in f^{-1}(y)} \min\{A_1(x_1), \ldots, A_n(x_n)\}, \ \forall\, y \in Y, \tag{5}$$

where $f^{-1}(y) = \{(x_1, \ldots, x_n) \in X_1 \times \ldots \times X_n : f(x_1, \ldots, x_n) = y\}$ and by definition $\sup \emptyset = 0$.

In case $f : X \to Y$, $\Omega \in \mathcal{F}(\mathbb{R})$, is an injection, the Zadeh's extension of f is $\widehat{f}(\Omega)(y) = \sup_{y=f(x)} \Omega(x)$. For each $y \in Im(f)$ there is only one $x \in Dom(f)$, such that $f(x) = y$. Hence,

$$\widehat{f}(\Omega)(y) = \Omega(f^{-1}(y)). \tag{6}$$

We denote $[\widehat{f}(\Omega)]_0 = [\widehat{f}(\Omega)_0^-, \widehat{f}(\Omega)_0^+]$ and $[\widehat{f}(\Omega)]_1 = [\widehat{f}(\Omega)_1^-, \widehat{f}(\Omega)_1^+]$. Besides that, if f increasing, then the following equivalences are valid

$$\Omega_0^- \le x \le \Omega_1^- \Leftrightarrow \widehat{f}(\Omega)_0^- \le f(x) \le \widehat{f}(\Omega)_1^-, \tag{7}$$
$$\Omega_1^+ \le x \le \Omega_0^+ \Leftrightarrow \widehat{f}(\Omega)_1^+ \le f(x) \le \widehat{f}(\Omega)_0^+. \tag{8}$$

Extending classical constraints, the constraint C is obtained, which can be written as $x \le B$, where $B \in \mathbb{R}_{\mathcal{F}}$. This approach was proposed by Barros et al. [3]. The set C can be rewritten in terms of set B. Note that the following equivalence holds

$$\Omega = \{x \in \mathbb{R}; x \le b\} \Leftrightarrow \Omega = \{x \in \mathbb{R}; \min\{x, b\} = x\}. \tag{9}$$

Let $B \in \mathcal{F}(\mathbb{R})$ with membership function $B(x)$, $\forall\, x \in \mathbb{R}$, with 0 and 1-cuts given by $[B]_0 = [b_0^-, b_0^+]$ and $[B]_1 = [b_1^-, b_1^+]$, respectively. A fuzzy set $\Omega = \{x \in \mathbb{R}; x \le B\}$ can be given as an extension of inequality $x \le b$.

Consider the cylindrical extension of B, denoted by $U = B \times \mathbb{R}$, in order to relate $x \in \mathbb{R}$ with $B \in \mathcal{F}(\mathbb{R})$. The membership function of U is given by

$$U(x, y) = B(x), \forall\, (x, y) \in \mathbb{R} \times \mathbb{R}. \tag{10}$$

Then, due to equivalence (9), the set Ω will be constructed by taking the Zadeh extension of the function $\varphi : \mathbb{R} \times \mathbb{R} \to \mathbb{R}$, defined by $\varphi(x, y) = \min\{x, y\}$, where $y \in B$, fixed. This extension, with respect to set U, is given by

$$\widehat{\varphi}(U)(z) = \sup_{(x,y)\in\varphi^{-1}(z)} U(x, y), \forall\, z \in \mathbb{R}. \tag{11}$$

The set $\varphi^{-1}(z)$ can be written as

$$\varphi^{-1}(z) = \underbrace{\{(z, y); z \le y\}}_{I} \cup \underbrace{\{(x, z); z \le x\}}_{II}. \tag{12}$$

1. If $(x, y) \in I$, then $U(x, y) = U(z, y) = B(z)$. The extension in (12) can be rewritten as

$$\widehat{\varphi}(U)(z) = \sup_{(x,y)\in I} U(x, y) = \sup_{(z,y), z\le y} B(z) = B(z).$$

2. If $(x, y) \in II$, then $U(x, y) = U(x, z) = B(x)$. In this case the extension (12) is

$$\widehat{\varphi}(U)(z) = \sup_{(x,y)\in(IV)} U(x, y) = \sup_{(x,z), z\le x} B(x) = \begin{cases} 1, \text{ if } z < b_1^+ \\ B(z), \text{ if } b_1^+ \le z < b_0^+ \\ 0, \text{ if } z \ge b_0^+ \end{cases} \tag{13}$$
$$= \Omega(z). \tag{14}$$

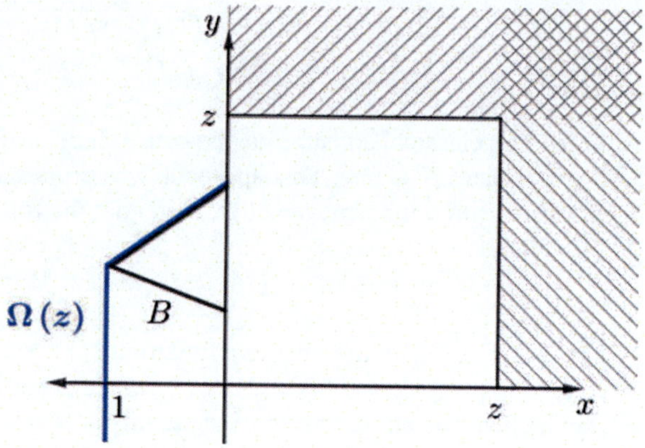

Fig. 2 Hatched region represents the set $\phi^{-1}(z)$ and fuzzy set in blue is the extension $\widehat{\phi}(U)(z) = \Omega(z)$

Therefore $\widehat{\varphi}(U)(z) = B(z) \vee L_{\Omega}(z) = \Omega(z)$, or yet, $\widehat{\phi} = \Omega$. This analysis is illustrated in Fig. 2.

Let be, in (1), $f : U \to \mathbb{R}$, where U is the domain. In the classical sense, the element $\overline{x}$ is the global maximum of (1) if there is $\delta > 0$ such that

$$f(\overline{x}) \geq f(x), \forall\, x \in \Omega. \tag{15}$$

Indeed, $\overline{x}$ can be seen as a point such that $f(\overline{x}) = \max f(\Omega)$.

Since Ω is a fuzzy set, we need to compare $f(\overline{x})$ with $\widehat{f}(\Omega)$, then we take the cylindrical extension $V = A \times \mathbb{R}$ with $A = \widehat{f}(\Omega)$:

$$V(f(x), f(y)) = A(f(x)) = \widehat{f}(\Omega)(w),$$

where $(f(x), f(y)) \in \mathbb{R} \times \mathbb{R}$, and $w = f(x)$.

The inequality in maximum definition (15) can be rewritten as $f(\overline{x}) \geq f(x) \Leftrightarrow \min\{f(\overline{x}), f(x)\} = f(x)$, hence we consider the function

$$m(f(x), f(y)) = \min\{f(x), f(y)\},$$

that represents the set of all $x \in \mathbb{R}$ such that $f(x)$ is smaller or equal than to the maximum value $f(\overline{x})$. We extend the function m by Zadeh extension principle:

$$\widehat{m}(V)(w) = \sup_{(f(x), f(y)) \in m^{-1}(w)} V(f(x), f(y)), \forall\, z \in \mathbb{R}. \tag{16}$$

Resembling what was done with $\varphi^{-1}(z)$ in (12), we have

$$m^{-1}(w) = \underbrace{\{(w, f(y));\, w = f(x) \text{ and } w \leq f(y)\}}_{III} \cup \underbrace{\{(f(x), w);\, w = f(y) \text{ and } w \leq f(x)\}}_{IV}.$$

If $(f(x), f(y)) \in III$, then $V(f(x), f(y)) = V(w, f(y)) = A(w) = A(f(x))$. From (16),

$$\begin{aligned}\widehat{m}(V)(w) &= \sup_{(f(x),f(y))\in m^{-1}(w)} V(f(x), f(y)) \\ &= \sup_{(w,y),w=f(x),w\leq f(y)} A(w) \\ &= \sup_{w=f(x),w\leq f(y)} \widehat{f}(\Omega)(w) \\ &= \widehat{f}(\Omega)(w).\end{aligned}$$

If $(f(x), f(y)) \in IV$, then $V(f(x), f(y)) = V(f(x), w) = A(f(x))$, and the extension in (16) becomes

$$\begin{aligned}\widehat{m}(V)(w) &= \sup_{(x,w),w=f(y),w\leq f(x)} \widehat{f}(\Omega)(w) \\ &= \begin{cases} 1, \text{ if } f(x) \geq (\widehat{f}(\Omega))_1^- \\ \widehat{f}(\Omega)(z), \text{ if } (\widehat{f}(\Omega))_1^- > f(x) \geq (f(\Omega))_0^- \\ 0, \text{ if } f(x) < (f(\Omega))_0^- \end{cases} \\ &= O(f(x)).\end{aligned}$$

Therefore $\widehat{m}(z) = \widehat{f}(\Omega)(z) \vee O(f(x)) = O(f(x))$. Note that $\widehat{f}(\Omega)$ is contained in $\widehat{m}$.

In classical case the value $f(\overline{x})$ is given by the intersection $m \cap \overline{m^C}$, where $m = \{\overline{x} \in \mathbb{R}; f(\overline{x}) \leq f(x), \forall x \in \mathbb{R}\}$:

$$f(\overline{x}) = m \cap \overline{m^C}. \tag{17}$$

Then it is necessary to search for $\overline{x}$ such that (17) is satisfied for $\widehat{m}$ a fuzzy set, that is,

$$f(\overline{x}) = \widehat{m} \cap \overline{\widehat{m}^C}. \tag{18}$$

Consider the set

$$O(x) = \overline{\widehat{m}^C}(f(x)).$$

The above equality means that the membership degree of x in O is calculated by taking the its image $f(x)$ and considering its membership degree in fuzzy set $\overline{\widehat{m}^C}$. This process is depicted in Fig. 3.

Precisely, the membership function of O is given by

$$O(x) = \begin{cases} 0, \text{ if } f(x) \geq (\widehat{f}(\Omega))_1^- \\ 1 - \widehat{f}(\Omega)(f(x)), \text{ if } (\widehat{f}(\Omega))_1^- > f(x) \geq (f(\Omega))_0^- \\ 1, \text{ if } f(x) < (f(\Omega))_0^- \end{cases}.$$

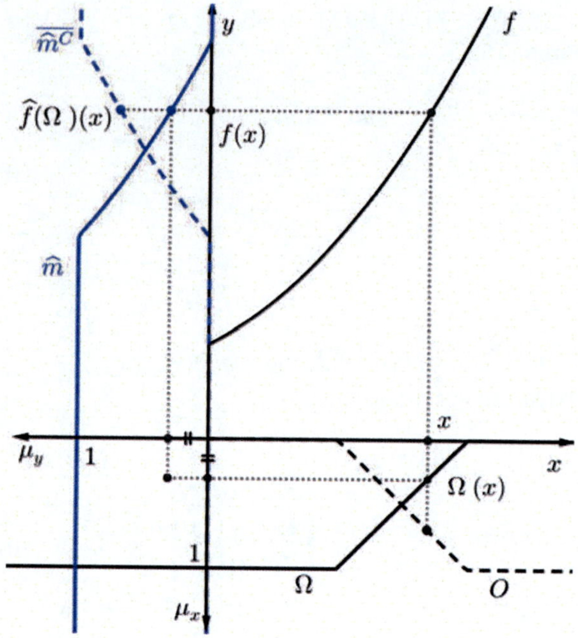

Fig. 3 The solid blue curve represents the membership function of $\widehat{m}$. In dashed blue curve represents the fuzzy set $\overline{\overline{x}}$. The sets Ω and O are represented by the solid and dashed black, respectively

If f is an increasing injection, then a combination of (7) and (8) leads to

$$O(x) = \begin{cases} 0, \text{ if } x \geq \Omega_1^- \\ 1 - \Omega(x), \text{ if } \Omega_1^- > x \geq \Omega_0^- \\ 1, \text{ if } x < \Omega_0^- \end{cases} . \quad (19)$$

In other words, if f is increasing and injective objective function, and constraint set in form $\Omega = \{x \in \mathbb{R}; x \leq B\}$ with $B \in \mathcal{F}(\mathbb{R})$, then the objective set O boils down to Ω: $O = \Omega^C$. Moreover, this definition of O coincides with the one given by Zimmermann [19, 20].

From (2) and from Bellman-Zadeh decision principle [4] in (3), the maximum $\overline{x}$ is given by

$$\overline{x} = \text{argmax } D(\overline{x}) = \text{argmax}(O(x) \cap \Omega(x)).$$

Therefore it is possible to redefine the notion of maximum.

Definition A point $\overline{x}$ is the global maximum of (1) if it satisfies

$$D(\overline{x}) > 0 \text{ and } D(\overline{x}) \geq D(x), \forall\, x \in \mathbb{R}^n. \quad (20)$$

A point $\overline{x}$ is the local maximum if there is $\delta > 0$ such that

$$D(\overline{x}) > 0 \text{ and } D(\overline{x}) \geq D(x), \forall\, x \in B(\overline{x}, \delta). \tag{21}$$

Expression (20) means that $\overline{x}$ must satisfies

$$\begin{cases} \Omega(\overline{x}) \cap O(\overline{x}) > 0 \\ \Omega(\overline{x}) \cap O(\overline{x}) \geq \Omega(x) \cap O(x), \forall\, x \in \mathbb{R}^n \end{cases}.$$

And expression (21) means that $\overline{x}$ must satisfies

$$\begin{cases} \Omega(\overline{x}) \cap O(\overline{x}) > 0 \\ \Omega(\overline{x}) \cap O(\overline{x}) \geq \Omega(x) \cap O(x) \cap B(\overline{x}, \delta) \end{cases}.$$

If f is increasing and injective and Ω has right end point $\Omega_\alpha^+ : U \to [0, 1]$ continuous with respect to α, then $\Omega(x) \leq 0.5 \Rightarrow 1 - \Omega(x) \geq 0.5$ and $\Omega(x) \geq 0.5 \Rightarrow 1 - \Omega(x) \leq 0.5$. Hence, from (19), we obtain

$$\min\{O(x), \Omega(x)\} = \min\{1 - \Omega(x), \Omega(x)\} \leq 0.5.$$

Therefore $D(\overline{x}) = 0.5$.

With these concepts at hand, let us return to the ETS problem.

4 Carbon Market as a Constraint

In terms of forecasting, policy makers may impose a country – Germany, for example – to emit at most 760 $MtCO_2$. This target T depends on the region/ sector and the decision-makers must choose how much less than T will be really emitted. This "how much" is uncertain and we model this linguistic term by a triangular fuzzy number $B = (L; E; T)$. So we create the constraint on emissions: $x \leq B$, where x is a measure of CO_2 emission.

We consider the logarithmic utility function [7] as the objective function in the problem. Our focus is only on the relation emissions *versus* utility, and not in the utility format function. Hence, we consider the utility as a function that only depends on x, the emission. Therefore, the goal is to maximize $f(x) = \log(x)$ subject to $x \leq B$. The problem (1) boils down to (22).

$$\begin{aligned} &\max\ \log(x) \\ &\text{s. t. } x \leq (L; E; T), \end{aligned} \tag{22}$$

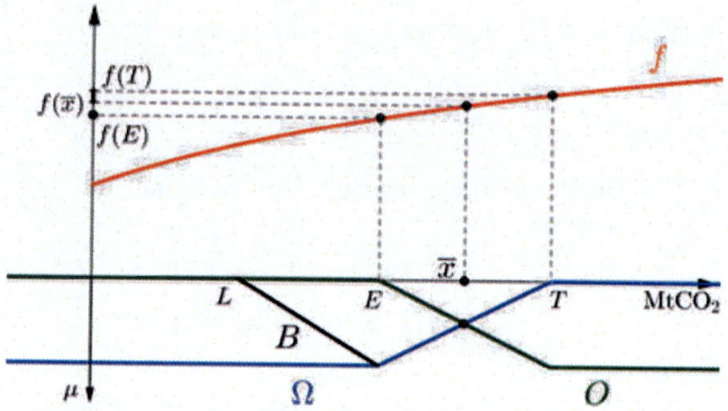

Fig. 4 The membership function of $B = (L; E; T)$. The utility function f is represented by the red curve. The constraint set Ω is represented by the blue curve. The goal O is represented by the green curve. The decision scheme leads to $\overline{x}$ as a solution for problem (22)

Following the idea proposed in Sect. 3, the objective O and constraint Ω sets have respectively, membership functions

$$O(x) = \begin{cases} 0, \text{ if } x \leq E \\ \frac{x-E}{T-E}, \text{ if } E < x \leq T \\ 1, \text{ if } x > T \end{cases}$$

and

$$\Omega(x) = \begin{cases} 1, \text{ if } x \leq E \\ \frac{T-x}{T-E}, \text{ if } E < x \leq T \\ 0, \text{ if } x > T \end{cases} .$$

The solution, represented in Fig. 4, is given by max $O \cap \Omega$, which corresponds to the point $\overline{x}$.

The lack $T - \overline{x}$ is the chosen quantity to be sell in carbon market and $f(T) - f(\overline{x})$ is the lost in utility (objective) function. Even though E would be the best decision in terms of reduction of emissions, $f(E)$ may not be the best decision in terms of utility decision. Depending on the value of tCO_2 in carbon market, it is desirable to sell more and more tons of carbon, and the lost $f(T) - f(\overline{x})$ causes less in the company revenue.

By choosing different forms of B, it is possible to put more importance on less emission. For B gaussian fuzzy set in form

$$B(x) = e^{-\frac{(x-E)^2}{\sigma^2}}, \tag{23}$$

with $\sigma = \frac{T-E}{2}$, for example. The σ could be any fraction of $T - E$. By choosing smaller σ, B would be narrower to E, modelling a mayor intention to attain E $MtCO_2$ emission. The solution of (22) is $\overline{\overline{x}}$. The decision scheme is represented in Fig. 5.

In practical terms, considering the carbon allowance price in 2019 at \$30 per carbon dioxide tonne [1]. The raised quantity obtain in cap and trade system is then $\$30(T - \overline{x})10^6$ in the first case and $\$30(T - \overline{\overline{x}})10^6$ in the second.

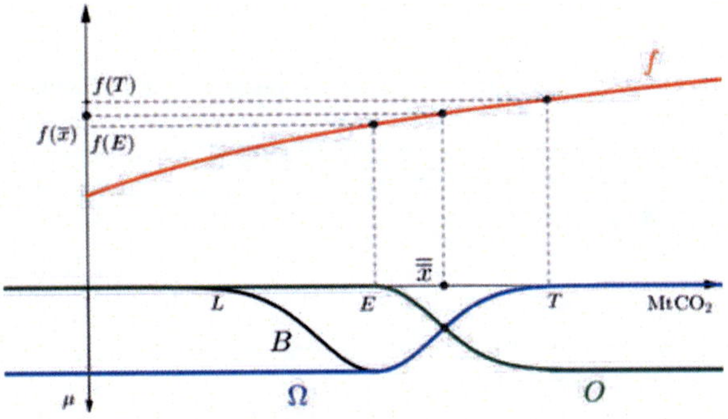

Fig. 5 The membership function of $B = (L; E; T)$.The utility function f is represented by red curve. The constraint set Ω is represented by the blue curve. The goal O is represented by the green curve. The decision scheme leads to $\bar{x}$ as a solution for problem (22) with B given in (23)

Comparing the two cases, the first solution led to more emission, with less gain in carbon market, but with little lost in the utility function. On the other hand, the second led to a solution with less emission and with more gain in carbon market and more lost in the utility function.

Furthermore, the trend is to raise carbon allowances price. So this study can be mixed with risk study on ETS. Also, this study may be done not in terms of utility function, but instead in marginal abatement cost, in order to compare if the profits on emissions trending are greater than the investment in reducing carbon dioxide emissions.

5 Final Remarks

Carbon market arose in a context of attempt to decelerate climate changes and impacts. Here we explained briefly the dynamics of carbon market and the subjective decision between pushing economy towards growing and slowing down the pollution in the whole world. We translated the emission target as a fuzzy constraint, and applied the Bellman-Zadeh scheme in order to find a solution for the problem. We extended the notion of solution by Zadeh's extension principle. By this incipient approach it is possible to estimate numerically the intention to emits less carbon and maintain good levels of welfare state. This solution not only ensures low emissions of CO_2, but also shows how many millions tons of carbon could be sold in carbon market. In terms of plan ahead policies on reduction of carbon emissions the next step is adopt this method to marginal abatement cost curves, the tool used to estimate the tax price on emissions of carbon dioxide.

Acknowledgements Pinto, N. J. B. thanks to CAPES for Financial support under grant no 1691227, Esmi, E. thanks FAPESP under grant no 2016/26040-7 and Barros, L. C. thanks CNPq under grant no 306546/2017-5.

References

1. J.E. Aldy, G. Gianfrate, Future-Proof your climate strategy. Harv. Bus. Rev. 86–97 (2019)
2. L.C. Barros, R.C. Bassanezi, W.A. Lodwick, *A First Course in Fuzzy Logic, Fuzzy Dynamical Systems, and Biomathematics* (Springer, Berlin, Heidelberg, 2017)
3. L. C. Barros, N. J. B. Pinto, E. Esmi, On fuzzy optimization foundation, in *Fuzzy Techniques: Theory and Applications. IFSA/NAFIPS 2019 Proceedings*. Advances in Intelligent Systems and Computing, vol. 1000 (Springer, Cham, 2019), pp. 148–156
4. R.E. Bellman, L.A. Zadeh, Decision-making in a fuzzy environment. Manag. Sci. **17**(4), B141–B164 (1970)
5. Y. Dafermos, M. Nikolaidi, G. Galanis, Climate change, financial stability and monetary policy. Ecol. Econ. **152**, 219–234 (2018)
6. Gilfillan et al., United Nations framework convention on climate change. *Statistical Review of World Energy* (2019). http://www.globalcarbonatlas.org/en/CO2-emissions. Accessed 26 Jan 2020
7. M. Golosov, J. Hassler, P. Krusell, A. Tsyvinski, Optimal taxes on fossil fuel in general equilibrium. Econometrica **82**(1), 41–88 (2014)
8. International Bank for Reconstruction and Development/The World Bank: Carbon Pricing Watch 2016 (2016). https://openknowledge.worldbank.org/bitstream/handle/10986/24288/CarbonPricingWatch2016.pdf?sequence=4&isAllowed=y. Accessed 22 Jan 2020
9. International Bank for Reconstruction and Development/The World Bank: Executive Summary, in *The First International Research Conference on Carbon Pricing* (World Bank Publications, New Delhi, 2019), pp. 11–15. https://static1.squarespace.com/static/54ff9c5ce4b0a53decccfb4c/t/5d9e0acf686073520c537675/1570638558624/Research+Conference+Report.2019+final.pdf. Accessed on 23 Jan 2020
10. P. Järvensivu, A post-fossil fuel transition experiment: exploring cultural dimensions from a practice-theoretical perspective. J. Clean. Prod. **169**, 143–151 (2017)
11. J. Meadowcroft, What about the politics? Sustainable development, transition management, and long term energy transitions. Policy Sci. **42**(4), 323–340 (2009)
12. I. Monasterolo, A. Roventini, T.J. Foxon, Uncertainty of climate policies and implications for economics and finance: an evolutionary economics approach. Ecol. Econ. **163**, 177–182 (2019)
13. R. Muradian, U. Pascual, Ecological economics in the age of fear. Ecol. Econ. **169**, 106498 (2020)
14. S.C. Newbold, C. Griffiths, C. Moore, A. Wolverton, E. Kopits, A rapid assessment model for understanding the social cost of carbon. Clim. Change Econ. **4**(1), 1350001-1-1350001–40 (2013)
15. United Nations Climate Change: Emissions Trading. https://unfccc.int/process/the-kyoto-protocol/mechanisms/emissions-trading. Accessed 22 Jan 2020
16. World Meteorological Organization. United In Science: High-level synthesis report of latest climate science information convened by the Science Advisory Group of the UN Climate Action Summit 2019 (2019). https://ane4bf-datap1.s3-eu-west-1.amazonaws.com/wmocms/s3fs-public/ckeditor/files/United_in_Science_ReportFINAL_0.pdf?XqiG0yszsU_sx2vOehOWpCOkm9RdC_gN. Accessed 22 Jan 2020
17. L.A. Zadeh, Fuzzy sets. Inf. Control **8**(3), 338–353 (1965)
18. L.A. Zadeh, The concept of a linguistic variable and its application to approximate reasoning-I. Inf. Sci. **8**(3), 199–249 (1975)
19. H.-J. Zimmermann, Description and optimization of fuzzy systems. Int. J. Gen. Syst. **2**, 209–215 (1976)
20. H.-J. Zimmermann, Fuzzy programming and linear programming with several objective functions. Fuzzy Sets Syst. **1**, 45–55 (1978)

Optimization of Neural Network Models for Estimating the Risk of Developing Hypertension Using Bio-inspired Algorithms

Patricia Melin, Ivette Miramontes, Oscar Carvajal, and German Prado-Arechiga

Abstract Nowadays, the use of intelligent systems can help in achieving a quick and timely diagnosis, with the aim of avoiding or controlling some diseases. In this case, the general goal of this work is to provide an intelligent model capable of solving a real life health problem, such as the risk of developing hypertension. For this reason, a new computational model is proposed using a neural network that has the ability to estimate the risk of developing high blood pressure in the next four years, which is optimized using the Flower Pollination Algorithm and Ant Lion Optimizer. The neural network model has seven inputs that are: age, gender, body mass index, systolic pressure, diastolic pressure, if the patient smokes, and if the patient has parents with hypertension, and one output, which is the risk of developing hypertension in the next 4 years. Simulation results show the advantage of the proposed approach.

Keywords Blood pressure · Hypertension · Optimization · Flower Pollination Algorithm (FPA) · Ant Lion Optimizer (ALO)

1 Introduction

Optimization can be understood as the mathematical process for finding the best solution to a problem [1]. Intelligent models have been previously optimized using bio-inspired algorithms [2–5]. One of the most important characteristic related to metaheuristic algorithms is that most of them benefit from stochastic operators that allow them to avoid local solutions much easier than deterministic algorithms. The No Free Lunch theorem allows researchers to propose and use bio-inspired algorithms because one algorithm can be very efficient in solving certain types of problems, but ineffective on other types of problems. The main contribution of this paper is the creation of an intelligent model applied to real life health problems. The model is able to obtain the risk of developing hypertension in a period of 4 years with good accuracy.

P. Melin (✉) · I. Miramontes · O. Carvajal · G. Prado-Arechiga
Tijuana Institute of Technology, Tijuana, BC, México
e-mail: pmelin@tectijuana.mx

B. Bede et al. (eds.), *Fuzzy Information Processing 2020*,
Advances in Intelligent Systems and Computing 1337,
https://doi.org/10.1007/978-3-030-81561-5_19

The Framingham Heart Study, which is a well-known study in the medical area, is able to calculate different risks; it is then taken as the basis for the study to design the proposed artificial neural network. The seven inputs to the neural network are: age, gender, body mass index, systolic pressure, diastolic pressure, if the patient smokes and if patient has parents with hypertension, corresponding to different risk factors that lead to the development associated with the disease and, and one output, the percentage derived from the risk that a patient has in developing hypertension in a period of 4 years.

Initially, the architecture was experimentally found, but to improve the obtained results, the architecture of the neural network was optimized. In this case, the necessary adjustments to the architecture of the neural network, this means, that varying the number of hidden layers and the number of neurons in the hidden layers. In this case, a database with 500 patients is used to train the neural network in each experiment, and this database has the risk factors mentioned before. For achieving this, two bio-inspired algorithms are used, the first one is the Flower Pollination algorithm and the second one is the Ant Lion Optimizer algorithm.

In addition, a Simple Enumeration Method (SEM) was used for comparison purposes, even if the SEM has all the possible combinations of neurons per hidden layer and number of hidden layers, it does not guarantee the best possible solution due to the fact that the weights related to each neuron are generated randomly in the training of the neural network, so, for this reason, we use bio-inspired algorithms.

This paper has been organized as follows: in Sect. 2 the literature review is presented, in Sect. 3 the proposed method is presented, in Sect. 4 the results and discussions are presented and in Sect. 5 the conclusions obtained after carrying out the tests based on the optimization of the neural network are offered.

2 Literature Review

2.1 Flower Pollination Algorithm

The Flower Pollination Algorithm [6] was originally introduced by Xin-She Yang in 2012, and this metaheuristic is inspired by using the process of plant pollination. The method uses both biotic and abiotic pollination, which can be computationally related to local and global search respectively. In addition, the method uses other parts of the pollination process to find the best solution to the problem to solve.

2.2 Ant Lion Optimizer

The Ant Lion Optimizer was originally proposed by Mirjalili in 2015 [7], which imitates the hunting mechanism derived from antlions in nature. The method uses five steps, that are used prior to hunting, and they are the random walk of the ants, the construction of traps, the trapping of ants in traps, the capture of prey and the reconstruction of traps.

2.3 Blood Pressure and Hypertension

Blood pressure is defined as the pressure associated with the blood within the arteries, and it is produced by the contraction of the heart muscle, and with this, the oxygen and nutrients are transferred to the whole body [8, 9]. The blood pressure has two components: Systolic pressure (the highest number), which occurs when the heart contracts and Diastolic pressure (the lowest number) is when the heart relaxes and refills with blood, and it is measured in millimeters of mercury (mmHg) [8].

Based on the European Guidelines for management of hypertension, the normal blood pressure is below 139 mmHg in systolic pressure and below 89 mmHg in diastolic pressure [10]. High blood pressure or hypertension is the sustained elevation of blood pressure above normal limits [11], and according to the European Guidelines for management of hypertension, it has three grades:

- Grade 1: is 140–159 mmHg in systolic pressure or 90–99 mmHg in diastolic pressure.
- Grade 2: is 160–179 mmHg in systolic pressure or 100–109 mmHg in diastolic pressure.
- Grade 3: is 180 or higher mmHg in systolic pressure or 110 or higher mmHg in diastolic pressure.

In addition, there is another classification that is called Isolated Systolic Hypertension, and this occurs when the Systolic pressure is higher than 140 mmHg, but the Diastolic pressure is lower than 90 mmHg [10].

2.4 Framingham Heart Study

The Framingham Heart Study began in 1948 [12], directed by the National Cardiac Institute, which is carried out to identify risk factors that have been the cause for the development of cardiovascular diseases.

To obtain the diagnosis, the patient information is used, such as: Age, Systolic Blood Pressure (SBP), Diastolic Blood Pressure (DBP), Body Mass Index (BMI), Sex, Smoking, Parental Hypertension.

2.5 Neural Networks

An artificial neural network is an information processing system, which has certain performance characteristics in common with biological neurons. Artificial neural networks have been developed as generalizations of mathematical models of human cognition [13]. A neural network consists of many simple processing elements called neurons or nodes. Each neuron is connected to other neurons through directed communication links, each with an associated weight, and this represents the information used by the network to solve a problem. Neural networks can be applied to different problems, such as robotics [14], function approximation, and time series prediction [15] or the search for solutions to optimization problems with restrictions [16]. The internal state that each neuron has is known as activity or activation, which is a function of the inputs it has received. Typically, a neuron sends its activation as a signal to several other neurons, and this has the capacity for sending only one signal at a time, even though the signal is transmitted to several other neurons [17].

3 Proposed Method

The proposed method is presented in summarized form in Fig. 1, as can be noticed a neural network based on [18–21] is related to the objective function of the bioinspired algorithms, and the block diagram shows the optimization process; as in any other optimization method, the first thing to do is the creation of the initial population and the fitness calculation associated with that one, after that the iterative process starts obtaining different solutions.

The individuals represent the number of hidden layers (HL), the number of neurons in the first hidden layer (H1), and the numbers of neurons in the second layer (H2). In Fig. 2, the search space associated with the FPA and ALO is presented. The number of hidden layers, due to the complexity derived from the problem was limited to 2, the number of neurons per hidden layer, which were limited to 30, this is due to the information with which we are working. 30 experiments were carried out, where the parameters of the algorithms were varied.

The weights from the neural network are not represented in the individual, due to the fact that they are generated randomly in the training process associated with the neural network, they are naturally stochastic, and if we consider the weights of the neural network the dimensions related to each algorithm increase and this is not desired; once the neural network is trained, the Mean Square Error is obtained.

The individuals, which in the case of FPA are the pollen, and in the case of ALO will be the ants and antlions, make the necessary changes in the architecture to obtain the one that provides the lower error when simulating the information, for which the mean square error (MSE) is used as an objective function, in order to reduce the error in the training, which is found in (1):

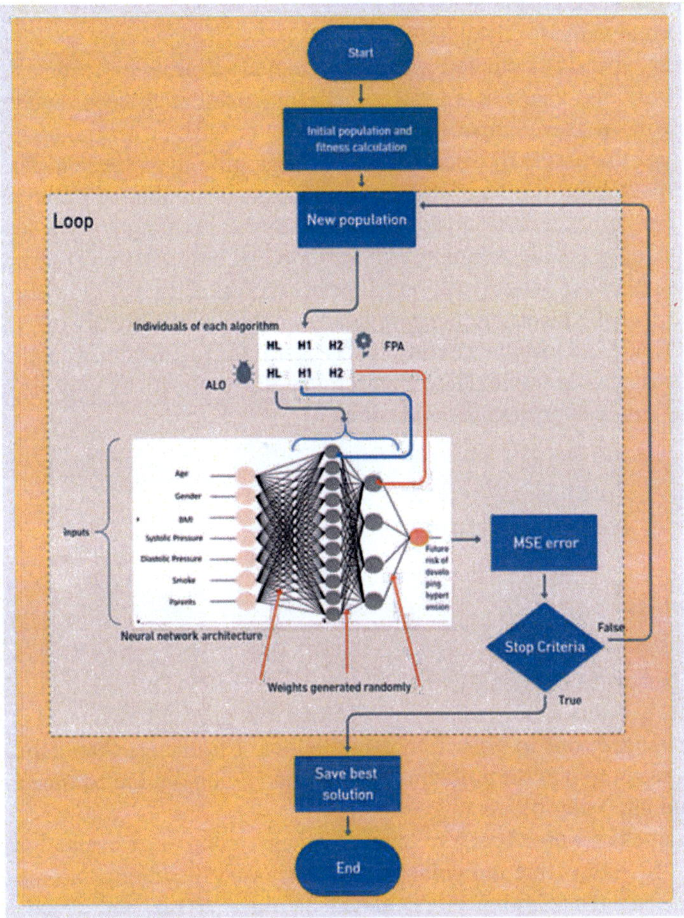

Fig. 1 Proposed method

Neural Network Parameters			
H. Layers		Neurons H.L1	Neurons H.L 2
1	2	1 - 30	1 – 30

Fig. 2 Search space of FPA and ALO

$$MSE = \frac{1}{2}\sum_{i=1}^{n}\left(\hat{Y}_i - Y_i\right)^2 \quad (1)$$

where:
n = number of data points,

$\hat{Y}$ = output value by the model,
Y = the real value for data point i.

This MSE error is calculated with the predicted information associated with the neural network and a new set of real patient information. As mentioned before the neural network has seven inputs, which represent the risk factors related to a person and an output that is the risk percentage of developing hypertension. The stopping criteria derived from the algorithm is the maximum number of iterations for this particular problem, the MSE error can be used as the stopping criteria depending of the problem, once the algorithm breaks the loop, the best solution is saved, this step is important, because not only we saved the best solution derived from the algorithm but also we save the whole architecture related to the neural network this includes all the weights and biases of the neurons, this is useful for implementation purposes, the neural network can be implemented in useful applications, for example in embedded systems, or in a web or mobile application.

4 Results and Discussion

4.1 Results

To test the performance related to the algorithm and analyze its operation, 30 different experiments are performed, where, in each one, the parameters related to the algorithm are changed, this in order to obtain the ideal parameters that help to solve in an optimal way the present problem. Tables 1 and 2 contain the parameters used in the FPA and ALO algorithms respectively.

The parameters were chosen to obtain the same number of evaluations for both algorithms. In Table 1 the individual size is increased by adding two more in each experiment, and the iteration is calculated dividing 930 by the actual individual size, as 930 is the desired number of evaluations which is the number related to the evaluations of the Simple Enumeration Method. The p parameter associated with the FPA is important due to the fact that it controls the algorithm exploration and exploitation, while higher is p the algorithm performs exploration, otherwise when p is small the exploitation is performed. It can be observed that the first experiment of the FPA was the one with the best fitness value.

In Table 2, the 30 experiments derived from ALO and SE are summarized. In the case of ALO, the number of Agents or individuals are increasing by adding two to the value associated with the Agents of the first experiment, and the number of iterations is calculated dividing 930 by the number of Agents in the actual experiment. This calculation guarantees the same number of evaluations with the others two methods, and the comparison is equivalent for the three methods, because if any of them has more iterations for example, it means more evaluations and probably a best average of the MSE error. In the case of ALO, the best fitness value is achieved in experiment 24 and experiment 22 for the SE. In the case of ALO, the exploration is guaranteed

Table 1 Parameters used in FPA for each experiment

FPA				
Exp	Ind	Iteration	p	MSE error
1	10	93	0.2	***5.78E-04***
2	12	78	0.7	1.21E-03
3	14	66	0.6	1.38E-03
4	16	58	0.4	1.69E-03
5	18	52	0.1	1.01E-03
6	20	47	0.8	1.20E-03
7	22	42	0.9	1.58E-03
8	24	39	0.3	8.41E-04
9	26	36	0.5	1.17E-03
10	28	33	0.7	9.20E-04
11	30	31	0.2	1.89E-03
12	32	29	0.8	1.41E-03
13	34	27	0.4	8.61E-04
14	36	26	0.5	1.18E-03
15	38	24	0.6	8.23E-04
16	40	23	0.8	1.66E-03
17	42	22	0.6	1.75E-03
18	44	21	0.4	1.15E-03
19	46	20	0.9	2.83E-03
20	48	19	0.2	1.73E-03
21	50	19	0.3	1.07E-03
22	52	18	0.7	1.10E-03
23	54	17	0.8	1.32E-03
24	56	17	0.1	1.20E-03
25	58	16	0.6	1.38E-03
26	60	16	0.2	1.56E-03
27	62	15	0.8	1.72E-03
28	64	15	0.4	2.20E-03
29	66	14	0.5	1.42E-03
30	68	14	0.3	1.73E-03
Average				1.39E-03
Standard deviation				4.568E-04

Table 2 Parameters used in ALO for each experiment

ALO				SE
Exp	Iterations	Agents	MSE error	MSE error
1	52	18	1.785E-03	1.491E-03
2	47	20	1.291E-03	1.927E-03
3	42	22	9.701E-04	4.243E-03
4	39	24	1.429E-03	5.790E-03
5	36	26	1.539E-03	2.631E-03
6	33	28	2.065E-03	1.444E-03
7	31	30	1.900E-03	4.671E-03
8	29	32	1.151E-03	5.277E-03
9	27	34	1.271E-03	4.300E-03
10	26	36	1.507E-03	3.769E-03
11	24	38	1.129E-03	2.799E-03
12	23	40	1.467E-03	1.383E-03
13	22	42	2.014E-03	2.782E-03
14	21	44	1.742E-03	5.922E-03
15	20	46	4.200E-03	4.974E-03
16	19	48	1.203E-03	2.498E-03
17	19	50	1.502E-03	1.463E-03
18	18	52	1.641E-03	3.412E-03
19	17	54	1.393E-03	1.762E-03
20	17	56	1.040E-03	4.338E-03
21	16	58	9.467E-04	2.758E-03
22	16	60	1.591E-03	***1.047E-03***
23	15	62	1.049E-03	4.915E-03
24	15	64	***7.341E-04***	5.657E-03
25	14	66	3.213E-03	6.103E-03
26	14	68	1.329E-03	3.298E-03
27	13	70	7.825E-04	3.669E-03
28	13	72	1.333E-03	5.492E-03
29	13	74	1.591E-03	4.612E-03
30	12	76	1.525E-03	2.309E-03
Average			1.544E-03	3.558E-03
Standard deviation			6.875E-04	1.571E-03

in the selection using the Roulette Wheel, and the exploitation is changed while the iteration is getting higher due to the fact that the search space is reduced. The simple enumeration method is the one with all the possible combinations related to the search space of the bio-inspired algorithms are used.

In Fig. 3, a comparison of the best experiments obtained by both methods is performed, for the FPA 10 individuals with 93 iterations were used and an MSE of 5.78E-04 was obtained. In the case of ALO, 64 agents with 15 iterations are used and an MSE of 7.34E-04 was obtained.

In Fig. 4, the comparison of the fastest convergences of each method is presented. For the FPA it was obtained in experiment 15, in this case 38 individuals and 24 iterations were used, obtaining a MSE of 8.23 E-04. For ALO, experiment 20 was

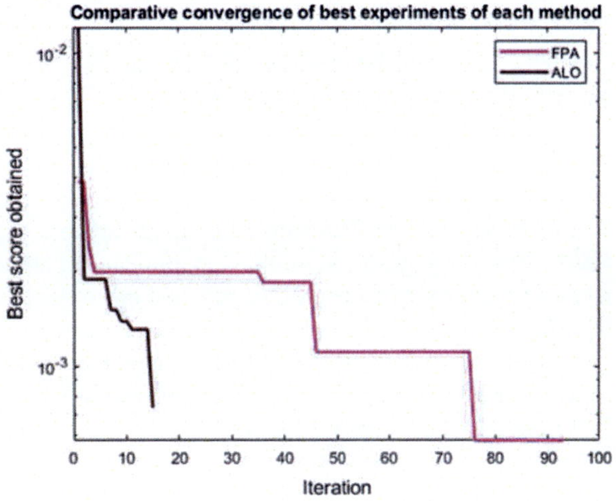

Fig. 3 Convergence of the best experiments

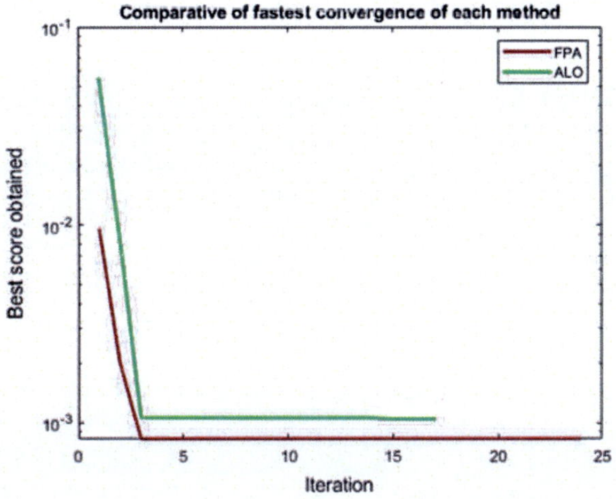

Fig. 4 Convergence of the best experiment

Table 3 Comparison of means of all methods

MSE	FPA	ALO	SEM
Average	0.001386461	0.001544412	0.0035578
Standard deviation	0.000457444	0.000687466	0.0015705
Experiments	30	30	30

the one that converged faster, using 56 agents with 17 iterations, obtaining an MSE of 9.47 E-04. To compare and verify which of the different methods yields the lower error, a statistical test is performed, using the Z test, which uses the following expression (2):

$$Z = \frac{(\overline{x}_1 - \overline{x}_2) - (\mu_1 - \mu_2)}{\sigma_{\overline{x}_1 - \overline{x}_2}} \tag{2}$$

where $\overline{x}_1 - \overline{x}_2$is the observed difference, $\mu_1 - \mu_2$ is the expected difference and $\sigma_{\overline{x}_1 - \overline{x}_2}$is the standard error of the differences.

For these cases, the null hypothesis states that the errors obtained by the Bioinspired methods (FPA and ALO) are greater than or equal to the errors obtained by the simple enumeration method (SEM). While the alternative hypothesis states that the errors obtained by the Bioinspired methods are less than the errors obtained by the SEM. Table 3 illustrates a comparison of means of all methods with 30 experiments; it can be observed that the bioinspired methods obtained a lower error compared with the SEM.

Table 4 presents the statistical parameters for this test.

Since it is observed that z test statistic value is Z = −6.432 for FPA and Z = −7.251 for ALO, is lower than the critical value $Z_c = -1.64$, it is then concluded that the null hypothesis is rejected and the alternative hypothesis is accepted. So, it can be concluded that there is sufficient evidence with a 5% level of significance to support the claim of the errors obtained by the bio-inspired method are lower than the errors obtained by the SEM, in Table 5 the mentioned results are presented.

Table 4 Statistical parameters

Parameters for statistical Z-test	
Critical value (Z_c)	−1.64
Significance level (α)	0.05
H0	$\mu 1 \geq \mu 2$
H_a (Claim)	$\mu 1 < \mu 2$
Experiments	30

Table 5 Results of the Z-test

Bioinspired methods	Not Bioinspired method	Z test	Evidence
FPA	SE	−6.432	Significant
ALO	SE	−7.251	Significant

4.2 Discussion

Previous works have focused on using neural networks for time series prediction for simulation and forecasting [22], pattern recognition. For example, the face and ear biometric measurement [23], classification problems in the medical area like arrhythmias classification [24] and using bio-inspired algorithms for optimizing fuzzy controllers to obtain the trajectory of an autonomous mobile robot using the Flower Pollination Algorithm [25]. Other works on optimizing neural networks with Genetic Algorithm are for plate recognition [26]. The findings of this paper agree with that of previous works that using soft computing achieves good results in each particular topic; another previous research was by J. Ben Ali et al. in 2018 focused on a neural network for blood glucose level prediction of Type 1 Diabetes [27]. The originality of the work in this paper is that the neural network not only makes a binary classification, it estimates the percentage of risk in a range of 1–100 percent for the next four years and also ALO and FPA are used to find the best architectures of the neural network, and it performs a 100% of risk detection of high blood pressure with simulated and real patients tested and compared with the Framingham Heart Study. The results in Table 5 suggest that by using optimization methods we have the capacity for achieving better performance than by not using them.

5 Conclusions and Future Work

The methodology proposed in this work achieves high-performance results in risk prevention of developing hypertension. ALO and FPA are helpful algorithms for optimizing the architecture of the neural networks. The relative small number of parameters, to take into account, in both algorithms compared with other metaheuristics in the literature allows simplicity in the optimization of the neural network architecture. The two algorithms obtain competitive results in optimizing models in medical problems when compared with other methods in the literature. In this case, 30 experiments were carried out with both algorithms and a simple enumeration method for comparing their performance, and based on the results obtained through the statistical tests of each metaheuristic, it can be concluded that the bio-inspired algorithms provide better performance and even less computational cost than using a simple enumeration method. The development of new models in the medical area are helpful for observing the behavior of health in incoming years, so this methodology can be applied not only with a hypertension health problem, it can also be used to other health problems or even in areas that are not from medicine, such as robotics, industrial applications, etc. In future research we envision extending this work by using this methodology to other heart problems or even other different health problems in the human body. In addition, other applications could be considered, like in granularity [28, 29] or in control [30, 31].

References

1. P. Jain, P. Kar, Non-convex optimization for machine learning. Found. Trends Mach. Learn. **10**, 142–336 (2017)
2. I. Miramontes, C.J. Guzman, P. Melin, G. Prado-Arechiga, Optimal design of interval type-2 fuzzy heart rate level classification systems using the bird swarm algorithm. Algorithms **11**(12) (2018). https://doi.org/10.3390/a11120206
3. J.C. Guzmán, I. Miramontes, P. Melin, G. Prado-Arechiga, Optimal genetic design of type-1 and interval type-2 fuzzy systems for blood pressure level classification. Axioms **8**(1) (2019). https://doi.org/10.3390/axioms8010008
4. P. Melin, G. Prado-Arechiga, I. Miramontes, J.C. Guzman, Classification of nocturnal blood pressure profile using fuzzy systems. J. Hypertens. **36**, e111–e112 (2018)
5. J.C. Guzman, P. Melin, G. Prado-Arechiga, Design of an optimized fuzzy classifier for the diagnosis of blood pressure with a new computational method for expert rule optimization. Algorithms **10**(3) (2017). https://doi.org/10.3390/a10030079
6. X.S. Yang, M. Karamanoglu, X. He, Flower pollination algorithm: a novel approach for multiobjective optimization. Eng. Optim. **46**(9), 1222–1237 (2014)
7. S. Mirjalili, The ant lion optimizer. Adv. Eng. Softw. **83**, 80–98 (2015)
8. V. Papademetriou, E.A. Andreadis, C. Geladari, *Management of Hypertension* (Springer International Publishing AG, Cham, 2019)
9. M. Paul et al., Measurement of Blood Pressure in Humans: a scientific statement from the American heart association. Hypertension **73**(5), e35–e66 (2019)
10. A. Zanchetti et al., 2018 ESC/ESH guidelines for the management of arterial hypertension. Eur. Heart J. **39**(33), 3021–3104 (2018)
11. G.L. Bakris, M.J. Sorrentino, *Hypertension, A Companion to Braunwald's Heart Disease, 3rd Edn.* (Elsevier, Philadelphia, 2018)
12. Framingham Heart Study (2019). Accessed 15 July 2019. https://www.framinghamheartstudy.org/risk-functions/hypertension/index.php
13. G. Cain, *Artificial Neural Networks: New Research* (Nova Science Publishers, Incorporated, New York, 2017)
14. L. Jin, S. Li, J. Yu, J. He, Robot manipulator control using neural networks: a survey. Neurocomputing **285**, 23–34 (2018)
15. J. Saadat, P. Moallem, H. Koofigar, Training echo estate neural network using harmony search algorithm. Int. J. Artif. Intell. **15**(1), 163–179 (2017)
16. G. Villarrubia, J.F. De Paz, P. Chamoso, F. De la Prieta, Artificial neural networks used in optimization problems. Neurocomputing **272**, 10–16 (2018)
17. C.C. Aggarwal, *Neural Networks and Deep Learning: A Textbook*, 1st edn. (Springer International Publishing, Cham, 2018)
18. P. Melin, G. Prado-Arechiga, I. Miramontes, M. Medina-Hernandez, Hybrid intelligent model based on modular neural network and fuzzy logic for hypertension risk diagnosis. J. Hypertens. **34**, e153 (2016)
19. I. Miramontes, G. Martínez, P. Melin, G. Prado-Arechiga, A hybrid intelligent system model for hypertension diagnosis, in *Nature-Inspired Design of Hybrid Intelligent Systems*. ed. by P. Melin, O. Castillo, J. Kacprzyk (Springer International Publishing, Cham, 2017), pp. 541–550

20. P. Melin, I. Miramontes, G. Prado-Arechiga, A hybrid model based on modular neural networks and fuzzy systems for classification of blood pressure and hypertension risk diagnosis. Expert Syst. Appl. **107**, 146–164 (2018)
21. J.C. Guzmán, P. Melin, G. Prado-Arechiga, Neuro-fuzzy hybrid model for the diagnosis of blood pressure, in *Nature-Inspired Design of Hybrid Intelligent Systems*. ed. by P. Melin, O. Castillo, J. Kacprzyk (Springer International Publishing, Cham, 2017), pp. 573–582
22. J. Soto, P. Melin, O. Castillo, A new approach for time series prediction using ensembles of IT2FNN models with optimization of fuzzy integrators. Int. J. Fuzzy Syst. **20**(3), 701–728 (2018)
23. P. Melin, D. Sánchez, Multi-objective optimization for modular granular neural networks applied to pattern recognition. Inf. Sci. (Ny) **460–461**, 594–610 (2018)
24. J. Amezcua, P. Melin, Classification of arrhythmias using modular architecture of LVQ neural network and type 2 fuzzy logic, in *Nature-Inspired Design of Hybrid Intelligent Systems*, 1st edn., ed. by P. Melin, O. Castillo, J. Kacprzyk (Springer International Publishing, Cham, 2017), pp. 187–194
25. O.R. Carvajal, O. Castillo, J. Soria, Optimization of membership function parameters for fuzzy controllers of an autonomous mobile robot using the flower pollination algorithm. J. Autom. Mob. Robot. Intell. Syst. **12**(1), 44–49 (2018)
26. J. Tarigan, Nadia, R. Diedan, Y. Suryana, Plate recognition using backpropagation neural network and genetic algorithm. Procedia Comput. Sci. **116**, 365–372 (2017)
27. J. Ben Ali, T. Hamdi, N. Fnaiech, V. Di Costanzo, F. Fnaiech, J.-M. Ginoux, Continuous blood glucose level prediction of type 1 diabetes based on artificial neural network. Biocybern. Biomed. Eng. **38**(4), 828–840 (2018)
28. M.A. Sanchez, O. Castillo, J.R. Castro, P. Melin, Fuzzy granular gravitational clustering algorithm for multivariate data. Inf. Sci. **279**, 498–511 (2014)
29. D. Sanchez, P. Melin, Optimization of modular granular neural networks using hierarchical genetic algorithms for human recognition using the ear biometric measure. Eng. Appl. Artif. Intell. **27**, 41–56 (2014)
30. O. Castillo, Type-2 fuzzy logic in intelligent control applications (Springer, 2012).
31. E. Ontiveros-Robles, P. Melin, O. Castillo, Comparative analysis of noise robustness of type 2 fuzzy logic controllers. Kybernetika **54**(1), 175–201 (2018)

Toward Improving the Fuzzy KNN Algorithm Based on Takagi–Sugeno Fuzzy Inference System

Eduardo Ramírez, Patricia Melin, and German Prado-Arechiga

Abstract In this paper we present a new approach in order to improve the performance of the Fuzzy K-Nearest Neighbor algorithm (Fuzzy KNN algorithm). We propose to use a Takagi–Sugeno Fuzzy Inference System with the Fuzzy KNN algorithm to improve classification accuracy. Also, we have used different measures to calculate the distance between the neighbors and the vector to be classified, such as the: Euclidean, Hamming, cosine similarity and city block distances. These distances represent the inputs for the Takagi–Sugeno Fuzzy Inference System. Simulation results with a classification problem show the potential of the proposed approach.

Keywords Fuzzy KNN algorithm · Takagi–Sugeno fuzzy inference system

1 Introduction

The K-Nearest Neighbor algorithm (KNN) is a widely used method to solve different classification problems [1, 11, 12]. Several variants based on Fuzzy Logic have been proposed, such as Type-1 Fuzzy Sets, Type-2 Fuzzy Sets, Possibilistic Methods, Intuitionistic Fuzzy Sets, Fuzzy Rough Sets, and Preprocessing methods via data reduction, that consider the aspects of member, distance, voting, independence of k, preprocessing, and center based [11].

Intuitionistic fuzzy sets have been used to improve the Fuzzy KNN classifiers. In [33] the IFSKNN intuitionistic classifier was proposed. This algorithm is based on the concept of the non-membership. A value of membership is computed for each instance as the distance to the mean of the class. The non-membership is computed by the distance to the nearest mean to the rest of the classes. The classification is

E. Ramírez · P. Melin (✉)
Tijuana Institute of Technology, Tijuana, BC, Mexico
e-mail: pmelin@tectijuana.mx

G. Prado-Arechiga
Tijuana Institute of Technology, 22010 Tijuana, BC, Mexico

B. Bede et al. (eds.), *Fuzzy Information Processing 2020*,
Advances in Intelligent Systems and Computing 1337,
https://doi.org/10.1007/978-3-030-81561-5_20

based on using both membership and non-membership to represent the distances computed by Fuzzy KNN algorithm.

Interval Type-2 fuzzy sets were proposed for a Fuzzy KNN algorithm in [5]. This approach consists as an alternative way of discarding of the necessity to setup the parameter of K nearest neighbor called *KInt*. This method was achieved by introducing interval type-2 fuzzy sets to represent the membership computed considering different values of the parameter *KInt*. Finally, the type reduction operation is performed to obtain the final result.

In [34] an approach based on Fuzzy C-Means was presented to obtain the membership weights of prototypes generated in an iterative way. Other example is presented in [35], with a PFKNN method, this consists, firstly to build a set of prototypes representing the border points of different clusters from the data discarding the non-relevant prototype in final phase.

It is well known that in different applications the implementation of fuzzy logic obtains a better [14–16].

In the mentioned approaches of the KNN algorithm only use a single measure to represent the distance between the unclassified vector to be classified and its nearest neighbors, this is the Euclidean distance. We believe that when considering different measures and not only the Euclidean distance, we are adding important information, which we can interpret as uncertainty. We propose the idea of implementing a Takagi–Sugeno Fuzzy Inference System to improve the performance of the original Fuzzy K-Nearest Neighbor algorithm offering a new method to improve the performance of the fuzzy KNN algorithm to be used on complex classification problems.

We used the MIT-BIH arrhythmia database to validate the proposed method [24, 25]. This database contains 48 half hour excerpts of two channel ambulatory electrocardiograms recordings belonging to 47 patients. The heartbeats are segmented and preprocessed [13, 18–20, 22, 23, 26, 27].

The rest of the paper is organized as follows. In Sect. 2, we present the problem statement and the proposed method. In Sect. 3 we describe a general review of the Fuzzy KNN algorithm in order to introduce the new approach of the Fuzzy KNN algorithm that we propose, as well as the important equations to calculate the measures of the distances that are used. In Sect. 4, we present details of the performed experiments and simulation results. Finally, conclusions and future work are presented in Sect. 5.

2 Problem Statement and Proposed Method

The proposed approach of the Fuzzy KNN algorithm using a Takagi–Sugeno Fuzzy Inference System, called TSFISKNN, is new because instead of calculating the inverse of the Euclidean distance to represent the membership degree of the unknown sample, we calculate the Euclidean, Hamming, cosine similarity and city block measures of the distances [28, 29]. The TSFISKNN uses these distances measures

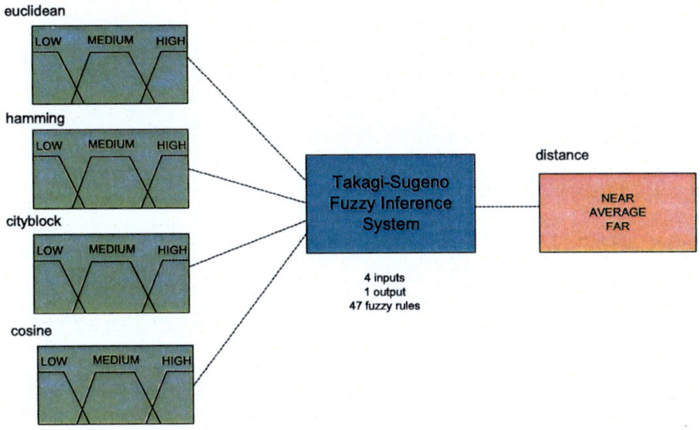

Fig. 1 Takagi–Sugeno fuzzy inference system

as inputs, and through the use of a set of fuzzy rules define the final distance that will replace the Euclidean distance used in the original Fuzzy KNN algorithm.

The TSFISKNN uses 4 inputs with three trapezoidal functions (Low, Medium and High), 47 rules, and 1 output with constant values (Near, Average and Far), and weighted average as defuzzification method, see Fig. 1. The rules for the TSFISKNN are presented below. The output of the TSFISKNN is used in the Fuzzy KNN algorithm. The rest of the steps of the Fuzzy KNN are executed as they are mentioned in the above section.

At this stage, we have not yet implemented Interval Type-2 Fuzzy Logic in the proposed method; however we are convinced that it could help to achieve our goal of improving the Fuzzy KNN algorithm.

The parameters for the Euclidean distance input are:

- Low = [−6.471 −0.275 2.511 7.817].
- Medium = [1.27 6.18 10.93 15.2].
- High = [9.027 12.69 16.78 22.98].

The parameters for the Hamming distance input are:

- Low = [0.855 0.895 0.9171 0.945].
- Medium = [0.905 0.935 0.965 0.995].
- High = [0.955 0.9844 1 1.04].

The parameters for the city block distance input are:

- Low = [−66.8 −1.73 15.37 79.57].
- Medium = [14.5 64.3 98.12 161].
- High = [95.8 146.6 177.242].

The parameters for the similarity distance input are:

- Low = [−0.559 −0.0621 0.09298 0.559].
- Medium = [0.0621 0.432 1.007 1.18].

- High = [0.6831.0991.31.8].

The parameters for the output are:

- Near = [0.879].
- Average = [6.55].
- Far = [30.28].

The rules of TSFISKNN are listed as follows:

1. **IF** (euclidean is LOW) **and** (hamming is LOW) **and** (cityBlock is LOW) **and** (similarity is LOW) **THEN** (distance is NEAR).
2. **IF** (euclidean is MEDIUM) **and** (hamming is MEDIUM) **and** (cityBlock is MEDIUM) **and** (similarity is MEDIUM) **THEN** (distance is AVERAGE).
3. **IF** (euclidean is HIGH) **and** (hamming is HIGH) **and** (cityBlock is HIGH) **and** (similarity is HIGH) **THEN** (distance is FAR).
4. **IF** (euclidean is LOW) **and** (hamming is LOW) **and** (cityBlock is LOW) **and** (similarity is MEDIUM) **THEN** (distance is NEAR).
5. **IF** (euclidean is LOW) **and** (hamming is LOW) **and** (cityBlock is LOW) **and** (similarity is HIGH) THEN (distance is NEAR).
6. **IF** (euclidean is LOW) **and** (hamming is LOW) **and** (cityBlock is MEDIUM) **and** (similarity is LOW) **THEN** (distance is NEAR).
7. **IF** (euclidean is LOW) **and** (hamming is LOW) **and** (cityBlock is HIGH) **and** (similarity is LOW) **THEN** (distance is NEAR).
8. **IF** (euclidean is LOW) **and** (hamming is MEDIUM) **and** (cityBlock is LOW) **and** (similarity is LOW) **THEN** (distance is NEAR).
9. **IF** (euclidean is LOW) **and** (hamming is HIGH) **and** (cityBlock is LOW) **and** (similarity is LOW) **THEN** (distance is NEAR).
10. **IF** (euclidean is MEDIUM) **and** (hamming is LOW) **and** (cityBlock is LOW) **and** (similarity is LOW) **THEN** (distance is NEAR).
11. **IF** (euclidean is HIGH) **and** (hamming is LOW) **and** (cityBlock is LOW) **and** (similarity is LOW) **THEN** (distance is NEAR).
12. **IF** (euclidean is LOW) **and** (hamming is LOW) **and** (cityBlock is MEDIUM) **and** (similarity is MEDIUM) **THEN** (distance is AVERAGE).
13. **IF** (euclidean is LOW) **and** (hamming is MEDIUM) **and** (cityBlock is LOW) **and** (similarity is MEDIUM) **THEN** (distance is AVERAGE).
14. **IF** (euclidean is MEDIUM) **and** (hamming is LOW) **and** (cityBlock is LOW) **and** (similarity is MEDIUM) **THEN** (distance is AVERAGE).
15. **IF** (euclidean is LOW) **and** (hamming is MEDIUM) **and** (cityBlock is MEDIUM) and (similarity is LOW) THEN (distance is AVERAGE).
16. **IF** (euclidean is MEDIUM) **and** (hamming is LOW) **and** (cityBlock is MEDIUM) **and** (similarity is LOW) **THEN** (distance is AVERAGE).
17. **IF** (euclidean is MEDIUM) **and** (hamming is LOW) **and** (cityBlock is LOW) **and** (similarity is LOW) **THEN** (distance is AVERAGE).
18. **IF** (euclidean is LOW) **and** (hamming is LOW) **and** (cityBlock is HIGH) **and** (similarity is HIGH) **THEN** (distance is FAR).

19. **IF** (euclidean is LOW) **and** (hamming is HIGH) **and** (cityBlock is LOW) **and** (similarity is HIGH) **THEN** (distance is FAR).
20. **IF** (euclidean is MEDIUM) **and** (hamming is MEDIUM) **and** (cityBlock is HIGH) **and** (similarity is MEDIUM) **THEN** (distance is FAR).
21. **IF** (euclidean is HIGH) **and** (hamming is HIGH) **and** (cityBlock is LOW) **and** (similarity is HIGH) **THEN** (distance is FAR).
22. **IF** (euclidean is LOW) **and** (hamming is HIGH) **and** (cityBlock is LOW) **and** (similarity is HIGH) **THEN** (distance is FAR).
23. **IF** (euclidean is HIGH) **and** (hamming is HIGH) **and** (cityBlock is LOW) **and** (similarity is HIGH) **THEN** (distance is FAR).
24. **IF** (euclidean is MEDIUM) **and** (hamming is MEDIUM) **and** (cityBlock is MEDIUM) **and** (similarity is LOW) **THEN** (distance is FAR).
25. **IF** (euclidean is MEDIUM) **and** (hamming is MEDIUM) **and** (cityBlock is MEDIUM) **and** (similarity is HIGH) **THEN** (distance is AVERAGE).
26. **IF** (euclidean is HIGH) **and** (hamming is HIGH) **and** (cityBlock is HIGH) **and** (similarity is MEDIUM) **THEN** (distance is FAR).
27. **IF** (euclidean is MEDIUM) **and** (hamming is MEDIUM) **and** (cityBlock is LOW) **and** (similarity is MEDIUM) **THEN** (distance is AVERAGE).
28. **IF** (euclidean is HIGH) **and** (hamming is HIGH) **and** (cityBlock is LOW) **and** (similarity is FAR) **THEN** (distance is HIGH).
29. **IF** (euclidean is MEDIUM) **and** (hamming is MEDIUM) **and** (cityBlock is LOW) **and** (similarity is MEDIUM) **THEN** (distance is AVERAGE).
30. **IF** (euclidean is HIGH) **and** (hamming is HIGH) **and** (cityBlock is LOW) **and** (similarity is HIGH) **THEN** (distance is FAR)
31. **IF** (euclidean is MEDIUM) **and** (hamming is MEDIUM) **and** (cityBlock is HIGH) **and** (similarity is MEDIUM) **THEN** (distance is AVERAGE).
32. **IF** (euclidean is HIGH) **and** (hamming is HIGH) **and** (cityBlock is LOW) **and** (similarity is HIGH) **THEN** (distance is FAR).
33. **IF** (euclidean is MEDIUM) **and** (hamming is LOW) **and** (cityBlock is MEDIUM) **and** (similarity is MEDIUM) **THEN** (distance is AVERAGE).
34. **IF** (euclidean is HIGH) **and** (hamming is LOW) **and** (cityBlock is HIGH) **and** (similarity is HIGH) **THEN** (distance is FAR).
35. **IF** (euclidean is MEDIUM) **and** (hamming is HIGH) **and** (cityBlock is MEDIUM) **and** (similarity is MEDIUM) **THEN** (distance is AVERAGE).
36. **IF** (euclidean is HIGH) **and** (hamming is LOW) **and** (cityBlock is HIGH) **and** (similarity is HIGH) THEN (distance is FAR).
37. **IF** (euclidean is LOW) **and** (hamming is MEDIUM) **and** (cityBlock is MEDIUM) **and** (similarity is MEDIUM) **THEN** (distance is AVERAGE).
38. **IF** (euclidean is MEDIUM) **and** (hamming is HIGH) **and** (cityBlock is HIGH) **and** (similarity is HIGH) **THEN** (distance is FAR).
39. **IF** (euclidean is MEDIUM) **and** (hamming is MEDIUM) **and** (cityBlock is HIGH) **and** (similarity is HIGH) **THEN** (distance is FAR).
40. **IF** (euclidean is MEDIUM) **and** (hamming is HIGH) **and** (cityBlock is MEDIUM) **and** (similarity is HIGH) **THEN** (distance is FAR).

41. **IF** (euclidean is HIGH) **and** (hamming is MEDIUM) **and** (cityBlock is HIGH) **and** (similarity is MEDIUM) **THEN** (distance is FAR).
42. **IF** (euclidean is LOW) **and** (hamming is HIGH) **and** (cityBlock is HIGH) **and** (similarity is LOW) **THEN** (distance is FAR).
43. **IF** (euclidean is MEDIUM) **and** (hamming is HIGH) **and** (cityBlock is MEDIUM) **and** (similarity is HIGH) **THEN** (distance is FAR).
44. **IF** (euclidean is HIGH) **and** (hamming is LOW) **and** (cityBlock is HIGH) **and** (similarity is LOW) **THEN** (distance is FAR).
45. **IF** (euclidean is LOW) **and** (hamming is MEDIUM) **and** (cityBlock is LOW) **and** (similarity is MEDIUM) **THEN** (distance is AVERAGE).
46. **IF** (euclidean is HIGH) **and** (hamming is MEDIUM) **and** (cityBlock is HIGH) **and** (similarity is MEDIUM) **THEN** (distance is FAR).
47. **IF** (euclidean is HIGH) **and** (hamming is HIGH) **and** (cityBlock is MEDIUM) **and** (similarity is MEDIUM) **THEN** (distance is FAR).

The fuzzy rules were designed thinking about combining the measures of the distances from the different perspectives provided by each metric with respect to the vector of unknown classification. For example, if the Euclidean, Hamming, city block and similarity measures provide the same perspective, then the consequent will be the same perspective. If the three measures have the same perspective, then will be that perspective as the consequent. In others words, we consider the majority perspective of the measures to represent the consequent. We have not considered all the possible options of the fuzzy rules that can be created, instead of this, the most important fuzzy rules are considered.

The pseudocode of the Fuzzy KNN algorithm using the TSFISKNN approach is as follows:

```
BEGIN
     Input x, of unknown classification
     Set K, 1 <= K <= n.
     Initialize i=1
     DO UNTIL (K-nearest neighbors to x found)
        Compute distances: Euclidean, Hamming, cosine
         similarity and city block from x to xi.
         Input the distances in TSFISKNN and obtain an
         output that represent the distances from x to xi.
        IF (i <= K) THEN
            Include xi in set of K-nearest neighbor
        ELSE IF (xi closer to x than any previous nearest)
             THEN
               Delete the farthest of the K-nearest neighbors
               Include xi in the set of K-nearest neighbors
         END IF
     END DO UNTIL
     Initialize i=1
     DO UNTIL (x assigned membership in all classes
            Compute ui(x) (1))
     END DO UNTIL
END
```

3 Review of Important Concepts

In this section, we describe important concepts, such as nearest neighbor methods based on fuzzy sets, Interval type-2 fuzzy sets, Possibilistic methods, Intuitionistic methods, fuzzy rough sets, preprocessing approaches via data reduction, as well as we present a general review of the original Fuzzy K-Nearest Neighbor as well as the equations to calculate the measures of the distances used in the proposed Takagi–Sugeno Fuzzy Inference System for the new approach of the Fuzzy KNN algorithm.

3.1 Nearest Neighbor Methods Based on Fuzzy Sets

The first Fuzzy KNN classifier was proposed by Jowik in 1983 [3]. This method is better than the standard KNN. It is based on a learning scheme of class memberships, providing each training instance with a membership array, which defines its fuzzy membership to each class. The final classification is performed similarly to the standard KNN algorithm; every neighbor uses its membership array for the voting rule, instead of just giving one vote as in the crisp KNN [2, 4].

3.2 K-NN Methods Based on Interval Type-2 Fuzzy Sets

Type-2 fuzzy sets have been the basis of a fuzzy nearest neighbor approach [5], the Interval Type-2 Fuzzy K-Nearest Neighbor (IT2FKNN) was proposed as an alternative way of discarding the necessity of setting up the parameter *K-Init* in the original definition of Fuzzy KNN. This is achieved by introducing interval type-2 fuzzy sets to represent the memberships computed by considering distinct choices of the fuzzy *K-Init* parameter. The fuzzy *K-Init* parameter is used to manage several numbers of initial *K* values for consequent to control the uncertainly. Type-2 fuzzy sets are built considering all the different computed memberships, and then a type reduction operation is performed to obtain a final, combined value, representative of all choices initially considered.

3.3 Possibilistic K-NN Methods

Possibilistic classification extends fuzzy classification in the sense that the set of assigned memberships to every instance is not constrained, that is, in most of the fuzzy classification approaches the sum of the membership degrees to every class of each instance must be 1. This property does not hold in possibilistic classification, which means than non-canonical situations can be represented using this paradigm: Using a possibilistic model an instance could belong to two classes simultaneously [6, 7].

3.4 Intuitionistic K-NN Methods

Intuitionistic fuzzy sets have also been used to develop Fuzzy Nearest Neighbor classifiers. By incorporating the concept of non-membership, it is possible to model some additional situations in an effort to characterize the classification problems as accurately as possible. In the Intuitionistic Fuzzy Sets applied to the KNN method (IFSKNN) [8] classifier this idea was proposed. In this algorithm, a membership value is computed for each instance, as the distance to the mean of the class. Then, the non-membership value is computed in a similar way, as the distance to the nearest mean of the rest of the classes. This enables typical instances to be represented with a high degree of membership whereas noisy instances will be assigned with a high degree of non-membership. The classification is completed using memberships and non-memberships to modify the distances computed by a KNN classifier.

3.5 K-NN Methods Based on Fuzzy Rough Sets

Recently, several approaches to nearest neighbor classification based on fuzzy rough sets have been proposed. Most of them aim at improving the quality of the classification performed with the combined support of the rough sets theories.

The first proposed Fuzzy Rough Nearest Neighbor Approach (FRNNA) [9], was presented based on these ideas. This classifier incorporates the lower and upper approximations of the memberships to the decision rule, in an effort to deal with both fuzzy uncertainties and rough uncertainties.

3.6 K-NN Methods Based on Fuzzy Rough Sets

Preprocessing methods have become an effective way for enhancing the performance of general nearest neighbor classifiers. Among them, prototype selection and prototype generation fields have inspired the first preprocessing methods for Fuzzy Nearest Neighbor classifiers [10].

Regarding prototype selection, the Edited Fuzzy KNN (EFKNN) classifier is based on Wilson's editing rule for KNN [17], all instances in the training set are checked and those whose classification by the Fuzzy KNN rule does not agree with its original class are removed. The Condensed Fuzzy KNN (CFKNN) is also inspired by a classic method [10].

In the following section, we present a brief review of the important concepts used in this work, such as the Fuzzy KNN algorithm as well as the equations to calculate the distances: Euclidean, Hamming, similarity and city block. Finally, we are also introducing the concepts of electrocardiograms and arrhythmias, due to the database that we have used to perform our experiments. Finally, in Sect. 5 we offer the conclusions.

3.7 Fuzzy K-Nearest Neighbor Algorithm

The fuzzy algorithm is similar to the crisp K-Nearest Neighbor. The Fuzzy K-Nearest Neighbor algorithm assigns class membership to a sample vector rather than assigning the vector to a particular class. The basis of the Fuzzy KNN algorithm is to assign membership as a function of the vector's distance from its K-Nearest Neighbors and those neighbors' memberships in the possible classes. The Fuzzy KNN algorithm is as follows:

```
BEGIN
        Input x, of unknown classification
        Set K, 1 <= K <= n.
        Initialize i=1
        DO UNTIL (K-nearest neighbors to x found)
           Compute distances Euclidean from x to xi
           IF (i <= K) THEN
               Include xi in set of K-nearest neighbor
           ELSE IF (xi closer to x than any previous nearest)
             THEN
               Delete the farthest of the K-nearest neighbors
               Include xi in the set of K-nearest neighbors
          END IF
        END DO UNTIL
        Initialize i=1
        DO UNTIL (x assigned membership in all classes (1))
                Compute ui(x)
        END DO UNTIL
END
```

$$u_i(x) = \frac{\sum_{j=1}^{k} u_{ij}(1/\|x - x_j\|^{2/(m-1)})}{\sum_{j=1}^{k}(1/\|x - x_j\|^{2/(m-1)})} \quad (1)$$

3.8 Euclidean Distance

The Euclidean distance between points p and q is the length of the line segment connecting them (pq).

In Cartesian coordinates, if p = (p_1, p_2 …, p_n) and q = (q_1, q_2,…, q_n) are two points in Euclidean n-space (d) from p to q, or from q to p is given by the Pythagorean equation, then:

$$\begin{aligned} d(p, q) &= d(q, p) \\ &= \sqrt{(q_1 - p_1)^2 + (q_2 - p_2)^2 + \cdots = (q_n - q_n)^2} \\ &= \sqrt{\sum_{i=1}^{n} (q_i - p_i)^2} \end{aligned} \tag{2}$$

3.9 Hamming Distance

The Hamming distance is a metric expressing the distance between two objects by the number of mismatches among their pairs of variables.

$$D_H = \sum_{i=1}^{k} |x_i - y_i| \tag{3}$$

Where

$$x = y \Rightarrow d = 0$$

$$x \neq y \Rightarrow d = 1$$

3.10 Cosine Similarity Distance

The cosine similarity between two vectors is a measure that calculates the cosine of the angle between them. This metric is a measurement of orientation and not magnitude.

$$\vec{a} \bullet \vec{b} = \|\vec{a}\| \|\vec{b}\| \cos\theta$$

$$\cos\theta = \frac{\vec{a} \bullet \vec{b}}{\| \vec{a} \| \| \vec{b} \|} \tag{4}$$

3.11 City Block Distance

The city block distance between two points, a and b, with k dimensions is calculated as:

$$d = \sum_{j=1}^{k} \left| a_j - b_j \right| \tag{5}$$

3.12 Electrocardiograms, Arrhythmias

An electrocardiogram represents the electrical activity of the heart, and contains important information that can help in the medical diagnosis process. The cardiac arrhythmias are alterations in the activity of the heart rhythm, such as amplitude, duration or shape of the rhythm [21].

4 Experiments

In this section, we describe details for the MIT-BIH Arrhythmia database, the experimental setup, and simulation results.

For this work, we used the MIT-BIH Arrhythmia Database to validate the proposed method. A set of samples of heartbeat records from this database are extracted. The heartbeats are segmented and transformed in the preprocessing stage [13]. The samples of heartbeats are normal beat (NB), left bundle branch block beat (LBBB), right bundle branch block beat (RBBB), premature ventricular contraction beat (PVC), fusion paced and normal beat (FPN), atrial premature beat (APB), aberrated atrial premature beat (AAPB), fusion of ventricular and normal beat (FVNB), ventricular escape beat (VEB) and paced beat (PB). In Fig. 2, some examples of the MIT-BIH arrhythmia database are presented.

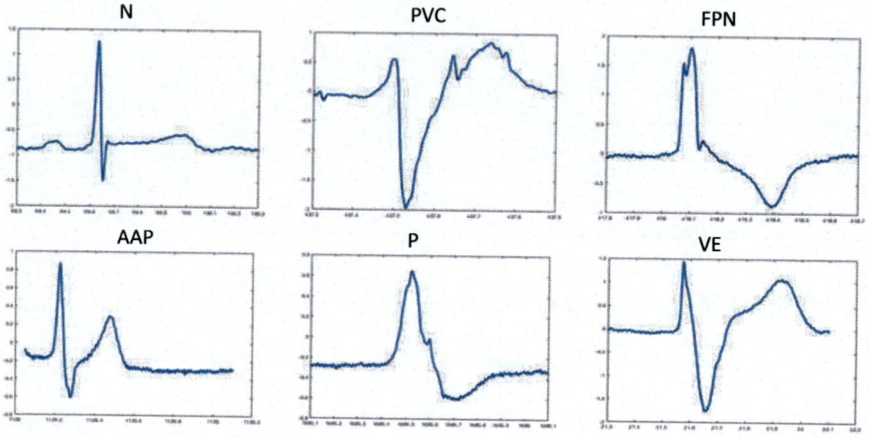

Fig. 2 Examples of the MIT-BIH Arrhythmia database

4.1 MIT-BIH Arrhythmia Database

The MIT-BIH arrhythmia database contains 48 half-hour excerpts of two channel ambulatory electrocardiogram recordings belonging to 47 patients, 22 women and 25 men aged between 23 and 89 years. The electrocardiograms were obtained from a Holter device in Beth Israel Hospital. The annotations were made separately by different cardiologists; the discrepancies were verified and resolved by consensus. We extracted 1,000 heartbeats, 100 samples per each class. The algorithm classified the 10 above mentioned classes using samples of each class. We applied tenfold cross validation to validate our work. The parameters used of the TSFISKNN are k = 3, 4 and 5 Nearest Neighbors. The results for tenfold cross validation of Fuzzy KNN algorithm and TSFISKNN are summarized in Tables 1, 2 and 3. Also, we present the results obtained using Interval Type-2 Fuzzy Logic IT2FISKNN in previous work [20].

4.2 Results

4.2.1 Results for Cross Validation Tenfold. Classifiers: Fuzzy KNN Algorithm, IT2FISKNN and TSFISKNN

In the performed experiments, we can observe that although we still do not have better results than the Fuzzy KNN algorithm. However, we are not far from achieving it, given the additional information that is provided to the Fuzzy Inference Systems, IT2FISKNN and TSFISKNN regarding distances with different perspectives we could find an optimal structure that can take advantage of this information and

Table 1 Results for cross validation with k = 3

	k = 3		
	Fuzzy KNN	IT2FISKNN	TSFISKNN
Fold 1	91	89	80
Fold 2	95	94	87
Fold 3	90	90	85
Fold 4	90	86	82
Fold 5	86	85	78
Fold 6	90	88	84
Fold 7	92	89	83
Fold 8	91	93	88
Fold 9	87	85	77
Fold 10	90	87	85

Table 2 Results for cross validation with k = 4

	k = 4		
	Fuzzy KNN	IT2FISKNN	TSFISKNN
Fold 1	92	93	82
Fold 2	95	95	92
Fold 3	91	91	84
Fold 4	90	87	81
Fold 5	87	85	78
Fold 6	89	89	86
Fold 7	92	91	82
Fold 8	93	93	88
Fold 9	87	87	77
Fold 10	93	90	88

improve the Fuzzy KNN algorithm. In Table 1, with k = 3 nearest neighbors, we obtained 82.9% of classification rate for TSFISKNN. In Table 2, with k = 4 nearest neighbors 83.8% of classification rate, and 84.5% of classification rate using k = 5 compared to 90.2%, 90.9% see Table 3, and 90% in the Fuzzy KNN algorithm respectively.

5 Conclusions

Most of the mentioned approaches of the KNN algorithm only use a single measure to represent the distance between the unclassified vector to be classified and its nearest neighbors, and this is usually the Euclidean distance. In our proposed method, we

Table 3 Results for cross validation with k = 5

	k = 5		
	Fuzzy KNN	IT2FISKNN	TSFISKNN
Fold 1	91	89	81
Fold 2	95	93	94
Fold 3	90	91	87
Fold 4	86	83	84
Fold 5	86	86	76
Fold 6	89	89	86
Fold 7	92	90	84
Fold 8	92	90	88
Fold 9	93	89	90
Fold 10	90	88.3	84.5

have considered more than one measure as a metric to determine how near or far the unclassified vector to be classified and its nearest neighbors are, such as the Euclidean, Hamming, cosine similarity, and city block distances. We believe that when considering different measures and not only the Euclidean distance, we are adding important information, which we can interpret as uncertainty, as well as knowledge, which can also be used with the implementation of fuzzy logic, through a fuzzy inference system, in this case Takagi–Sugeno type, and that it can help the original algorithm to gain advantage over the rest of the mentioned approaches. A disadvantage in the proposed method is how to find the optimal characteristics or attributes that the fuzzy inference system must have in order to adequately represent the problem to be solved. Therefore, it is necessary to use an additional method that helps adjusting the appropriate attributes for the fuzzy inference system, such as genetic algorithms or some heuristic method for optimization.

In this work, we have proposed a new approach of the Fuzzy K-Nearest Neighbor algorithm using a Takagi–Sugeno Fuzzy Inference System and several measures to calculate the distances between the neighbors and the vectors to classify. In the performed experiments using cross validation we can find some folds with better classification rate than Fuzzy KNN algorithm. Although until now we have not obtained a better global classification rate using TSFISKNN that exceeds the Fuzzy KNN algorithm. We continue to perform experiments and making changes to the Takagi–Sugeno Fuzzy Inference System to improve the results in the classification rate.

As future works, we will apply some techniques in order to optimize the Takagi–Sugeno Fuzzy Inference System (TSFISKNN), such as genetic algorithms, as well as create the TSFISKNN using Interval Type-2 Fuzzy Inference System. In addition, we will perform more experiments with different benchmark databases to perform a comparison with them toward improving the Fuzzy KNN algorithm. We can also try other types of applications, like in recognition [30–32] or control [36, 37].

References

1. D.W. Aha, D. Kibler, M.K. Albert, Instance-based learning algorithms. Mach. Learn. **6**, 37–66 (1991)
2. J.M. Keller, M.R. Gray, J.A. Givens, A fuzzy k-nearest neighbor algorithm. IEEE Trans. Fuzzy Syst., Man, Cybern. **15**, 580–585 (1985)
3. A. Jowik, A learning scheme for fuzzy k-NN rule. Pattern Recognit. Lett. **1**, 287–289 (1983)
4. M. Yang, C. Chen, On strong consistency of the fuzzy generalized nearest neighbor rule. Fuzzy Sets Syst. **60**, 273–281 (1993)
5. F. Chung-Hoon, C. Hwang, An interval type-2 fuzzy k-nearest neighbor, in *Proceedings of the 12th IEEE International Conference on Fuzzy Systems (IEEE FUZZY'03)* (St. Louis, Missouri, USA), pp. 802–807. May 25–29
6. T. Denoeux, A k-nearest neighbor classification rule based on Dempster-Shafer theory. IEEE Trans. Syst. Man Cybern. **25**, 804–813 (1995)
7. S. Destercke, An k-nearest neighbors method based on imprecise probabilities. Soft. Comput. **16**, 833–844 (2012)
8. S. Hadjittodorov, An intuituionistic fuzzy sets applicaion to the k-NN method. Notes Intuitionistic Fuzzy Sets **1**, 66–69 (1995)
9. H. Bian, L. Mazlack, Fuzzy-rough nearest neighbor classification approach, in *Proceedings of the 22th International Conference of the North American Fuzzy Information Processings Society (NAFIPS'03)* (Chicago, Illinois, USA, 2003), pp. 500–505. July 24–26
10. J.H. Zhai, N. Li, M.Y. Zhai, The condensed fuzzy k-nearest neighbor rule based on sample fuzzy entropy, in *Proceedings of the 2011 International Conference on Machine Learning and Cybernetics (ICMLC'11)* (Guilin, China), pp. 282–286. July 10–13
11. S. Garcia, J. Derrac, J.R. Cano, F. Herrera, Prototype selection for nearest neighbor classification: taxonomy and empirical study. IEEE Trans. Pattern Anal. Mach. Intell. **34**, 417–435 (2012)
12. J. Derrac, S. Garcia, F. Herrera, Fuzzy nearest neighbor algorithms: taxonomy, experimental analysis and prospects. Inf. Sci. **260**, 98–119 (2014)
13. O. Castillo, P. Melin, E. Ramirez, J. Soria, Hybrid intelligent system for cardiac arrythmia classification with fuzzy k-nearest neighbors and neural networks combined with a fuzzy system. Expert Syst. Appl. **39**, 2947–2955 (2012)
14. J.M. Mendel, Explaining the performance potential of rule-based fuzzy systems as a greater sculpting of the state space. IEEE Trans. Fuzzy Syst. 2199–2216 (2017)
15. P. Melin, O. Castillo, A review on the applications of type-2 fuzzy logic in classification and pattern recognition. Expert Syst. Appl. **40**(13), 5413–5423 (2013)
16. O. Castillo, L. Amador-Angulo, J.R. Castro, M. Garcia-Valdez, A comparative study of type-1 fuzzy logic systems, interval type-2 fuzzy logic systems and generalized type-2 fuzzy logic systems in control problems. Inf. Sci. **354**, 257–274 (2016)
17. M. Yang, C. Cheng, On the edited fuzzy k-nearest neighbor rule. IEEE Trans. Syst. Man Cybern. B Cybern. **28**, 461–466 (1998)
18. P. Melin, J. Amezcua, F. Valdez, O. Castillo, A new neural network model based on the LVQ algorithm for multi-class classification of arrhythmias. Inf. Sci **279**, 483–497 (2014)
19. E. Ramirez, O. Castillo, J. Soria, Hybrid system for cardiac arrhythmia classification with fuzzy K-nearest neighbors and neural networks combined by a fuzzy inference system. Soft Comput. Recognit. Based Biom., Stud. Comput. Intell. **312**, 37–53. ISBN 978-3-642-15110-1. (Springer, 2010)
20. P. Melin, E. Ramirez, G. Prado-Arechiga, A new variant of fuzzy K-nearest neighbor using interval type-2 fuzzy logic. FUZZ-IEEE, 1–7 (2018)
21. R. Jayasinghe, *ECG Workbook* (Churchill Livingstone, Elsevier, 2012)
22. E. Ramirez, P. Melin, G. Prado-Arechiga, Hybrid model based on neural networks, type-1 and type-2 fuzzy systems for 2-lead cardiac arrhythmia classification. Expert. Syst. Appl., Elsevier **126**, 295–307 (2019)

23. E. Ramirez, P. Melin, G. Prado-Arechiga, *Hybrid Model Based on Neural Networks and Fuzzy Logic for 2-Lead Cardiac Arrhythmia Classification* (Hybrid Intelligent Systems in Control, Pattern Recognition and Medicine, 2020), pp. 193–217
24. G. Moody, R. Mark, The impact of the MIT-BIH arrhythmia database. IEEE Eng. Med. Biol. **20**(3), 45–50 (2001)
25. A. Goldberger, L. Amaral, L. Glass, J. Hausdorff, P. Ivanov, R. Mark, J. Mietus, G. Moody, C. Peng, H. Stanley, PhysioBank, physiotoolkit, and physionet: components of a new research resource for complex physiologic signals. Circulation **101**(23), e215–e220 (2003)
26. J. Amezcua, P. Melin, O. Castillo, A new classification method based on LVQ neural networks and fuzzy logic. NAFIPS/WConSC, 1–5 (2015)
27. J. Martindale, D. Brown, A visual guide to ECG interpretation, 2nd edn. (Wolters Kluwer, 2017). ISBN 978-1-4963-2153-4
28. Z. Takac, H. Bustince, J. Pintor, C. Marco-D, Width-based interval-value distances and fuzzy entropies. IEEE Access **7**, 14044–14057 (2019)
29. H. Salles, I. Couso, B. Bedregal, Z. Takac, M. Minarova, A. Asiain, E. Barrenechea, H. Bustince, Similarity measures, penalty functions and fuzzy entropy form new subsethood measures. Int. J. Intell. Syst. **34**(6), 1281–1302 (2019)
30. P. Melin, D. Sánchez, O. Castillo, Genetic optimization of modular neural networks with fuzzy response integration for human recognition. Inf. Sci. **197**, 1–19 (2012)
31. M.A. Sanchez, O. Castillo, J.R. Castro, P. Melin, Fuzzy granular gravitational clustering algorithm for multivariate data. Inf. Sci. **279**, 498–511 (2014)
32. D. Sanchez, P. Melin, Optimization of modular granular neural networks using hierarchical genetic algorithms for human recognition using the ear biometric measure. Eng. Appl. Artif. Intell. **27**, 41–56 (2014)
33. S. Hadjitodorov, An intuitionistic fuzzy sets application to the k-NN method. Notes Intuitionistic Fuzzy Sets **1**, 66–69 (1995)
34. N. Gayar, F. Schwenker, G. Palm, A study of the robustness of kNN classifiers trained using soft labels, in *Proceedings of the Second IAPR Workshop on Artificial Neural Networks in Pattern Recognition (ANNPR 2006)* (Ulm, Germany), pp. 67–80. August 31–September 2
35. M. Arif, M.U. Akram, F.A. Afsar, F.A.A. Minhas, Pruned fuzzy k-nearest neighbor classifier for beat classification. J. Biomed. Sci. Eng. **3** (2010)
36. O. Castillo, *Type-2 Fuzzy Logic in Intelligent Control Applications* (Springer, 2012)
37. E. Ontiveros-Robles, P. Melin, O. Castillo, Comparative analysis of noise robustness of type 2 fuzzy logic controllers. Kybernetika **54**(1), 175–201 (2018)

Random Fuzzy-Rule Foams for Explainable AI

Akash Kumar Panda and Bart Kosko

Abstract A random foam trains several fuzzy-rule-foam function approximators and then combines them into a single rule-based approximator. The foam systems train independently on bootstrapped random samples from a trained neural classifier. The foam systems convert the neural black box into an interpretable set of rules. The fuzzy rule-based systems have an underlying probability mixture structure that gives rise to an interpretable Bayesian posterior over the rules for each input. A rule foam also measures the uncertainty in its outputs through the conditional variance of the generalized probability mixture. A random foam combines the learned additive fuzzy systems by averaging their throughputs or rule structure. A random foam is also interpretable in terms of its rules, its posterior of its rules, and its conditional variance. Thirty 1000-rule foams trained on random subsets of the MNIST digit data set. Each such foam system had about 93.5% classification accuracy. The random foam that averaged throughputs achieved 96.80% accuracy while the random foam that averaged only their outputs achieved 96.06% accuracy. The throughput-averaged random foam also slightly outperformed a standard random forest that output-averaged 30 classification trees. Thirty 1000-rule foams also trained on a deep neural classifier that had 96.26% accuracy. The random foam that averaged these foam throughputs was itself 96.14% accurate. The random foam that averaged their outputs was just 95.6% accurate. The appendix proves a Gaussian combined-foam version of the uniform approximation theorem for additive fuzzy systems.

1 Explaining Neural Black Boxes with Fuzzy Rule Foams

Deep neural classifiers are black-box function approximators that map pattern inputs to class labels through multiply layers of neurons. They do not explain their predictions or the inner workings of their black box. So they do not yield explainable

A. K. Panda · B. Kosko (✉)
Department of Electrical and Computer Engineering, University of Southern California, Los Angeles, CA, USA
e-mail: kosko@usc.edu

B. Bede et al. (eds.), *Fuzzy Information Processing 2020*,
Advances in Intelligent Systems and Computing 1337,
https://doi.org/10.1007/978-3-030-81561-5_21

AI or XAI systems [1, 5, 6, 18]. Nor do they give a confidence measure for their input-output predictions. Yet some predictions should be more credible than others given the nature of the pattern inputs and given what the neural system has learned or forgotten in prior bouts of training. These XAI problems tend to get only worse as the network gets deeper.

We recently showed that an additive fuzzy rule-based approximator can learn from a trained neural classifier and thereby produce an interpretable system [16]. The rules' if-part sets vary in size in analogy with bubbles in a foam as in Fig. 1. Varying the size of the rules controls the granularity of the approximators and helps mitigate the fuzzy system's inherent exponential rule explosion.

Rule foams have a natural representation as a generalized probability mixture $p(y|x)$ [12]. A set of m fuzzy if-then rules $R_{A_1 \to B_1}, \ldots, R_{A_m \to B_m}$ defines such a generalized probability mixture $p(y|x) = p_1(x)\ p_{B_1}(y|x) + \cdots + p_m(x)\ p_{B_m}(y|x)$ so long as the fuzzy system $F : \mathbb{R}^n \to \mathbb{R}^p$ is additive [12]. So a rule foam converts a sampled neural network into an explanatory probabilistic system by way of the rules. The mixture's likelihood functions $p_{B_1}(y|x)$ absorb the structure of the fuzzy rules. The foam's mixture gives rise to a Bayes theorem that in turn gives a complete posterior probability description of the rule firings. This Bayesian posterior gives an input-by-input description of the inner workings of the fuzzy-system proxy for the sampled deep neural network. The mixture's second moment further gives a conditional variance that describes the uncertainty or confidence of the foam system's output for each given pattern input. So the rule foam defines an XAI system that is modular and self-descriptive and does so completely in probabilistic terms.

This paper introduces a *random* rule foam in rough analogy to a random forest of decision trees [3] that arise from sampling with replacement from a data set. Figure 1 shows the random-foam training process. A deep neural classifier first trains on a set of pattern data. Then we form several rule foams by randomly sampling *with replacement* from the trained neural classifier and thus perform bootstrap sampling from the neural network itself (or more generally from any data set). We then combine the rule foams by directly combining their rules or their throughput structure because the foams are additive fuzzy systems. Simulations show that such throughput combination tends to outperform simply combing the foam outputs.

The random construction tends to reduce the output conditional variance. Figure 2 shows how this output condition variance helps with the classification process. The random foam maintains its mixture structure because mixing mixture gives a mixture. The additive fuzzy structure goes further and gives a simple double-sum form for the overall mixture. This sum structure allows throughput combination in addition to the simpler output combination.

The random foams substantially outperformed the individual rule foams that comprised them. Simulations showed that random foams slightly outperformed random forests for the same number of combined systems. Random forests lack the self-explanation structure of the random foams. The random foam closely approximated the neural network that it trained on and achieved a similar classification accuracy.

Figure 3 shows the Bayesian posterior over the rules for a given input image and observed classification output. This information also helps explain the classification

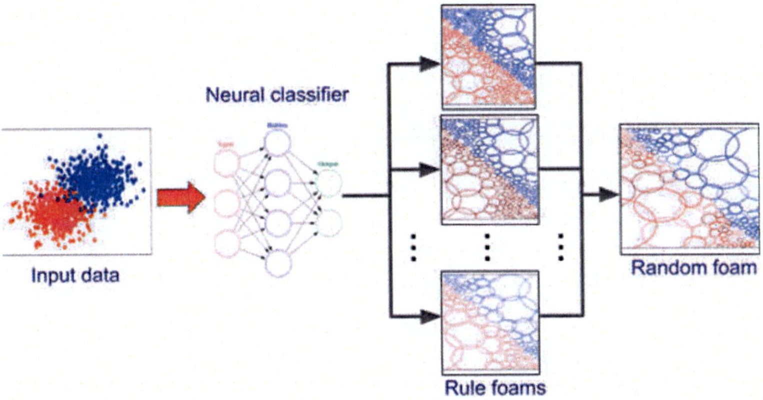

Fig. 1 Random foam structure: The deep neural classifier first learns from the data. Then several rule foams randomly sample from the trained network and independently learn from those random samples. The random foam combines these foams by additively combining or mixing rules

process. Figure 1 shows the bubble-like detail of rule foam. Figure 4 shows how classification advantage of using random foams. The appendix presents a multi-foam Gaussian generalization of the original fuzzy approximation theorem for additive systems [10, 11, 13]. The next section presents the pertinent mixture structure of additive fuzzy systems.

2 Standard Additive Model (SAM) Fuzzy Systems

A standard additive model (SAM) fuzzy system approximates the function $f : \mathbb{R}^d \to \mathbb{R}$ using m "if then" fuzzy rules. The j-th rule has the fuzzy if-part set $A_j \subset \mathbb{R}^d$ and the fuzzy then-part set $B_j \subset \mathbb{R}$. The fuzzy sets A_j and B_j have their respective unit-interval-valued indicator functions $a_j : \mathbb{R}^d \to [0, 1]$ and $b_j : \mathbb{R} \to [0, 1]$ such that $a_j(x) = \text{Degree}(x \in A_j)$ and $b_j(y) = \text{Degree}(y \in B_j)$. The jth fired then-part set $B_j(x)$ has set function $b_j(y|x) = a_j(x)b_j(y)$ because the system is standard. The rule-based system's total fired then-part set $B(x)$ sums the fired rules by summing the fired then-part sets: $B(x) = \sum_{j=1}^{m} w_j a_j(x) B_j$ for rule weight w_j. Then the fuzzy rules define the generalized probability mixture $p(y|x)$ [12]:

$$p(y|x) = \frac{b(y|x)}{\int b(y|x)dy} = \sum_{j=1}^{m} \frac{w_j a_j(x) V_j}{\sum_{j=1}^{m} w_j a_j(x) V_j} \frac{b_j(y)}{V_j} = \sum_{j=1}^{m} p_j(x) p_{B_j}(y) \quad (1)$$

where a_j and b_j are the jth rule's respective if-part and then-part fuzzy set functions. The volume $V_j = \int b_j(y)dy > 0$ is the volume of then-part set B_j. The SAM structure $b_j(y|x) = a_j(x)b_j(y)$ gives the likelihood as $b_j(y|x) = b_j(y)$.

The fuzzy system $F : \mathbb{R}^d \to \mathbb{R}$ arises naturally as the first moment of $p(y|x)$:

$$F(x) = E[Y|X = x] = \int y \; p(y|x) \; dy = \sum_{j=1}^{m} p_j(x) c_j \tag{2}$$

where c_j is the j-th then-part centroid. The system output $F(x)$ also arises as $F(x) = \text{Centroid}(B(y|x))$. Mixing just two normal bell curves can always exactly represent any bounded non-constant real function f such that $f(x) = E[Y|X = x]$ for curves centered at the infimum α and supremum β of f [12]: $p(y|x) = w(x)N(y|\alpha, \sigma_\alpha^2) + (1 - w(x))N(y|\beta, \sigma_\beta^2)$ for any variances σ_α^2 and σ_β^2 if $w(x) = \frac{\beta - f(x)}{\beta - \alpha}$.

2.1 *Fuzzy Approximation Theorem for Gaussian SAMs*

A SAM fuzzy system is a universal function approximator [10]. It can uniformly approximate any continuous function f defined on a compact set using finite number of rules m. This approximation may need m that grows exponentially with the number of dimensions d leading to rule explosion.

Consider a SAM with m Gaussian rules that approximates the continuous function f on a compact set. Let ε be the approximation error. The function f is uniformly continuous. So we can find some positive δ so that

$$||x - \hat{x}_j|| < \delta \implies |f(x) - f(\hat{x}_j)| < \varepsilon/2. \tag{3}$$

Theorem 1 in the appendix shows that the Gaussian SAM can uniformly approximate f with approximation error ε and with m rules if $m \propto \frac{1}{\delta^d}$. The fuzzy rule foam mitigates this rule explosion by controlling the bubble size.

2.2 *Conditional Variance*

The SAM measures the uncertainty in its output through the conditional variance

$$V[Y|X = x] = \sum_{j=1}^{m} p_j(x)\sigma_{B_j}^2 + \sum_{j=1}^{m} p_j(x)(c_j - F(x))^2 \tag{4}$$

where σ_{B_j} is the j-th rule's then-part dispersion. The second term in Eq. (4) imposes an interpolation penalty on the system for guessing with respect to the given set of rules. We are confident in the system's output if the variance is low and we do not trust the system's output when the variance is high.

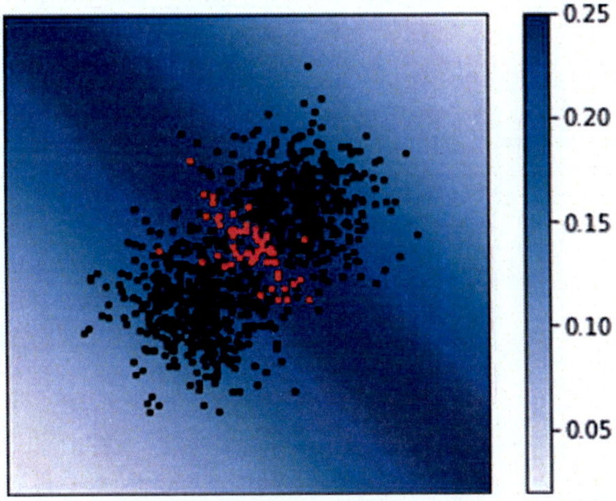

Fig. 2 Output uncertainty of the rule-foam classifier. The correctly classified points are in black. The misclassified points are in red. The background color shows the conditional variance. The color bar gives the value of the conditional variance $V[Y|X = x]$ that corresponds to the color: It is highest where the pattern classes overlap

Consider a SAM that approximates the simple function f that represents an ideal classifier. A simple function maps an input space to a finite number of values [17]. The classifier's output has lower conditional variance in the class interior and has higher conditional variance at the class boundary. The misclassification rate is higher in regions of high conditional variance. Neural classifiers lack a conditional variance or indeed any intrinsic measure of their system uncertainty. SAMs that train on a neural network's output can measure the uncertainty in the network's classification.

Figure 2 shows the conditional variance of a 2-class fuzzy rule based classifier. The conditional variance is high near the class boundary. The misclassification rate is also high close to the class boundary.

2.3 *Bayesian Posteriors of Subsystems and Rules*

The generalized mixture $p(y|x)$ in (1) has the same form as the elementary theorem on total probability. The mixing weights $p_j(x)$ define generalized rule priors and the $p_{B_j}(y)$ define generalized rule likelihoods. This total-probability structure gives at once a posterior probability density over the rules from Bayes theorem:

$$p(j|y,x) = \frac{P(Z=j, Y=y|X=x)}{p(y|x)} = \frac{p_j(x)p_{B_j}(y)}{p(y|x)} = \frac{p_j(x)p_{B_j}(y)}{\sum_{j=1}^{m} p_j(x)p_{B_j}(y)} \tag{5}$$

for the j-th rule firing given observed input x and output y. These posteriors also give the contribution of each rule to the final output.

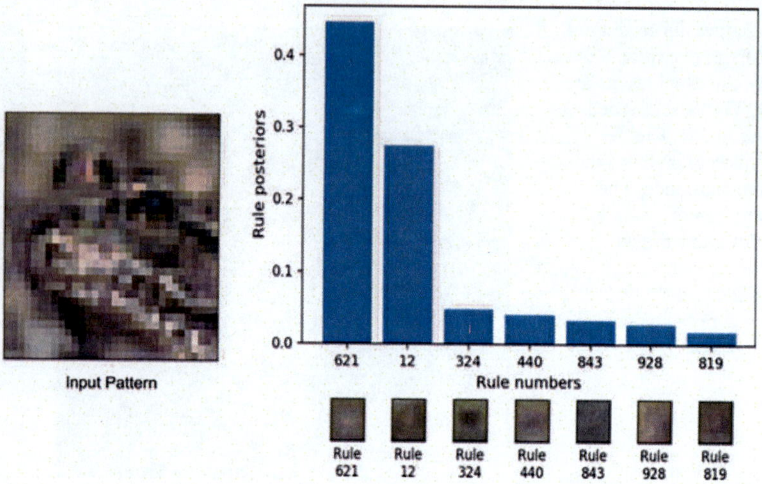

Fig. 3 Bayesian posterior $p(j|y, x)$ over the rule firings when input x produces output y. The input image is on the left. The 7 largest rule posteriors for the input are on the right. The if-part centroids of these rules are also on the right. The rule foam had 1000 rules. The x-axis lists the rule number. The y-axis lists the corresponding posterior probability of rule firing. The posterior distribution when the foam correctly classified an input from class 'Frog'. The 621st rule ($j = 621$) contributed most to the classification of this input pattern

Figure 3 shows a snapshot of the rule posteriors for a CIFAR-10 image input. The x-axis lists the rule numbers that correspond to the 7 highest posteriors. The 621st rule contributed most to the classification of the input image.

Consider a SAM classifier that misclassified the input x. The SAM identifies the rule responsible for the mistake. This makes the fuzzy rule-based system interpretable. A fuzzy system that learns from a neural classifier can interpret the neural black box. Users can also use these posteriors to later prune or modify the rule base.

2.4 SAM Combination of Throughput Rules

SAM systems combine by taking sum of their then part sets. Consider q SAMs that each approximate the function f. The k-th SAM uses m_k rules and has the weight v^k. This helps us combine knowledge from multiple experts and also allows us to combine closed-form equations [19] with the softer knowledge of rules. The rules of the SAM combination also give the underlying mixture density:

$$p(y|x) = \sum_{k=1}^{q} \sum_{j=1}^{m_k} \frac{v^k w_j^k a_j^k(x) V_j^k}{\sum_{k=1}^{q} \sum_{j=1}^{m_k} v^k w_j^k a_j^k(x) V_j^k} \frac{b_j^k(y)}{V_j^k} = \sum_{k=1}^{q} \sum_{j=1}^{m_k} p_j^k(x) p_{B_j^k}(y). \quad (6)$$

The function approximation is just the conditional expectation of the mixture:

$$F(x) = E[Y|X = x] = \int y\, p(y|x)\, dy = \sum_{k=1}^{q} \sum_{j=1}^{m_k} p_j^k(x) c_j^k. \tag{7}$$

Here c_j^k is the k-th SAM's j-th then-part centroid. This SAM combination also measures the uncertainty in its output through its conditional variance $V[Y|X = x]$:

$$V[Y|X = x] = \sum_{k=1}^{q} \sum_{j=1}^{m_k} p_j^k(x) \sigma_{B_j^k}^2 + \sum_{k=1}^{q} \sum_{j=1}^{m_k} p_j^k(x)(c_j^k - F(x))^2 \tag{8}$$

where $\sigma_{B_j^k}$ is the dispersion of k-th SAM's j-th then-part fuzzy set B_j^k. The generalized mixture $p(y|x)$ in (6) also gives a finer-grained Bayesian posterior

$$p(j, k|y, x) = \frac{p_j^k(x) p_{B_j^k}(y)}{\sum_{k=1}^{q} \sum_{j=1}^{m_k} p_j^k(x) p_{B_j^k}(y)} \tag{9}$$

of the k-th SAM's j-th rule firing. Summing over m gives the posterior over the constitutive foams.

3 Fuzzy Rule Foam

A SAM fuzzy system's rules resemble bubbles in the input-output product space. The bubbles or rule patches cover the graph of the function they approximate. The SAM's graph-cover structure leads to rule explosion. Rule foam mitigates rule explosion and allows fuzzy systems to approximate high-dimensional pattern classifiers.

A classifier foam's rules do not cover the input space equally. The rule foam concentrates its rules at the class boundaries. There are a few large rules covering the class interior and a lot of smaller rules covering the class boundary. This defines a foam like structure of the rule if-part set bubbles [16]. Figure 1 shows this structure. The circles represent the if-part sets of the rule foam. The radius of the circle represents the dispersion of the if-part set. The if-part dispersion characterizes the size of the rule. Rules are smaller close to the class boundary and are larger away from the boundary. The foam also avoids covering empty regions of input space through Adaptive Vector Quantization (AVQ).

3.1 Training with Adaptive Vector Quantization (AVQ)

Adaptive vector quantization (AVQ) is a sample-based scheme for estimating an unknown data distribution [2]. We use reinforcement version of AVQ to distribute rule if-part sets. AVQ is a from of k-means clustering [7, 15] or competetive learning [4] or self-organizing maps [8]. AVQ gives Quantization Vectors (QVs) $\{\hat{x}_j\}_{j=1}^{m}$ whose distribution approximates that of the data set $\{x_n\}_{n=1}^{N}$. The AVQ algorithm cycles through the data set every epoch. It finds the QV $\hat{x}_j$ closest to the input vector x and either rewards it or punishes it by moving it either towards or away from x. AVQ moves $\hat{x}_j$ closer to the x if they belong to the same class and moves $\hat{x}_j$ away from the x if they belong to different classes.

Let $\hat{x}_j^{(t)}$ denote the j-th QV after the t-th iteration. Let $\hat{x}_j^{(t)}$ be the closest QV to the data point x_n. Let $\hat{x}_j^{(t)}$ belong to the class C_j. AVQ updates $\hat{x}_j^{(t)}$ as

$$\hat{x}_j^{(t+1)} = \hat{x}_j^{(t)} + \eta_t(x_n - \hat{x}_j^{(t)})r_j(x_n) \tag{10}$$

for decreasing learning rates η_t. The bipolar reinforcement function r_j is the indicator difference [9]

$$r_j(x_n) = \mathbb{I}_{C_j}(x_n) - \sum_{C \neq C_j} \mathbb{I}_C(x_n). \tag{11}$$

It gives $r_j(x_n) = +1$ if $x_n \in C_j$ and $r_j(x_n) = -1$ otherwise. So the jth QV looks a little more like the current sample point x_n if $r_j(x_n) = 1$ and a little less like it if $r_j(x_n) = -1$.

The distribution $\hat{x}_j$'s approximates the distribution of the input vectors. So there are no $\hat{x}_j$'s in empty regions of input space. We center the foam's if-part sets around $\hat{x}_j$'s and avoid covering the empty input space.

4 Random Foams

A random foam combines several fuzzy rule foams that train on random subsets of a data set. This method resembles how random forest combines the output of several trees [3]. The random foam performs better than the individual constituent foams.

We can combine foams by simply combining or averaging their outputs or by combining their rules or the throughputs. The additive structure of SAM systems permits this richer form of combination through (6). Random forests do not permit such throughput combination.

Let the k-th foam F_k with m_k rule approximate f. Then (2) gives F_k as

$$F_k(x) = \sum_{j=1}^{m_k} \frac{w_j^k a_j^k(x) V_j^k}{\sum_{j=1}^{m_k} w_j^k a_j^k(x) V_j^k} c_j^k. \tag{12}$$

Averaging the foam *outputs* gives the random foam output $F_{avg}(x)$ as

$$F_{avg}(x) = \frac{1}{q} \sum_{k=1}^{q} F_k(x) \ . \tag{13}$$

We combine the throughputs of the q SAMs using Eq. (6). We choose weight each SAM equally by $u^k = 1/q$. Then the random foam output is

$$F_{com}(x) = \sum_{k=1}^{q} \sum_{j=1}^{m_k} p_j^k(x) c_j^k \tag{14}$$

where

$$p_j^k(x) = \frac{u^k w_j^k a_j^k(x) V_j^k}{\sum_{k=1}^{q} \sum_{j=1}^{m_k} u^k w_j^k a_j^k(x) V_j^k} = \frac{w_j^k a_j^k(x) V_j^k}{\sum_{k=1}^{q} \sum_{j=1}^{m_k} w_j^k a_j^k(x) V_j^k} . \tag{15}$$

5 Experiments on MNIST Data

We tested the random foam on the MNIST data set [14]. The MNIST data set consists of 60,000 28×28 gray-scale images of handwritten digits from 0 to 9. The random foam trained 30 rule foams on random subsets of MNIST data with bootstrap resampling. Each subset had 10,000 MNIST images. Each individual foam used 1000 rules was about 93.5% accurate. The random foam then combined these foams in both ways.

The output-averaged random foam was 96.06% accurate while the throughput-averaged random was was 96.80% accurate. We trained a random forest with 30 trees for comparison. The random forest was 96.55% accurate and thus less accurate than the throughput-combined random foam trained on the same MNIST data. Figure 4 compares the foam accuracies against the number of foams in the random foams. It also shows the performance of the random forest against the number of trees.

The random foam trained by sampling from a multilayer neural classifier that itself trained on the MNIST data. The neural network had 784 identity input neurons and 10 softmax output neurons. It had two hidden layers with 256 rectified-linear-unit (ReLU) neurons each. The network trainsed on the MNIST dataset for 100 epochs and had 96.26% accuracy. The 30 1000-rule foams trained by randomly sampling 10,000 input-output pairs each from this trained neural classifier. Each individual foam was about 94.47% accurate. The random foam then averaged their outputs and their throughputs. The throughput-averaged random foam was 96.14% accurate while the output-averaged random foam was only 95.6% accurate.

Figure 5 shows the accuracy of the output-averaged random foam $F_{avg}(x)$ and the throughput-averaged random foam $F_{com}(x)$ against the number of their constituent

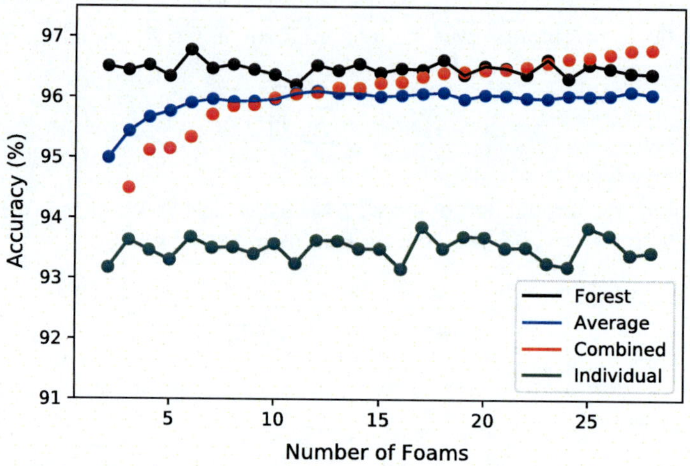

Fig. 4 Accuracy of the random foams: Random foam with throughput averaging performed best. Accuracy of the throughput-combined foam F_{com} in red. Accuracy of the output-averaged foam F_{avg} in blue. The accuracy of the individual foams F_k in green. Plot of random forest accuracy against number of trees in black

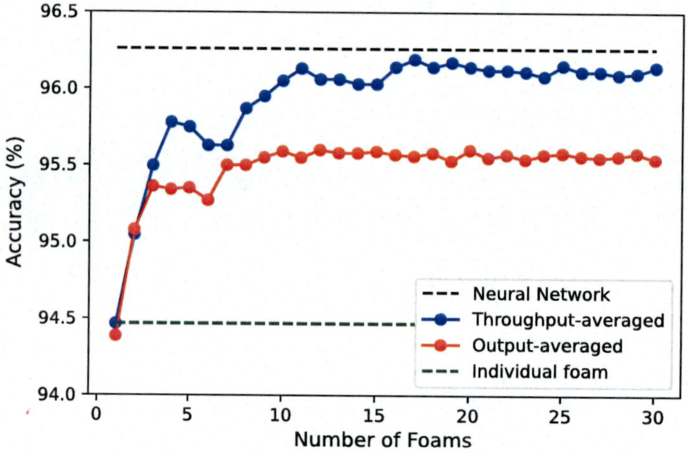

Fig. 5 Random foam trained on a neural classifier. The throughput-averaged random foam performed almost as good as the neural network that trained it. The accuracy of the neural network is in black. The accuracy of the throughput averaged random foam is in blue. The accuracy of the output-averaged random foam is in red. The accuracy of an individual constituent foam is in green

foams. It also shows the accuracy of the neural classifier that trained the random foam and shows the accuracy of an individual foam. Figure 5 shows that the random foam performed significantly better than its constituent foams. It also shows that averaging throughputs performed better than averaging outputs. The throughput-averaged random foam was nearly as accurate as the neural classifier that trained it.

6 Conclusions

Fuzzy rule foams are granular function approxiimators with interpretable rule structures and output uncertainties. They suffer from exponential rule explosion because their if-part fuzzy sets must cover the input pattern space. A random foam combines several independent fuzzy rule-foam approximators and performs better than the individual rule-foam approximators. The random foam acts as type of variance reduction because of the independent random sampling involved in the foam training. The random foam interprets its rules though its Bayesian posterior over the rules and measures its output uncertainty with the conditional variance of its underlying generalized mixture. Simulations showed that random foams performed comparably on the MNIST test set when it trains on random samples from a trained deep neural classifier. The random foam that trained on the MNIST data itself had slightly better accuracy than the neural classifier. Future random foams may find more efficient if-part rule covers that mitigate rule explosion as they achieve higher accuracy.

7 Appendix : Fuzzy Approximation Theorem for a Combination of Gaussian SAMs

Theorem 1 extends a version of the fuzzy approximation theorem [10] to the case of combined additive systems. The Gaussian structure of the if-part sets gives insight into the uniform approximation. The approximation may need ab exponentially growing total number of rules mq with respect to the number of dimensions n.

Theorem 1 *A combination of q Gaussian SAMs can uniformly approximate a continuous function f over a compact space.*

Proof We combine q SAMs to approximate $f(x)$ with $F(x)$ by mixing their then-part sets. We want to show that the approximation error obeys $|f(x) - F(x)| < \varepsilon$ for any $\varepsilon > 0$ using a finite number of rules. We use Gaussian if-part sets with set functions $a_j^k(x) = \exp\{-||x - \hat{x}_j^k||^2/(d_j^k)^2\}$. We assign equal weights, equal volumes, and equal dispersions to all the rules: $v^k = v$, $w_j^k = w$, $V_j^k = V$, and $d_j^k = d$. This reduces the generalized mixing weights $p_j(x)$ to softmax activation functions:

$$p_j^k(x) = \frac{vwa_j^k(x)V}{\sum_{k=1}^q \sum_{j=1}^m vwa_j^k(x)V} = \frac{\exp\{-||x-\hat{x}_j^k||^2/d^2\}}{\sum_{k=1}^q \sum_{j=1}^m \exp\{-||x-\hat{x}_j^k||^2/d^2\}}. \tag{16}$$

Multiply $f(x)$ by $\sum_{k=1}^q \sum_{j=1}^m p_j^k(x) = 1$ and then use (7):

$$|f(x) - F(x)| = \left|\left(\sum_{k=1}^q \sum_{j=1}^m p_j^k(x)\right) f(x) - \sum_{k=1}^q \sum_{j=1}^m p_j^k(x) c_j^k\right|. \tag{17}$$

Rearrange the terms and use the triangle inequality:

$$|f(x) - F(x)| \le \sum_{k=1}^q \sum_{j=1}^m p_j^k(x) \big| f(x) - c_j^k \big| = \sum_{k=1}^q \sum_{j=1}^m p_j^k(x) \big| f(x) - f(\hat{x}_j^k) \big| \tag{18}$$

because we center the then-part set around $f(\hat{x}_j^k)$. So $c_j^k = f(\hat{x}_j^k)$. The target function f is uniformly continuous because it is continuous and X is compact. So for all $\varepsilon > 0$ there is a $\delta > 0$ such that $|f(x) - f(\hat{x}_j^k)| < \varepsilon/2$ if $||x - \hat{x}_j^k|| < \delta$. This splits the rules into the two sets $J_1 = \{(j,k) : ||x - \hat{x}_j^k|| < \delta\}$ and $J_2 = \{(j,k) : ||x - \hat{x}_j^k|| \ge \delta\}$. We assume that $J_1 \neq \emptyset$. Then (18) becomes

$$|f(x) - F(x)| \le \sum_{(j,k)\in J_1} p_j^k(x) \big| (f(x) - f(\hat{x}_j^k)) \big| + \sum_{(j,k)\in J_2} p_j^k(x) \big| (f(x) - f(\hat{x}_j^k)) \big|. \tag{19}$$

Consider the first term on the right side of (19). The absolute differences obey $\big|(f(x) - f(\hat{x}_j^k))\big| < \varepsilon/2$ because $(j,k) \in J_1$implies that $||x - \hat{x}_j^k|| < \delta$. Add these terms over all $(j,k) \in J_1$:

$$\sum_{(j,k)\in J_1} p_j^k(x) \big| (f(x) - f(\hat{x}_j^k)) \big| < \sum_{(j,k)\in J_1} p_j^k(x) \frac{\varepsilon}{2} < \frac{\varepsilon}{2} \tag{20}$$

because $\sum_{(j,k)\in J_1} p_j^k(x) \le \sum_{k=1}^q \sum_{j=1}^m p_j^k(x) = 1$. Now pick a real number C large enough that $|f(x) - f(\hat{x}_j^k)| < C$ for all j and k. Pick $(j,k) \in J_2$ and pick $(j_1, k_1) \in J_1$. Then $||x - \hat{x}_j^k|| \ge \delta$ and $||x - \hat{x}_{j_1}^{k_1}|| < \delta$. Pick a real number D small enough that $||x - \hat{x}_j^k||^2 - ||x - \hat{x}_{j_1}^{k_1}||^2 \ge D > 0$. Choose the rule if-part dispersion d small enough that $\exp(D/d^2) > 2mqC/\varepsilon$. Then the definition of D gives

$$\frac{\exp(-||x - \hat{x}_{j_1}^{k_1}||^2/d^2)}{\exp(-||x - \hat{x}_j^k||^2/d^2)} = \exp\left(\frac{||x - \hat{x}_j^k||^2 - ||x - \hat{x}_{j_1}^{k_1}||^2}{d^2}\right) > \frac{2mqC}{\varepsilon}. \tag{21}$$

Add $\sum_{i\neq j_1} \sum_{l \neq k_1} \frac{\exp(-||x-\hat{x}_i^l||^2/d^2)}{\exp(-||x-\hat{x}_j^k||^2/d^2)}$ to both sides of the above inequality:

$$\frac{\sum_{j_1=1}^{m}\sum_{k_1=1}^{q}\exp(-||x-\hat{x}_{j_1}^{k_1}||^2/d^2)}{\exp(-||x-\hat{x}_j^k||^2/d^2)} > \frac{2mqC}{\varepsilon} + \frac{\sum_{i\neq j_1}\sum_{l\neq k_1}\exp(-||x-\hat{x}_i^l||^2/d^2)}{\exp(-||x-\hat{x}_j^k||^2/d^2)} > \frac{2mqC}{\varepsilon} \tag{22}$$

because $\sum_{i\neq j_1}\sum_{l\neq k_1}\frac{\exp(-||x-\hat{x}_i^l||^2/d^2)}{\exp(-||x-\hat{x}_j^k||^2/d^2)} > 0$. Equation (16) and the definition of C give

$$p_j^k(x)\big|f(x)-f(\hat{x}_j^k)\big| < \frac{\varepsilon}{2mqC}\big|f(x)-f(\hat{x}_j)\big| < \frac{\varepsilon}{2mq}. \tag{23}$$

Sum both sides over all $(j,k)\in J_2$:

$$\sum_{(j,k)\in J_2} p_j^k(x)|f(x)-f(\hat{x}_j^k)| < \sum_{(j,k)\in J_2}\frac{\varepsilon}{2mq} \leq mq\frac{\varepsilon}{2mq} = \frac{\varepsilon}{2} \tag{24}$$

because J_2 has fewer than mq elements. Then (20) and (24) add to give the approximation-error bound: $|f(x)-F(x)| \leq \varepsilon/2+\varepsilon/2 < \varepsilon$. So the combined SAM with mq rules uniformly approximates f. □

The proof gives insight into a foam's rule explosion. The proof assumes that $J_1 \neq \emptyset$. So for all $x \in X$ there is a $\hat{x}_j^k$ such that $||x-\hat{x}_j|| < \delta$. Let $N_\delta(\hat{x}_j^k) = \{x : ||x-\hat{x}_j^k|| < \delta\}$ denote the δ-neighborhood around $\hat{x}_j^k$. Then $\forall\, x \in X : \exists\, \hat{x}_j^k : x \in N_\delta(\hat{x}_j^k)$. This gives

$$X \subseteq \bigcup_{k=1}^{q}\bigcup_{j=1}^{m} N_\delta(\hat{x}_j^k). \tag{25}$$

X is bounded because X is compact by hypothesis. Let an n-dimensional sphere of radius R contain X. The volume of this sphere is cR^n for some constant c. Each $N_\delta(\hat{x}_j^k)$'s is an n-dimensional sphere with radius δ and volume $c\delta^n$. Use the δ-neighborhoods to cover the R-sphere that contains X. The $N_\delta(\hat{x}_j^k)$'s are not disjoint. So the volume of the R-sphere is less than the volume of the mq $N_\delta(\hat{x}_j^k)$'s combined: $cR^n \leq mqc\delta^n$. Then $mq \geq (R/\delta)^n$. This inequality shows that the required total number of rules grows exponentially with the dimension n.

References

1. A. Adadi, M. Berrada, Peeking inside the black-box: a survey on explainable artificial intelligence (XAI). IEEE Access **6**, 52138–52160 (2018)
2. A. Buzo, R.M. Gray et al., An algorithm for vector quantizer design. IEEE Trans. Commun. **28**(1), 84–95 (1980)
3. B. Efron, T. Hastie, *Computer Age Statistical Inference*, vol. 5 (Cambridge University Press, 2016)

4. S. Grossberg, Competitive learning: from interactive activation to adaptive resonance. Cogn. Sci. **11**(1), 23–63 (1987)
5. R. Guimerà, I. Reichardt, A. Aguilar-Mogas, F.A. Massucci, M. Miranda, J. Pallarès, M. Sales-Pardo, A Bayesian machine scientist to aid in the solution of challenging scientific problems. Sci. Adv. *6*(5) (2020)
6. F. Horst, S. Lapuschkin, W. Samek, K.-R. Müller, W.I. Schöllhorn, Explaining the unique nature of individual gait patterns with deep learning. Sci. Rep. **9**(1), 1–13 (2019)
7. A.K. Jain, Data clustering: 50 years beyond K-means. Pattern Recogn. Lett. **31**(8), 651–666 (2010)
8. T. Kohonen, The self-organizing map. Proc. IEEE **78**(9), 1464–1480 (1990)
9. B. Kosko, Stochastic competitive learning. IEEE Trans. Neural Netw. **2**(5), 522–529 (1991)
10. B. Kosko, Fuzzy systems as universal approximators. IEEE Trans. Comput. **43**(11), 1329–1333 (1994). November
11. B. Kosko, *Fuzzy Engineering* (Prentice-Hall, 1996)
12. B. Kosko, Additive fuzzy systems: from generalized mixtures to rule Continua. Int. J. Intell. Syst. **33**(8), 1573–1623 (2018)
13. V. Kreinovich, G.C. Mouzouris, H.T. Nguyen, Fuzzy rule based modeling as a universal approximation tool, in *Fuzzy Systems* (Springer, 1998), pp. 135–195
14. Y. LeCun, C. Cortes, MNIST handwritten digit database (2010)
15. J.B. MacQueen, Some methods for classification and analysis of multivariate observations. *Western Management Science Institute University*. Technical report, of California Working Paper (1966)
16. A.K. Panda, B. Kosko, Converting neural networks to rule foam, pp. 519–525 (2019)
17. W. Rudin, *Real and Complex Analysis* (McGraw-hill education, 2006)
18. W. Samek, *Explainable AI: Interpreting, Explaining and Visualizing Deep Learning*, vol. 11700 (Springer Nature, 2019)
19. F. Watkins, The representation problem for additive fuzzy systems, in *Proceedings of the International Conference on Fuzzy Systems (IEEE FUZZ-95)* (pp. 117–122) (1995)

An Approach for Solving Fully Interval Production Planning Problems

Juan Carlos Figueroa García and Carlos Franco

Abstract This paper shows a method for solving production planning problems involving interval-valued costs, standard times and constraints. An iterative method is used to solve the fully production planning problem using a global satisfaction degree and it is applied to an inventory/shortage example.

Keywords Interval linear programming · Interval uncertainty · Optimization

1 Introduction and Motivation

Interval uncertainty comes from a lack of certainty about the exact value of a variable/parameter, so it can be expressed as an interval set of possible values (often understood as a range of values). Some definitions and applications are given by Kreinovich (see [7, 10, 16, 20]) and R. Baker (see [13, 18]).

Interval Linear Programming (ILP) has been studied by Garajová et al. [9], Hladík [11], Hladík [12, 12], Černý, Hladík [2] and Lodwick & Jamison [17] who introduced theoretical conditions to obtain optimal solutions for ILPs.

Mixed Production Planning (MPP) is a popular decision making problem where different uncertainty sources can affect its parameters i.e. costs, technological parameters and constraints, so those uncertain parameters can be represented with intervals. A similar reasoning to that used by Zimmermann [23] and Verdegay [3] is implemented in this paper where the main idea is to involve interval-valued parameters and an additional variable that operates as an global satisfaction degree among the goal, technological parameters and constraints of the problem. Therefore, the proposed method finds a solution enclosed into the boundaries of the ILP instead of

J. C. Figueroa García (✉)
Universidad Distrital Francisco José de Caldas, Bogotá, Colombia
e-mail: jcfigueroag@udistrital.edu.co

C. Franco
Universidad del Rosario, Bogotá, Colombia
e-mail: carlosa.franco@urosario.edu.co

B. Bede et al. (eds.), *Fuzzy Information Processing 2020*,
Advances in Intelligent Systems and Computing 1337,
https://doi.org/10.1007/978-3-030-81561-5_22

computing them, so the question *How to select a particular solution enclosed into the interval set of possible choices?* is answered in this paper.

The paper is divided into five sections. Section 1 shows the introduction; Sect. 2 presents the interval-valued LP model; Sect. 3 presents the proposed approach. In Sect. 4, the fully interval MPP is presented; Sect. 5 presents an application example, and finally some concluding remarks are presented in Sect. 6.

2 The Interval LP Problem

2.1 Real-Interval Numbers

Let us define $\mathscr{I}(X)$ as the class of all intervals where $\mathbb{R} \subseteq \mathscr{I}(\mathbb{R})$. A real-interval (or simply interval) is denoted as $\boldsymbol{x} \in \mathscr{I}(\mathbb{R})$ and is a subset of real numbers i.e.

$$\boldsymbol{x} = \{[\underline{x}, \overline{x}] | \, \underline{x} < \overline{x}; \; (\underline{x}, \overline{x}) \in \mathbb{R}\}$$

Real-interval numbers are useful to represent uncertain parameters in decision making. Intervals $\mathscr{I}(\mathbb{R})$ are often used to represent generic uncertainty since they do not weight its elements in any particular way.

2.2 Interval LP Model

The ILP basic model is as follows (see Garajová et al. [9] and Hladík [11]):

$$\begin{aligned} \text{Max } z &= \sum_{j=1}^{n} \boldsymbol{c}_j x_j \\ & s.t. \\ \sum_{j=1}^{n} \boldsymbol{a}_{ij} x_j &\leqslant \boldsymbol{b}_i \quad \forall i \in \mathbb{N}_m \\ x_j &\geqslant 0 \end{aligned} \tag{1}$$

where $\boldsymbol{c}_j \in \mathscr{I}(\mathbb{R})$, $\boldsymbol{a}_{ij} \in \mathscr{I}(\mathbb{R})$, and $\boldsymbol{b}_i \in \mathscr{I}(\mathbb{R})$.

In the model $\text{Max}_x\{z = \boldsymbol{c}'x : \boldsymbol{A}x \leqslant \boldsymbol{b}, x \geqslant 0\}$ all its parameters $\boldsymbol{A} \in \mathscr{I}(\mathbb{R}^{mn})$, $\boldsymbol{b} \in \mathscr{I}(\mathbb{R}^m)$, $\boldsymbol{c} \in \mathscr{I}(\mathbb{R}^n)$ are intervals. There are different binary relations for intervals, so in this paper we use the binary relation $\leqslant$ defined as follows:

Definition 1 Let $\boldsymbol{x}, \boldsymbol{y} \in \mathscr{I}(\mathbb{R})$ be two intervals. Then it is said that $\boldsymbol{x} \leqslant \boldsymbol{y}$ only if $\underline{x} \leqslant \underline{y}$ and $\overline{x} \leqslant \overline{y}$.

In general, the solution of an ILP is an interval of solutions $\boldsymbol{z} = [\underline{z}, \overline{z}]$ (see Kreinovich and Kosheleva [15], Garajová et al. [9] and Hladík [11]):

$$\underline{z}(\boldsymbol{A}, \boldsymbol{b}, \boldsymbol{c}) = \inf\{z(A, b, c) : A \in \boldsymbol{A}, b \in \boldsymbol{b}, c \in \boldsymbol{c}\} \tag{2}$$

$$\overline{z}(\boldsymbol{A}, \boldsymbol{b}, \boldsymbol{c}) = \sup\{z(A, b, c) : A \in \boldsymbol{A}, b \in \boldsymbol{b}, c \in \boldsymbol{c}\} \tag{3}$$

where $\boldsymbol{A} \in [\underline{A}, \overline{A}]$, $\boldsymbol{b} \in [\underline{b}, \overline{b}]$, $\boldsymbol{c} \in [\underline{c}, \overline{c}]$ and usually $\underline{z}, \overline{z}$ come from two instances:

$$\underline{z} = \text{Max}\{z = \underline{c}'x : \overline{A}x \leqslant \underline{b}, x \geqslant 0\} \tag{4}$$

$$\overline{z} = \text{Max}\{z = \overline{c}'x : \underline{A}x \leqslant \overline{b}, x \geqslant 0\} \tag{5}$$

Necessary and sufficient conditions for the existence of feasible/optimal solutions for $\underline{z}, \overline{z}$ are discussed by Mráz [19] and Hladík [5]. The problem addressed in this paper is *how to select an optimal solution enclosed into* $\boldsymbol{z}$ instead (good works on this issue include Tuan et al. [21], Welteand et al. [22] and Alefeld et al. [1]).

3 The Fully Interval Mixed Production Planning Problem (IMPP)

MPP models look for optimal production quantities constrained to its demands the capacities of the system. In this paper, the goal of the IMPP is to maximize total utilities composed by the incomes (selling price) minus production, inventory and shortage costs under interval uncertainty. This way, the IMPP model is as follows:

$$\text{Max } \boldsymbol{z} = \sum_{j=1}^{n}\sum_{k=1}^{K}(\boldsymbol{Sp}_{jk} - \boldsymbol{Cp}_{jk})\, x_{jk} - \boldsymbol{h}_{jk}\, q_{jk} - \boldsymbol{o}_{jk}\, s_{jk} \tag{6}$$

s.t.

$$\sum_{j=1}^{n} \boldsymbol{ts}_{ijk}\, x_{jk} \leqslant \boldsymbol{Ac}_{ik} \;\; \forall \;\, i \in \mathbb{N}_m; k \in \mathbb{N}_K \tag{7}$$

$$x_{jk} + q_{j,k-1} + s_{jk} - q_{jk} - s_{j,k-1} \leqslant \boldsymbol{d}_{jk}^{(+)} \;\; \forall \;\, j \in \mathbb{N}_n; k \in \mathbb{N}_K \tag{8}$$

$$x_{jk} + q_{j,k-1} + s_{jk} - q_{jk} - s_{j,k-1} \geqslant \boldsymbol{d}_{jk}^{(-)} \;\; \forall \;\, j \in \mathbb{N}_n; k \in \mathbb{N}_K \tag{9}$$

$$x_{jk}; q_{jk}, s_{jk} \geqslant 0 \tag{10}$$

where x_{jk}, q_{jk}, s_{jk} are the quantities of product/inventory/shortage j to be manufactured/held/supported per period k; $\boldsymbol{Sp}_{jk}$ is the unitary interval-selling price, and $\boldsymbol{Cp}'_{jk}, \boldsymbol{h}_{jk}, \boldsymbol{o}_{jk}$ are the unitary interval-production cost, inventory and shortage costs for x_{jk}, q_{jk} and s_{jk}, $\boldsymbol{ts}_{ijk}$ is the unitary interval-production time for x_{jk}; $\boldsymbol{d}_{jk}^{(-)}, \boldsymbol{d}_{jk}^{(+)}$

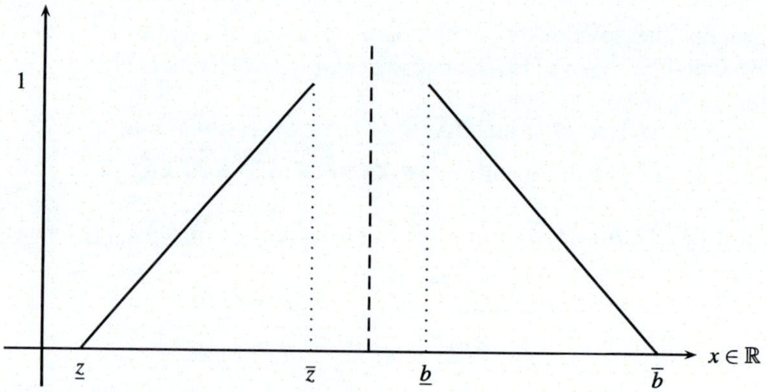

Fig. 1 Conflicting interests between $\tilde{z}$ and $\boldsymbol{b}$

are minimum/maximum interval-demands of x_{jk}, and $\boldsymbol{Ac}_{ik}$ is the available interval-capacity of stage i per period k (in hours).

4 The Proposed Approach

The same reasoning used by Zimmermann [23] and Verdegay [3] namely the Soft Constraints Method (SCM) is used here. The main idea of the SCM is to compose a fuzzy set of optimal solutions $\tilde{z} \in [\check{z}, \hat{z}]$ coming from fuzzy constraints $\boldsymbol{b}$ to then maximize a global satisfaction degree λ among them.

It is important to note that $\tilde{z}$ and the constraints $\boldsymbol{b}$ are conflicting each other since the goal is to maximize using the less possible resources (see Fig. 1). Then the SCM looks for the maximum satisfaction degree between the goal $\tilde{z}$ and its constraints $\boldsymbol{b}$ using an auxiliary variable λ where $\check{z}$, $\hat{z}$ come from an LP for $\underline{b}$, $\overline{b}$ respectively.

In order to use less resources and the highest possible income to maximize Eq. (6) (lesser usage of resources $\boldsymbol{A}$ and bigger profits $\boldsymbol{c}$ imply a bigger satisfaction degree) we define the shape of $\boldsymbol{ts}_{ijk}$, $\boldsymbol{AC}_{ik}$, $\boldsymbol{d}_{jk}^{(+)}$ as equal to $\boldsymbol{b}$ and the shapes of $\boldsymbol{Sp}_{jk}$, $\boldsymbol{Cp}_{jk}$, $\boldsymbol{h}_{jk}$, $\boldsymbol{o}_{jk}$ as equal to $\tilde{z}$, as a function of α:

$$\alpha = \frac{\overline{b} - b(\alpha)}{\overline{b} - \underline{b}} \rightarrow b(\alpha) = \overline{b} - \alpha \cdot (\overline{b} - \underline{b}) \tag{11}$$

$$\alpha = \frac{z(\alpha) - \underline{z}}{\overline{z} - \underline{z}} \rightarrow z(\alpha) = \underline{z} + \alpha \cdot (\overline{z} - \underline{z}) \tag{12}$$

The strategy proposed to solve the model (6) uses a new decision variable $\alpha \in [0, 1]$ to obtain real values for $\boldsymbol{ts}_{ijk}$, $\boldsymbol{AC}_{ik}$, $\boldsymbol{d}_{jk}^{(+)}$, $\boldsymbol{Sp}_{jk}$, $\boldsymbol{Cp}_{jk}$, $\boldsymbol{h}_{jk}$, $\boldsymbol{o}_{jk}$, then α is updated with the λ obtained by the SCM in three iterative steps: (i) computation of the boundaries for $\tilde{z} = [\check{z}, \hat{z}]$, (ii) computation of a global satisfaction degree λ and (iii) update α with λ^* till $\alpha \approx \lambda$, as shown as follows:

Iterative IMPP

1. SELECT an initial $\alpha_1 \in [0, 1]$ and SET $n = 1, \lambda_0^* = \alpha_1$
2. FOR n COMPUTE $Sp_{jk}(\alpha_n), Cp_{jk}(\alpha_n), h_{jk}(\alpha_n), o_{jk}(\alpha_n), ts_{ijk}(\alpha_n), d_{jk}^{(+)}(\alpha_n), d_{jk}^{(-)}(\alpha_n)$
3. COMPUTE $\check{z}$ as follows:

$$\check{z}_n = \text{Max} \sum_{j=1}^{n}\sum_{k=1}^{K}(Sp_{jk}(\alpha_n) - Cp_{jk}(\alpha_n)\, x_{jk} - h_{jk}(\alpha_n)\, q_{jk} - o_{jk}(\alpha_n)\, s_{jk}$$

$$\text{s.t.}$$

$$\sum_{j=1}^{n} ts_{jk}(\alpha_n)\, x_{jk} \leqslant \underline{Ac}_{ik} \quad \forall\ i \in \mathbb{N}_m; k \in \mathbb{N}_K$$

$$x_{jk} + q_{j,k-1} + s_{jk} - q_{jk} - s_{j,k-1} \leqslant \underline{d}_{jk}^{(+)} \quad \forall\ j \in \mathbb{N}_n; k \in \mathbb{N}_K$$

$$x_{jk} + q_{j,k-1} + s_{jk} - q_{jk} - s_{j,k-1} \geqslant \underline{d}_{jk}^{(-)} \quad \forall\ j \in \mathbb{N}_n; k \in \mathbb{N}_K$$

$$x_{jk}, q_{jk}, s_{jk} \geqslant 0$$

4. COMPUTE $\hat{z}$ as follows:

$$\hat{z}_n = \text{Max} \sum_{j=1}^{n}\sum_{k=1}^{K}(Sp_{jk}(\alpha_n) - Cp_{jk}(\alpha_n)\, x_{jk} - h_{jk}(\alpha_n)\, q_{jk} - o_{jk}(\alpha_n)\, s_{jk}$$

$$\text{s.t.}$$

$$\sum_{j=1}^{n} ts_{jk}(\alpha_n)\, x_{jk} \leqslant \overline{Ac}_{ik} \quad \forall\ i \in \mathbb{N}_m; k \in \mathbb{N}_K$$

$$x_{jk} + q_{j,k-1} + s_{jk} - q_{jk} - s_{j,k-1} \leqslant \overline{d}_{jk}^{(+)} \quad \forall\ j \in \mathbb{N}_n; k \in \mathbb{N}_K$$

$$x_{jk} + q_{j,k-1} + s_{jk} - q_{jk} - s_{j,k-1} \geqslant \overline{d}_{jk}^{(-)} \quad \forall\ j \in \mathbb{N}_n; k \in \mathbb{N}_K$$

$$x_{jk}, q_{jk}, s_{jk} \geqslant 0$$

5. COMPUTE λ_n^* as follows:

$$\lambda_n^* = \text{Max}\ \ \lambda$$

$$\text{s.t.}$$

$$\sum_{j=1}^{n}\sum_{k=1}^{K}(Sp_{jk}(\alpha_n) - Cp_{jk}(\alpha_n)\, x_{jk} - h_{jk}(\alpha_n)\, q_{jk} - o_{jk}(\alpha_n)\, s_{jk} - \lambda(\hat{z} - \check{z}) = \check{z}$$

$$\sum_{j=1}^{n} ts_{jk}(\alpha_n)\, x_{jk} + \lambda(\overline{Ac}_{ik} - \underline{Ac}_{ik}) \leqslant \overline{Ac}_{ik} \;\; \forall \;\; i \in \mathbb{N}_m; k \in \mathbb{N}_K$$

$$x_{jk} + q_{j,k-1} + s_{jk} - q_{jk} - s_{j,k-1} + \lambda(\overline{d}_{jk}^{(+)} - \underline{d}_{jk}^{(+)}) \leqslant \overline{d}_{jk}^{(+)} \;\; \forall \;\; j \in \mathbb{N}_n; k \in \mathbb{N}_K$$

$$x_{jk} + q_{j,k-1} + s_{jk} - q_{jk} - s_{j,k-1} - \lambda(\overline{d}_{jk}^{(-)} - \underline{d}_{jk}^{(-)}) \geqslant \underline{d}_{jk}^{(-)} \;\; \forall \;\; j \in \mathbb{N}_n; k \in \mathbb{N}_K$$

$$x_{jk}, q_{jk}, s_{jk} \geqslant 0, \;\; \lambda \in [0, 1]$$

6. IF $\lambda_n^* \approx \lambda_{n-1}^*$ then STOP, ELSE SET $n = n + 1$, $\alpha_n = \lambda_{n-1}^*$
7. END FOR
8. RETURN $\lambda_n^*, \hat{z}_n, \check{z}_n, x_{jk}^*, q_{jk}^*, s_{jk}^*$

After computing $Sp_{jk}(\alpha_n), Cp_{jk}(\alpha_n), h_{jk}(\alpha_n), o_{jk}(\alpha_n), ts_{ijk}(\alpha_n), d_{jk}^{(+)}(\alpha_n), d_{jk}^{(-)}(\alpha_n)$ (see Eqs. (12) and (11)) and solving Eq. (5) for an initial α, they are updated with λ^* (see Eq. (5)) until it converges to a stable solution (see Kreinovich and Figueroa-García [14], Figueroa-García et al. [4, 8] for further details).

The solution gets an equilibrium point $\alpha = \lambda^*$ between $\tilde{z}, \boldsymbol{b}$ and the values $Sp_{jk}(\alpha_n), Cp_{jk}(\alpha_n), h_{jk}(\alpha_n), o_{jk}(\alpha_n), ts_{ijk}(\alpha_n), d_{jk}^{(+)}(\alpha_n), d_{jk}^{(-)}(\alpha_n)$ for $\alpha = \lambda^*$. Note that $\alpha = \lambda^*$ leads to one of the possible choices $Sp_{jk}(\alpha_n), Cp_{jk}(\alpha_n), h_{jk}(\alpha_n), o_{jk}(\alpha_n), ts_{ijk}(\alpha_n), d_{jk}^{(+)}(\alpha_n), d_{jk}^{(-)}(\alpha_n)$ that leads to an optimal satisfaction degree among $\tilde{z}, \boldsymbol{AC}, \boldsymbol{d}^{(+)}, \boldsymbol{d}^{(-)}$. This way, $\alpha = \lambda^*$ answers the question *How to select a particular solution enclosed into the interval set of possible choices?* since it is an optimal satisfaction/equilibrium degree among $\tilde{z}, \boldsymbol{Sp}, \boldsymbol{Cp}, \boldsymbol{h}, \boldsymbol{o}, \boldsymbol{ts}, \boldsymbol{AC}, \boldsymbol{d}^{(+)}, \boldsymbol{d}^{(-)}$.

5 Application Example

The IMPP to solve is composed by 4 products manufactured in 3 stations during 4 periods. A production, inventory and backorder strategy is applied to a utility maximization objective. All interval parameters shown in Tables 1 and 2. Its shapes are defined in Eqs. (5), (12) and (11).

This IMPP can be solved using an initial value $\alpha \in [0, 1]$ which leads to a final λ^*. The obtained results for some instances are shown next.

5.1 *Obtained Results*

Five different experiments for $\alpha_1 = \{0.1, 0.4, 0.5, 0.6, 0.9\}$ were performed, all of them leading to an optimal $\alpha^* = \lambda^* = 0.51347$ in different iterations. The optimal values for z were $\underline{z} = -6445780$, $\overline{z} = 56556200$ (see Eqs. (4) and (5)), and the optimal values for $\tilde{z}$ were $\check{z} = 6516920.02$, $\hat{z} = 9252219.66$ and $z(\alpha^*) = 7921414.33$. Note that z includes negative values which mean that the worst possible scenario needs to pay much more in shortage to satisfy demands.

Table 1 Unitary interval-production times and capacities

Interval-production time $\boldsymbol{ts}_{ijk}\ \forall\{k = 1, 2, 3, 4\}$

i, j	$\underline{ts}_{i1k}$	$\underline{ts}_{i2k}$	$\underline{ts}_{i3k}$	$\underline{ts}_{i4k}$	$\overline{ts}_{i1k}$	$\overline{ts}_{i2k}$	$\overline{ts}_{i3k}$	$\overline{ts}_{i4k}$
$i = 1$	0	0.2	0.1	0.05	0.35	0.5	0.45	0.4
$i = 2$	0.1	0.025	0	0.05	0.4	0.175	0.25	0.125
$i = 3$	0.025	0.15	0.05	0.025	0.25	0.55	0.45	0.65

Interval-capacities $\boldsymbol{Ac}_{ik}$

i, k	$\underline{Ac}_{ik}$	$\overline{Ac}_{ik}$	i, k	$\underline{Ac}_{ik}$	$\overline{Ac}_{ik}$	i, k	$\underline{Ac}_{ik}$	$\overline{Ac}_{ik}$
1,1	7700	8300	2, 1	3200	4100	3, 1	1800	2400
1,2	5500	6400	2, 2	3100	3800	3, 2	1900	2100
1,3	5200	5800	2, 3	3600	4000	3, 3	2200	3200
1,4	6900	7800	2, 4	2100	3000	3, 4	2600	3100

Table 2 Interval-production, inventory, backorder costs and demands

	Unitary costs								Demands			
j, k	$\underline{Cp}_{jk}$	$\overline{Cp}_{jk}$	$\underline{Sp}_{jk}$	$\overline{Sp}_{jk}$	$\underline{h}_{jk}$	$\overline{h}_{jk}$	$\underline{o}_{jk}$	$\overline{o}_{jk}$	$\underline{d}^{(-)}_{jk}$	$\overline{d}^{(-)}_{jk}$	$\underline{d}^{(+)}_{jk}$	$\overline{d}^{(+)}_{jk}$
1, 1	125	245	150	395	10	12	320	350	3700	3900	9249	9723
1, 2	135	185	180	365	10	14	270	300	3200	3400	8128	8606
1, 3	120	200	140	340	8	13	220	280	2400	2700	6834	7593
1, 4	115	120	165	285	11	15	200	220	3400	3500	8107	8339
2, 1	115	220	155	375	10	12	300	320	2650	2800	6657	7014
2, 2	125	185	180	365	8	14	265	300	3100	3400	8390	9130
2, 3	110	190	175	365	9	13	225	270	1750	1900	4641	5007
2, 4	105	250	140	390	7	15	310	330	2600	2800	6785	7270
3, 1	137	70	165	235	10	12	175	190	1950	2100	5088	5451
3, 2	146	82	175	257	10	14	190	220	3700	3900	9249	9723
3, 3	147	118	180	298	7	13	220	250	2450	2600	6208	6566
3, 4	182	155	210	365	12	15	300	320	4000	4100	9456	9687
4, 1	190	175	230	405	6	12	320	350	2650	2800	6657	7014
4, 2	180	110	230	340	8	14	225	270	3300	3400	7882	8114
4, 3	145	90	195	285	9	13	185	210	1900	2100	5222	5719
4, 4	168	95	210	305	11	15	210	240	2400	2500	5859	6093

For instance, when starting by $\alpha_1 = 0.6$ the method gets the final stable solution $\lambda^* = 0.51347$ in six iterations, as shown in Table 3.

Now, the results of all six experiments get the same $\lambda^* = 0.5137$ (see Fig. 2).

Table 3 Obtained results starting from $\alpha_1 = 0.6$

$\check{z}$	$\hat{z}$	α	λ^*
8888225.64	11770084.09	0.6	0.51438
6539861.76	9275423.50	0.51438	0.51314
6507901.79	9243105.27	0.51314	0.51359
6519658.78	9254988.39	0.51359	0.51340
6515321.80	9250604.11	0.51340	0.51349
6516920.02	9252219.66	0.51349	0.51347

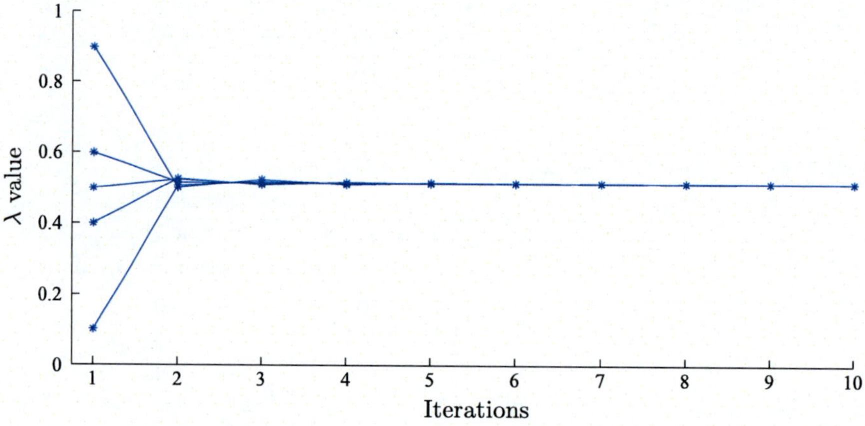

Fig. 2 Obtained results starting from $\alpha_1 = \{0.1, 0.4, 0.5, 0.6, 0.9\}$

6 Concluding Remarks

An iterative method to solve interval IMPPs using an auxiliary equilibrium variable λ is proposed. The method looks for an overall satisfaction/equilibrium degree among interval costs, production times and constraints. The experiments have shown that λ works well and finds a global solution no matter the initial value $\alpha_1 \in [0, 1]$ used to initialize the method. Thus, the question *How to select a particular solution enclosed into the interval set of possible choices?* has been answered using $\alpha = \lambda^*$ as an optimal satisfaction/equilibrium degree among the goal and constraints of the problem.

The presented example is a guide for solving basic interval IMPPs through iterative LPs. Bigger problems including other uncertain variables such as overtime, benchmarks, raw material consumption etc. can be solved as well. A natural extension of the proposed method is to solve fuzzy MMPs using a similar reasoning (see Figueroa-García [6]).

Finally, it is important to note that the presented IMPP is based on a utility maximization scheme. It can also be applied to a minimization problem using a similar reasoning for defining the shape of $\boldsymbol{ts}_{ijk}$, $\boldsymbol{AC}_{ik}$, $\boldsymbol{d}_{jk}^{(+)}$, $\boldsymbol{Sp}_{jk}$, $\boldsymbol{Cp}_{jk}$, $\boldsymbol{h}_{jk}$, and $\boldsymbol{o}_{jk}$.

Acknowledgements The authors would like to thank to Prof. Vladik Kreinovich for his interest and invaluable support, and a special thanks is given to all members of *LAMIC* research group of Universidad Distrital Francisco José de Caldas - Bogotá, Colombia.

References

1. G. Alefeld, V. Kreinovich, G. Mayer, On the solution sets of particular classes of linear interval systems. J. Comput. Appl. Math. **152**(1), 1–15 (2003)
2. M. Černý, M. Hladík, Optimization with uncertain, inexact or unstable data: linear programming and the interval approach, in *Proceedings of the 10th International Conference on Strategic Management and its Support by Information Systems*, ed. by R. Němec, F. Zapletal, VŠB - Technical University of Ostrava (2013), pp. 35–43
3. M. Delgado, J.L. Verdegay, M. Vila, A general model for fuzzy linear programming. Fuzzy Sets Syst. **29**(1), 21–29 (1989)
4. J.C. Figueroa-García, An iterative procedure for fuzzy linear programming problems, in *2011 Annual Meeting of the North American Fuzzy Information Processing Society (NAFIPS)* (IEEE, 2011), pp. 1–6
5. J.C. Figueroa-García, An approximation method for type reduction of an interval Type-2 fuzzy set based on α-cuts, in *IEEE Proceedings of FEDCSIS 2012*. IEEE (2012), pp. 1–6
6. J.C. Figueroa-García, Solving linear programming problems with interval type-2 fuzzy constraints using interval optimization, in *Proceeedings of NAFIPS 2013*. IEEE (2013), pp. 1–6
7. J.C. Figueroa-García, E.R. Lopez, C. Franco-Franco, A note about the (x, y) coordinates of the centroid of a fuzzy set, in *Proceeedings of WEA 2018*. Springer (2018), pp. 78–88
8. J.C. Figueroa-García, Y. Olarte, F. Camargo, Linear programming with fuzzy joint parameters: an iterative method (2010), pp 1–6
9. E. Garajová, M. Hladík, M. Rada, Interval linear programming under transformations: optimal solutions and optimal value range. Cent. Eur. J. Oper. Res. **27**(1), 601–614 (2019)
10. G. Heindl, V. Kreinovich, A.V. Lakeyev, Solving linear interval systems is NP-Hard even if we exclude Overflow and Underflow. Rel. Comput. **4**(4), 383–388 (1998)
11. M. Hladík, Optimal value range in interval linear programming. Fuzzy Opt. Dec. Making **8**(3), 283–294 (2009)
12. M. Hladík, Weak and strong solvability of interval linear systems of equations and inequalities. Linear Algebra Appl. **438**(11), 4156–4165 (2013)
13. R.B. Kearfott, V. Kreinovich, Beyond convex? Global optimization is feasible only for convex objective functions: a theorem. J. Global Opt. **33**(4), 617–624 (2005)
14. V. Kreinovich, J.C. Figueroa-Garcia, Optimization under fuzzy constraints: from a heuristic algorithm to an algorithm that always converges, in *Proceeedings of WEA 2018*. Springer (2018), pp. 3–16
15. V. Kreinovich, O. Kosheleva, The range of a continuous functional under set-valued uncertainty is always an interval. Rel. Comput. **24**, 27–30 (2017)
16. V. Kreinovich, C.W. Tao, Checking identities is computationally intractable (NP-Hard), so human provers will be always needed. Int. J. Intell. Syst. **19**(1), 39–49 (2004)
17. W.A. Lodwick, K.D. Jamison, Interval methods and fuzzy optimization. Int. J. Uncertain. Fuzziness Knowl.-Based Syst. **5**(3), 239–249 (1997)
18. R.E. Moore, R.B. Kearfott, M.J. Cloud, *Introduction to Interval Analysis* (SIAM, 2009)
19. F. Mráz, Calculating the exact bounds of optimal values in lp with interval coefficients. Ann. Oper. Res. **81**, 51–62 (1998)

20. H.T. Nguyen, V. Kreinovich, Computing degrees of subsethood and similarity for interval-valued fuzzy sets: fast algorithms, in *9th International Conference on Intelligent Technologies InTec'08*. IEEE (2008), pp. 47–55
21. T.A. Tuan, V. Kreinovich, T.N. Nguyen, Decision making under interval uncertainty: beyond hurwicz pessimism-optimism criterion, in *Beyond Traditional Probabilistic Methods in Economics*. IEEE (2019), pp. 176–184
22. A. Welteand, L. Jaulin, M. Ceberio, V. Kreinovich, Avoiding fake boundaries in set interval computing. J. Uncertain Syst. **11**(2), 137–148 (2017)
23. H.J. Zimmermann, Fuzzy programming and Linear Programming with several objective functions. Fuzzy Sets Syst. **1**(1), 45–55 (1978)

Enhanced Cascaded Genetic Fuzzy System for Counterfeit Banknote Detection

Anirudh Chhabra, Donghoon Kim, and Kelly Cohen

Abstract Counterfeit currency is a very serious issue from an economic point of view for any country. In rural communities around the world, where digital currency has not yet become the norm and the usage of banknotes is still dominant, the detection of counterfeit banknotes is very important. Authentication of currency is a classification problem and therefore a Fuzzy Inference System (FIS) is designed with its parameters optimized using a Genetic Algorithm. This FIS takes in features extracted from the images of banknotes and classifies the corresponding banknotes as genuine or fake. A cascaded approach is selected to design the FIS in order to reduce the number of parameters. In this study, an algorithm based on the Fuzzy C-means clustering algorithm is designed to define the best combination of inputs for the cascaded FIS. Also, another cascaded FIS is developed with an attempt at decreasing the number of parameters by introducing symmetry and other constraints. The results are compared with the Machine Learning (ML) approaches, and efforts are also made to get a highly reliable precision similar to that of ML methodologies.

1 Introduction

Smoothness in transactions is highly dependent on the effective detection of counterfeit banknotes. With newer techniques being developed every day to forge currency, the approach to authenticate banknotes should also be seamless and consistently upgraded. This is necessary to ensure that the economy of a state is not the victim of such unethical practices.

A. Chhabra (✉) · D. Kim · K. Cohen
Department of Aerospace Engineering & Engineering Mechanics, University of Cincinnati, Cincinnati, OH 45221, USA
e-mail: chhabrad@mail.uc.edu

D. Kim
e-mail: Donghoon.Kim@uc.edu

K. Cohen
e-mail: cohenky@ucmail.uc.edu

B. Bede et al. (eds.), *Fuzzy Information Processing 2020*,
Advances in Intelligent Systems and Computing 1337,
https://doi.org/10.1007/978-3-030-81561-5_23

Neural Networks (NNs) and other Machine Learning (ML) techniques have recently become the standard of classification problems which also involve banknote authentication. With precision rates of up to 100% in some cases, these methodologies have been widely used in classification problems. However, the issue is that these approaches are black boxes [6]. They give high precision at the cost of no explainability. The purpose of ML must be to augment human decision but these techniques are better at replacing humans. Also, with no explainability comes the fear of unexpected and unpredictable errors. Therefore, for sensitive applications, ML algorithms that provide explainability have been preferred over the traditional approaches such as NNs. Fuzzy Inference System (FIS) is one of the intelligent approaches that provide explainability [11]. The idea is to design a system capable of interpreting linguistic rules in decision making. Fuzzy systems basically provide membership to each category over the range of inputs. This process consists of multiple steps: fuzzification, rule-inference, and defuzzification. It basically creates an explainable mapping between the inputs and the outputs of a system. The parameters of the FIS used here are optimized using a Genetic Algorithm (GA) [5]. A supervised learning approach is used to tune these parameters and optimize the FIS. The use of a GA-optimized FIS is also known as a Genetic Fuzzy System (GFS) [1]. GFSs have been successful in a variety of applications in the past as demonstrated in [1, 4].

In this study, a GFS-based algorithm is developed to efficiently detect counterfeit banknotes and provide explainability in terms of the classification features. There are two cases in which the number of parameters used to optimize the FIS is different and the results are observed for both of them. The results have been then compared with other Artificial Intelligence (AI) approaches. This work is organized as follows: A dataset, FIS and the GA are described in Sect. 2; the methodology proposed is explained in Sect. 3; the simulation study results using the proposed method are explained in Sect. 4 and are also compared with the existing AI methodologies; and finally, Sect. 5 concludes this work.

2 Preliminaries

2.1 Banknote Authentication Dataset

The dataset used in this study, *Banknote Authentication Dataset* [8], is obtained from the University of California, Irvine - Machine Learning Repository. The dataset contains 5 attributes out of which 4 are input features and 1 is the output which is defined as the class of the banknote. Here is the attribute information:

- **Variance** of Wavelet Transformed Image (continuous)—a measure of the spread of a distribution about its mean.
- **Skewness** of Wavelet Transformed Image (continuous)—the direction of variation of the lack of symmetry.

- **Kurtosis** of Wavelet Transformed Image (continuous)—a measure of whether the data are heavy-tailed or light-tailed relative to a normal distribution.
- **Entropy** of Image (continuous)—the amount of information that must be encoded by a compression algorithm.
- **Class** (discrete)—1 means genuine. 0 means fake.

It is a discrete value with the assumption that 0 means fake, and 1 means genuine. The dataset contains 1,372 sets of values for each attribute. The following is the description of the dataset available at the source [8]: *Data were extracted from images that were taken from genuine and forged banknote-like specimens. For digitization, an industrial camera usually used for print inspection was used. The final images have* 400 × 400 *pixels. Due to the object lens and distance to the investigated object gray-scale pictures with a resolution of about 660 dpi were gained. Wavelet Transform tool was used to extract features from images.*

2.2 Fuzzy Inference System

The FIS consists of three sections: the Membership Functions (MFs), Rule-base & Inference, and Defuzzification. The MFs are specified using the parameters obtained from the chromosome optimized by the GA. These are updated in every generation of the GA (See Sect. 3.2). Each fuzzy set has a corresponding membership for any given variable, and these membership values lie in the range [0,1]. Then the rule-base consists of the logical operator 'AND' which is used to evaluate the rules. Then these rules are aggregated based on the parameters from the chromosome. When the fuzzified inputs have been evaluated into a single number, then this number is used to correlate the inputs to an output MF. The defuzzification technique used here is *Centroid Defuzzification* [7] with the scaling of the output MFs.

2.3 Genetic Algorithm

To obtain an optimal FIS, one needs to optimize its parameters. The process is to randomly select a solution set for all the parameters, and update it until an optimal set of parameters is achieved. In terms of GA, this initial set is known as the initial population. GA is a heuristic search algorithm based on the idea of Darwin's theory of natural evolution. It performs operations like crossover and mutation on the initial set of solutions just like in nature. A candidate solution is known as a chromosome and the parameters contained in it are known as genes. Each iteration is known as a generation. The GA implemented here also consists of elitism, which allows the preservation of the best solutions and make sure that natural processes do not remove the best chromosomes from the population. This allows for the population to improve itself over a number of generations.

3 Suggested Approach

3.1 Cascaded Fuzzy Inference System

A cascaded FIS (CFIS) is developed for this classification problem. Three interlinked FISs are designed with 2 inputs and 1 output each as shown in Fig. 1. The reason behind using a cascaded approach is to reduce the number of parameters that need to be optimized. In a single FIS with n inputs and m MFs for each input, the number of rules that exist is m^n. In this problem, there are 4 inputs with 5 MFs each. This leads to $5^4 = 625$ rules. Now, one needs 3 parameters for each of the 36 MFs (5 each for 4 inputs and 2 intermediate fuzzy sets, and 2 for the single output) resulting in a total of $(5 \times 6 + 2) \times 3 = 96$ parameters. This leads to a total of 721 parameters. This is a problem as it will take a long time for the GA to optimize these parameters.

In the cascaded approach, three FISs are used instead of one. This leads to 25 rules in each FIS which leads to 75 parameters required for the optimization of the rule-base for the three FISs. The number of parameters required for the fine-tuning of the MFs remains the same as before. This induces a total of $75 + 96 = 171$ parameters. This also requires a large amount of time to be optimized using the GA but it gives a significant reduction in computational burden compared to the single FIS approach. Furthermore, the relationship between variables can be studied to combine the more related inputs with each other in the CFIS.

Finding the Best Combination of Inputs
Some features are more closely related to each other than to other features depending on how well they can be clustered into the required output. In this study, this relationship of features is observed to find out the best combination of inputs for

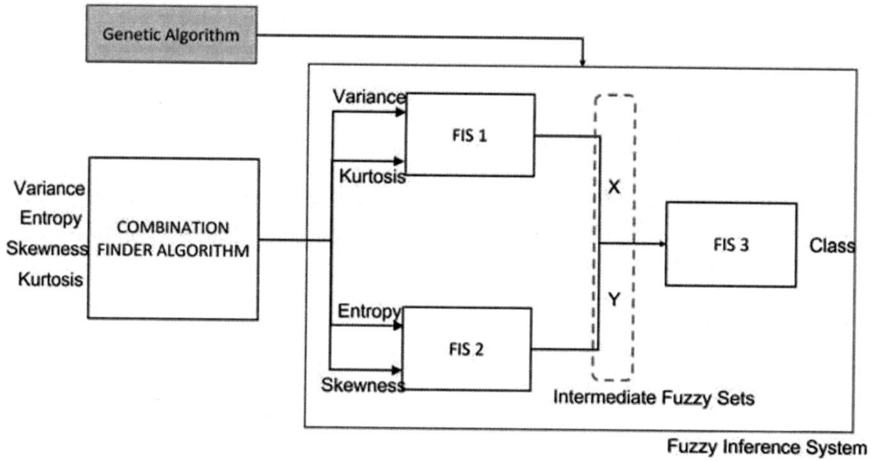

Fig. 1 Cascaded fuzzy inference system

the cascaded structure, and the *fuzzy C-means algorithm* [3] is used. It picks out the best combination when all the possible combinations perform well enough but none of them is a poor combination. This means that 2 combinations that perform above average are better than the group of a good combination and a bad combination. This approach is tested, and the results are provided for all the possible combinations.

Reducing the Number of Parameters
In addition, an attempt is made to reduce the number of parameters to be optimized by the GA. The approach is to equate the position of the left end of an MF to the center of the preceding MF. The extreme ends are allowed to be optimized according to the GA. Also, the center of the central MF is fixed at zero. The MFs of the output, *Class*, is fixed around 0 and 1. With this, the number of parameters required for fine-tuning of the MFs is reduced to 36 leading to the total number of parameters equal to 111. The results are compared with the original approach, and it is observed that suggested approach provides more clarity regarding the shape of the MFs making them easier to interpret.

3.2 GA-based Parameter Optimization

In this study, a proportional Roulette wheel selection algorithm is preferred over a Rank-based or a Tournament selection algorithm because of its simplicity for design and debug. It also offers quicker results than the other two approaches. For the purpose of the crossover of candidate solutions or chromosomes, a single-point crossover algorithm has been developed as a part of the GA. Figure 2 depicts the workflow of the GA, and the structure of each candidate solution (or chromosome) for both the cases in this study is shown in Figs. 3 and 4, respectively. As depicted in Fig. 2, the first step is to initialize a random population of chromosomes within the specified bounds for each parameter. Then the fitness value of each of these solutions is determined by passing them as the required set of parameters of the FIS. The data points from the training data are used to evaluate the resulting FIS. The best chromosomes with a higher fitness value are then selected using the specified selection algorithm. The selected chromosomes are known as parent chromosomes, which undergo the crossover operation to produce two children with a probability of crossover, P_c, which is determined by multiple trials. The best chromosomes then undergo some random mutation with a probability of mutation, P_m, which is kept very low to avoid mutation of a good solution into a bad solution. Then, the population undergoes elitism where a small percentage of the population with the highest fitness values is preserved in order to avoid unnecessary operations which might reduce the quality of the population. This complete process is carried out for 10,000 generations (or iterations). In the end, the population is sorted according to the fitness values and the best chromosomes are taken as the final set of trained parameters for the FIS.

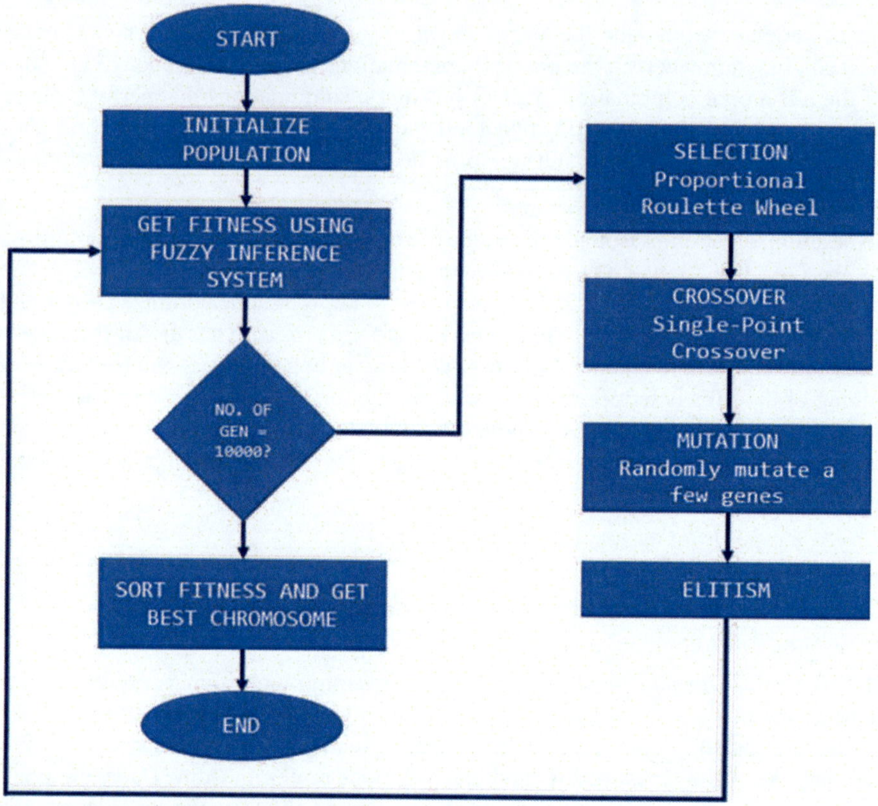

Fig. 2 Workflow diagram for the GA

MF(v)	MF(k)	MF(s)	MF(e)	MF(x)	MF(y)	MF(class)	Rule Base for FIS 1	Rule Base for FIS 2	Rule Base for FIS 3
0-15	16-30	31-45	46-60	61-75	76-90	91-96	97 - 121	122-146	147 - 171

Fig. 3 Chromosome structure for Case I

MF(v)	MF(k)	MF(s)	MF(e)	MF(x)	MF(y)	Rule Base for FIS 1	Rule Base for FIS 2	Rule Base for FIS 3
0-6	7-12	13-18	19-24	25-30	30-36	37-61	62-86	87-111

Fig. 4 Chromosome structure for Case II

4 Simulation Results

Note. The figure of merit applied here for the evaluation of the system is the mean absolute error, which is found as a measure of the absolute difference between expected and observed values. This has been used as the criteria in this classification problem as the absolute error is always one (fake-0 or genuine-1) and not a continuous value. Therefore, there is no need to measure the root mean square error as we don't have a variable error.

Initial conditions and stochastic parameters used for the two cases are listed in Table 1. Table 2 shows that the objective values found for the various combinations. The objective value correlates to the intra-cluster variance. Clustering is desirable when the objective value is minimum. Less intra-cluster variance corresponds to higher intra-cluster similarity. Therefore, it can be concluded that the best pair of combinations of inputs is *Variance & Kurtosis* and *Skewness & Entropy*. The optimization results of other combinations are also determined, and they offer less precision as predicted.

The MFs obtained for each of the fuzzy sets involved, namely: *Variance, Skewness, Kurtosis, & Entropy* along with the two intermediate fuzzy sets *X* & *Y* that take the outputs from Level 1 of the system as inputs to the Level 2 are shown in Figs. 5 and 6 for Cases I and II, respectively.

Table 1 Simulation parameters

Parameter	Case I	Case II
Length of chromosome, N	171	111
Population size, M	10	10
Probability of crossover, P_c	0.95	0.95
Probability of mutation, P_m	0.05	0.05
Elitism ratio, E_r	0.2	0.2
Maximum number of generations	10,000	10,000

Table 2 Objective values for the possible combinations

Combination	Objective value
Variance & Skewness	20,918.67
Kurtosis & Entropy	11,326.96
Variance & Kurtosis	**14,448.48**
Skewness & Entropy	**16,318.51**
Variance & Entropy	6,700.88
Kurtosis & Skewness	22,758.86

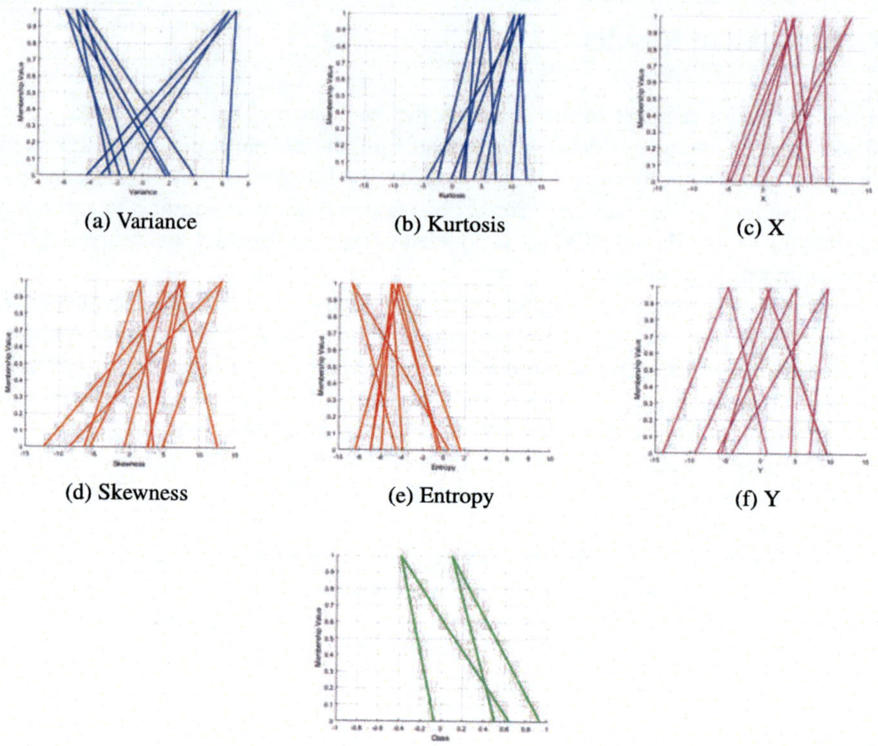

(a) Variance (b) Kurtosis (c) X

(d) Skewness (e) Entropy (f) Y

(g) Class

Fig. 5 Optimized MFs for Case I

The rule-base is defined using a vector optimized by the GA. It basically contains discrete values corresponding to the output MF of that FIS. This creates a linguistic mapping between the inputs and outputs of a FIS. 3 such vectors of 25 values each are obtained for each of the FISs. The rule-base (consisting of 75 rules) is used to infer the value of the output from the set of input data points. These rules are the basis of the explainability of the approach as they can be represented in linguistic terms. An example of linguistic representation of one such rule (obtained from Case II implementation) can be given as follows:

> If Variance is in (−1, 1), Kurtosis is in (−2, 7), Skewness is in (7, 13), and Entropy is in (−7, −3), then Class is "0" or "Fake".

The optimization process is carried out 10 times to cross-validate the results in order to ensure that the results obtained are not due to a specific set of training data used to validate the approach. This allows for a more true measure of precision.

Figure 7 shows the training and validation curves for the case one where the length of the chromosomes in 171. The precision observed at the end of the training stage is 98.2%. Each spike in the validation curve highlights an error in the validation stage.

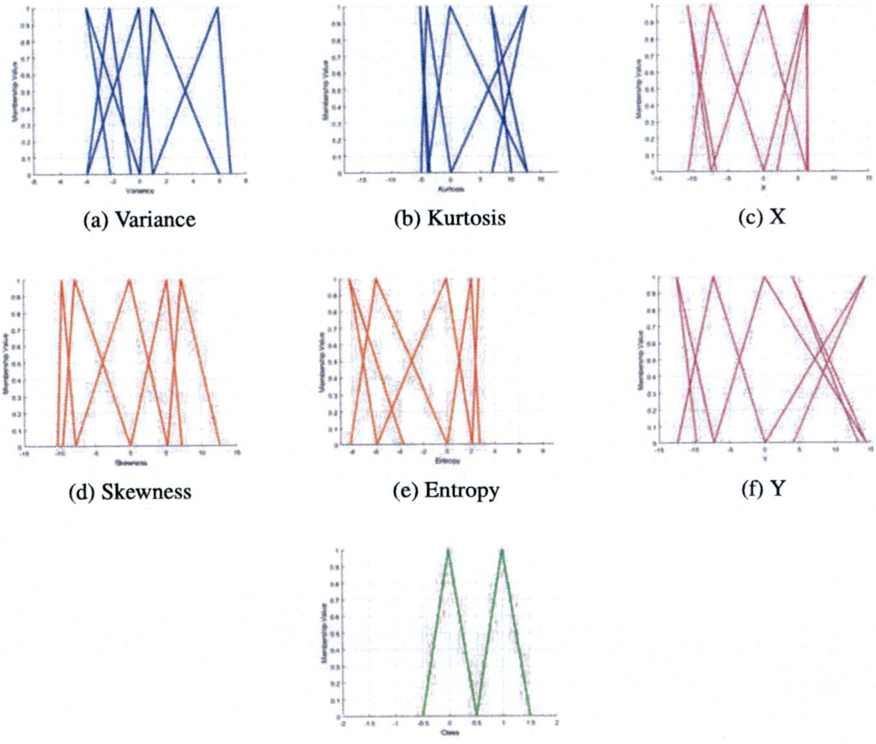

(a) Variance (b) Kurtosis (c) X

(d) Skewness (e) Entropy (f) Y

(g) Class

Fig. 6 Optimized MFs for Case II

8 spikes are observed in the curve so that there are 8 errors out of 372 validation data points leading to a validation precision of 97.85%. Figure 8 shows the training and validation curves for the case one where the length of the chromosomes in 111. The precision percentage observed at the end of the training stage is 99.70%. Each spike in the validation curve highlights an error in the validation stage. 10 spikes are observed in the curve corresponding to 9 errors out of 372 validation data points leading to a validation precision of 97.58%.

Some of the existing AI methodologies have been widely accepted as the benchmarks for the newer algorithms. For this reason, the results of this study are compared with some of the most common and acceptable AI approaches. These include Surface Vector Machines (SVMs) [2], Artificial NNs with Bayesian Regularization Backpropagation (BR Method) [10], and the Decision Trees classification learning approach [9]. Table 3 compares the results of this study with the existing AI methodologies, which are designed by the authors.

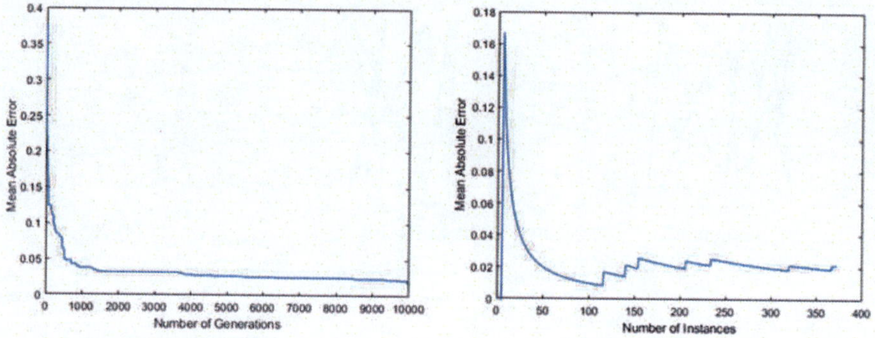

(a) Training Error (b) Validation Error

Fig. 7 Error curves for Case I

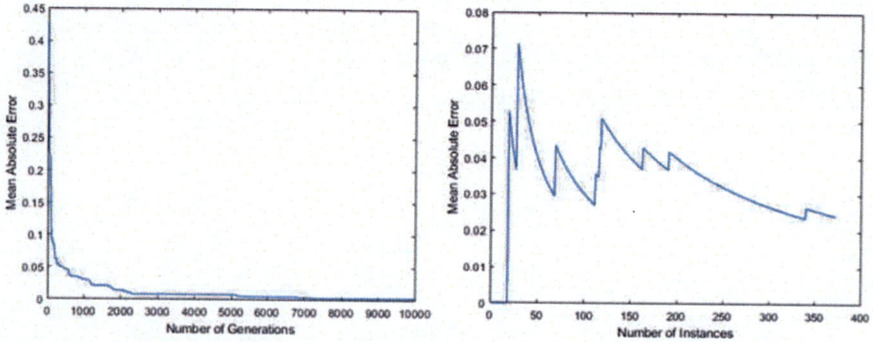

(a) Training Error (b) Validation Error

Fig. 8 Error curves for Case II

Table 3 Precision comparison results

Approach	Precision
NNs (BR Method)	99.91
Decision trees	98.3%
SVM	100%
GFS—Case I	97.85%
GFS—Case II	97.58%

5 Conclusion

The purpose of this study is to introduce explainability in the Machine Learning (ML) approaches and to confirm that Genetic Fuzzy Systems (GFSs) can be trusted for applications in classification problems. Comparable results to existing artificial intelligence approaches, which do not offer any explanation with regard to the inference of raw data, enables us to apply GFSs in areas that specifically demand explainability from the ML approaches due to sensitivity of operations. Reducing the number of parameters by using constraints leads to a lesser training error. With a larger data set, this can be an advantageous approach as it limits the search space of the genetic algorithm offering it more time to converge to the optimal solutions. Also, since the validation error is around the same, it is safe to assume that using constraints can be more desirable in some problems as they are also easier to interpret visually. From this study, it can also be concluded that clustering methodologies can be used to gain more insight into the original dataset which allows for better design on fuzzy inference systems.

References

1. O. Cordon, F. Herrera, F. Hoffmann, L. Magdalena, Genetic fuzzy systems: evolutionary tuning and learning of fuzzy knowledge bases **19**(2001). https://doi.org/10.1142/4177
2. C. Cortes, V. Vapnik, Support-vector networks. Mach. Learn. **20**(3), 273–297 (1995). https://doi.org/10.1007/BF00994018
3. J.C. Dunn, A fuzzy relative of the isodata process and its use in detecting compact well-separated clusters. J. Cybern. **3**(3), 32–57 (1973). https://doi.org/10.1080/01969727308546046
4. N. Ernest, K. Cohen, E. Kivelevitch, C. Schumacher, D. Casbeer, Genetic fuzzy trees and their application towards autonomous training and control of a squadron of unmanned combat aerial vehicles. Unmanned Syst. **03**(2015). https://doi.org/10.1142/S2301385015500120
5. D.E. Goldberg, *Genetic Algorithms in Search, Optimization and Machine Learning*, 1st edn (Addison-Wesley Longman Publishing Co., Inc, USA, 1989)
6. D. Heaven, Why deep-learning AIS are so easy to fool (2019). URL https://www.nature.com/articles/d41586-019-03013-5
7. W.V. Leekwijck, E.E. Kerre, Defuzzification: criteria and classification. Fuzzy Sets Syst. **108**(2), 159–178 (1999). https://doi.org/10.1016/S0165-0114(97)00337-0. http://www.sciencedirect.com/science/article/pii/S0165011497003370
8. V. Lohweg, H. Doerksen, UCI machine learning repository (2017). https://archive.ics.uci.edu/ml/datasets/banknote+authentication
9. L. Rokach, O. Maimon, *Data Mining With Decision Trees: Theory and Applications*, 2nd edn. (World Scientific Publishing Co., Inc, USA, 2014)
10. C.E. Shannon, J. McCarthy, *Automata Studies*, vol. AM-34 (Princeton University Press, Princeton, 2016). https://princetonup.degruyter.com/view/title/521931
11. L. Zadeh, Fuzzy sets. Inf. Control **8**(3), 338–353 (1965). https://doi.org/10.1016/S0019-9958(65)90241-X. http://www.sciencedirect.com/science/article/pii/S001999586590241X

ExTree—Explainable Genetic Feature Coupling Tree Using Fuzzy Mapping for Dimensionality Reduction with Application to NACA 0012 Airfoils Self-Noise Data Set

Javier Viaña and Kelly Cohen

Abstract This research presents an AI-based tool (ExTree) that provides high explainability in prediction problems that involve multiple continuous inputs. The algorithm uses an input coupling tree that gradually reduces the dimension of the system. The desired dimension reduction is achieved developing a network of fuzzy inference systems (FIS) wherein in each layer of the network, two inputs get combined to yield a single outcome. These outcomes are then submitted to the same procedure at the following layer until we arrive at a single output, thereby reducing the dimensionality of the problem in every step. Hence, large scale problems with more inputs require more layers. The final outcome is that we obtain a set of FIS nodes across the network, where each FIS may be characterized using an explainable control surface. The structure of the tree is optimized using a genetic algorithm that gets the best hierarchy of fuzzy features to minimize the dispersion of the final outcome. This tool has been benchmarked using NASA's wind tunnel testing database of NACA 0012 Airfoils. The results, demonstrating accurate validation, are of value not only from the perspective of a high performing AI-based algorithm, but also because of the substantial amount of interpretability and traceability that the algorithm offers.

Keywords ExTree · Input coupling tree · Genetic algorithm · Fuzzy feature mapping · Dispersion · Airfoils · Dimensionality reduction · Explainability

1 Introduction

The complexity of the data increases every day, requiring more features to characterize properly a meaningful problem. The curse of dimensionality makes it difficult, and sometimes impossible, to visualize the data. Therefore, the algorithms that make

J. Viaña (✉) · K. Cohen
University of Cincinnati, Cincinnati, OH 45219, USA
e-mail: vianajr@mail.uc.edu

K. Cohen
e-mail: cohenky@ucmail.uc.edu

B. Bede et al. (eds.), *Fuzzy Information Processing 2020*,
Advances in Intelligent Systems and Computing 1337,
https://doi.org/10.1007/978-3-030-81561-5_24

use of this information should provide more clarity. In other words, AI has the task of shedding light on the problem, providing traceability for every prediction made.

However, this is currently not the case. Instead of doing so, the tools obscure the process, jeopardizing the understanding from the human perspective. It has even reached a point where the term "black box" is used to define the functions that predict the results. The vast majority of the algorithms that are being used lack explainability.

In the aerospace sector, as in many others, a wise decision (as understood by the machine) is not enough for humans. It is also necessary to justify the reasoning that leads to that prediction [1]. Neural Networks, indeed, can be easily fooled [2]. Small modifications in the inputs of the system, typically imperceptible to humans, can produce a big error in the results. Thus, the outcomes of such an opaque reasoning require an update. Explainable Artificial Intelligence, XAI as identified by DARPA [3], seeks to find this clarification [4, 5].

The tool presented in this research seeks primarily explainability in the predictions of multidimensional problems. The proposed algorithm couples features in a tree fashion. The input features are grouped in pairs based on the dispersion they demonstrate in the target feature. The result of each coupling is a fuzzy output that ranges from 0 to 1. Therefore, it could be understood as a mapping of two features into a non-physical fuzzy feature that still contains the information of both its inputs. To optimize the structure of the tree, a Genetic Algorithm (GA) is used, which aims to minimize the dispersion of the fuzzy outputs obtained from each pair.

The tree is composed of layers. In each layer, the size of the problem can be reduced by half. In the last layer, all the information is contained in a single fuzzy feature. Therefore, the system is translated into a two-dimensional problem (the fuzzy variable produced in the last layer and the values of the target feature). The information of the mapping is recorded inside the tree in the form of control surfaces or matrices. These control surfaces relate the two input features to the fuzzy output feature produced, thus creating an Explainable Tree ("ExTree") with accessible visual data. The term "ExTree" will be used to refer to the whole explainable structure of couplings. The software developed offers the possibility of filtering the data in case there is a high presence of noise. For this, the dispersion of each individual is compared in relation to the dispersion shown by the average in the last layer of the tree as this contains all the information of the original features in a single variable.

This tool has been tested and validated with one of the NASA databases. Particularly, the self-noise results of an anechoic wind tunnel testing of several NACA 0012 airfoil blade sections [6] provided by the UCI Machine Learning Repository [7]. The results obtained have demonstrated that the algorithm not only has a high explainability in the decision making, but also accuracy in the prediction of values.

2 NASA Airfoil Self-Noise Data Set

This data set was obtained from a series of aerodynamic and acoustic tests carried out by NASA of two and three-dimensional airfoil blade sections conducted in an anechoic wind tunnel. The experiments were conducted in an anechoic wind tunnel with the observer's position and the span of the airfoil being constant through the whole testing. The data acquired was multivariable, formed by 1503 instances, with no missing values, and six continuous real features. The five input features are vibration frequency of the airfoil (measured in [Hz]), angle of attack (measured in [°]), chord length of the airfoil (measured in [m]), wind tunnel's free-stream velocity (measured in [m/s]) and suction side displacement thickness (measured in [m]) which refers to the maximum displacement in the suction side. The only target feature is noise produced due to the vibration, measured with the scaled sound pressure level ([dB]). Other authors [8, 9] have already studied this data set in depth using Regression tools, Random Forests, Neural Networks, Boosting or Bagging [10, 11]. Between those prediction models, Bagging and Random Forests outperformed the others, obtaining RMSEs of 0.18 and 0.05 respectively (in non-normalized noise units), with the only inconvenient of the lack of explainability. The regression method on the other hand was able to keep its inherent transparency at the cost of getting a result of 5.72 in the RMSE. In that regard, the method proposed in this research seeks a trade of in traceability and accuracy.

3 Objectives of the Research

The first objective is dimensionality reduction, showing the ability of this tool to gather all the information of a multidimensional problem in a visual bidimensional plot. The second objective is explainability, creating a structure of control surfaces that map the fuzzy features together (two to one) in a tree shape, the ExTree. Furthermore, the tool has an expanding property, which fills any missing combination of features with inferred values. The final objective is validation, proving the aforementioned objectives using the NASA Airfoil Self-Noise Data Set while still getting high efficiency in the prediction.

4 Methodology

4.1 Dimensionality Reduction, Structure of the Tree

For the dimensionality reduction, the input features $\{X_1, X_2, \ldots, X_n\}$ will be grouped in pairs. From every pair, a fuzzy outcome feature will be created $\{A_1, A_2, \ldots, A_{n/2}\}$ (if n is even, if n is odd then there will be $(n+1)/2$ fuzzy features). A control surface,

CS_κ, will relate both input features, X_i and X_{i+1}, to the fuzzy feature, A_j. This control surface will depend on both the dispersion of the variables (D_κ),

$$D_\kappa = f(X_i, X_{i+1}) \tag{1}$$

and the values of the target feature (Y),

$$Y = f(X_1, X_2, \ldots, X_i, X_{i+1}, \ldots, X_n) \tag{2}$$

such that

$$CS_\kappa = f(D_\kappa, Y) \tag{3}$$

When two new values of the two features that define D_κ are considered, X'_i and X'_{i+1}, the resulting fuzzy feature, A'_j, can be extracted mapping X'_i and X'_{i+1} in CS_κ,

$$A'_j = CS_\kappa(X'_i, X'_{i+1}) \tag{4}$$

In such a way that the data can be understood as dependent on the new fuzzy features instead of using the original features (2). Thus, obtaining a reduction of dimensionality. The process is then repeated considering $\{A_1, A_2, \ldots, A_{n/2}\}$ as the new input features; a new set of fuzzy features $\{B_1, B_2, \ldots, B_{n/4}\}$ is obtained (again, if n/2 is even). Eventually, iterations will lead to a single output fuzzy feature, Z_1. This last feature will thus contain the information of all the inputs $\{X_1, X_2, \ldots, X_n\}$ and the dispersion of each coupling. Therefore, the multidimensional data could then be represented in two dimensions as,

$$Y = f(A_1, A_2, \ldots, A_{n/2}) \tag{5}$$

$$Y = f(Z_1) \tag{6}$$

For the case of the NASA Airfoil Self-Noise Data Set, the ExTree is as described in Fig. 1.

4.2 *Fine Tuning, Genetic Algorithm*

A GA will be used to select accordingly the couples. Other authors [12] have already used Gas for camber control morphing of airfoils depending on the external conditions. But in this case, the chromosomes will refer to the order of the features in the ExTree algorithm. Each gene will contain the position of the feature to which it is

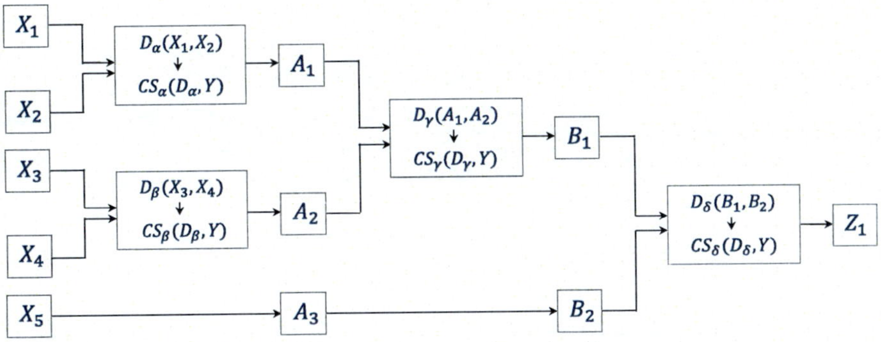

Fig. 1 ExTree structure for 5 input features, NASA airfoil self-noise data set

referring. Therefore, it could be understood as a similar optimization to the travelling salesman problem [13]. However, the fitness function in this case will reward those chromosomes that lead to a structure of Dispersion matrices with low values, an ExTree with low randomness.

4.3 Expansion, Filling Unknown Entries

The control surface that arises from the combination of two features, CS_κ, is a matrix. However, the individuals of the data set might not cover all the entries of this matrix. Thus, once the known individuals are incorporated in the calculation of CS_κ, an inference process begins. Here, the main purpose is to obtain a matrix with all entries filled. The algorithm for the inference of those unknown combinations uses an interpolation of the surrounding known information.

5 Results

5.1 Genetic Algorithm

For the optimization problem, 60 generations of chromosomes were considered with a population of 20 individuals. The probability of crossover and mutation were set to 0.8 [–] and 0.1 [–] respectively. Additionally, elitism was considered with a ratio of 0.9 [–]. The fittest chromosome of the last population after running the GA is the order for which the minimum dispersion in the results. The result obtained was the following: X_1 suction side displacement thickness, X_2 frequency, X_3 angle of attack, X_4 free-stream velocity and X_5 chord length. Additionally, other parameters were optimized, such as weights for the ponderations used in the inference.

5.2 Control Surfaces

The structure of control surfaces created by the algorithm for the case studied is shown below. For each of the four couplings, a set of six plots are represented. The first two plots of every coupling (Fig. 2 for coupling 1) are scattered representations of all the individuals of the data set. The green data points refer to the real data. The blue, purple and pink data points refer to the amount of real data points that are contained within each cell of the three-dimensional space. The lighter the color and the bigger in size, the more concentrated that cell of the workspace. For the plots all the features have been normalized.

After finding the fuzzy feature A_j, of each couple (X_i, and X_{i+1},), the results are presented in a two-dimensional matrix. For the case of the coupling 1, the first matrix shown in Fig. 3 represents the preliminary version of the control surface. It has some black areas, for which there is no known information. The yellow color represents the highest value of the CS_α, whereas the dark blue has the lowest values. Higher values of CS_α imply higher dispersion on the values of Y. The algorithm is then able to expand the information to those NaN (Not a Number) areas of the matrix for which there is no prior information (second matrix of Fig. 3). This will be very helpful when a certain new value of a feature is encountered for which there is no

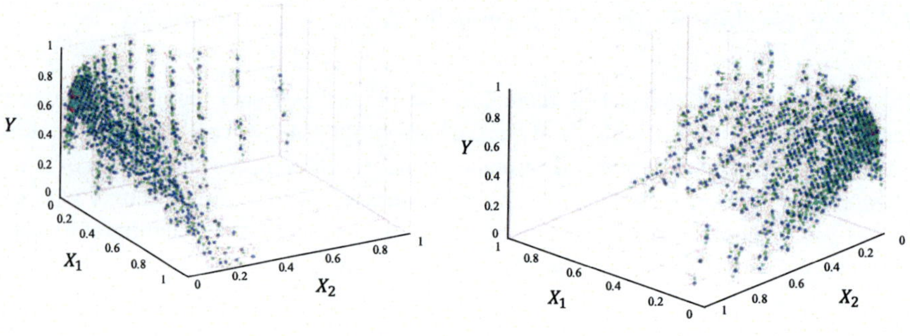

Fig. 2 Original data set plotted using X_1 and X_2 as independent features and Y as dependent feature

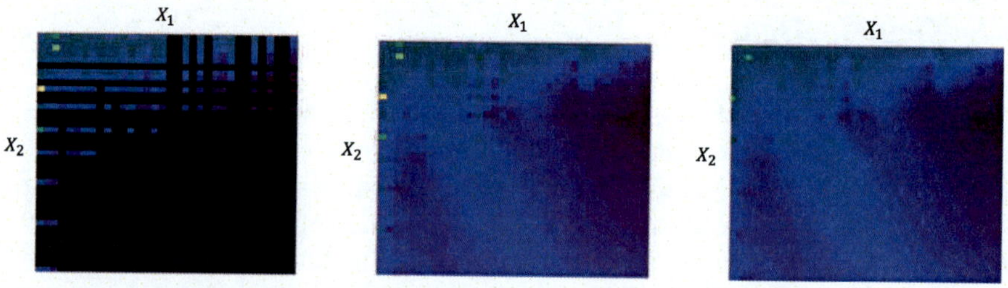

Fig. 3 Inference process to obtain the control Surface CS_α for the original data set

similar test in the data set. Finally, the matrix is interpolated to have a continuous solution of the CS_α (third matrix of Fig. 3).

After substituting for the different values of the data set in CS_α, the output fuzzy feature A_1 can be obtained (Fig. 4). In this figure, the red line provides the separation of the data according to the dispersion ranges of each set. Thus, the most useful part is contained within the first interval, and those data points that refer to higher randomness are constrained in the second. The algorithm allows the possibility of modifying the number of intervals and their limits depending on the user's commands.

The process is then repeated for all the different layers; the results can be seen below for each of the remaining three couplings of the system (Figs. 5, 6, 7, 8, 9, 10, 11, 12, 13).

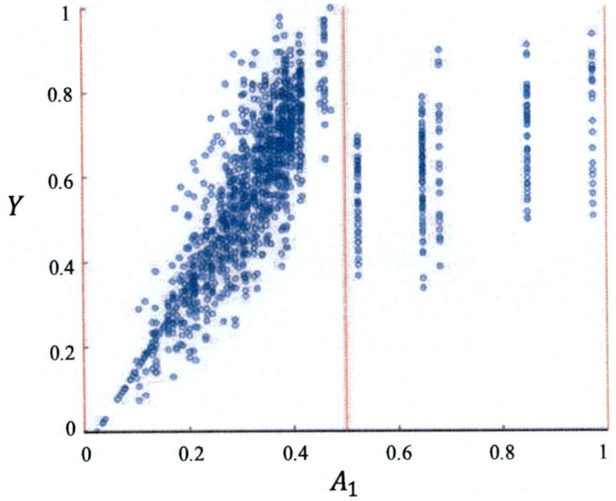

Fig. 4 Plotting of the fuzzy output variable A_1 against Y

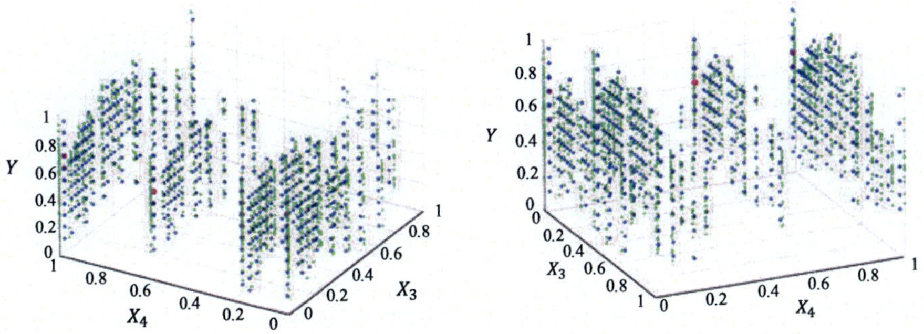

Fig. 5 Original data set plotted using X_3 and X_4 as independent features and Y as dependent feature

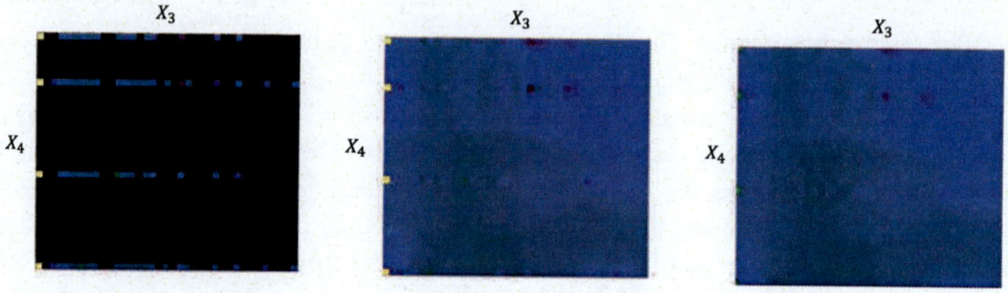

Fig. 6 Inference process to obtain the control Surface CS_β for the original data set

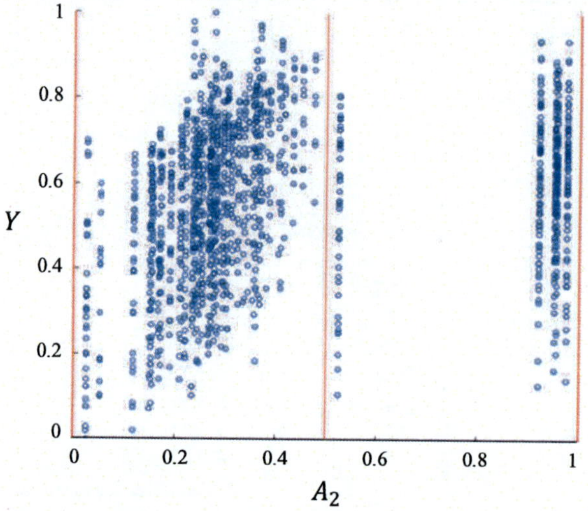

Fig. 7 Plotting of the fuzzy output variable A_2 against Y

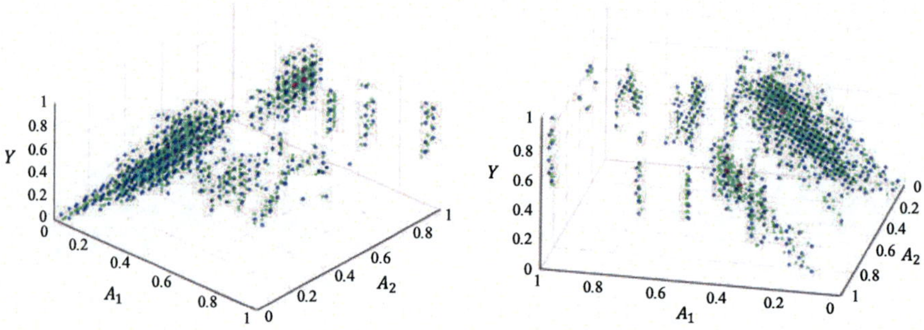

Fig. 8 Original data set plotted using A_1 and A_2 as independent features and Y as dependent feature

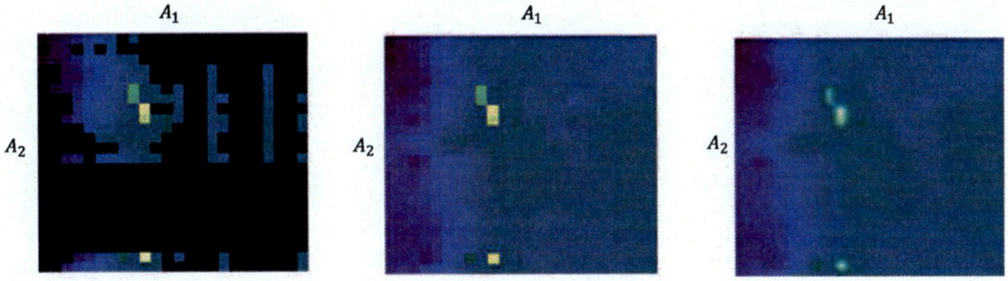

Fig. 9 Inference process to obtain the control surface CS_Y for the original data set

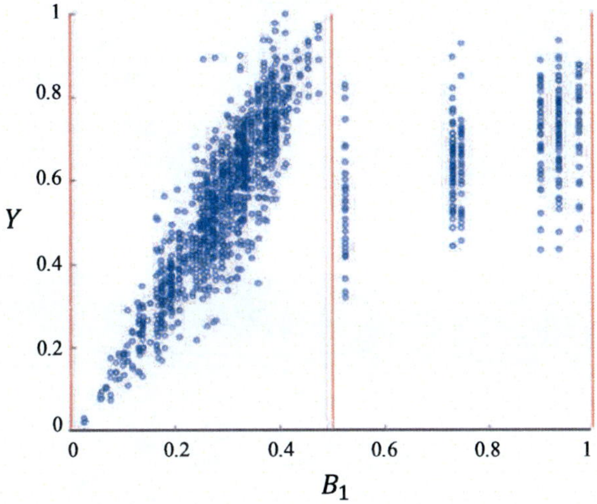

Fig. 10 Plotting of the fuzzy output variable B_1 against Y

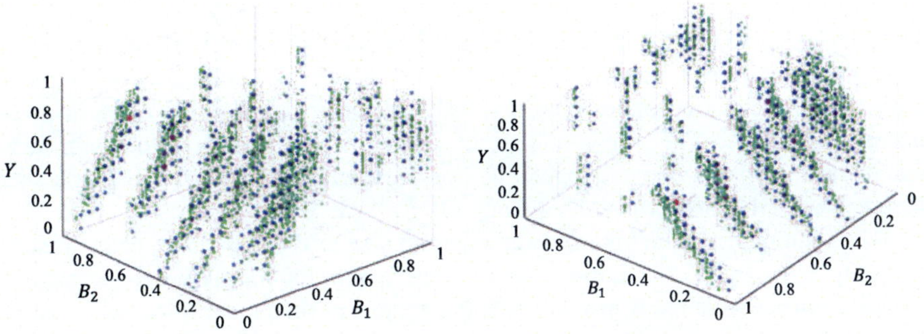

Fig. 11 Original data set plotted using B_1 and B_4 as independent features and Y as dependent feature

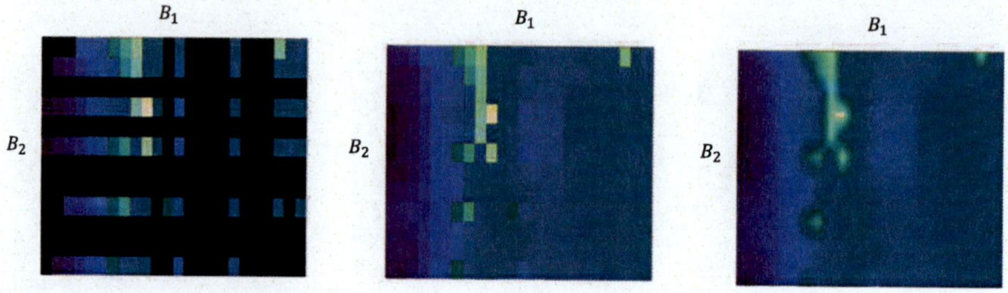

Fig. 12 Inference process to obtain the control surface CS_8 for the original data set

Fig. 13 Plotting of the fuzzy output variable Z_1 against Y

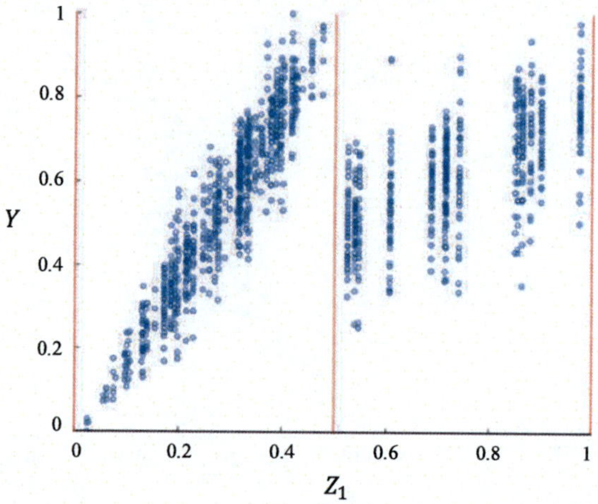

5.3 *Predicted Values*

The data was divided in 80 [%] training and 20 [%] testing. Figure 14 shows the difference in the results between predicted and real values. The average RMSE obtained after the normalization of the data is 0.0838 using cross validation (2.86 in non-normalized values). The computational time required for the training to obtain such result was less than 1 min in a regular computer (Intel® Core i5-10210U Processor, 4 cores, 4.20 GHz Max Turbo Frequency, DDR4 8 Gb RAM).

This solution has been obtained for a certain size of the workspace cells of each coupling; however, it can be improved by properly refining the lattice. Nevertheless, the case considered already provides a better prognosis than models like Regression (5.72 RMSE), and Trees (3.34 RMSE), while having a significantly low computational cost. In addition, it allows extrapolating the information for unknown cases with high uncertainty (as previously mentioned for the process of matrix extension), proving accurate results. But above all, it reduces the dimensionality of the problem increasing significantly the transparency of the algorithm.

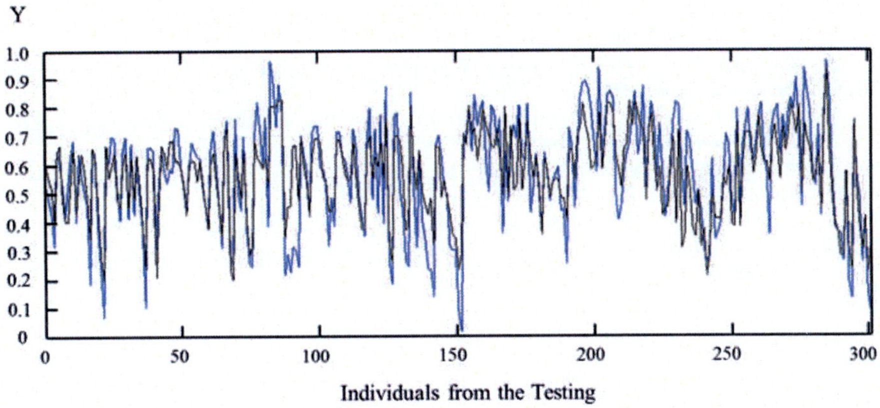

Fig. 14 Predicted in black versus real in blue

6 Conclusion

The tool proposed in this study, ExTree, allows the visualization of the data for any n-dimensional problem. Therefore, contributes significantly in the explainability of data prediction. The algorithm uses multiple inferencing systems in every layer, producing a tree shape structure with intermediate plots that provide transparency and clarity in the decision making. For the optimization problem, a GA correctly obtains the most effective combination of features for the coupling process, minimizing the dispersion in the output fuzzy features of the tree. The preliminary results obtained proved that this methodology can still obtain accurate solutions while avoiding the use of any "black box".

References

1. B. Galitsky, Customers' retention requires an explainability feature in machine learning systems they use, in *Proceedings of the 2018 AAAI Spring Symposium Series, Palo Alto* (2018)
2. D. Heaven, Why deep-learning AIs are so easy to fool. Nature **574**, 163–166 (2019)
3. D. Gunning, Explainable artificial intelligence (xai), in *Defense Advanced Research Projects Agency, DARPA* (2017)
4. A. Adadi, M. Berrada, Peeking inside the black-box: a survey on explainable artificial intelligence (XAI). IEEE Access **6**, 52138–52160 (2018)
5. J.M. Alonso, From Zadeh's computing with words towards eXplainable artificial intelligence, in *Fuzzy Logic and Applications 2018*, ed. by R. Fullér, S. Giove, F. Masulli, WILF, vol. 11291. (Springer, Cham, 2019)
6. T.F. Brooks, D.S. Pope, A.M. Marcolini, Airfoil self-noise and prediction. Technical report, NASA RP-1218 (1989)
7. D. Dua, C. Graff, UCI machine learning repository. University of California, School of Information and Computer Science (2019). http://archive.ics.uci.edu/ml

8. K. Lau, A neural networks approach for aerofoil noise prediction. Department of Aeronautics. Imperial College of Science, Technology and Medicine, London (2006)
9. R. Lopez, Neural networks for variational problems in engineering. Technical University of Catalonia, Barcelona (2008)
10. Y. Freund, R.E. Schapire, Experiments with a new boosting algorithm, in *Proceedings of the 13th International Conference on Machine Learning* (San Francisco, 1996), pp. 148–156
11. L. Breiman, Bagging predictors. Mach. Learn. **24**(2), 123–140 (1996)
12. C. Lafountain, *Matlab-Based Development of Intelligent* (Systems for Aerospace Applications. University of Cincinnati, College of Engineering and Applied Science, 2015)
13. H. Braun, On solving travelling salesman problems by genetic algorithms, in *Parallel Problem Solving from Nature*, pp. 129–133 (1991)

Fast Training Algorithm for Genetic Fuzzy Controllers and Application to an Inverted Pendulum with Free Cart

Javier Viaña and Kelly Cohen

Abstract The classical control theory cannot be applied in those systems whose complexity is too high to be analytically modeled. In these cases, mathematical methods with more degrees of freedom are used because they provide better adaptation. One method widely used in control problems is the Fuzzy Inferencing Systems. However, the process of calibration of the parameters required may involve a high computational cost. Among them, Genetic Algorithms have demonstrated great convergence towards ideal solutions. As the dimensions of the control problem (input features) increase, the optimization process requires much more time. Therefore, the present work proposes a gradual search and parameter update criteria for Genetic Fuzzy Controllers that improves several orders of magnitude the processing time. The algorithm developed has been applied to the control problem of the Inverted Pendulum with Free Cart. The results obtained demonstrate an effective parameter calibration in seconds, while the traditional method of tuning for the same problem takes more than 2 h. Currently, many of the mechanical systems of the different industries undergo sudden changes in their properties during use, therefore an instant effective recalibration of the controllers is necessary. This method allows fast adaptation and also guarantees the same performance in the control process.

Keywords Fast tuning · Chained genetic algorithms · Fuzzy inferencing system · Symmetric mechanical systems · Gradual update

J. Viaña (✉) · K. Cohen
University of Cincinnati, Cincinnati, OH 45219, USA
e-mail: vianajr@mail.uc.edu

K. Cohen
e-mail: cohenky@ucmail.uc.edu

B. Bede et al. (eds.), *Fuzzy Information Processing 2020*,
Advances in Intelligent Systems and Computing 1337,
https://doi.org/10.1007/978-3-030-81561-5_25

1 Introduction

1.1 Problem

The use of a single Genetic Algorithm (GA) that optimizes the parameters of an entire Fuzzy Inference System (FIS) in a problem with several degrees of freedom is computationally expensive.

1.2 Main Purpose

Offer a substantially faster training method for the FIS of a control problem with symmetry, while proving high precision in the testing.

1.3 Approach

Instead of applying a single GA for the exploratory search of the parameters for the control, the following steps are proposed.

Divide the problem into small independent blocks (workspace regions of the FIS).

Apply different chained Gas to some of the blocks (strategically chosen) to optimize these regions.

Finally, extrapolate the information to the remaining blocks (symmetrically equivalent) without the use of Gas.

In the end, a fully trained FIS is achieved in a faster and elegant way rather than using "brute force" in the optimization.

1.4 Nature of the Case Study

The case chosen to benchmark this technique is the Inverted Pendulum with Free Cart (IPFC). The System is composed of a homogeneous bar with a mass M_2 and a length L_2 (known moment of inertia I_2), joined with a linkage that allows its rotation to the top of the cart. The free cart, of mass M_1, glides without friction along a horizontal track. A unique actuator exerts a horizontal force, F, in the center of gravity of the cart, G_1.

The IPFC problem has 2° of freedom and thus it should have 2 actuators. Nevertheless, the configuration considered includes a single actuator to increase the complexity of the problem. Thus, proving the efficiency of the algorithm developed in the control of highly unstable mechanical systems.

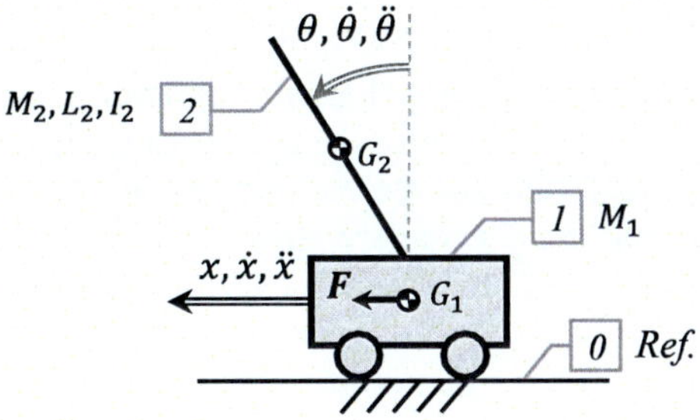

Fig. 1 Schematics of the inverted pendulum with free cart (IPFC)

Three input variables have been chosen to control the system, θ,$\ddot{\theta}$,$\ddot{x}$ angle of the rod, angular acceleration and linear acceleration of the base of the cart respectively. There is no control over $\dot{\theta}$ (angular velocity of the rod) or x (linear displacement of the cart). As said, F (horizontal force applied to the cart) will be the control variable on the actuator. Thus, the problem is 4-dimensional with 3 inputs and 1 output (see Fig. 1).

For the given configuration, the differential equations of the motion of the IPFC with no friction are defined as,

$$F = (M_1 + M_2)\ddot{x} + M_2\frac{L}{2}\ddot{\theta}cos(\theta) - M_2\frac{L}{2}\dot{\theta}^2sin(\theta) \tag{1}$$

$$\left(I_{2_G} + M_2\frac{L^2}{4}\right)\ddot{\theta} = M_2g\frac{L}{2}sin(\theta) - M_2\frac{L}{2}\ddot{x}cos(\theta) \tag{2}$$

These can be easily obtained isolating each of the rigid bodies of the system and removing the linkage forces.

2 Related Work

For the control of this nonlinear system (IPFC), many and very diverse strategies have been used previously. In [1] such objective is obtained with the Lyapunov function and applying LaSalle's invariance theorem.

In [2] a network of the Fuzzy Logic Controllers (FLCs) with different weights and a GA for the tuning is used. The control of the double inverted pendulum is also an interesting problem, similar to the one considered in this study, where a Genetic Fuzzy System proves to be a good solution [3].

A comparative study of the LQR and PID control strategies is introduced in [4]. In [5] the passivity properties of Lagrangian and Hamiltonian systems are used to find the best static feedback controller.

Other authors have used GAs for the optimization of the parameters for the FIS of the inverted pendulum and cart system but at a high computational cost, [6].

Particle Swarm Optimization (PSO) algorithms or Ant Colony System optimization techniques are viable substitutes to replace the use of GA for the tuning of the PID or the LQR [7, 8]. Alternative inferencing rules considering the Takagi-Sugeno were introduced in [9] for the control of the inverted pendulum.

Nevertheless, the training time of the inferencing rules for those methodologies is still high. There are already algorithms which use neuro-fuzzy logic controllers with the aim to diminish such computational cost [10].

3 Theoretical Background

The concept of fuzzy refers to a scale of values between 0 and 1, where the analyzed variable can belong to more than one set. In this way, the transition from one group of values to another is blurred. Thus, a gradation in the contours that limit the membership of each set is obtained. In short, it is intended to resolve the sudden changes that crisp sets imply.

In engineering control systems, a fuzzy data processing approach significantly improves the accuracy. That is, by increasing the degrees of freedom, the control surface (input to output graphical representation), or hypersurface for more than 2 dimensions, becomes smooth, with no sudden jumps. A great ability to adjust is achieved regardless of the problem considered.

In the vast majority of aerospace systems these precision and adaptation properties are essential, given the complex nature of the infrastructure. However, this technique presents a clear drawback when compared to the classical control theory; it requires the optimization of a greater amount of values.

Therefore, an efficient optimization method is needed to perform an exploratory search until the closest combination of parameters is found. Genetic Algorithms (GAs) have been one of the biggest bets for the search of these parameters. This algorithm is based on the principles of natural evolution. A group of chained functions perform the basic processes that are observed in the biological mechanisms for development; selection, mutation, crossover and elitism. In each generation / iteration, the individuals (chromosomes) of the population change, progressively improving their fitness. At all times, chromosomes are evaluated with a fitness function that determines how valid the individual is. A certain degree of randomness is also incorporated in the evolutionary process. This is done to consider alternative solutions, which although in the initial generations are not very adequate, in the future they will lead to better results.

The IPFC problem involves an exploration of continuous values. Therefore, the training process using a single GA (in this study referred to as the traditional approach) can take a long time if the number of parameters to be optimized is high.

Therefore, the authors have focused their work on improving the Genetic Fuzzy (GF) system so that the computational cost for training is significantly lower than the one required in the traditional approach.

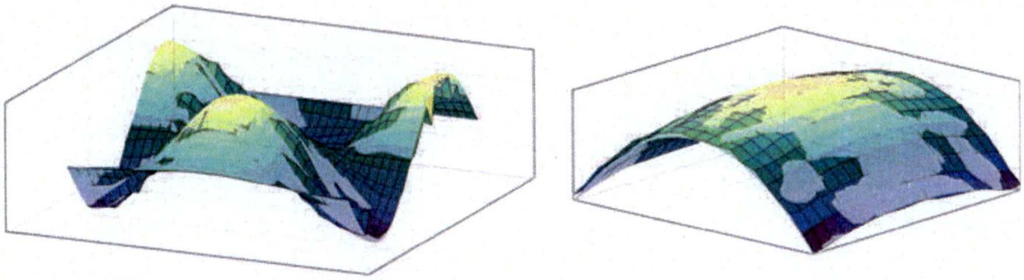

Fig. 2 3D genetic fuzzy fitting with a modified mamdani approach (real solution in transparent, and approximation in color)

One of the techniques developed is the Modified Mamdani [11] that allows solving the problem of 3D fuzzy fitting in seconds (see Fig. 2). This simulates the adaptation of a Neural Network with step functions by making individual approximations in each section of the surface.

The technique proposed in this paper seeks also this property of fast training. The physical properties of many industrial systems are subject to sudden modifications. Thus, an immediate recalibration process should be used in order to minimize the loss of efficiency. The recalibration and maintenance operations of the tools are still a big problem in many automated manufacturing processes.

For this preliminary development, the primary goal was to obtain a successful fast learning algorithm. In the next stage, this fast adaption skill will be incorporated in the controller of an IPFC prototype.

The system will have a health monitoring systems integrated (accelerometers and gyroscopes will be used for the measurements), which will evaluate the efficiency of the controller. When the IPFC shows a decrease in performance (Identification Procedure) the fast learning algorithm will recalibrate automatically the coefficients of the FIS to obtain a tuned system (Self Adaption Procedure). This procedure is described in the Fig. 3.

At every time step, the performance is evaluated, in case it is not satisfactory, the system starts the adaption sequence until the system is tuned for the new conditions.

4 Methodology

First, a preprocessing of the variables has been carried out, being standardized with values between 0 and 1. As initial input, it has been decided to divide the 4 variables into 5 equally spaced regions. Identifying the letters L, H, Z, N and P as Low, High, Zero, Negative and Positive respectively.

In the proposed algorithm, membership functions with symmetrical triangular shape have been used. The two functions of the ends are therefore divided in half, generating right triangles. The problem is thus reduced to the choice of the values (5 possible) of each combination in the three-dimensional mesh of the if-then-rules.

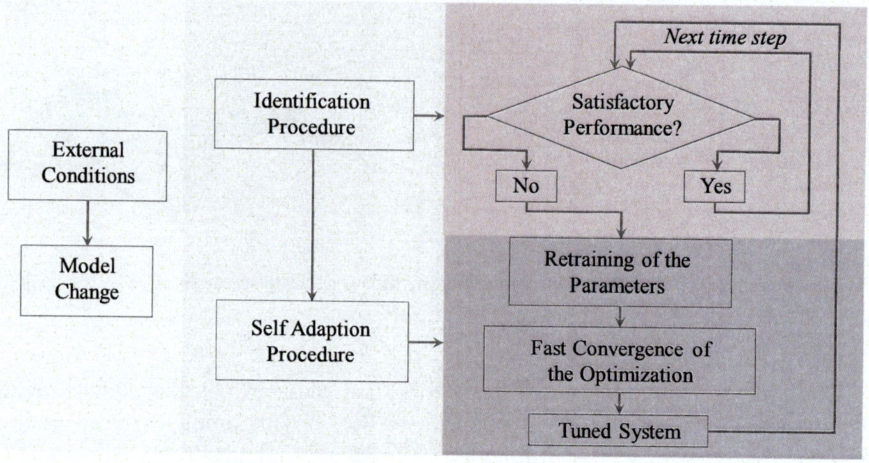

Fig. 3 Block diagram for self adaptation of the IPFC

Figure 4 shows the structure or 3D tower of if-then rules along with the three input variables and their associated membership functions. The form of the membership

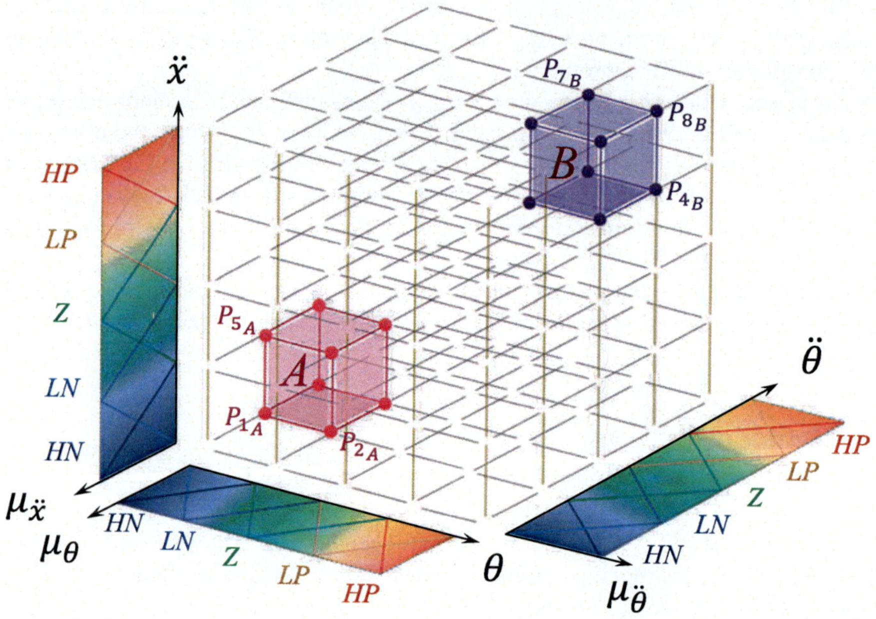

Fig. 4 Tower or grit of if-then rules representing all the possible combinations of the fuzzy sets from the input variables

functions of the output variable is not represented in Fig. 4, however, it is the same as the other three.

Every point in the 3D grit represents an if-then rule. Thus, the total number of if-then rules is 5^3, 125 parameters for the optimization problem. In the traditional approach (single GA for the optimization of all the parameters) if the form of the memberships is not defined, the number of degrees of freedom due to the membership functions should also be counted. Since there are 4 variables, each with 5 membership functions, and each membership function has 3 edges, there are 60 degrees of freedom. Thus, making a total of 185 values for the optimization problem. In a regular computer (Intel® Core i5-10210U Processor, 4 cores, 4.20 GHz Max Turbo Frequency, DDR4 8 Gb RAM) this traditional approach requires more than 2 h of computation. On the other hand, the training of the algorithm presented requires less than 10 s (using the same machine).

For the obtention of the membership value for the target variable the geometric mean is calculated using the three variables' memberships. Then, a defuzzification process is carried out. The final value is proportional to the areas enclosed under the membership functions of the output force. The fitness function used is such that it seeks to maintain vertical the pendulum ($\theta \to 0$), minimizing its angular acceleration ($\ddot{\theta} \to 0$) and linear acceleration ($\ddot{x} \to 0$). Thus, the formula of the fitness function is the inverse of the error to those desired values.

Figure 4 also represents with different colors two subsets (A and B) that contain eight if-then rules each. Due to the symmetry of the problem, these subsets are complementary, such that each point of the subset A has its exact complementary inside B. As defined below.

$$P_{1A} \equiv \overline{P_{8B}} \tag{3}$$

$$P_{2A} \equiv \overline{P_{7B}} \tag{4}$$

$$P_{5A} \equiv \overline{P_{4B}} \tag{5}$$

For the optimization problem, several recursive functions are used. The ultimate goal is to find which of the membership functions from the output variable, F, corresponds to each of the points in the 3D lattice.

The tuning is then performed inside each subset. It starts from one corner of the 3D lattice, where a single GA is applied to search for the eight parameters of the cell in the lattice. After the generations converge to the best chromosome, the corners of that subset are updated with the values of the genes. The first exploration process finishes, and a new GA is applied in the following subset that lies next to the current one. The only difference is that the length of the chromosomes is reduced to the half. Thus, convergence is much faster.

This process is repeated for each subset of the 3D tower, updating the inferencing rules as the chained GAs provide the best solution for each subset. The optimization is not the same for all the cells in the workspace, thus the dimensionality of the problem plays a key role to determine the most efficient updating protocol.

Additionally, as it has been pointed out before, due to the symmetry, the computational time of the problem can be reduced by half. But in fact, this is not the condition that improves significantly the computational cost. The key to minimize the processing time is the discretization of the workspace into subsets, and the gradual update of the lattice by using several GAs that pass the information forward, creating a chain. In the end, the exploratory problem is reduced tremendously, proving that is more efficient the application of many "small" GAs rather than a single "big" GA in the tuning of the FIS.

5 Preliminary Results

The functions of this algorithm have been developed in MATLAB. For this computation, the physical parameters of the system have been set to the followings values: Mass of the free cart, (M_1), 0.2 [kg], mass of the rod, (M_2), 1 [kg], length of the rod, (L_2), 1 [m]. For the specified conditions, the tower's training process took 7 s. Unlike the traditional approach which training took more than 2 h.

For the validation phase, in each case studied, a set of initial conditions have been considered;$\theta_0,\dot{\theta}_0,\ddot{\theta}_0,x_0,\dot{x}_0,\ddot{x}_0$, F_0. Both methodologies are able to keep the inverted pendulum upright in the operating range for which they have been trained. In addition, the two methods converge towards their stationary states in a similar way. Below are four cases (Figs. 5, 6, 7, 8) in which the pendulum control is more complex; where $\ddot{x}_0<0$ and $\ddot{\theta}_0>0$.

In addition to keeping the pendulum in an upright position, it is observed how the applied force is such that a tradeoff between $\ddot{\theta}$ and $\ddot{x}$ is obtained so that θ,$\ddot{\theta}$ and $\ddot{x}$ are minimal. These do not become exactly zero because of the granularity

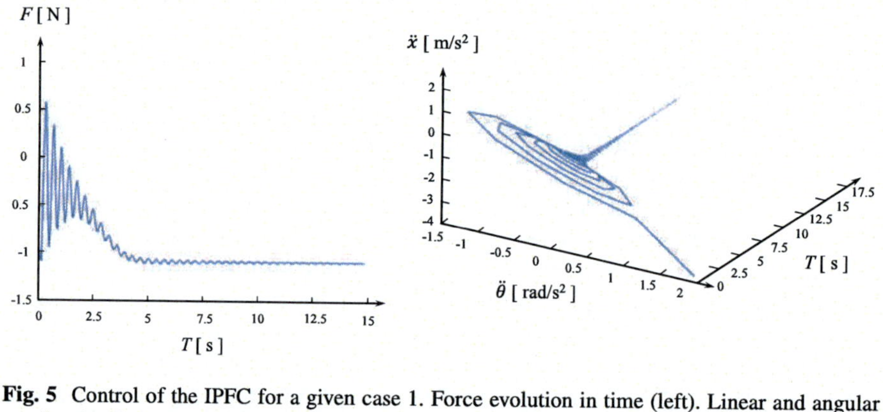

Fig. 5 Control of the IPFC for a given case 1. Force evolution in time (left). Linear and angular accelerations' convergence in time (right)

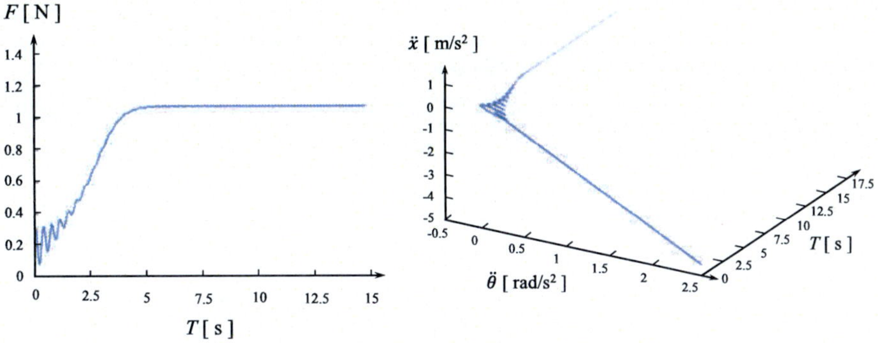

Fig. 6 Control of the IPFC for a given case 2. Force evolution in time (left). Linear and angular accelerations' convergence in time (right)

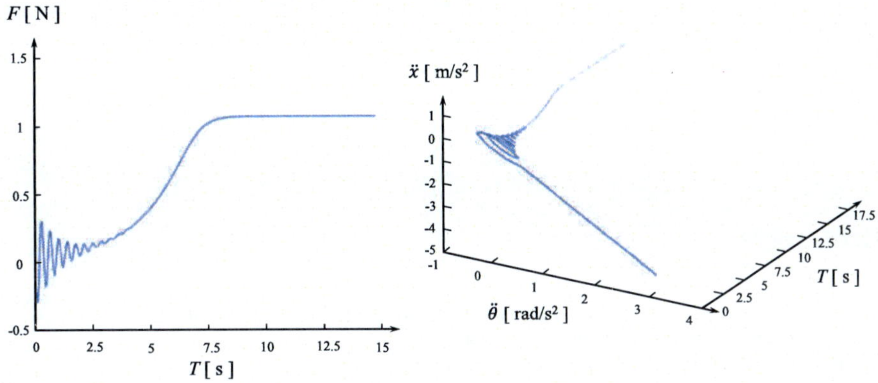

Fig. 7 Control of the IPFC for a given case 3. Force evolution in time (left). Linear and angular accelerations' convergence in time (right)

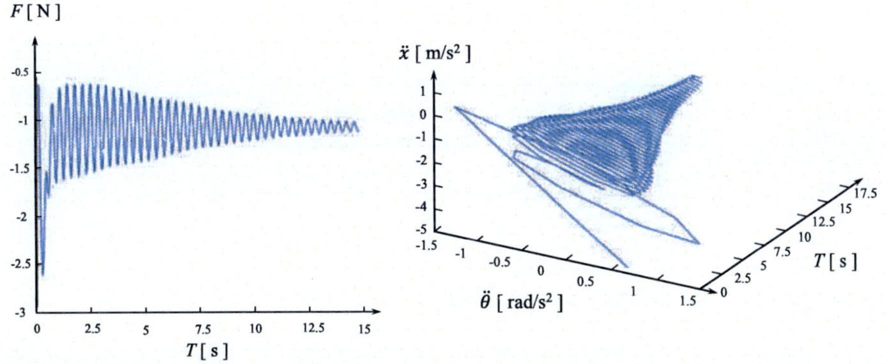

Fig. 8 Control of the IPFC for a given case 4. Force evolution in time (left). Linear and angular accelerations' convergence in time (right)

of the problem considered. Increasing the number of membership functions would make the values in the steady state closer to zero. On the other hand, it would also increase the computational cost. However, to verify the algorithm proposed in this study (which is the main objective) is sufficient.

6 Conclusions

A fast training algorithm for the control of symmetric systems has been proposed. Its effectiveness has been proven with the Inverted Pendulum + Free Cart system. The basis of the algorithm is a Fuzzy Inferencing System whose parameters have been optimized using different chained Genetic Algorithms. The training time is substantially improved by dividing the workspace into blocks and gradually transmitting the information obtained to the following phases of the optimization process. Using a single GA for the optimization of the entire system has a computational cost of more than 2 h, whereas this method required less than 10 s. Furthermore, such benefit is achieved while still proving a successful and efficient result competitive with the traditional approach.

References

1. C.A. Ibañez, O.G. Frias, M.S. Castañón, Lyapunov-based controller for the inverted pendulum cart system. Nonlinear Dyn. **40**, 367–374 (2005)
2. K. Belarbi, F. Titel, Genetic algorithm for the design of a class of fuzzy controllers: an alternative approach. IEEE Trans. Fuzzy Syst. **8**(4), 398–405 (2000)
3. A. Sathyan, K. Cohen, Development of a genetic fuzzy controller and its application to a noisy inverted double pendulum, in *Fuzzy Logic Based in Optimization Methods and Control Systems and Its Applications* (IntechOpen, 2018)
4. A.N.K. Nasir, M.A. Ahmad, M.F. Rahmat, Performance comparison between LQR and PID controllers for an inverted pendulum system, in *AIP Conference Proceedings*, vol. 1052, p. 124 (2008)
5. A. Shiriaev, A. Pogromsky, H. Ludvigsen, O. Egeland, On global properties of passivity-based control of an inverted pendulum. Int. J. Robust Nonlinear Control **10**, 283–300 (2000)
6. J. Kim, Y. Moon, B.P. Zeigler, Designing fuzzy net controllers using genetic algorithms. IEEE Control Syst. Mag. **15**(3), 66–72 (1995)
7. T.O.S. Hanafy, Stabilization of inverted pendulum system using particle swarm optimization, in *8th International Conference on Informatics and Systems (INFOS)* (IEEE, Cairo, 2012), pp. 207–210
8. A. Jacknoon, M.A. Abido, Ant colony based LQR and PID tuned parameters for controlling inverted pendulum, in *International Conference on Communication, Control, Computing and Electronics Engineering (ICCCCEE)* (IEEE, Khartoum, 2017), pp. 1–8
9. F. Song, S.M. Smith, A Takagi-Sugeno type fuzzy logic controller with only 3 rules for a 4 dimensional inverted pendulum system, in *2000 IEEE International Conference on Systems, Man and Cybernetics Conference Proceeding* (IEEE, Nashville, 2000), pp. 3800–3805

10. G.R.C. Tatikonda, V.P. Battula, V. Kumar, Control of inverted pendulum using adaptive neuro fuzzy inference structure (ANFIS), in *Proceedings of 2010 IEEE International Symposium on Circuits and Systems* (Paris, 2010), pp. 1348–1351
11. E.H. Mamdani, S. Assilian, An experiment in linguistic synthesis with a fuzzy logic controller. Int. J. Man Mach. Stud. **7**(1), 1–13 (1975)

Genetic Fuzzy Systems: Genetic Fuzzy Based Tetris Player

Lynn Pickering and Kelly Cohen

Abstract Tetris is a single player game, the objective being to place four-piece blocks and clear as many rows of blocks as possible. The game requires quickness and flexibility in its decision making, which makes it a good candidate for Fuzzy Logic decision making. To train the Fuzzy Logic Tetris Player (FLTP), a genetic algorithm is used. The genetic algorithm trains the input and output membership functions, along with the rules of the fuzzy logic system. Previously the FLTP was created by playing the game repeatedly, creating input functions and rules and iterating upon these parameters by identifying faulty decisions from previous runs. The genetic algorithm is able to search a bigger solution space, exploring this complex problem and the many parameters to tune; therefore, achieving higher game scores than this previous FLTP not trained by a genetic algorithm.

1 Introduction

The highest score possible in Tetris is 999, 999. The Classic Tetris World Championship cites 36 players as having reached this score [1]. Tetris requires a quickness and flexibility in decisions, which makes it a good application for fuzzy logic [2]. A Fuzzy Logic Tetris Player (FLTP) is created using expert knowledge. The application of an evolutionary algorithm to a fuzzy system has many advantages [3]. To improve upon the performance of the FLTP a genetic algorithm is used to train the entire player and then only the input membership functions of the player.

L. Pickering (✉)
PhD Candidate, Department of Aerospace Engineering & Engineering Mechanics, University of Cincinnati, 2600 Clifton Ave., Cincinnati, USA
e-mail: pickerln@mail.uc.edu

K. Cohen
Brian H. Rowe Chair and Interim Head, Department of Aerospace Engineering & Engineering Mechanics, University of Cincinnati, 2600 Clifton Ave., Cincinnati, USA
e-mail: cohenky@ucmail.uc.edu

B. Bede et al. (eds.), *Fuzzy Information Processing 2020*,
Advances in Intelligent Systems and Computing 1337,
https://doi.org/10.1007/978-3-030-81561-5_26

2 Methodology

2.1 *Fuzzy Logic Tetris Player Expert Training*

A fuzzy logic controller was initially designed using expert knowledge, and a three input, one output controller is used. The Tetris game was played many times to identify the top three identifiers of a good move in Tetris [4, 5]. These identifiers are the number of holes created by a move, the number of rows cleared by a move, and the balance of the board created by a move. The first input, holes, is the number of holes created for each spin of the block and each position of the block. The membership functions of this input are shown in Fig. 1, domain [0, 201]. The second input, balance, is the height the blocks will reach for each spin of the block and each position of the block. The membership functions of this input are shown in Fig. 2, domain [0, 23]. The final input, rows, is how many rows will be cleared for each spin of the block and each position of the block. The membership functions of this input are shown in Fig. 3, domain [0, 5].

A falling Tetris block can be rotated four times and positioned across the board in 10 locations, so there are 40 possible options for a block placement each time. The fuzzy system takes the three inputs described, and calculates the membership value on the shapes shown in the figures for each input. The membership functions are limited to triangles and trapezoids for the input membership functions, and the number of membership functions is limited to three. This provides sufficient granularity, while keeping the controller calculations simple, and reducing the number of genes in the genetic algorithm chromosomes. The output in this system is simply a score, on how

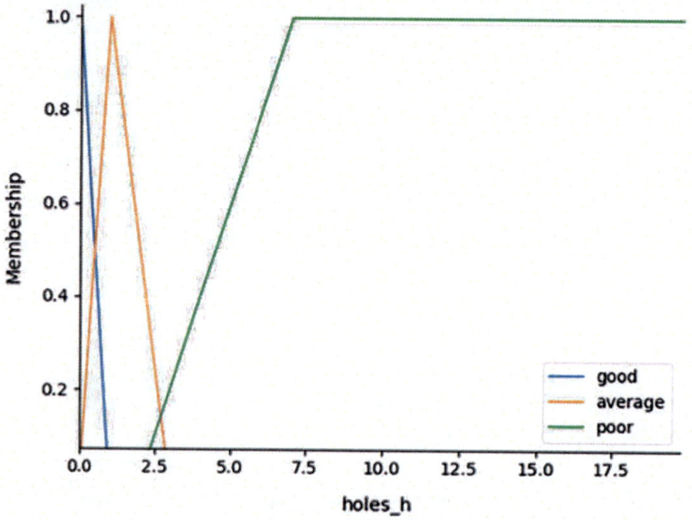

Fig. 1 Membership functions for Input 1: Holes

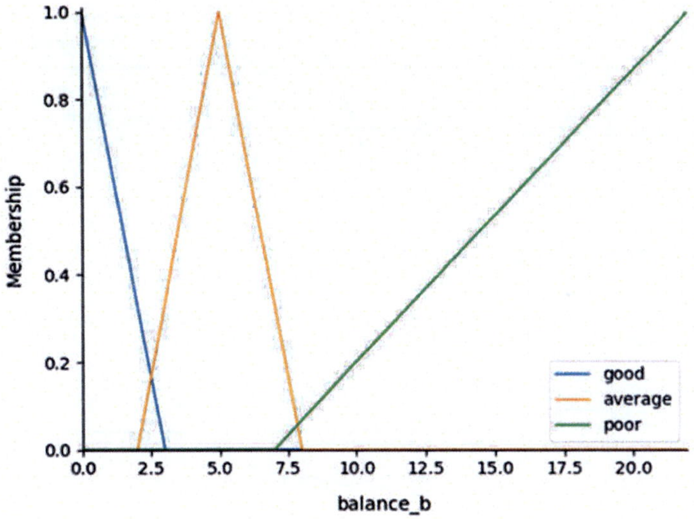

Fig. 2 Membership functions for Input 2: Balance

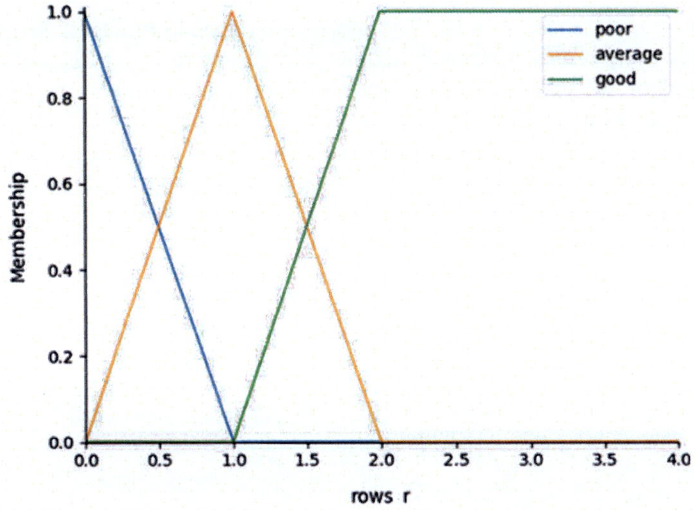

Fig. 3 Membership functions for Input 3: Rows

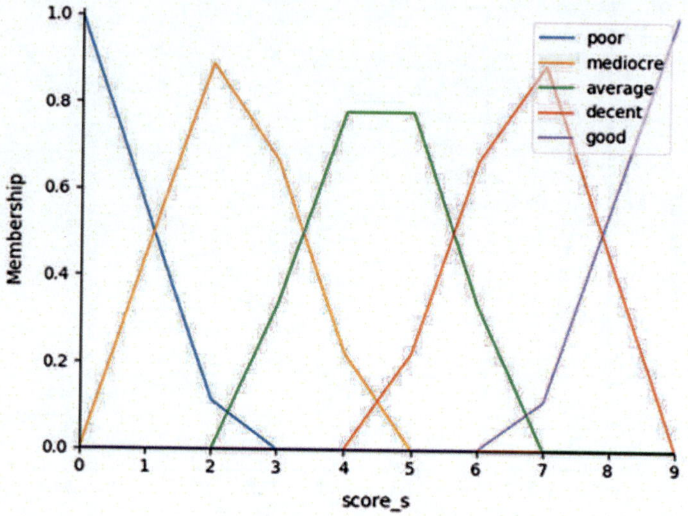

Fig. 4 Membership functions for Output: Score

good a move is determined to be. The output membership functions are shown in Fig. 4. These output membership functions were automatically generated using the skfuzzy toolbox [6] such that the membership functions at any vertical line sum to one. The center of area method is used as the defuzzification method [7].

A great strength of fuzzy logic is that the rules are in linguistic form, allowing for the easy comprehension of the rules and understanding of how decisions are made by the system [8]. The rule base of the system was created by trial and error, using expert knowledge. The rules that define a good score are:

IF 'holes' is good AND 'balance' is average AND 'rows' is good OR
'holes' is good AND 'balance' is good AND 'rows' is average OR
'holes' is good AND 'balance' is good AND 'rows' is good THEN
'score' is good.

A block diagram of this system is shown in Fig. 5.

2.2 *Fuzzy Logic Tetris Player with Genetic Algorithm Training*

Using the same 3 inputs and output as in the FLTP created using expert knowledge, the FLTP is now trained using a genetic algorithm [9]. A block diagram of this system is shown in Fig. 6. The expert knowledge has been replaced with the genetic algorithm. The genetic algorithm trains 51 aspects(chromosomes) of the system, which are stored in a gene. Each input and output has three membership functions,

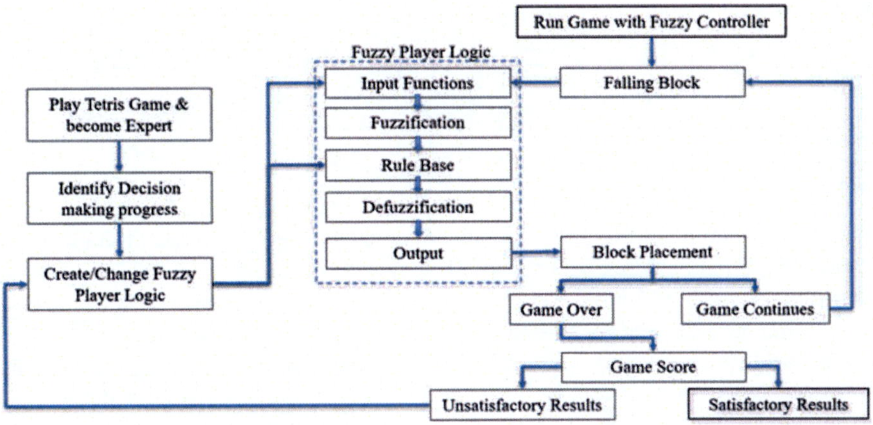

Fig. 5 Block diagram of system

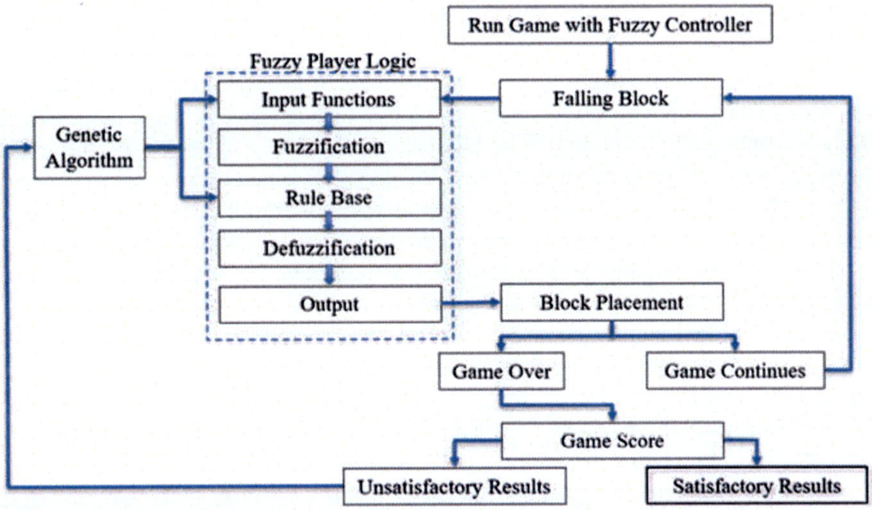

Fig. 6 Block diagram of system

each membership function defined by the center of the triangle and the width of the triangle. The remaining 27 chromosomes in the gene define the rule system of the FLTP. There are 9 membership functions total, so the total number of rules created once subtracting those rules that combine membership functions of the same input, is 27 rules. Each possible two membership function combination is covered, and the GA trains the output that is assigned of each of these combinations. An example rule is:

IF balance is small AND holes is medium THEN score is large.

Table 1 Genetic algorithm stochastic parameters

Probability of crossover	Elitism	Probability of mutation
0.95	0.20	0.40

The genetic algorithm used to train the fuzzy logic system has 10 chromosomes in each population [10]. The stochastic parameters used were the probability of crossover, elitism, and the probability of mutation, summarized in Table 1. Elitism helps the genetic algorithm converge onto a solution faster, because it keeps the best genes from the previous generation without altering them. Elitism was set at 0.2 so that the best 2 chromosomes were kept without mutation in the next population. The probability of crossover was set at 0.95. The probability of crossover determines how likely it is that crossover between two parents will happen. As in biology, where crossover must almost always occur to create offspring, the probability is very high. A high probability means that it is very likely that the crossover children will move to the next generation instead of the parents. A probability of crossover of 1, means parents will never move to the next generation, only the children will. A probability of 0 means the parents will always move to the next generation and there will be no crossover. A low crossover creates a global search space, but one that is not likely to converge to a solution. The higher the probability of crossover, the more focused the search space becomes. The probability of mutation was set at 0.4 and is the probability that, after crossover, a gene will undergo mutation, and become a random number in the range of the boundaries. Though mutation is often lower in biology, a high mutation was needed to more effectively cover the search space. A mutation of 1 means the genes will be mutated every time, and the search space will be very global and random. A mutation of 0 means the genes will never be mutated, and the search space will be limited to within the box created by the initial population.

Each chromosome in a population will play 3 random Tetris games, with each of these random games being the same for the chromosomes in that population. Because of the randomness of Tetris games, this creates a better estimate of the strength of a chromosome, and therefore that FLTP.

2.3 Fuzzy Logic Tetris Player with Genetic Algorithm Training (Membership Functions)

This system is the same as the system described above, other than the rules of the fuzzy logic system were set to the same rules as defined in the expert knowledge trained system. This allowed for a simplification of the gene to a gene that is 24 long, instead of 51. This reduces the complexity for the training of the fuzzy logic system by the genetic algorithm.

Table 2 The expert training test runs

Run	1	2	3	4	5	6	7	8	9	10
Score	5240	5365	695	749	585	2172	2717	1490	969	17611
Level	5	5	1	1	1	3	4	2	2	7
Run	11	12	13	14	15	16	17	18	19	20
Score	1231	511	487	642	11979	648	2239	707	663	760
Level	2	1	1	1	7	1	3	1	1	1

3 Results

To compare the various trained FLTPs, the same evaluation method is used across each training method. 20 sets of 100 random combination of blocks are created. If the player reaches the end of the 100 blocks, the loop starts again. In this way, each player plays the same 20 games, and the score is then an average across these games. The score of the Tetris game increases throughout as follows in Eq. 1:

$$Add\ To\ Score = Blocks\ Cleared \times Level, \tag{1}$$

where

$$Blocks\ Cleared = 0, 40, 100, 300, 1200, \tag{2}$$

dependant on the number of blocks cleared. Single points are given throughout the game as well for manual movements of the blocks by the player, and the level is increased as the lines cleared become greater than 6 times the level number.

3.1 *Results: Fuzzy Logic Tetris Player Expert Training*

The FLTP created using expert knowledge is evaluated using the set of test games. The average score is evaluated to be 2873.0, and the average level achieved is 2.5. The full results on the 20 test games are shown in Table 2.

3.2 *Results: Fuzzy Logic Tetris Player with Genetic Algorithm Training*

The genetic algorithm to train the FLTP is run for 2000 generations. The total time to run the 2000 generations was 24 h. The convergence of the genetic algorithm is shown in Fig. 7.

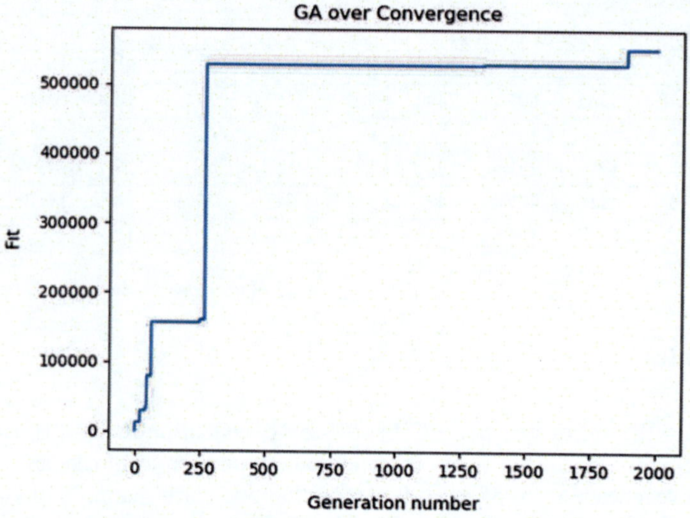

Fig. 7 Convergence of the genetic algorithm

Table 3 The GA training test runs

Run	1	2	3	4	5	6	7	8	9	10
Score	2060	29171	438	974	33637	3065	10782	658	17925	5655
Level	3	12	1	2	14	4	8	1	11	5
Run	11	12	13	14	15	16	17	18	19	20
Score	1624	3661	1069	4269	5876	59678	578	16405	5892	27223
Level	3	4	2	5	6	17	1	9	6	12

The average score is evaluated to be 11532.0, and the average level achieved is 6.3. The full results on the 20 test games are shown in Table 3.

The membership functions as trained by the genetic algorithm are shown in Figs. 8, 9, and 10. The trained output membership functions are shown in Fig. 11.

3.3 *Results: Fuzzy Logic Tetris Player with Genetic Algorithm Training (Membership Functions)*

The genetic algorithm to train the FLTP is run for 2000 generations. The total time to run the 2000 generations was 48 h. The convergence of the genetic algorithm is shown in Fig. 12. The fitness value of each FLTP is the score of the Tetris game, as previously defined in Eqs. 1 and 2.

The average score is evaluated to be 16122.05, and the average level achieved is 5.6. The full results on the 20 test games are shown in Table 4.

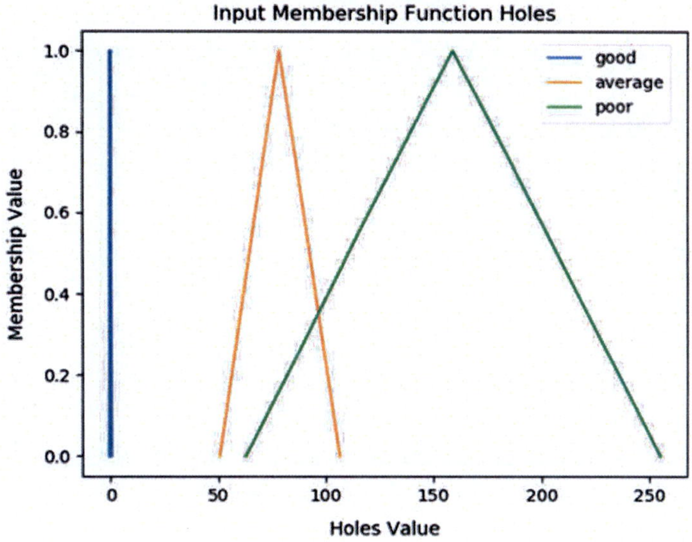

Fig. 8 Trained membership functions for Input 1: Holes

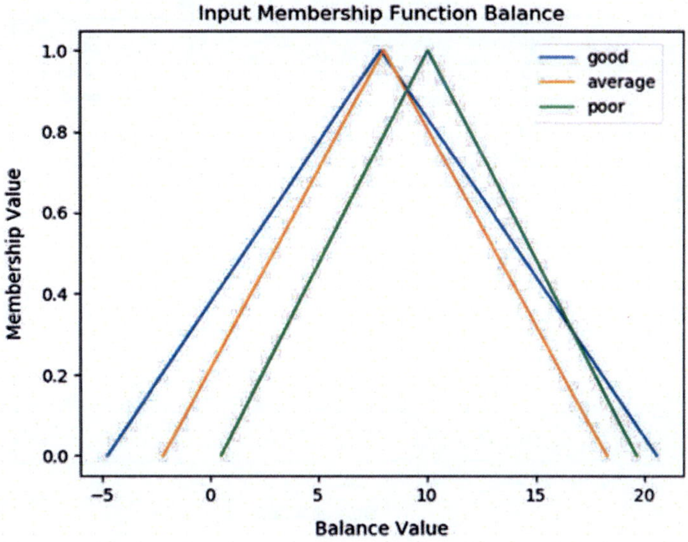

Fig. 9 Trained membership functions for Input 2: Balance

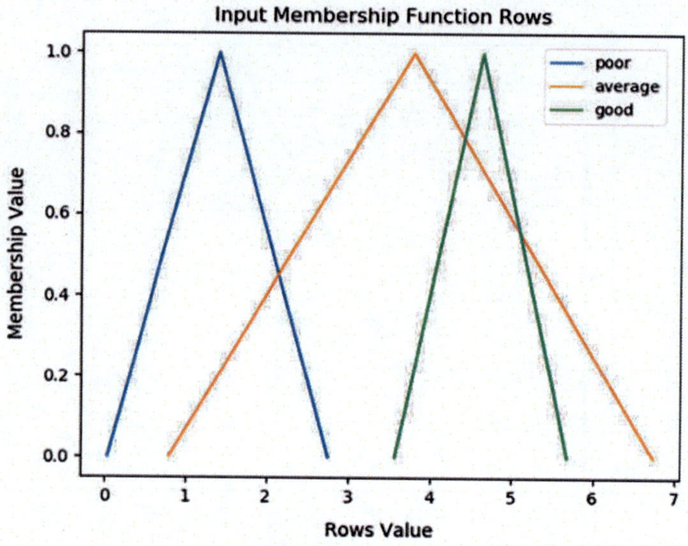

Fig. 10 Trained membership functions for Input 3: Rows

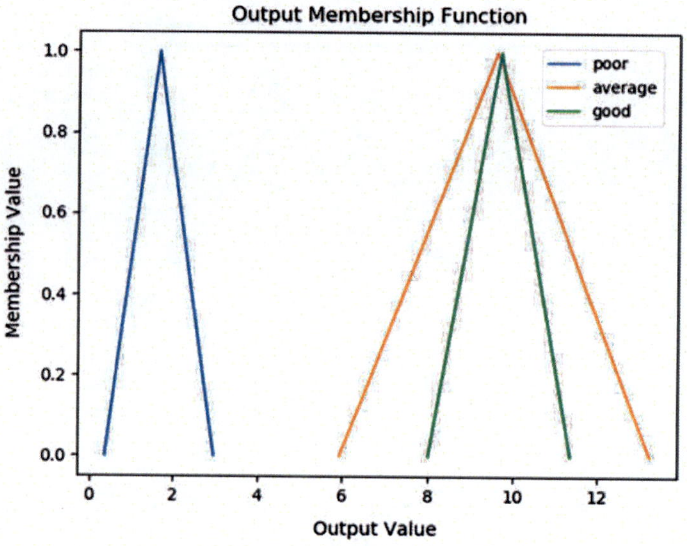

Fig. 11 Trained membership functions for the Output

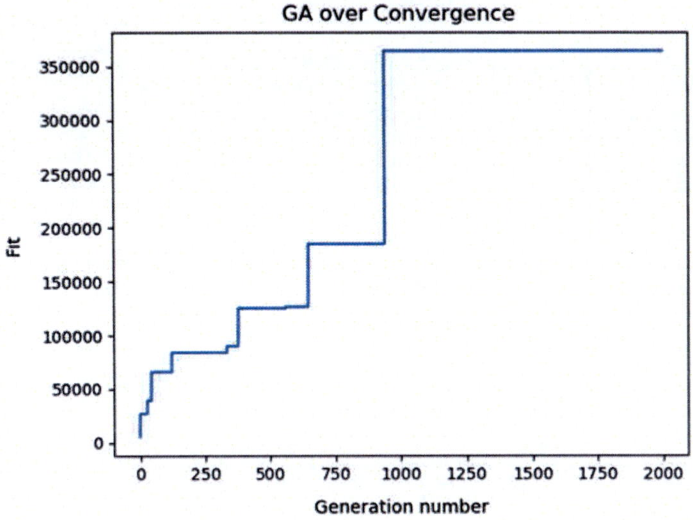

Fig. 12 Convergence of the genetic algorithm

Table 4 The GA training (MF) test runs

Run	1	2	3	4	5	6	7	8	9	10
Score	20116	24280	524	1214	8477	2051	5650	1708	2165	2733
Level	9	9	1	2	5	3	4	3	3	3
Run	11	12	13	14	15	16	17	18	19	20
Score	1204	629	1265	8263	11162	136900	793	84532	8149	626
Level	2	1	2	7	7	24	1	18	7	1

The membership functions as trained by the genetic algorithm are shown in Figs. 13, 14, and 15. The rules and the output membership functions were fixed to be the same as those in the FLTP trained by an expert.

4 Discussion

The best FLTP was the FLTP trained by the genetic algorithm, with the rules and output membership functions fixed, with an average score of 16122.05 and the highest score on the test data achieved as 136900. The second best FLTP was the FLTP completely trained by the genetic algorithm, with an average score of 11532.0 and the highest score on the test data achieved as 59678. The worst FLTP was the FLTP trained by an expert, with an average score of 2873.0 and a high score of 17611 achieved on the test data.

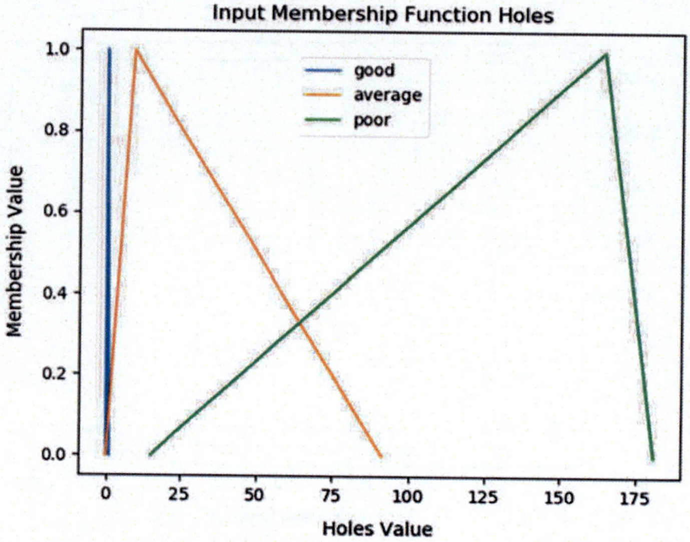

Fig. 13 Trained membership functions for Input 1: Holes

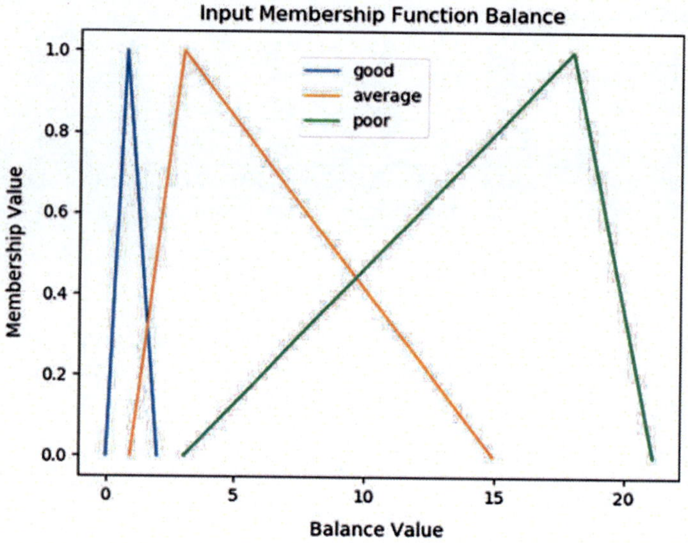

Fig. 14 Trained membership functions for Input 2: Balance

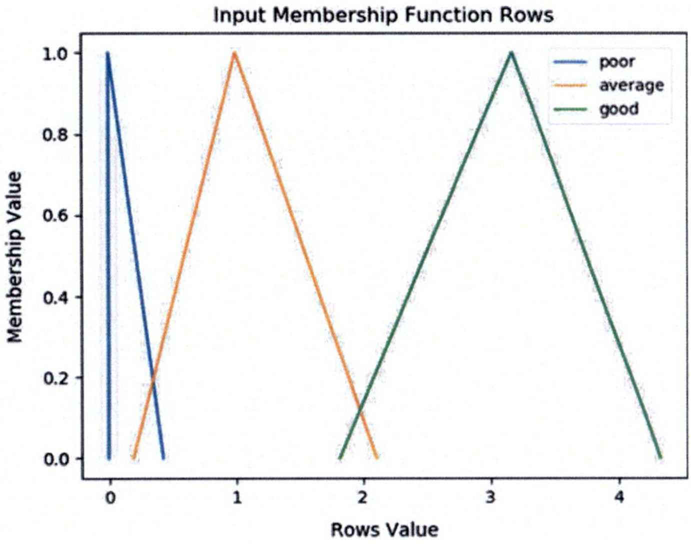

Fig. 15 Trained membership functions for Input 3: Rows

It is expected that the FLTPs trained by the genetic algorithm do better than the FLTP trained by a human expert. It is interesting that the FLTP that kept the rules of the expert trained player scores higher than the FLTP trained completely by the genetic algorithm. This is most likely due to the number of generations run. With a greater number of generations run, the FLTP trained completely by the genetic algorithm would most likely perform better than the FLTP with the rules already set.

It is interesting to observe the similarities and differences across the membership functions developed by the different training methods. The 'good' membership function for the holes input becomes much narrower in both FLTPs trained by the genetic algorithm. It is notable that there is a space not covered by membership functions in the holes input space, which may be addressed in the future by adding another membership function to this input. The membership functions for the other inputs are relatively similar across the FLTP trained by the expert and the FLTP with membership functions trained by the genetic algorithm. Further notable is that the output space of the FLTP trained completely by the genetic algorithm also has space not covered by membership functions.

It is clear, looking at the tables of the scores attained on the testing tetris games, that the scores can vary quite greatly across the games. A human playing tetris would have a much more even scoring level reached across any tetris game. This shows that the order of bricks falling can have a great affect on the scores.

5 Summary and Conclusions

A FLTP has been created using expert knowledge, a genetic algorithm to train the entire FLTP, and a genetic algorithm to train only the input membership functions of the FLTP. It was shown that the FLTP trained by the genetic algorithm for its input membership functions only performed the best. Additional research is needed to train the FLTP with the genetic algorithm in both situations for many more generations, potentially allowing the FLTP trained only by the genetic algorithm to perform better. Human behavior emerged from the genetic algorithm as evident in many of the rules it trained. Techniques must be considered to increase the speed of the genetic algorithm, so that this is possible [11, 12]. Further modifications would be to create more diversity in the training data so that the genetic algorithm trains for a broader range of tetris games, and therefore performs better overall, achieving a more even range of scores on the test data.

Acknowledgements Without the help and guidance of my mentor, Dr. Kelly Cohen, and the support from the Ohio Space Grant Consortium, this project would have not been possible. Thank you. Thank you also to my family and friends for their support along the way.

References

1. "Leaderboard." Classic Tetris World Championship, Classic Tetris World Championship, http://thectwc.com/leaderboard/
2. L.A. Zadeh, Fuzzy logic. Computer **21**(4), 83–93 (1988). (April)
3. A. Fernandez, F. Herrera, O. Cordon, M. Jose del Jesus, F. Marcelloni, Evolutionary fuzzy systems for explainable artificial intelligence: why, when, what for, and where to? IEEE Comput. Intell. Mag. **14**(1), 69–81 (2019). https://doi.org/10.1109/MCI.2018.2881645 (Feb.)
4. L.C. Chabowski, Tetris, github (2010). http://hi-im.laria.me/progs/tetrispyexefied.zip
5. Pajitnov, A., *Tetris—The game, The Tetris Company*, June 1984. https://tetris.com/
6. "API Reference" API Reference—Skfuzzy v0.2 Docs, Scikit-Fuzzy Development Team (2020). http://pythonhosted.org/scikit-fuzzy/api/api.html
7. T.J. Ross, *Fuzzy Logic with Engineering Applications* (Wiley, 2017)
8. L.A. Zadeh, Fuzzy Logic = Computing with Words. In: Zadeh L.A., Kacprzyk J. (eds) *Computing with Words in Information/Intelligent Systems 1. Studies in Fuzziness and Soft Computing*, vol. 33. (Physica, Heidelberg, 1999)
9. O. Cordon et al., *Genetic Fuzzy Systems: Evolutionary Tuning and Learning of Fuzzy Knowledge Bases* (World Scientific, 2001)
10. S. Mirjalili, *Introduction to Genetic Algorithms: Theory and Applications* (Online course offered by Udemy, 2020). https://www.udemy.com/course/geneticalgorithm/
11. R. Tanese, J.H. Holland, Q.F. Stout, Distributed genetic algorithms for function optimization. Ph.D. Dissertation. University of Michigan, USA. Order Number: AAI9001722 (1989)
12. G.R. Harik, F.G. Lobo, D.E. Goldberg, The compact genetic algorithm. IEEE Trans. Evol. Comput. **3**(4), 287–297 (1999). (Nov.)

A Dynamic Hierarchical Genetic-Fuzzy Sugeno Network

Owen Macmann and Kelly Cohen

Abstract A class of predictive algorithms that use genetic-fuzzy trees has risen to prominence in recent years. These algorithms are defined by three features: fuzzy logic, cascading decision trees, and genetic optimization. This paper presents an algorithm in this class that uses a cascading tree of Takagi-Sugeno systems, where the structure of the tree and the order of the elements within the tree are calculated procedurally by genetic algorithm. The resulting structure is unlike a tree and more akin to a honeycomb. This algorithm is shown herein to compare favorably with similar classes of predictive algorithms.

1 Introduction

Genetic-fuzzy logic systems have become distinguished in recent years for two reasons: their excellent ability to extrapolate, and their relative explainability as predictive algorithms. The model of this success has been the genetic-fuzzy tree, presented in the work of Ernest [1]. The genetic-fuzzy tree (GFT) describes a network of fuzzy inference systems (FIS) arranged in a cascading fashion, where the earlier layers process mainly the system inputs, and the later layers process both the fuzzy outputs of previous layers and any remaining inputs. In this way, the inputs are spread across a handful of FISes and appraised at different levels of abstraction. The primary advantage of such an approach over, say, one very large FIS is that the rule base is significantly reduced in size with no reduction in overall performance. This reduced-size rulebase can then be optimized by a genetic algorithm much more rapidly.

The GFT is distinguished from earlier, myriad usages of "fuzzy network", such as that first proposed by Hanebeck and Schmidt in 1996 [2] and the still-different fuzzy network of Gegov's in 2010 [3, 4]. In Hanebeck and Schmidt's original concept of the

O. Macmann (✉) · K. Cohen
University of Cincinnati, 2600 Clifton Ave, Cincinnati, OH 45221, USA
e-mail: macmanob@mail.uc.edu

K. Cohen
e-mail: cohenky@ucmail.uc.edu

B. Bede et al. (eds.), *Fuzzy Information Processing 2020*,
Advances in Intelligent Systems and Computing 1337,
https://doi.org/10.1007/978-3-030-81561-5_27

fuzzy network, they proposed an alternative model of a single FIS which resembles a radial neural network. The GFT instead maintains a network of two-inputs FISes, thus minimizing the total number of rules, and allowing for hybrid implementation of fuzzy models such as the Mamdani or Takagi-Sugeno models.

It would be possible, however, to arrange a genetic-fuzzy tree such that each FIS element—for all intents and purposes "fuzzy neurons"—can be arranged in a cascading fashion according to a hierarchical index for each element. This index determines the place a particular FIS element has in the hierarchy of inferences, and how it relates to the anterior and posterior inference layers: which inputs it takes, and which neurons take its outputs. A network of Sugeno inference systems, in this way, resembles an artificial neural network where the neurons are Sugeno gateways and the structure of the data cascade is procedurally generated. This structure cannot be calculated through backpropagation, necessitating the use of a genetic algorithm. Rather than a web or a tree, the structure becomes like a honeycomb, being filled with "honey" (data) by "bees" (the connection between cells) who collect "pollen from flowers" (inputs).

The purpose of this paper is to demonstrate that a genetic fuzzy network with an adaptively-determined hierarchy of fuzzy neurons, a genetic fuzzy honeycomb, can be used to perform unsupervised learning on regression and classification problems, and do so achieving higher accuracy than can be achieved with non-genetic approaches in neural networks that rely on Levenberg-Marquardt backpropagation or Bayesian regularization.

2 Methodology

The genetic-fuzzy honeycomb is ultimately defined by a vector of parameters consisting of its *gene*, or genetic code. This code determines the parameters of each Sugeno FIS or "cell" within the honeycomb. These parameters configure the Sugeno model in each cell, and assign a position in the honeycomb hierarchy to that cell.

2.1 *Takagi-Sugeno Fuzzy Model*

The fuzzy model used for the honeycomb cells is known as the Sugeno fuzzy model, also called the Takagi-Sugeno model or the TSK model, which was first proposed by Takagi and Sugeno in 1985 [5, 6]. The Sugeno model uses a fuzzy rule of the form in Eq. 1.

$$\text{if } x \text{ is } A \text{ and } y \text{ is } B \text{ then } z = f(x, y) \tag{1}$$

where A and B are antecedent fuzzy sets, and $z = f(x, y)$ is a consequent function with a crisp output. When $f(x, y)$ is a first-order polynomial, i.e. $f(x, y) = Ax +$

By, the fuzzy model is called a *first-order Sugeno fuzzy model*; by this token, if $f(x, y) = C$ where C is a constant, it's a *zero-order Sugeno fuzzy model*, and when it is a second-order polynomial, a *second-order Sugeno fuzzy model*, and so on.

In a second-order genetic fuzzy honeycomb, $f(x, y)$ is a second-order polynomial for each cell, of the form in Eq. 2.

$$z(x, y) = A_2x^2 + A_1x + B_2y^2 + B_1y + C \tag{2}$$

In this second-order polynomial, five real parameters need to be defined for $f(x, y)$, with additional parameters for defining the membership functions of x and y. These functions can be calculated in any way; they can even be ignored, assuming $z(x, y)$ incorporates "membership" through the design of its parameters, if the designer wishes. Otherwise, there will be a unique set of parameters for $z(x, y)$ for each membership function antecedent. Assuming two rules, the resulting output for a cell with two rules is demonstrated in Eq. 3, where w is the firing strength of the rule. Each honeycomb cell is therefore a 2-input Sugeno FIS of this generic form.

$$z = \frac{w_1z_1 + w_2z_2}{w_1 + w_2} \tag{3}$$

2.2 *Hierarchical Network*

The hierarchical network is the structure organizing the cells of the genetic-fuzzy honeycomb described in Sect. 2.1. The "hierarchy" refers to the design of the network which orders the cells from anterior to posterior and assigns them to successive layers of cells. Anterior layers pass their outputs to posterior layers as further inputs; ultimately, the terminal cell produces the final output. It's possible to configure a hierarchical network like this with two or more terminal cells, as well.

Some genetic parameters are used to construct this hierarchy. Being genetic parameters, the structure is therefore generated non-deterministically based on the holistic solution to the problem. The two parameters are called *Rank* and *Salient*.

Rank is a real scalar value, and is compared directly to the Ranks of other cells. The higher the Rank, the higher the position in the hierarchy—the further towards the posterior of the network it is. *Salient* is a network-wide positive integer that determines the rate that cells in the honeycomb "funnel" to a final output. A higher salient means a longer "tunnel" section, where cascaded inputs may continue to be added network; a lower salient means a shorter tunnel section, eventually shrinking entirely to a section only as large as half of the number of inputs, rounded up, only one layer deep. The "funnel" section then contains a number of cells calculable by Gauss's method, as in Eq. 4.

$$CellsInFunnel = \frac{\frac{inputs}{2}(\frac{inputs}{2} + 1)}{2} \tag{4}$$

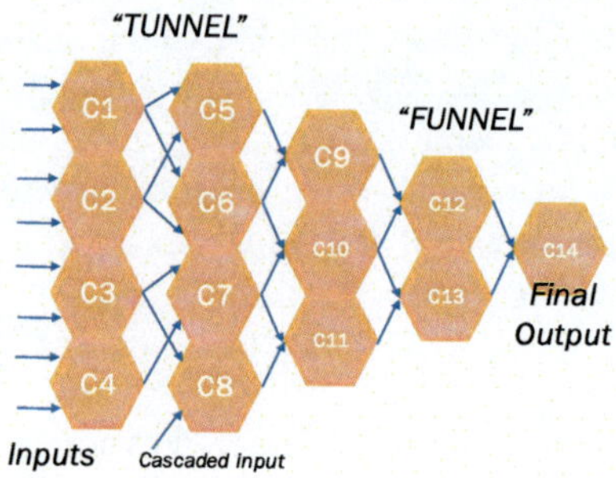

Fig. 1 The transition from the tunnel section of the structure to the funnel section is dictated by the *salient* parameter. Here, a salient of 2 is shown, across 8 inputs. The cells are ordered in the structure according to their rank: C14 has the highest rank, and C1 has the lowest

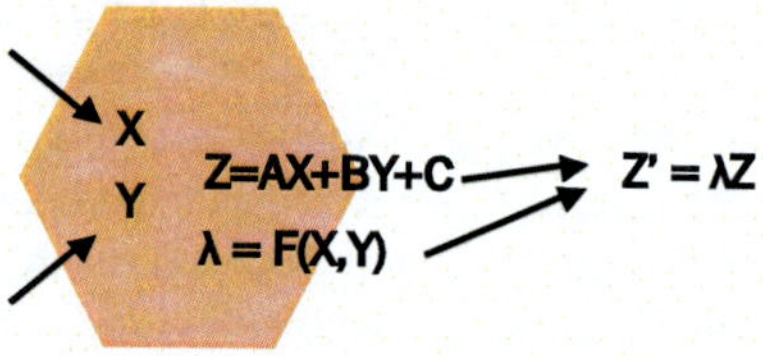

Fig. 2 A singular honeycomb cell, containing a basic first-order Sugeno fuzzy model

A graphical representation of this salient is shown in Fig. 1. Take note that the total number of cells and, therefore, genes remains constant even if the salient changes. A sufficiently high salient will have a shorter tunnel section: care must be taken the salient is not set so high that cells begin dropping outputs. An output that does not continue to the terminal cell is wasted.

Finally, we note, again, that each cell within this hierarchy is simply a Takagi-Sugeno system, of first- or second-order, with a membership λ decided by some membership function for X and Y. A diagram of the elementary structure of the first-order cell is shown in Fig. 2.

2.3 *Genetic Algorithm*

In order to tune the parameters and allow the fuzzy-genetic honeycomb to converge on a final solution, one that closely predicts the true or ideal solution, a genetic algorithm is deployed, using the *gene* of the honeycomb mentioned at the beginning of Sect. 2 as the string of genetic variables. A genetic algorithm finds a solution for a fitness function that incorporates the *gene* as a vector of parameters for that function.

In the case of a honeycomb, for just two cells, as an example, that vector may be (1.23, 4.56, −8.9, 2.7, 99.5, −16.5, 4.11, 13.8, 17.7, 6. 09, 2); these values assigning, in sequence, $A_{2_1} = 1.23$, $A_{1_1} = 4.56$, $B_{2_1} = -8.9$, and so on. Then, a genetic algorithm generates an initial population of possible vectors, which can be called *individuals*, abiding by some upper and lower limits that can be specified (to reduce the search space, some domains can be omitted if they are thought to be trivial). Each member of this population is evaluated for fitness, and some percentage of the fittest individuals are preserved and used to seed a second generation of the population. The seed is then "mutated", offering random variations to any elements within the genetic vector for any individual), and fit individuals are crossed over with other fit individuals to generate "offspring," which retain some of the same genes (sans mutation) that its "parents" had. In this manner is a new generation of the population created, which is then tested for fitness, with the fittest individuals selected to seed another generation; and so on, ad infinitum.

The convergence criteria for this process is based on some tolerance for stalling in the growth of the population's best fitness, or sometimes the number of generations to be tested, or the amount of simulation time to be allocated. Therefore, no absolute basis for deterministic convergence on a guaranteed optimum solution exists. Nevertheless, the breadth of the genetic algorithm search space and the possibility of continuous reiterative training can allow it to ferret solutions from uncooperative problem spaces. In the case of the genetic-fuzzy tree, uncommonly difficult and non-linear problems like intelligent flight direction and aerial combat strategy are well within the capability of the system to decide [7].

2.4 *Artificial Neural Networks*

A broad class of algorithm, the artificial neural network (ANN) can be used to describe any network of logic gates that take in inputs and produce at least one final output. The standard artificial neural network uses a Sigmoidal neuron, which produces some output on a Sigmoid function bound between an absolute minimum and an absolute maximum.

A standard method of optimizing the gains in a neural network is known as *Levenberg-Marquardt backpropagation*, which is based on the Levenberg-Marquardt algorithm [8, 9], also called the damped least-squares method. The algorithm is used for solving non-linear least squares problems, which when applied to backpropagation in a neural network allows it to tune the gains towards some minimum value. This approach is related to the Gauss-Newton algorithm. It is a mainstay method for optimizing a neural network.

An entirely different approach, also towards tuning the gains of a neural network, is known as *Bayesian Regularization*. Unlike Levenberg-Marquardt, and like genetic algorithms, Bayesian regularized artificial neural networks (BRANN) have no deterministic criteria for convergence. A BRANN may be continuously trained until it stops improving entirely (whatever the reason) [10].

In this study, these methods are compared with the genetic-fuzzy honeycomb for their potential to achieve effective classification and regression. For the Levenberg-Marquardt ANN (LMANN), the BRANN, and the genetic-fuzzy honeycomb, the training and test data were randomly and evenly selected; the RMSE's shown in Tables 1, 2, and 3 are just for the test data set. Two LMANNs will be considered: one with 10 neurons, and another with 36 neurons, to demonstrate how the consistency of ANN performance is only loosely connected to the number of internal data nodes.

Table 1 Root-Mean-Squared-Error for the Forest Fire problem: 8 inputs, 517 samples

Algorithm	RMS error	Computation time (s)
ANN Levenberg-Marquardt backpropagation (10 neurons)	62.6915	1.66
ANN Levenberg-Marquardt backpropagation (36 neurons)	71.0210	2.53
Bayesian Regularization (10 neurons)	62.9668	15.67
Genetic-Fuzzy Honeycomb (10 cells)	54.7198	96.13

Table 2 Root-Mean-Squared-Error for the O-ring problem: 3 inputs, 23 samples

Algorithm	RMS error
ANN Levenberg-Marquardt backpropagation (10 neurons)	0.6015
ANN Levenberg-Marquardt backpropagation (36 neurons)	0.5605
Bayesian Regularization (10 neurons)	0.5274
Genetic-Fuzzy Honeycomb (6 cells)	0.5357

Table 3 Root-Mean-Squared-Error for the QSAR fish toxicity problem: 6 inputs, 908 samples

Algorithm	RMS error	Computation time
ANN Levenberg-Marquardt backpropagation (10 neurons)	0.9249	1.93 s
ANN Levenberg-Marquardt backpropagation (36 neurons)	0.9309	2.15 s
Bayesian Regularization (10 neurons)	0.8241	16.48 s
Genetic-Fuzzy Honeycomb (12 cells)	1.6147	113.52 s
Genetic-Fuzzy Honeycomb (64 cells)	0.5976	1593.19 s

3 Results

This study compares the performance of the three algorithms for three data problems obtained from the UC Irvine Machine Learning Repository, accessible online. [11] Each problem tests either the ability of the algorithm to either calculate a non-linear regression between the input data and the output for that problem, or to classify the input correctly according to the classifications given as the output.

The primary figure of merit is the root-mean-squared error. Calculation time for the forest-fire problem is given to provide some comparison between the efficiency of the methods. Calculations were performed on a 2.8 GHz 4-core processor.

3.1 Forest Fires

The forest fires data set was collected by Cortez and Morais in 2007 [12]. This data set collects 517 instances of data on indices and measurements of forest in northwestern Portugal with an indication of whether or not a fire occurred, and how widespread the burning was. In this study, the final anticipated output is a twofold response: a classification of whether or not a fire occurred, and how large the fire was. The performance of the three algorithms are recorded in Table 1.

We note that a significant improvement in the regression is achieved over the LMANNs and the BRANN for the Genetic-Fuzzy Honeycomb for the same number of cells. Increasing the number of neurons available to the LMANN does not increase its performance; rather, the opposite.

We also note that the LMANN and the BRANN have superior computational time to the genetic-fuzzy honeycomb. It is likely that the genetic-fuzzy honeycomb can be optimized to function similarly.

3.2 O-Rings

The O-rings problem, based on data associated with the Challenger space shuttle disaster collected by David Draper in 1993 [13], has a very small number of samples (23) and only 3 inputs, which must somehow be correlated to the number of O-rings that are endangered by any particular flight. While heuristically it is simple to understand, by reading Draper's paper, the exact nature in which the Challenge risk was unquantified, that being that it was simply untested for the flight conditions, in terms of pure data it provides an interesting challenge about extrapolating from a very small data set. The results are shown in Table 2.

We note that, as before, the genetic-fuzzy honeycomb achieves a superior RMS error to the LMANNs with fewer cells. However, the BRANN achieves a lower RMS error than the genetic-fuzzy honeycomb yet. It's possible a greater number of cells would improve this figure.

3.3 QSAR Fish Toxicity

The QSAR fish toxicity problem refers to a quantitative structure-activity relationship model devised in a study by Cassotti et al. towards the prediction of acute aquatic toxicity in the fish *Pimephales promelas* over a set of 908 chemicals [14]. With six molecular descriptors, the data they collected seeks to establish a relationship between those six inputs with a quantitative response called LC50, the negative logarithm of mols per liter necessary to kill 50% of test fish over a test duration of 96 h. Ultimately, this is a simple regression problem. The results are recorded in Table 3.

We note that the 12-cell configuration for the genetic-fuzzy honeycomb produces a very meager correlation compared to the LMANNs and the BRANN. Reasoning that perhaps the search space was too small with only 12-cells, the genetic-fuzzy honeycomb was tested again with a 64-cell correlation, thus allowing the algorithm to take full advantage of the *salient* parameter. The response was vastly improved, though at the cost of an increased computation time.

4 Conclusion

In this paper we have presented a new model of genetic-fuzzy network, that builds on the concept of genetic-fuzzy trees and allows for the creation of procedurally-generated fuzzy network architectures. These fuzzy networks can deploy hybrid methods towards the inclusion of other fuzzy models, such as Sugeno, and are optimized genetically. The resulting networks have been shown in this paper to be capable and comparable in regression capabilities to traditional artificial neural networks. These genetic-fuzzy networks have been termed genetic-fuzzy honeycombs, due to their non-ordinal structure.

Where we identify space for improvement, and the development of more advanced algorithms in this class, is in the optimization of the algorithms' computational time and in further refining of the concept of the configurable architecture. Although the *salient* variable allows for the network to assume different shapes, there is no tool at present for allowing this architecture to assume *radically* different shapes, with non-discrete layers and semi-recursive loops. Additionally, a fully recursive version of this architecture may be able to incorporate a genetic fitness feature in real time, and thus self-regulate through a constant series of internal pressures. This capability would allow the architecture to be truly multiform.

This version of the genetic-fuzzy honeycomb was implemented in MATLAB, and there may be ways to further refine and optimize the algorithm by eliminating unnecessary steps or combining those steps together. The ANNs it was compared with were all highly refined and optimized versions of their respective algorithms. Additionally, the computer hardware on which these tests were being conducted was relatively modest.

In future work, this concept can be expanded to inform a variety of different methods towards unsupervised learning using fuzzy Sugeno networks, such as for unsupervised classification or clustering algorithms. More importantly, the capability of fuzzy systems to serve as "explainable AI," even when trained by a genetic algorithm, makes a system like this potentially very valuable in the field of verifiable and validatable machine learning algorithms, should the method be outlined that converts a fuzzy Sugeno network into an equivalent series of verbal logic chains. It is clear the genetic-fuzzy honeycomb offers new potential to the algorithm class known as the genetic-fuzzy tree.

References

1. N. Ernest, *Genetic Fuzzy Trees for Intelligent Control of Unmanned Combat Aerial Vehicles* (University of Cincinnati, 2015). (Electronic Thesis or Dissertation)
2. U. Hanebeck, G. Schmidt, Genetic optimization of fuzzy networks. Fuzzy Sets Syst. **79**, 59–68 (1996)
3. A. Gegov, *Fuzzy Networks for Complex Systems: A Modular Rule Base Approach* (Springer, Berlin, 2010)
4. A. Gegov, *Fuzzy Rule Based Networks* (2011)
5. T. Takagi, M. Sugeno, Fuzzy identification of systems and its applications to modeling and control. IEEE Trans. Syst. Man Cybern. **SMC–15**, 116–132 (1985)
6. M. Sugeno, G.T. Kang, *Structure Identification of Fuzzy Model* (Department of Systems Science, Tokyo Institute of Technology, 1988)
7. N. Ernest, D. Carroll, C. Schumacher, M. Clark, K. Cohen, G. Lee, Genetic fuzzy based artificial intelligence for unmanned combat aerial vehicle control in simulated air combat missions. J. Def. Manag. **06** (2016)
8. K. Levenberg, A method for the solution of certain non-linear problems in least squares. Quart. Appl. Math. **2**(2), 164–168 (1944)
9. D. Marquardt, An algorithm for least-squares estimation of nonlinear parameters. SIAM J. Appl. Math. **11**(2), 431–441 (1963)
10. F. Burden, D. Winkler, Bayesian regularization of neural networks, in *Artificial Neural Networks, Methods in Molecular Biology*, ed. by D.J. Livingstone, vol. 458 (Humana Press, 2008)
11. D. Dua, C. Graff, *UCI Machine Learning Repository* (University of California, School of Information and Computer Science, 2019). http://archive.ics.uci.edu/ml
12. P. Cortez, A. Morais, A data mining approach to predict forest fires using meteorological data, in *New Trends in Artificial Intelligence, Proceedings of the 13th EPIA 2007—Portuguese Conference on Artificial Intelligence*, pp. 512–523 (2007). http://www.dsi.uminho.pt/~pcortez/fires.pdf
13. D. Draper, Assessment and propagation of model uncertainty, in *Proceedings of the Fourth International Workshop on Artificial Intelligence and Statistics*, pp. 497–509 (1993)
14. M. Cassotti, D. Ballabio, R. Todeschini, V. Consonni, A similarity-based QSAR model for predicting acute toxicity towards the fathead minnow (Pimephales promelas). SAR QSAR Environ. Res. **26**, 217–243 (2015)

Obstacle Avoidance and Target Tracking by Two Wheeled Differential Drive Mobile Robot Using ANFIS in Static and Dynamic Environment

Dhruv Patel and Kelly Cohen

Abstract This paper proposes an idea for navigation and target acquiring for an autonomous robot in static as well dynamic environment using a tuned and trained Sugeno-type fuzzy inference system. This term for such system is Adaptive Neuro Fuzzy Inference System (ANFIS). ANFIS incorporates benefits from both fuzzy logic controller and neural networks and therefore is a robust tool to solve the stated problem. The input/output training data used to tune the system parameters is intuitively chosen and can be considered expert knowledge. Simulation results are achieved using in-build MATLAB functions. The architecture of the entire system consists of four ANFIS controllers. Two of them are used to control the wheel velocities of left and right wheel of the robot in order to get to the target defined by the user. The remaining two ANFIS controllers are utilized to successfully avoid obstacles encountered on the way to target.

Keywords Obstacle avoidance · Fuzzy logic controller · ANFIS · Navigation

1 Introduction

A two wheeled differential drive robot is the most rudimentary type of autonomous vehicle which finds its application ranging from a simple backyard project to industrial applications such as monitoring, transportation and countless more. The most essential behavior we seek in robots is obstacle avoidance and energy efficiency. Energy efficiency can also be defined as finding an optimal route to target position. The path planning and navigation are one of central design problem and also an area of current ongoing research. Considerable efforts have been done in the last decade to deal with such problems. The autonomous must be capable to gather and extrapolate

D. Patel (✉)
Department of Mechanical Engineering, University of Cincinnati, Cincinnati, OH, USA
e-mail: patel4db@mail.uc.edu

K. Cohen
Department of Aerospace Engineering, University of Cincinnati, Cincinnati, OH, USA
e-mail: cohenky@ucmail.uc.edu

B. Bede et al. (eds.), *Fuzzy Information Processing 2020*,
Advances in Intelligent Systems and Computing 1337,
https://doi.org/10.1007/978-3-030-81561-5_28

information from the surrounding using sensors. The acquired data from all onboard sensors is fed to a controller to plan and execute its mission within its environment without human intervention [1].

Primarily, there are two ways a path planning problem can be classified: the local planning and the global planning [2]. In the local planning, the environment in which the robot is deployed is either only partially known or completely unknown. Whereas, in the global planning, complete environment is priori-known and the components of the environment (i.e. obstacles) should be static [3]. The robot must use sensors to perceive its surroundings and plan the motion accordingly.

In the current literature, there have been many techniques used to solve the dilemma of the navigation and path planning. Some of the most popular methods are the Visibility Graph algorithm [4], Voronoi Diagram [5], Bug Algorithm and the Potential Field Method. Just to give an example, in Potential field method, there are two opposite forces at work—an attractive potential pulling the object towards the goal point while a repulsive force pushing the object away from the obstacles present in the environment. Nevertheless, this method is susceptible to becoming stuck at a local search minimum.

In recent years, due to the advent of modern computing techniques coupled with computational resources has made solving the path planning and navigation problem undemanding. These techniques include genetic algorithm [6, 7]. In [8], a control algorithm for obstacle avoidance for mobile robots based on fuzzy controller was presented. The environment information surrounding the robot detected by the ultrasonic sensors is fuzzified, and then used as an input into a fuzzy control system. The output obtained from the fuzzy control system is used to drive the robot. The simulation results show that the algorithm can help the robot to avid obstacles safely and is certainly feasible. In [9], fuzzy logic controllers having different membership functions are developed, tested and then used to navigate robots. Firstly, a fuzzy controller was used with 4 types of input members, two types of output members and three parameters each. Upon further development, two types of fuzzy controllers were developed having similar input and output membership functions but with five parameters each rather than three. Each of the robots were using ultrasonic sensor array to measure the distances of obstacles around them and an infrared sensor for detecting heading of the target. In [10], path planning of a mobile robot in an unknown environment was designed by utilizing a fuzzy logic system. Similar to [9], ultrasonic sensors were deployed to detect the distance and positions of obstacles. Using this fuzzy logic system, angular velocities of left and right wheel for the two wheeled robots were controlled. This method also required another new rule table that was required considering the distances to obstacles and the angle between the robot and the goal.

Chi and Less [11], showed an artificial neural network which was applied to learn the environment from sensory data for moving along a collision free course. The FIS was trained offline by using suitable datasets and the obtained parameters were fed into the robot. The training and testing of the FIS were done with the help of the ANFIS toolbox of MATLAB. Singh [12], did an analysis on an adaptive neuro-fuzzy technique for the navigation of a mobile robot in an unknown environment. The results improved the validity of the proposed technique to determine the optimal time and create a collision free trajectory. A neuro-fuzzy based system for the behavior-based control of a mobile robot for reactive navigation was touched upon in [13], by Joshi and Zaveri.

From the literature survey, it can be seen that there are still some areas that have not been addressed. Such as the online sensing information from the 3 directions (front, right, left) during the navigation and the complexity of the navigation environment. All those areas have been included in this paper to formulate a new problem definition for solving the path planning problem in a complex static environment containing multiple obstacles having different sizes and shapes.

The overall goal of this proposed idea can be seen as the implementation of ANFIS model which has the ability to avoid becoming stuck at any local minimum and generates a smooth trajectory from the start point to goal point in unknown environment. Simply put, ANFIS helps in guiding the mobile robot to reach its final goal position while efficiently and successfully avoiding obstacles through the workspace.

The entire process of developing a fuzzy-based controller that can successfully avoid collisions and reach a target in an efficient manner has been divided into two main steps: firstly, the obstacle avoidance controller and then the target acquiring controller. The obstacle avoidance controller constitutes two ANFIS controllers. Each of them receives the same three inputs, which are the front, right and left distances acquired from the sensors mounted on the robot. The output of the first ANFIS is used to control the right wheel's angular velocity while the other ANFIS is used for angular velocity control of the left wheel. The target acquiring controller is implemented, similar to the obstacle avoidance controller, using two ANFIS controllers. Instead of using three inputs, these controllers use only one input each which is the angle difference between the autonomous ground vehicle's direction and the target's angle. Again, the output of first ANFIS controller is used to control the right wheel velocity while the output of second ANFIS is used to control the wheel velocity of the left wheel of the robot.

2 The Architecture and Design of the Mobile Robot's ANFIS Controller

In this section, the proposed ANFIS controllers are discussed in depth. There are four ANFIS controllers which have been designed to accomplish the navigation task, which can be grouped under two groups. The first group is the one that oversees the obstacle avoidance mission while the other group has the oversight on the target acquisition task.

As both controller groups are giving output of the same nature (i.e. left and right wheel velocities), a decision has to be made upon which controller output to choose in order to get an overall intended behavior. This is resolved through a switch block structure and it decides on which controller will be activated. The decision is quite intuitive also. The obstacle avoidance is the primary goal of the robot regardless of target approaching efficiency. Therefore, it should always exhibit collision avoidance behavior and when there are no obstacles in the vicinity it should shift to target tracking behavior. Moreover, while the robot is in target approaching behavior and suddenly it detects an obstacle, it then should fall back to its basic primitive behavior. Such behavioral switching is done with the help of the switch block. The information from the sensors (front distance, left distance, and right distance) can be combined in order to generate a Boolean signal namely, 'Obstacle Sensing Signal' abbreviated as OS. If the value of OS is True (1) then the bot will fall back to obstacle avoidance and if the value is False (0) then the set behavior is of target approaching.

3 Target Reaching ANFIS Controller

As stated previously, there are two dedicated ANFIS controllers to drive the right and left wheel with a given angular velocities. The input for this pair of controllers is the angle difference, AD. However, when both the target point and mobile robot are assumed to be a point entity then there is no requirement to calculate the AD. As, the robot can be manipulated to move in the line of sight when there is no obstacle around. However, the final aim of this project is to move beyond point entities and solve the path planning problem for a two-dimensional body. The output of the first controller is the angular velocity for the right wheel and for the second controller, the returned value is the angular velocity for the left wheel. The training for the target acquiring controller is done using a total of 40 sets of data. The training data for input and output ranges are given in Table 1. The purpose of this training is to adjust the membership function parameters to obtain the required model.

Figures 1 and 2 depicts the training instance for left and right wheel velocities. The training is based upon the training set discussed in Table 1. The number of membership functions were kept three for the angle difference input. This number gives a reasonable amount of error as shown in the Figure. The type of membership function is triangular. The relationship of surface view between angle difference and the output for the first and second ANFIS controllers is illustrated in Figs. 3 and 4 as follows.

4 Obstacle Avoidance ANFIS Controller

The most focal behavior in the entire deployment of the mobile robot is finding a path with zero collisions with obstacles. The third and fourth ANFIS controllers are utilized to steer away the robot when it is in close proximity with obstacles. The structure of both FIS files is similar and each ANFIS has three inputs and one output. Inputs are namely Front Distance, Left Distance and Right Distance while the output

Table 1 Training Data for Target Acquiring ANFIS Controller

Dataset index	Angle difference	Right angular velocity	Left angular velocity
1	0	80	80
2	−44.007	40.882	−37.352
3	−29.214	54.031	2.093
4	−14.226	67.353	42.061
5	−4.319	76.16	68.48
6	0.505	78.652	79.55
7	1.975	74.731	78.243
8	1.809	75.173	78.391
…	…	…	…
…	…	…	…
33	−37.315	46.83	−19.507
34	−22.376	60.11	20.33
35	−22.843	59.694	19.082
36	−15.563	66.165	38.496
37	−7.999	72.888	58.66
38	−2.812	77.5	72.5
39	0.2973	79.207	79.735
40	4.682	70.152	72.25

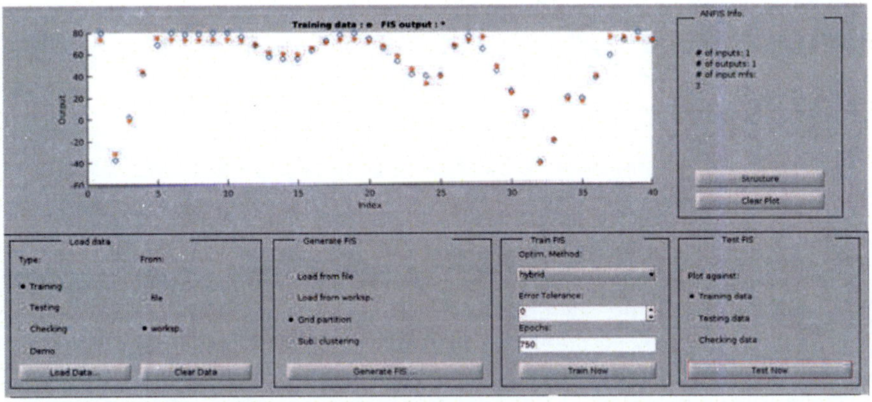

Fig. 1 Training quality for left velocity—output data before and after training

is left or right wheel's angular velocity. To train the ANFIS for obstacle avoidance behavior, a dataset consisting 20 entries was selected for the training purpose. This training dataset is given in Table 2. Such training decides the optimum membership function limits, resulting in the required model. Figures 5 and 6 denotes the training instance for the fuzzy inference system for left and right wheel velocity respectively.

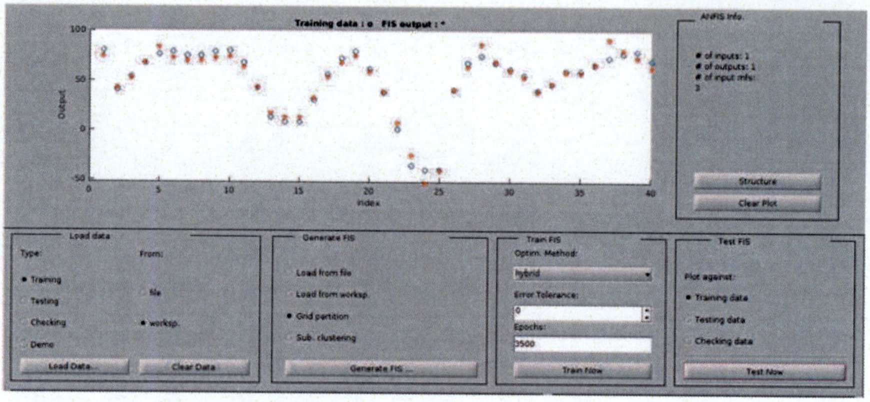

Fig. 2 Training quality for right velocity—output data before and after training

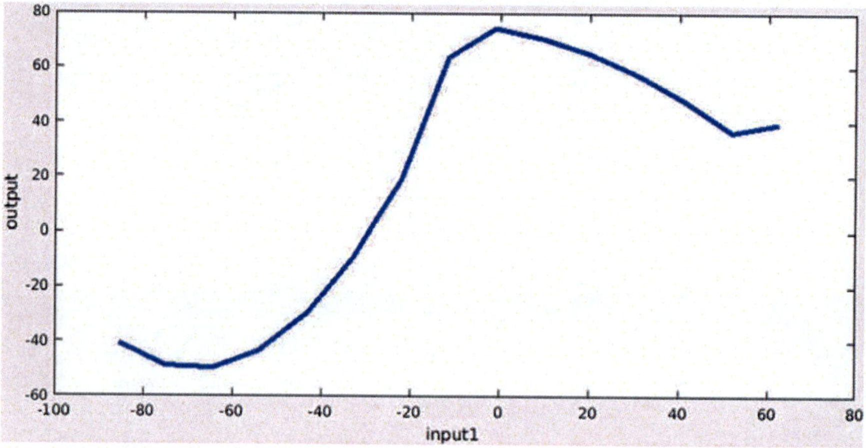

Fig. 3 The surface view for target acquiring ANFIS: first ANFIS controller (Left Wheel)

It can be seen that, there is a considerable amount of error in the model even after the training. However, the number of membership functions chosen for the ANFIS were three and for all inputs the type of all membership functions was kept triangular. Keeping this in mind, the error in the model is reasonable.

The surface view between inputs and the output for the given ANFIS controllers are outlined in Figs. 7, 8, 9, 10, 11, 12.

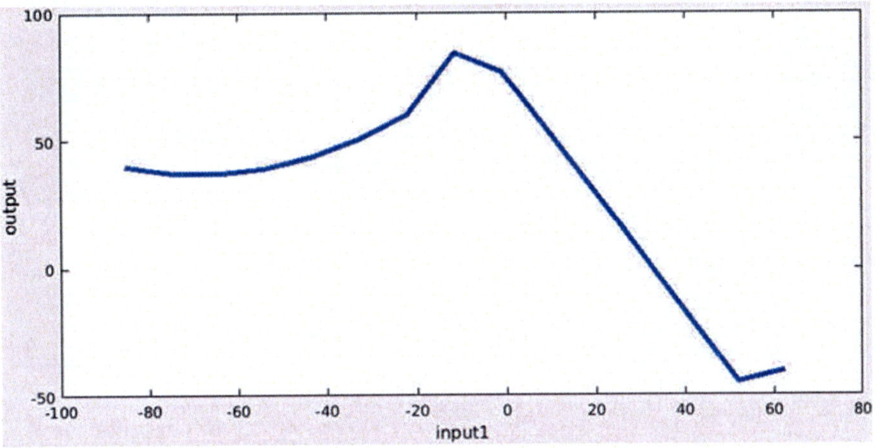

Fig. 4 The surface view for target acquiring ANFIS: second ANFIS controller (Right Wheel)

Table 2 Training Data for Target Acquiring ANFIS Controller

Dataset index	Front distance	Right distance	Left distance	Right angular velocity	Left angular velocity
1	0.1	0.1	0.1	−40	−40
2	0.8	0.8	0.8	80	80
3	0.8	0.399	0.8	79.99	79.99
…	…	…	…	…	…
16	0.8	0.8	0.304	22.418	60.806
17	0.8	0.8	0.304	22.803	60.934
18	0.8	0.8	0.335	41.084	67.028
19	0.8	0.8	0.395	77.33	79.11
20	0.8	0.8	0.455	80	80

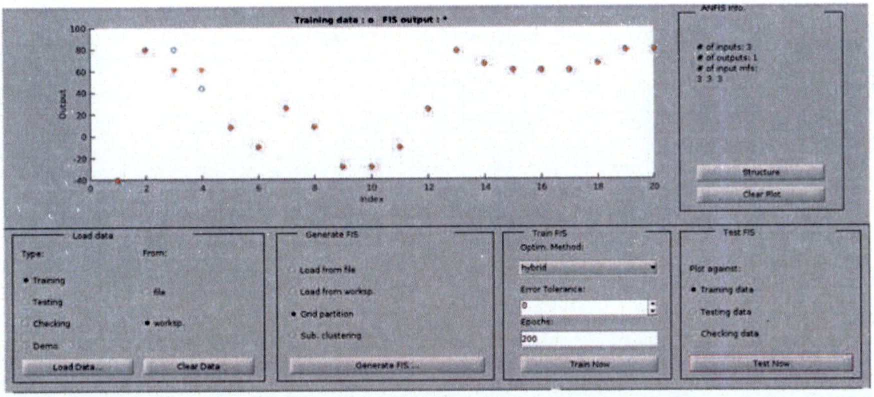

Fig. 5 Training quality for left velocity—output data before and after training

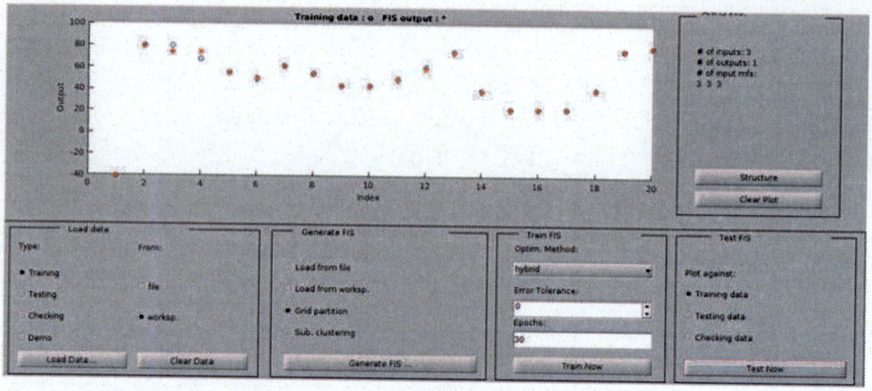

Fig. 6 Training quality for right velocity—output data before and after training

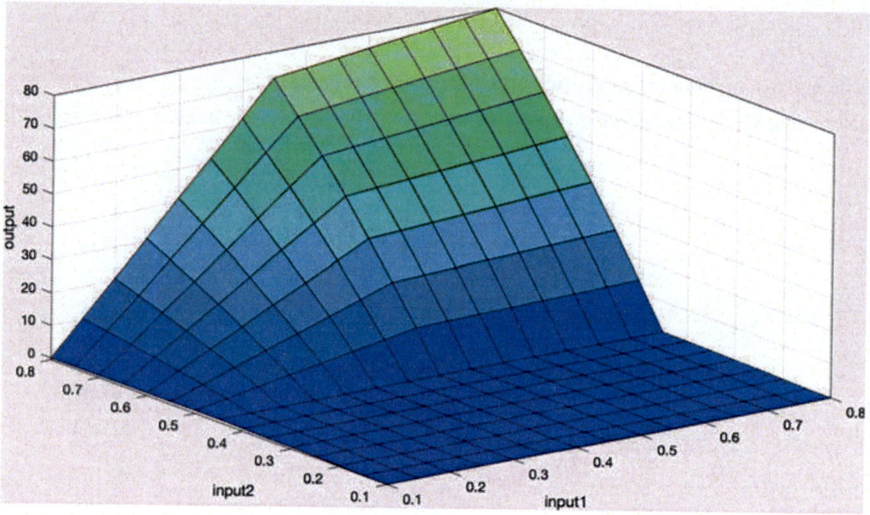

Fig. 7 The surface view for obstacle avoidance ANFIS—between input 1, input 2 and output (left wheel velocity)

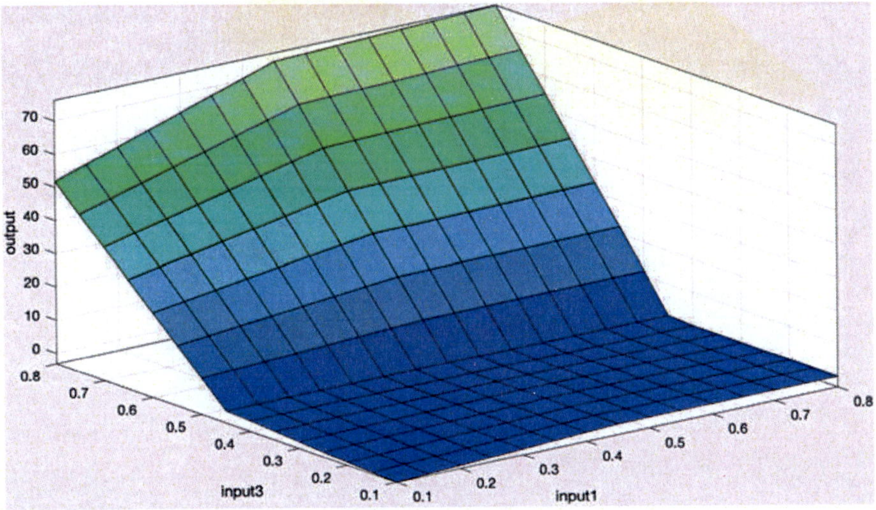

Fig. 8 The surface view for obstacle avoidance ANFIS—between input 1, input 3 and output (left wheel velocity)

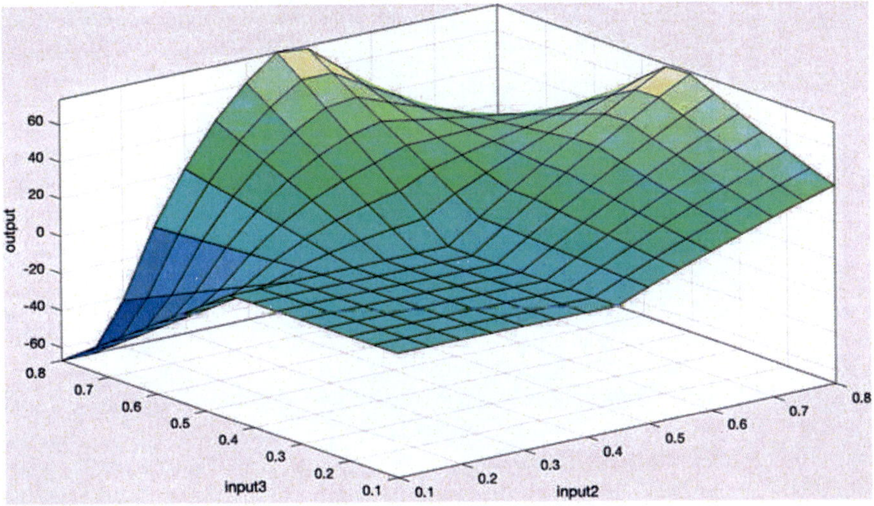

Fig. 9 The surface view for obstacle avoidance ANFIS—between input 2, input 3 and output (left wheel velocity)

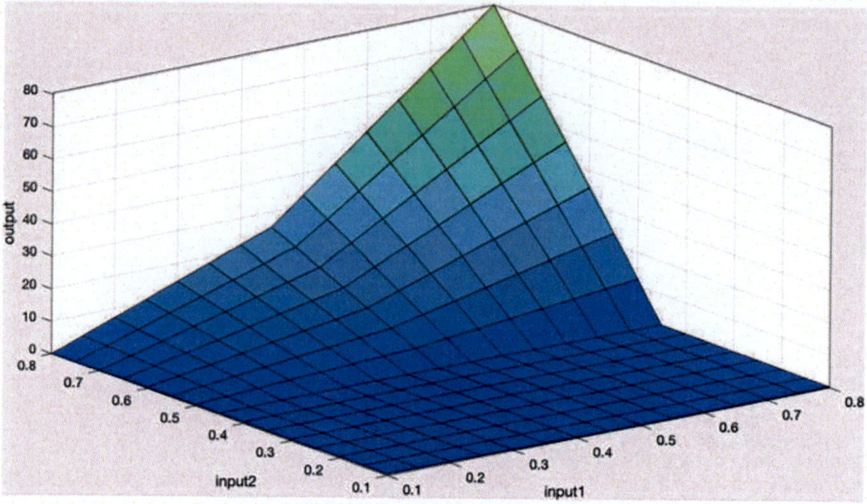

Fig. 10 The surface view for obstacle avoidance ANFIS—between input 1, input 2 and output (right wheel velocity)

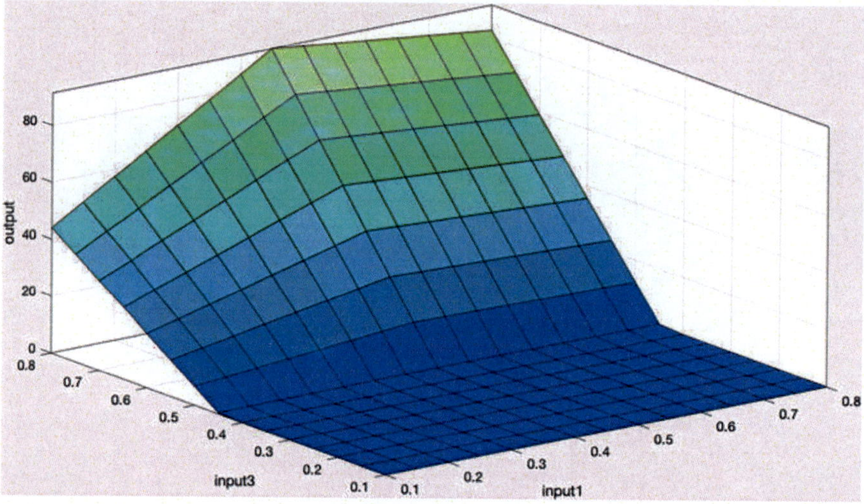

Fig. 11 The surface view for obstacle avoidance ANFIS—between input 1, input 3 and output (right wheel velocity)

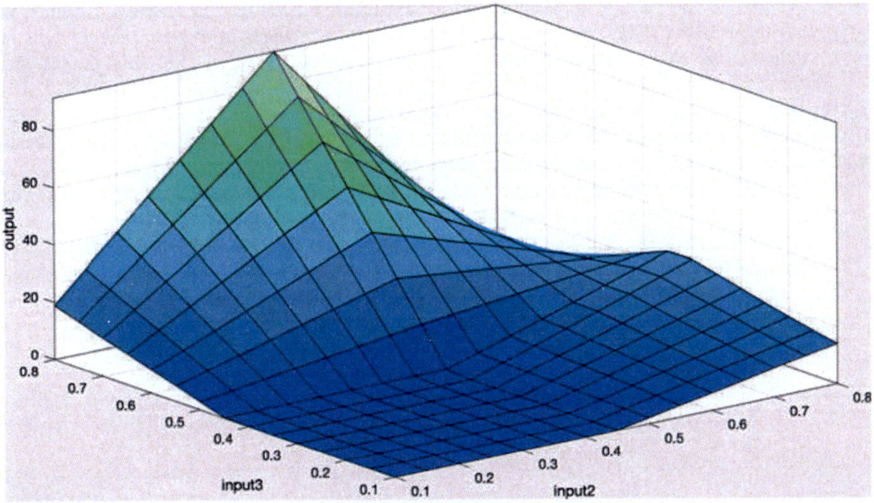

Fig. 12 The surface view for Obstacle Avoidance ANFIS—between input 2, input 3 and output (right wheel velocity)

References

1. K. Nonami, M. Kartidjo, K.-J. Yoon, A. Budiyono, *Autonomous Control Systems and Vehicles: Intelligent Unmanned Systems* (Springer Tokyo Heidelberg: Tokyo, Japan, 2013), p. 306
2. K.H. Sedighi, K. Ashenayi, T.W. Manikas, R.L. Wainwright, Autonomous local path planning for a mobile robot using a genetic algorithm, in *Proceedings of the IEEE Congress on Evolutionary Computation* (Portland Marriott Downtown, Portland, OR, USA, 2004), pp. 1338–1345. 19–23 June 2004
3. M. Wang, J.N.K. Liu, Fuzzy logic based robot path planning in unknown environment, in *Proceedings of the IEEE International Conference on Machine Learning and Cybernetics* (Guangzhou, China, 2005), vol. 1, pp. 813–818. 18–21 August 2005
4. O.S.M. De Berg, M. van Kreveld, M. Overmars, *Computational Geometry: Algorithms and Applications* (Springer, Berlin, Germany, 2000)
5. S. Garrido, L. Moreno, D. Blanco, P. Jurewicz, Path planning for mobile robot navigation using Voronoi diagram and fast marching. Int. J. Robot. Autom. **2**, 42–64 (2011)
6. I.A. L-Taharwa, A. Sheta, M. Al-Weshah, A mobile robot path planning using genetic algorithm in static environment. J. Comput. Sci. **4**, 341–344 (2008)
7. M. Samadi, M.F. Othman, Global path planning for autonomous mobile robot using genetic algorithm, in *Proceedings of the IEEE International Conference on Signal-Image Techonology & Internet-Based Systems* (Kyoto, Japan, 2013), pp. 726–730. 2–5 Dec 2013
8. S. Cui, X. Su, L. Zhao, Z. Bing, G. Yang, Study on ultrasonic obstacle avoidance of mobile robot based on fuzzy controller, in *Proceedings of the IEEE International Conference on Computer Application and System Modeling* (TaiYuan, China, 2010), pp. 233–237. 22–24 Oct 2010
9. S.K. Pradhan, D.R. Parhi, A.K. Panda, Fuzzy logic techniques for navigation of several mobile robots. Appl. Soft Comput. **9**, 290–304 (2009)
10. X. Li, B. Choi, Design of obstacle avoidance system for mobile robot using fuzzy logic systems. Int. J. Smart Home **7**, 321–328 (2013)
11. K.H. Chi, M.-F.R. Lee, Obstacle avoidance in mobile robot using neural network, in *Proceedings of the IEEE International Conference on Consumer Electronics, Communication and Network* (XianNing, China, 2011), pp. 5082–5085. 16–18 April 2011

12. M.K. Singh, D.R. Parhi, J.K. Pothal, ANFIS approach for navigation of mobile robots, in *Proceedings of the IEEE International Conference on Advances in Recent Technologies in Communication and Computing* (Kottayam, Kerala, India, 2009), pp. 727–731. 27–28 Oct 2009
13. R. Fierro, F.L. Lewis, Control of a nonholonomic mobile robot using neural networks. IEEE Trans. Neural Networks **9**, 589–600 (1998)
14. M.M. Joshi, M.A. Zaveri, Neuro-fuzzy based autonomous mobile robot navigation system, in *Proceedings of the 11th International Conference Control, Automation, Robotics and Vision* (Singapore, 2010) pp. 384–389. 7–10 Dec 2010
15. The Table 1 and Table 2 data has been taken from Adaptive Neuro-Fuzzy Technique for Autonomous Ground Vehicle Navigation by Auday Al-Mayyahi, William Wang and Phil Birch. This stated article is an open access article distributed under the terms and conditions of the Creative Commons Attribution License (https://creativecommons.org/licenses/by/4.0/)

An Investigation into the Impact of System Transparency on Work Flows of Fuzzy Tree Based AIs

Nick Ernest, Brandon Kunkel, and Timothy Arnett

Abstract Genetic Fuzzy Trees (GFTs) have demonstrated high performance in a variety of AI-style applications ranging from air to air combat to bipolar disorder drug treatment effectiveness predictions. Throughout most of these, the most frequent alternative approaches that have been employed are considered "black box" methods; that is, the internal operations cannot be fully understood by a human, if at all. Several of the primary benefits leading to the certifiability of the GFT are largely due to the transparency, or "white box" nature of these systems. However, the increased safety and mission critical assurances of the GFT are not the focus of this study. Instead, this work seeks to demonstrate the benefit of transparency in terms of work-flow and man-hours, rather than the verifiability and certifiability of the methodology. This is inevitably difficult to quantify empirically but is especially noticed in problems with specific behavior requirements that are difficult to fully encapsulate within an objective function.

1 Introduction

Within the domain of mission-critical and safety-critical control systems, Artificial Intelligence (AI) applicability is highly reliant upon the trust and certifiability capabilities of the underlying methodology. Many of the top performing AI methodologies today can be considered as "black boxes" to some degree; that is, the inner workings of the control action are not human understandable. While many parties are researching methods to mitigate this for some of the most popular machine learning approaches, it is a major weakness for these types of AI applications.

N. Ernest (✉) · B. Kunkel · T. Arnett
Thales USA, 312 Walnut St, 45202 Cincinnati, OH, USA
e-mail: nick.ernest@psibernetix.com

B. Kunkel
e-mail: brandon.kunkel@psibernetix.com

T. Arnett
e-mail: tim.arnett@psibernetix.com

B. Bede et al. (eds.), *Fuzzy Information Processing 2020*,
Advances in Intelligent Systems and Computing 1337,
https://doi.org/10.1007/978-3-030-81561-5_29

Derived from fuzzy logic controllers oft touted for their transparency, the Genetic Fuzzy Tree (GFT) methodology is consequently entirely transparent [1]. This, among other benefits of fuzzy logic, has significant safety and certifiability implications which are not the focus of this study [2]. Instead, a benefit of this technology that is often overlooked will be investigated; that this transparency allows for the simpler development of the system to adhere to design requests that are difficult to encapsulate in an objective function or training data. While the impact of this ease of development clearly depends quite heavily on the specifics of the application, the difference in man-hours and CPU-hours can be quite significant. These work-flow issues, as well as specific difficulties which the transparency of these type of high-performance AI systems can avoid, are discussed and an experiment with a sample application area is utilized to provide a sample quantification of this difference.

The remainder of the paper is organized as follows. Section 2 describes the problem statement in detail that was addressed by the experiment. Section 3 describes the set up of the simulation environment. Section 4 details the proposed experimentation that will measure the impact of transparency within this example problem. This is followed by a thorough description of the approach for both variants of the AI system in Sect. 5. The results of the experimentation will be discussed in Sect. 6, with a conclusion in Sect. 7.

2 Problem Statement

Suppose a client is funding development of a centralized multi-agent control system whereby this system's goal is to collect a maximal amount of resources within the environment. This client provides a set of design requirements outlining how this system should control the autonomous agents at the beginning of the program. The engineers opt to use machine learning to build and optimize the behavior of this system to meet the needs of the client.

After weeks of design, model training, and testing, the engineers produce a system that meets the client's original design requirements. However, the client acknowledges that they have more requirements or a different system behavior than what was delivered that were never explicitly discussed in the beginning. While the engineering team produced a system with various search path geometries shown in Fig. 1a, the client wants the system to behave in a different manner, with concentrically shaped search paths similar to Fig. 1b. The client states that while the algorithm should generally utilize a concentric search path, slight deviations in order to maximize performance will be allowed.

The nature of software design is inherently iterative, and the algorithmic design approach matters significantly as projects grow and change. This experiment investigated the impacts of how selecting different algorithm architectures may impact the work load costs and performance to develop and modify a centralized multi-agent controller for a system of foraging bees. This bee foraging behavior is the topic of many studies by biologists and computer scientists [3] to understand how organisms can learn and adapt behavior to their environment.

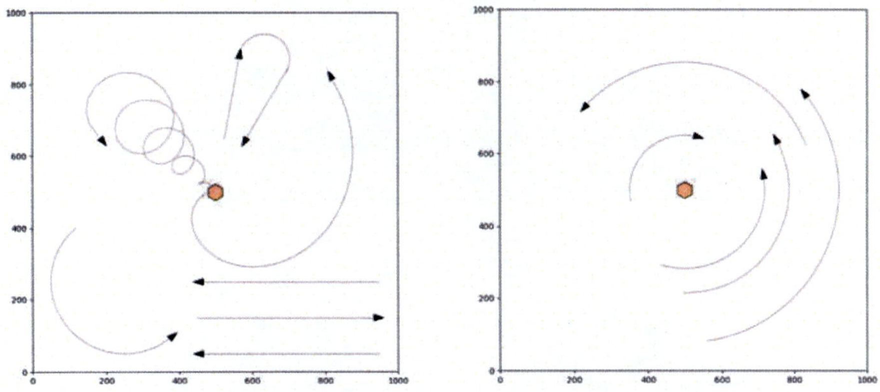

(a) Various search path geometries.

(b) Sample concentric flight paths.

Fig. 1 Comparison of search path shapes

With a primary strength of the GFT methodology being complete system transparency, both mathematically and linguistically, this example experiment serves the purpose of demonstrating this effect on engineering team work flows. While this example is quite simple and notional, many GFTs developed for complex use-cases contain elements that are quite similar. One such potential example is a GFT for air-to-air combat purposes [4], wherein the client wants the system to behave more humanistic, or perform in a particular manner, with insufficient data traces to automate ensuring the system adheres to such a requirement. GFTs have been utilized in a wide variety of markets thus far, from defense applications such as the above, to biomedical [5].

3 Experimental Setup

For the purposes of this experiment, a custom low fidelity simulation environment was developed to model bee foraging behavior. The environment contains a 2-dimensional area that represents a 1 km by 1 km field from which a swarm of robotic bees must search for and collect nectar from flowers. Bees take off from a singular depot, or hive, which is located within the center of this region, and the bees have a limited carrying capacity before needing to deposit their findings into the hive. Figure 2 shows a still image that represents the state of the simulation environment at one point in its evaluation.

The field may be filled with a plurality flowers, each acting as a point source for "smell" which propagates throughout the field in an inverse relationship to distance. Each flower object's smell magnitude relates to its nectar value, each having a maximum nectar normalized to 100%. The flowers' nectar will regenerate over time at a set rate until the maximum value is reached.

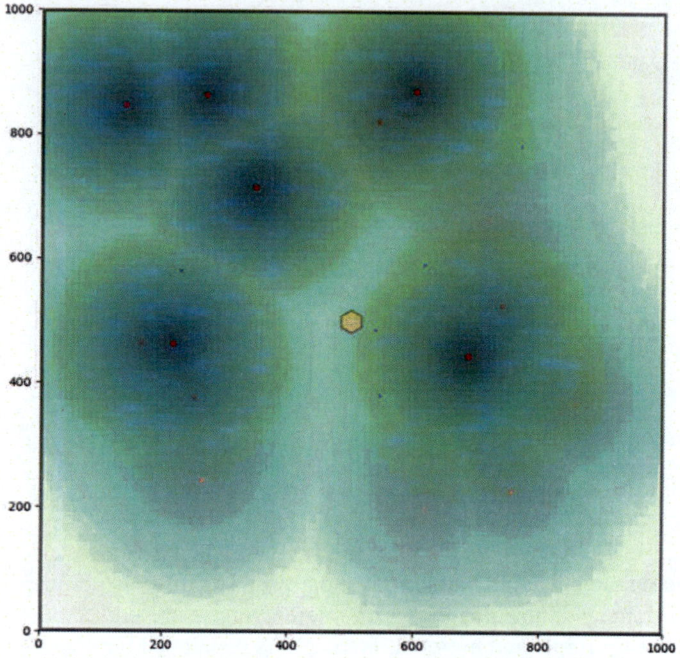

Fig. 2 Still image of simulation environment showing hive and bees located within a field of flowers

Each bee records its recent state history up to the last 4 time steps, as well as the position information of all other bees in the swam. Additionally, each bee has one external sensory input which is the intensity of smell at their current position. The total sensory input to a bee cannot exceed 100%, even if multiple flowers are nearby, though their distributions overlap. Each bee has a maximum turn rate dependent upon speed, and once landed on a flower or the hive, must harvest or deposit the nectar over a duration of time.

By interpreting these state and sensory histories and states relative to the other bees, the controller algorithms are able to control the flower's heading and speed. Various metrics were used to evaluate the performance of the developed controllers over various environment starting states, also referred to as "scenarios".

4 Proposed Solution

In this experiment, a generic GFT solution was assumed given, and then two GFT derivative systems were developed, one modeling after transparent "white box" algorithmic approaches, through which all of the data flow is clear to the developers and another modeling after the opaque "black box" algorithmic approach whereby.

For the sake of consistency in the experiment, both “white box” and “black box” approaches used the same genetic algorithm for optimization and the same fuzzy logic controller type. While the “black box” system is indeed utilizing transparent fuzzy logic, the developers will be enforced to leave variables within their GFT as generic, and solely utilize optimization of pre-processing and the objective function in order to obtain the desired behavior.

Additionally, all bees share the same instantiation of this algorithm and their behaviors will differ based on their own state relative to the rest of the system. Therefore, while the GFT approach is indeed a “white box”, no differences in results will be caused by algorithmic reasons, and instead will purely be a function of work flow and design by the human developers. This investigation, similar to other analyses between “black box” and “white box” approaches, will demonstrate fuzzy logic’s, and more specifically, the GFT’s strengths within these areas [6–8].

5 Work Flow Analysis

Both design approaches were used to develop algorithms which can control and modify the bees’ behavior to reach the design requirements that the hypothetical client desires, as outlined in the following:

1. All bees must attempt to adhere to concentric flight paths.
2. Bees may deviate slightly from their concentric flight paths if they are near a flower.
3. Once a bee is at max nectar carrying capacity, it can fly directly back to the hive, ignoring any concentric requirements, until it has fully unloaded its nectar to the hive.
4. While adhering to other requirements, maximize the amount of nectar collected such that the average nectar collection for scenarios is within 80% of the original multi-path GFT system.

In the following subsections, each design approach is described in detail. These design approaches were closely monitored and their time costs were tracked on the basis of computer and developer time.

5.1 White Box Process

The “white box” approach utilizes the base GFT system that was developed and exploits the benefit of transparency within the system. As such, the design requirements can be directly encoded within the system, included within the objective function, or some combination of aspects of both. The “white box” algorithm design process is detailed below:

1. Develop a "white box" concentric flight path algorithm that assigns the bees to radial distance channels. The developer manually selects some parameters and leaves others to be optimized via objective function.
2. Integrate GFTs to modify the behavior intended in the concentric flight path algorithm by updating the heading and landing behaviors.
3. Train the model over a specified objective function.
4. Analyze training results.
5. Modify any aspect of the transparent algorithm.
6. Repeat Steps 3–5 until system is observed to perform desired behaviors.

The primary strength of this approach is that the objective function and pre-processing functions are not the sole targets of development and optimization. By being able to directly work with the design requirements via the transparency of the system, it is expected that overall effort levels would be reduced.

5.2 *Black Box Process*

The "black box" approach is meant to mimic that of popular modern machine learning approaches which use a learning agent to tune a universal approximator to map an input set to an output set based on an objective function. By careful selection and tuning of optimization algorithms, agent algorithm structure/architecture, data pre-processing, and an objective function, machine learning techniques are capable of solving most problems.

In a similar structure to the "white box" approach, the design process for the "black box" algorithm is detailed below.

1. Develop a "black box" concentric flight path algorithm that pre-processes all parameters that may be used by the GFT.
2. Integrate GFTs to directly control heading and landing behaviors.
3. Train the model over a specified objective function.
4. Analyze training results.
5. Modify objective function via selection of parameters and their weights.
6. Modify solely the "black box" visible portions of the algorithm.
7. Repeat Steps 3–6 until system is observed to perform desired behaviors.

The key aspect of the "black box" approach is that the internal workings on the solution are not able to be directly modified upon analysis of training results, and instead, solely the objective function is altered until the desired behavior is found.

6 Results Comparison

To compare the two design approaches in terms of their adherence once completed, visual inspections of the search paths, as well as an analysis of the average aspect angles of the bees to the hive was performed. Additionally, while not required to surpass the initial algorithm in terms of nectar collection, both algorithms were expected

Table 1 Validation scenarios Bee and flower count

Scenario	Flowers	Bees
1	6	2
2	8	2
3	10	3
4	12	3
5	12	4
6	15	4
7	15	5
8	20	5
9	20	5
10	20	6

Table 2 Concentric search aspect error (degrees) of search path

Scenario	Original multi-path search	White box concentric	Black box concentric
1	27.3	2.7	8.9
2	25.4	2.9	8.1
3	20.1	5.9	4.6
4	20.2	5.8	4.7
5	15.4	4.2	2.5
6	14.4	4.7	2.4
7	11.4	4.5	1.8
8	12.1	4.9	2.1
9	9.2	3.7	1.7
10	9.1	3.8	1.6
Average	16.46	4.31	3.84

to be near-peers in terms of overall mission performance. The two algorithms were completed to these standards, and then evaluated over a set of scenarios which were different from the training scenarios. The scenarios were generated randomly with a seeded random number generator to ensure consistency between evaluations.

Table 1 lists the number of bees and flowers over 10 validation scenarios that were utilize to examine the output of each approach.

Table 2 shows the average per-bee deviation from 90° aspect for each algorithm per each scenario. Essentially, if each bee is performing a concentric orbit of the hive, the absolute value of the aspect should be roughly 90°, unless the bee has collected nectar and is flying towards to the hive. As such, if the bee is returning to the hive, its aspect error was not included. This serves as an ideal metric beyond the eye test for determining the search path meets the client's requirements.

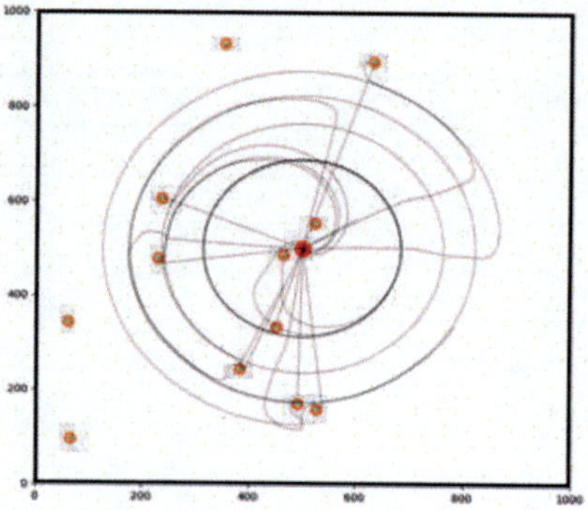

(a) Resultant path for black box approach on Scenario 5.

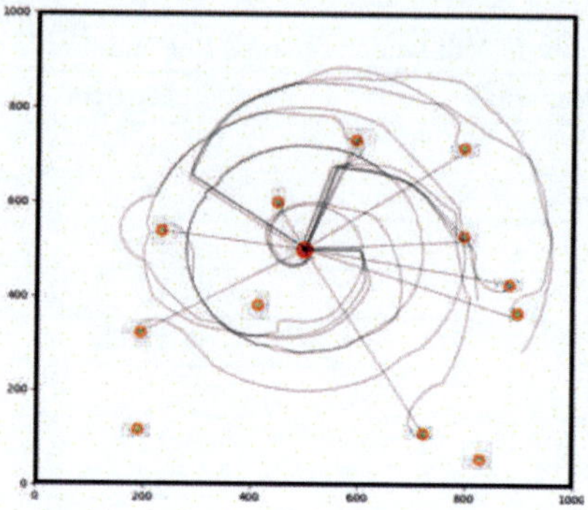

(b) Resultant path for white box approach on Scenario 5.

Fig. 3 Comparison of search path routes for trained algorithms

Additionally, two sample figures are provided which show a comparison of route path history flown by bees controlled by both algorithms. Both algorithms fly concentrically, though as can be seen, the "black box" algorithm had to be heavily enforced via objective function to meet such path. The "white box" algorithm, which had an unaltered objective function, deviated some from the concentric path in order to maximize overall mission performance through nectar collection. However, graphical analysis of their search paths, combined with aspect error was determined to exceed the minimum benchmarks for adherence to the theoretical client's requirements (Fig. 3).

Beyond adherence to the concentric search requirement, there was additionally a requirement to be within an acceptable range of performance (nectar collection) as compared to the initial draft algorithm. Table 3 displays the nectar collected by each algorithm over each validation scenario. Both the "white box" and "black box" are considered to have met the minimum requirements, however the "white box" algorithm was not only the superior of the two new approaches, but surpassed the original multi-search draft algorithm on average. This difference is very likely attributable to the fact that the "white box" approach's objective function was primarily interested in performance, with adherence to the design requirements loosely enforced via the direct development of the fuzzy logic controllers. In comparison, the "black box" approach had to heavily weight adherence to design requirements within the objective function.

The time spent on different phases in the design process were recorded for each design approach. Table 4 shows the recorded engineering and computer time used to iterate through the algorithm design. The "white box" design approach used significantly less time in both the development, training and design iteration in comparison to the "black box" approach. The "white box" design approach was much faster in terms of design iterations to achieve the desired behavior. In fact, the "white box" algorithm met the desired performance characteristics in just two training cycles,

Table 3 Concentric search aspect error (degrees) of search path

Scenario	Original multi-path search	White box concentric	Black box concentric
1	4.0	0.0	5.0
2	4.0	0.0	5.0
3	5.0	5.0	6.0
4	5.0	5.0	8.0
5	8.0	7.0	10.0
6	12.0	7.0	9.0
7	13.0	12.0	11.0
8	13.0	13.0	18.0
9	15.0	14.0	18.0
10	16.0	8.0	14.0
Total	95.0	71.0	104.0
Average	9.5	7.1	10.4

Table 4 Time duration (in hours) used to perform tasks in the white box and black box design approaches

Cycle	Task	White box approach	Black box approach
0	Initial development	3.25	1.50
1	Training	3.00	13.00
	Training analysis	0.25	0.50
	Design iteration	3.50	6.00
2	Training	4.00	13.00
	Training analysis	0.25	0.25
	Design iteration		1.50
3	Training		10.00
	Training analysis		0.25
	Design iteration		0.50
Total developer time (hrs)		7.25	10.50
Total computer time (hrs)		7.00	36.00
Total time (hrs)		14.25	46.50

while the "black box" algorithm took three training cycles to achieve the desired behaviors.

As expected, the "white box" approach had a 217% higher start up cost in terms of developer time in the initial instantiation. To exploit the transparency benefit of fuzzy logic does involve manual labor beyond that of "black box" approaches typically. However, the "black box" approach had a 145% increase in overall developer time cost. In terms of computational cost, the "black box" approach was also 514% higher.

7 Conclusion

This experiment investigated the cost of selecting different architectures; namely a traditional "black box" approach for machine learning compared to the "white box" GFT approach. While this experiment was performed at a small scale it highlights important factors to consider when selecting and implementing machine learning algorithms. This experiment showcases that "white box" algorithms compare very favorably to "black box" machine learning algorithm development when being judged on merits engineering and computing costs to build each system for applications that are difficult to objectively quantify within an objective function.

The measurements found within this small-scale experiment would certainly be compounded in larger-scale and more complex applications. The objective function for this experiment was heavily optimized in C++, and quite simple. More computationally costly objective functions would have an immense negative impact on the "black box" approach. Additionally, this would be further compounded by a plurality of design requirements. The adherence to specific and difficult requirements being enforced within the objective function also places a lesser weight on overall system performance, as can be seen in terms of nectar collection within this example application. This implies that for certain applications, not only do "black box" approaches require more developer and computational time, but will be impacted more negatively in terms of overall performance by specific requirements that are difficult to encapsulate within an objective function.

Relying on iterative objective function and algorithm modifications can be unpredictable and add significant costs to training times and costs when systems need to be updated over their functional life span. This cost scales with system complexity and time to evaluate the objective function and can be particularly expensive for reinforcement learning. A vast majority of industry uses can benefit from the maintainability and transparency of using white box solutions vs black box solutions.

As such, the benefits of this fuzzy logic based approach such as formal verification and complete transparency do not merely lend themselves towards certifiability, but also have a significant impact on real-world application development cycles.

References

1. N. Ernest, D. Carroll, C. Schumacher, M. Clark, K. Cohen, G. Lee, Genetic fuzzy based artificial intelligence for unmanned combat aerial vehicle control in simulated air combat missions. J. Def. Manag. **6**(1), 0374–2167 (2016)
2. T. Arnett, *Iteratively Increasing Complexity During Optimization for Formally Veriffable Fuzzy Systems*. University of Cincinnati. Ph.D. Dissertation, July 2019
3. Y. Niv, D. Joel, I. Meilijson, E. Ruppin, Evolution of reinforcement learning in uncertain environments: a simple explanation for complex foraging behaviors. Adapt. Behav. **10**(1), 5–24 (2002). https://doi.org/10.1177/1059-712302-010001-01

4. N. Ernest, K. Cohen, E. Kivelevitch, C. Schumacher, D. Casbeer, Genetic fuzzy trees and their application towards autonomous training and control of a squadron of unmanned combat aerial vehicles. Unmanned Syst. **3**, 185–204 (2013)
5. D. Fleck, N. Ernest, C.M. Adler, K. Cohen et al., Prediction of lithium response in first-episode mania using the LITHium Intelligent Agent (LITHIA): Pilot data and proof-of-concept. Bipolar Disord. **19**(4), 259–272 (2017)
6. B. Delibašić, M. Vukićević, M. Jovanović, M. Suknović, White-Box or Black-Box decision tree algorithms: which to use in education?. in *IEEE Transactions on Education*, vol. 56, no. 3, pp. 287–291, Aug 2013
7. Howar, F., Jonsson, B., Vaandrager, F.: Combining black-box and white-box techniques for learning register automata, in *Computing and Software Science*, ed. by Steffen B., Woeginger G. Lecture Notes in Computer Science, vol. 10000 (Springer, Cham, 2019)
8. A. Sablayrolles et al., White-box versus black-box: Bayes optimal strategies for membership inference (2019). arXiv:1908.11229

Formal Verification of a Genetic Fuzzy System for Unmanned Aerial Vehicle Navigation and Target Capture in a Safety Corridor

Timothy Arnett, Nicholas Ernest, Brandon Kunkel, and Hugo Boronat

1 Introduction

There are a number of methods for advanced control of Unmanned Aerial Vehicles (UAV) in the literature. These range from linear control for small-scale problems to advanced Artificial Intelligence (AI) methods for large-scale, complex, coordinated control [1–3]. For simpler methods, analysis and development methods exist that can guarantee certain properties of the system *a priori*. When utilizing more advanced control and AI methods such as Neural Networks (NN) and Genetic Fuzzy Trees (GFT) [4, 5], traditional analytical methods for guaranteeing aspects of the system's behavior are not applicable. Typically other methods of verification are needed such as numerical evaluations through Monte Carlo methods and simulation. Recently, Formal Methods have been utilized to verify aspects of behavior for more complex systems [6–9] in order to greatly increase the confidence in their correctness. Being able to increase confidence in these systems is of the utmost importance in order to show a path towards implementation in widespread, and particularly safety-critical, applications.

T. Arnett (✉) · N. Ernest · B. Kunkel
Thales USA, 312 Walnut St, 45202 Cincinnati, OH, USA
e-mail: tim.arnett@psibernetix.com

N. Ernest
e-mail: nick.ernest@psibernetix.com

B. Kunkel
e-mail: brandon.kunkel@psibernetix.com

H. Boronat
Thales AVS, Toulouse, France
e-mail: hugo.boronat@fr.thalesgroup.com

B. Bede et al. (eds.), *Fuzzy Information Processing 2020*,
Advances in Intelligent Systems and Computing 1337,
https://doi.org/10.1007/978-3-030-81561-5_30

In this work, a safety corridor is constructed that can be viewed as vehicle trajectory boundaries. It is desired that the vehicle stay within this corridor at all times while also maneuvering to capture multiple points of interest (POI) throughout the environment. Due to the constraints imposed on the vehicle, safety specifications are created that can be evaluated with formal verification methods. The results of the evaluations show that counterexamples can be found for all but one of the specifications. However, having this capability is important as it allows definitive evidence that a specification will hold, or identify cases where it does not.

The remainder of the paper is organized as follows. Section 2 describes the problem in detail along with background information. Section 3 details the proposed solution along with simulation and training results. This is followed by a description of the verification procedures in Sect. 4. The results of the verification methods are presented and compared in Sect. 5. Finally, conclusions and possible extensions to future work is discussed in Sect. 6.

2 Background and Problem Statement

The problem driving this work is navigation of a UAV through a "safety corridor" while also capturing points of interest (POI). The UAV has minimal sensing of the boundaries of the corridor, consisting of five distance sensors arranged around the front and sides of the vehicle. Although not the focus of this paper, more detail on the environment and models can be provided upon request to the authors. A visualization of the environment with the vehicle, POIs, corridor, and distance sensors are shown in Fig. 1.

As it is not known *a priori* what behavior is necessary in order to achieve the goal, a type of reinforcement learning will need to be utilized. GFTs have shown their ability to be implemented in these situations with little or no knowledge of optimal behavior. However, an issue with this is that the resulting control system needs to be verified against behavioral specifications to show that it will operate safely. There are many methods for analyzing the adherence of a system to its specifications, including numerical and more rigorous, formal techniques.

A common method for numerical verification is the Monte Carlo technique. It can be extremely useful when there are complex probability distributions involved and a distribution of behavior is desired. However, it relies on a large number of evaluations of the system in an attempt to cover the space. There are several issues with doing this. The first is the curse of dimensionality. Monte Carlo and other numerical sampling techniques suffer from the increasing hyperspace with the number of dimensions involved. This grows rapidly and quickly becomes untenable for complex systems where extremely high levels of confidence are desired. The second issue is related to this in that it is unclear how much sampling is enough before you can say that the system is verified. It may lead to higher confidence that the system will perform as intended, but will not guarantee correctness. That leads to the final issue. Numerical evaluations cannot definitively guarantee that the system is always correct. They can

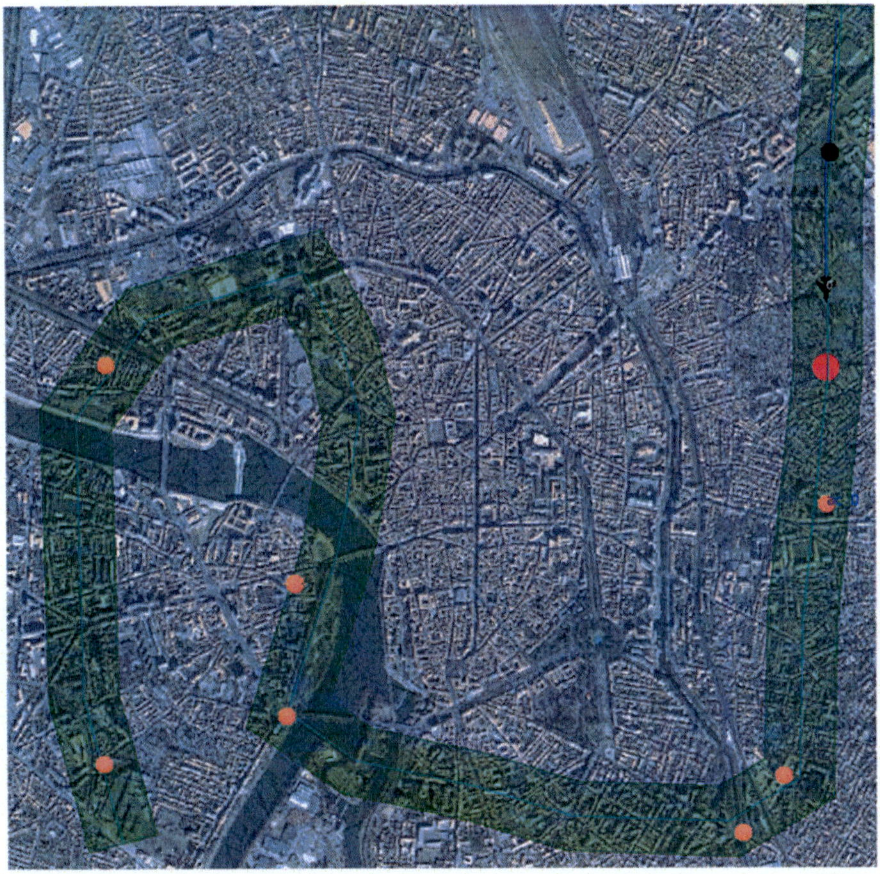

Fig. 1 Example scenario with safety corridor, points of interest, and vehicle

only identify cases where the specification fails. In order to show higher levels of correctness, Formal Methods can be utilized.

Formal Methods are mathematically rigorous methods and techniques for the design, specification, and verification of systems. There are a multitude of areas that fall under Formal Methods but for the purposes of this paper the focus will be on Satisfiability Modulo Theories (SMT) problems. These are problems that deal with the satisfiability of First-Order logic (FOL) formulas that describe the system's behavior. The specifications that are desired to hold over all system states, or are *valid*, are *safety* specifications. Due to this, they can be formulated as a quantified FOL formula, or a temporal logic formula with predicates. If appropriate specifications about the system's behavior can be formulated along with an encoding of the model with a sufficient level of fidelity, much higher levels of confidence in the correctness of the system can be achieved than purely numerical techniques.

It has already been shown that certain types of Fuzzy Inference Systems (FIS) [7, 10–12] can be evaluated using Formal Methods tools. In this work, the same constraints are applied as described in the referred works and include: triangular input membership functions, Ruspini partitioning, input normalization and bounding, singleton output membership functions, product T-norm for conjunction and implication, and weighted average defuzzification.

3 Proposed Solution

In order to develop a GFT for the problem, a structure must be defined in order to take in the known information and output the necessary information. For this problem, the state information available to the controller is limited to the values from the five distance sensors, or "whiskers", the target distance and angle, the distance to the center of the corridor, and the vehicle's speed. The five whiskers are distributed uniformly across the front half of the vehicle and are labeled as d_{-90}, d_{-45}, d_0, d_{45}, d_{90} from left to right, respectively. The subscript labels refer to the angle in degrees from the centerline of the vehicle. The distance to the center of the corridor is d_c. The target distance and angle are labeled as d_{target} and ψ_{target}, respectively. Vehicle speed is simply v. It should be noted that all of these values are normalized to the range of $[-1, 1]$ before being passed to the FISs.

These values are fed as inputs to a series of FISs in order to generate two outputs, vehicle bank angle and acceleration. These are then used to calculate the new vehicle states based on 2D level flight dynamics. In order to account for dependencies in the state variables, they are grouped together in order to produce intermediate outputs that convey the information in reduced dimensionality. Additionally, any symmetry that can be defined beforehand can be used in order to reduce the number of parameters that the system must learn.

The first information used in the structure is the whisker values. These are arranged such that a single FIS takes in the 90, 45, and 0° whiskers. This allows the same 3-input, 1-output FIS to be used for both the left and ride sides. The outputs of these are then combined with the vehicle speed in order to generate an avoid score. This is shown in Fig. 2. The avoid score is used to compare to a "capture" score which decides which mode the vehicle is currently in. While in the avoid mode, the vehicle is avoiding obstacles. While in capture mode, the vehicle is attempting to reach POI.

In order to calculate the capture score, the target information is needed. The target distance and angle are fed into a FIS for calculating the capture score. This is shown in Fig. 3. The capture and avoid scores are then compared and the maximum value determines the vehicle mode.

Avoid

While the vehicle is in avoid mode, a number of other FISs are activated. The structure of these is shown in Figs. 4 and 5. As different structures are needed for commanded acceleration and bank angle, two separate subtrees are used.

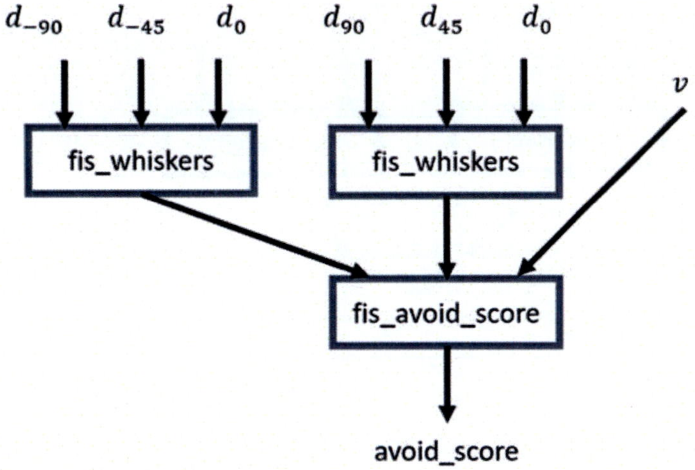

Fig. 2 Structure for calculating wall avoidance score

Fig. 3 Structure for calculating POI capture score

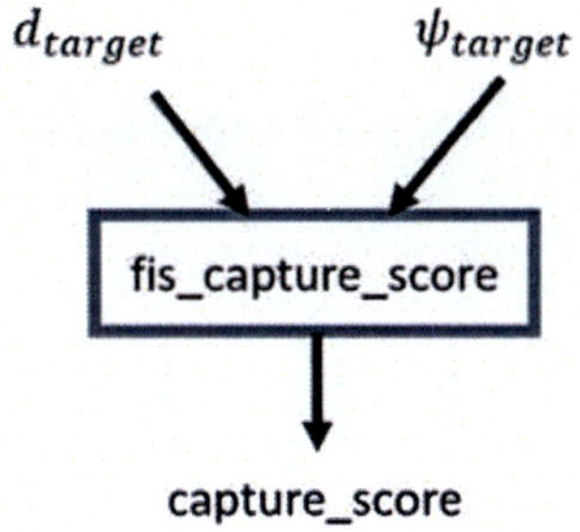

Capture
When the capture mode is activated, the vehicle attempts to reach the POI. The structure used for calculating bank and acceleration ratios are shown in Fig. 6. Note that $sgn()$ refers to the sign function.

Training
After creating the structure, the system was trained through multiple training and testing scenarios similar to the one shown in Fig. 1. The system was trained with a type of reinforcement learning with a Genetic Algorithm. An objective function was created to reward the vehicle for capturing POI and avoiding crash conditions. As the focus of this work is only on the verification aspect, the details of this process are omitted.

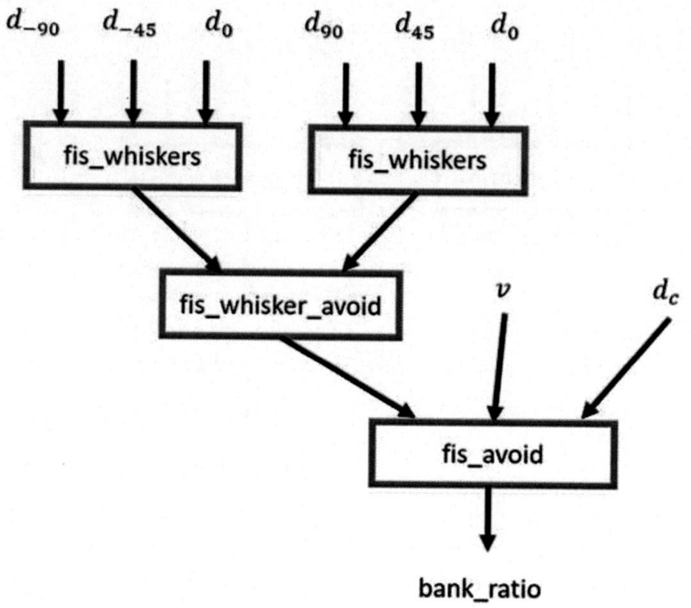

Fig. 4 Structure for calculating avoidance bank angle

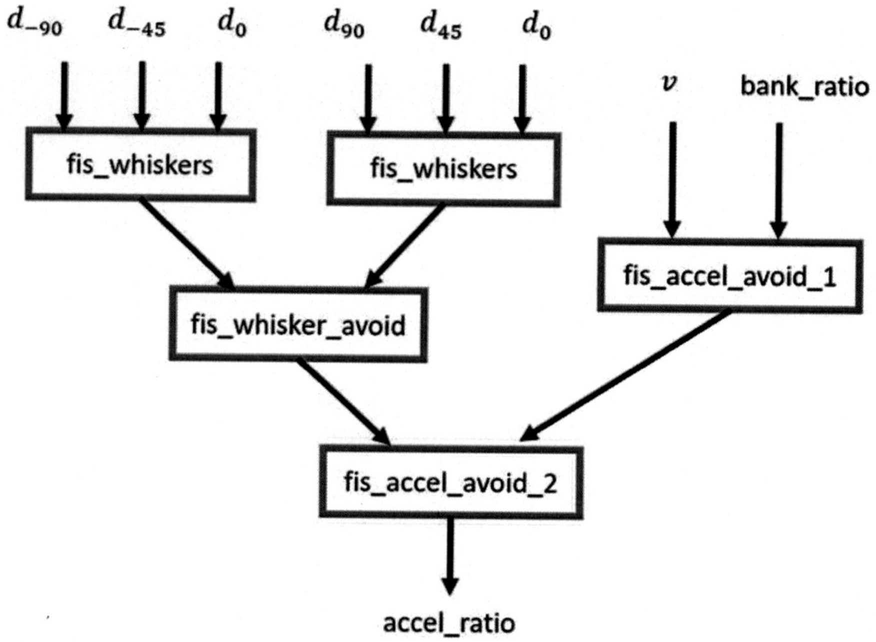

Fig. 5 Structure for calculating avoidance acceleration

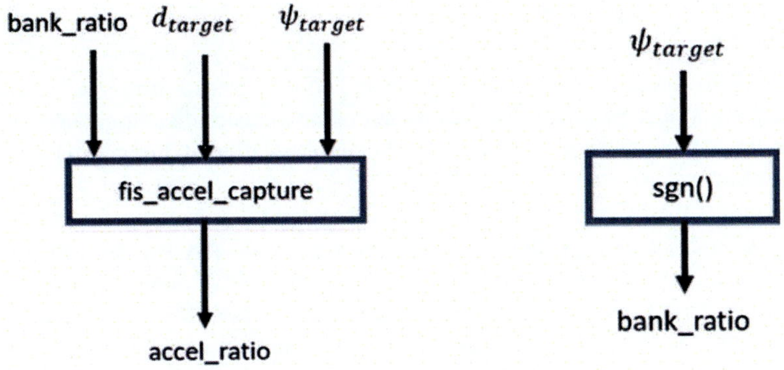

Fig. 6 Structure for calculating capture acceleration (left) and bank (right) ratios

4 Verification

After developing the system, it is necessary to ensure that it adheres to specifications about its behavior. Verification is a way to do this by comparing what a system *should do* to what it *actually does*. This requires formal specifications describing the desired behavior along with decision procedures that can show whether they are, or can be, violated or not. For this system, specifications that refer to the output of the controller can be formulated as *constraint satisfaction problems* and evaluated with a number of Formal Methods tools. Specifically, the specifications can be formulated as either First-Order or temporal logic formulas with predicates. For this particular problem, the specifications created were such that the designer expects them to lead to unsafe conditions if they are reached. For example these unsafe conditions could be a vehicle collision, leaving the flight envelope, violating a regulatory requirement, etc. The specifications are as follows:

Spec 1: If the vehicle is at minimum or maximum speed, do not give an acceleration command that will force it to leave the flight envelope.

Spec 2: If the front whisker distance is less than the turn radius at the current vehicle state +10%, then the chosen action should be to avoid.

Spec 3: If a 90° whisker reads less than 50m, the bank angle command should be such that the vehicle turns away from the wall.

Spec 4: If the vehicle is flying straight and near the middle of the corridor and the corridor is somewhat straight and the vehicle is not in capture mode, the commanded bank ratio should be small.

4.1 Formal Verification

In contrast to numerical techniques such as Monte Carlo, Formal Verification is rigorous in that the specifications are created using mathematical logic and checked against a model of the system. In this case, the representation of the system is such that the operations occurring within the FIS are represented within the tool. The tool used in this work is the infinite-state model checker JKind [13]. It allows for checking of FOL specifications along with temporal properties. To do this, the English language specifications are translated to formal specifications using Linear Temporal Logic (LTL) along with FOL predicates. Due to the fact that the FISs being used have an explicit output in the form of an elementary symmetric polynomial [12], they can be evaluated with SMT using theories for nonlinear arithmetic. The specifications after being converted to this form are shown below. Note that the first specification was split into two cases. Specification three was not because of the symmetry of the whisker FIS as mentioned previously.

Spec 1a: $\square((v = 1.0) \rightarrow (accel\ ratio \leq 0.0))$
Spec 1b: $\square((v = -1.0) \rightarrow (accel\ ratio \geq 0.0))$
Spec 2: $\square((front\ dist < 1.1 * turn\ radius) \rightarrow (avoid\ score \geq capture\ score))$
Spec 3: $\square((d_{-90} < -0.667) \rightarrow (bank\ ratio \geq 0.0))$
Spec 4: $\square\big(((|d_{-90} - d_{90}| \leq 0.1) \wedge (|d_c| \leq 0.1) \wedge (avoid\ score \geq capture\ score)) \rightarrow (|bank\ ratio| \leq 0.1)\big)$

The specifications, GFT, and input variable relationships were then translated and encoded into the Lustre synchronous language for evaluation in JKind utilizing the Z3 SMT solver [14].

5 Results

5.1 Specification 1

The first specification (both a and b) were checked against the avoidance acceleration structure only due to the fact that the capture structure does not take vehicle speed into account. After evaluation however, counterexamples were found showing that the specifications failed. The conditions that generated these failures are shown in Table 1. Note that whiskers left and right are the outputs from "fis whiskers" for the left and right whiskers, respectively.

Although technically this is a failure of the system (as currently constructed) to adhere to safety specifications, a simple fix is to introduce a saturation filter after the acceleration output that simply limits it to be 0 if the commanded output would make the vehicle leave the flight envelope. Additionally, more training may be beneficial as the vehicle was not trained over situations where it started at speeds much closer to the envelope. The result is that the GFT may have effectively learned how to keep the vehicle from reaching these portions of the state space.

Table 1 Specification 1a and 1b counterexample values

Variable	1a value	1b value
Bank ratio	0.402	0.330
v	1	−1
Whiskers left	0	0
Whiskers right	0.5	0
Accel ratio	0.095	−0.408

5.2 *Specification 2*

Checking the second specification also lead to a generated counterexample. It is shown in Table 2.

Although this specification fails also, this one is essentially reflecting what the GFT learned. It seems to be that in this situation, the target location is such that it has learned to prioritize capturing the target over performing an aggressive maneuver to avoid crashing. It is still also possible that it may not crash. Additionally, the specification may itself be refined in order to rule out situations like this where it is not absolutely clear if the system should be performing avoid or capture.

Table 2 Specification 2 counterexample values

Variable	Value
Bank ratio	0.238
v	0
d_{target}	0
ψ_{target}	0
d_0	0
d_{-90}	0
d_{-45}	0.980
d_{90}	−0.875
d_{45}	0.967
front dist	150
turn radius	1408.93
avoid score	−0.393
capture score	0.713

Table 3 Specification 3 counterexample values

Variable	Value
Left wall distance	1
v	0.5
d_0	1
d_{-90}	−0.993
d_{-45}	−0.990
d_{90}	1
d_{45}	1
d_c	−0.993
Whisker avoid	−0.900
Bank ratio	−0.232

5.3 Specification 3

The third specification also produced a counterexample. This one however is more complicated than it appears at first glance. This is due to the size of the corridor, the sensing range of the whiskers, and the relationships between the input variables to the whisker FIS. In order to best represent the system, the wall and how it is affecting the whisker values together would need to be modeled. In this case, some relationships were implemented in the form of constraints, while others were set such that it assumes the vehicle is traveling parallel to the wall. The counterexample is shown in Table 3.

Again an example of a failed specification that may not actually lead to crash conditions. However, in this case it very well might due to the speed that the vehicle is traveling (0.5 normalized speed corresponds to approximately 190 kph). If it were to bank left at that speed and only 1m from the wall, it is highly likely to trigger a crash condition. More training could potentially alleviate this, along with previously mentioned saturation filters that explicitly prevent it from turning left in these situations.

5.4 Specification 4

The final specification is similar to the third specification in that it also uses the relationships between the whisker values and the distance to the corridor centerline. This specification is essentially dealing with stability of the bank commands when in avoid mode. After evaluation of this specification, the result was that this specification is *valid*. That is, it always holds for the inputs as shown in the FOL representation. This is an important capability compared to numerical methods such as Monte Carlo.

Those techniques cannot guarantee a property will always hold, but only for the points sampled. This gives much more confidence that undesired behaviors will not occur after GFT training.

6 Conclusion

In conclusion, a method for applying formal verification techniques for a trained GFT system was demonstrated. Although most of the specifications failed, there are some valuable takeaways. First is that counterexamples were generated that identify a point in the space where the specification fails. This can definitively show that either the system lacks crucial safety properties or that the specification itself does not fully encapsulate the possible behaviors that occur in unsafe conditions. This also aids the designer in adjusting behaviors in order to meet safety specifications. Secondly, this demonstrates the capability to implement a number of FISs simultaneously in tools for definitive, formal verification that they adhere to specifications from a designer. Lastly, the final specification being shown to be valid highlights the advantage of this method over numerical techniques that cannot make such guarantees. The combination of these traits represents an important capability in order to increase confidence that AI systems such as GFTs are correct and will behave as intended.

Future Work

There are several rich opportunities for future work. Perhaps the highest priority item is to include idealized dynamics into the formal verification process. This can help address two concerns: insufficient specification coverage, and input relationships between dependent variables. Some other potential avenues include expanding the list of tools and methodologies for formal verification for GFTs. Additionally, several opportunities exist for inclusion of Formal Verification and correction measures within the optimization loop itself.

References

1. N. Ernest, K. Cohen, *Fuzzy Logic Based Intelligent Agents for Unmanned Combat Aerial Vehicle Control* (2015). https://doi.org/10.4172/2167-0374.1000139
2. A. Sathyan, N. Ernest, K. Cohen, An efficient genetic fuzzy approach to UAV Swarm routing. Unmanned Syst. **04**(02), 117–127 (2016). https://doi.org/10.1142/S2301385016500011
3. N. Ernest, D. Carroll, C. Schumacher, M. Clark, K. Cohen, G. Lee, Genetic fuzzy based artificial intelligence for unmanned combat aerial vehicle control in simulated air combat missions. J. Def. Manag. **6.1**, 2167–0374 (2016). https://doi.org/10.4172/2167-0374.1000144
4. N. Ernest, K. Cohen, E. Kivelevitch, C. Schumacher, D. Casbeer, *Genetic Fuzzy Trees and their application Towards Autonomous Training and Control of a Squadron of Unmanned Combat Aerial Vehicles*, vol. 03, No. 03, pp. 185–204 (2015). https://doi.org/10.1142/S2301385015500120
5. N. Ernest, *Genetic Fuzzy Trees for Intelligent Control of Unmanned Combat Aerial Vehicles*. Ph.D. Dissertation, University of Cincinnati (2015)

6. G. Katz, C. Barret, D. Dill, K. Julian, M. Kochenderfer, Reluplex: an efficient SMT solver for verifying deep neural networks. Comput. Aided Verif. **2017**, 97–117 (2017)
7. T. Arnett, *Verification of Genetic Fuzzy Systems*. MS thesis, University of Cincinnati (2016)
8. T. Arnett, B. Cook, M. Clark, K. Rattan, Fuzzy logic controller stability analysis using a satisfiability modulo theories approach, in *19th AIAA Non-Deterministic Approaches Conference*, p. 1773 (2017)
9. B. Cook, T. Arnett, K. Cohen, A fuzzy logic approach for separation assurance and collision avoidance for unmanned aerial systems. Ch. 12, in *Modern Fuzzy Control Systems and its Applications*, pp. 225–256 (2017)
10. M. Clark, K. Rattan, Piecewise affine hybrid automata representation of a multistage fuzzy PID controller, in *2014 AAAI Spring Symposium Series* (2014)
11. M. Clark, K. Rattan, Hybrid representation of rule-based systems, in *18th International Conference on Hybrid Systems: Computation and Control*, pp. 287–288 (2015)
12. T. Arnett, *Iteratively Increasing Complexity During Optimization for Formally Verifiable Fuzzy Systems*. Ph.D. Dissertation, University of Cincinnati (2019)
13. A. Gacek, J. Backes, M. Whalen, L. Wagner, E. Ghassabani, *The JKind Model Checker* (2017). arXiv:1712.01222
14. L. De Moura, N. Bjorner, Z3: an efficient SMT solver, in *International Conference on Tools and Algorithms for the Construction and Analysis of Systems*, pp. 337–340 (2008)

How to Reconcile Randomness with Physicists' Belief that Every Theory Is Approximate: Informal Knowledge Is Needed

Ricardo Alvarez, Nick Sims, Christian Servin, Martine Ceberio, and Vladik Kreinovich

Abstract In this paper, we show that physicists' intuition about randomness is not fully consistent with their belief that every theory is only approximate. We also prove that there is no formal way to reconcile these two intuitions, this reconciliation has to be informal. Thus, there are fundamental reasons why informal knowledge is needed for describing the real world.

1 Randomness: An Algorithmic Description (A Brief Reminder)

From the computational viewpoint, what sequences of measurement results can we expect: case of a fixed physical theory. The usual analysis of algorithms and computability takes into account what we can compute "from scratch", by using only the given input and performing appropriate computational steps; see, e.g., [4].

In principle, in addition to the given input, we could also use the results of physical experiments. We normally do not use these results when performing unrelated computations, since we do not expect the results of physical experiments to help

R. Alvarez · C. Servin · M. Ceberio · V. Kreinovich (✉)
Department of Computer Science, University of Texas at El Paso, 500 W. University, 79968 El Paso, TX, USA
e-mail: vladik@utep.edu

R. Alvarez
e-mail: ralvarezlo@miners.utep.edu

C. Servin
e-mail: cservin1@epcc.edu

M. Ceberio
e-mail: mceberio@utep.edu

N. Sims
Computer Science and Information Technology Systems Department, El Paso Community College (EPCC), 919 Hunter Dr.m, El Paso, TX 79915-1908, USA
e-mail: nsims2@my.epcc.edu

B. Bede et al. (eds.), *Fuzzy Information Processing 2020*,
Advances in Intelligent Systems and Computing 1337,
https://doi.org/10.1007/978-3-030-81561-5_31

us solve complex mathematical problems. It would be very strange to find, e.g., some radioactive material whose emissions will solve a given NP-hard problem like propositional satisfiability. We may get some help in solving this problem from other people, maybe from signals sent to us by an alien civilization—in short, from someone with intelligence—but we do not expect such help from an (unintelligent) physical process.

This intuitive idea has been formally described in [2] in terms of the important notion of *Kolmogorov complexity* (see, e.g., [3]). Let us briefly recall the corresponding formalization.

Kolmogorov complexity: a brief reminder. To understand this notion, we need to recall that while computers can process many different types of objects—arrays, graphs, images, videos, etc.—in a computer, every object is represented as a sequence of 0s and 1s, i.e., to use a computer science terminology, as a *binary sequence*.

The notion of Kolmogorov complexity was invented to formalize another intuitive idea—that of a random binary sequence. From the purely mathematical viewpoint, if we consequently flip a coin and write down all heads as 1s and all tails as 0s, then all length-n sequences of 0s and 1s have the exact same probability 2^{-n}. From this viewpoint, they are equally probable and thus, seem to have an equal right to be called random. In particular, for a large n, a sequence that we will actually get after flipping the coin n times has the same probability as the sequence 0101...01, in which the sequence 01 is repeated many times. However, intuitively, the seemingly lawless sequence that we will actually get after flipping a coin looks random, while the sequence 0101...01 does not look random—and it would be very surprising if such a sequence indeed appears as a result of flipping a coin. Moreover, if something like this sequence will be observed in a casino, after a while we will be absolutely sure that cheating is taking place, and that the corresponding slot machine is not truly random (as it is supposed to be).

Kolmogorov noticed that there is a difference between sequences which are intuitively random and sequences which are not intuitively random: namely, the reason why we do not believe that the sequence 0101...01 is truly random is that it can be generated by a reasonably short program, in which we print 01 in a loop. On the other hand, if we consider a real sequence of 0s and 1s obtained by flipping a coin, we do not expect to find any regularity there, so the only way to generate this sequence is to print it bit-by-bit.

In other words, a sequence like 0101...01 can be generated by a very short program, while to generate the binary sequence 01... corresponding to the actual flipping results, we cannot use any program which is much shorter than a straightforward bit-by-bit printing program `print`(01...) whose length is practically equal to the length of the binary string. To describe this difference in precise terms, Kolmogorov introduced the notion of Kolmogorov complexity $C(x)$—the shortest length of a program (in some fixed programming language) that generates the binary string x. In these terms, if the Kolmogorov complexity $C(x)$ is much smaller than the length $\text{len}(x)$ of the sequence, then this sequence is not random. On the other hand, if $C(x) \approx \text{len}(x)$—or, to be more precise, if $C(x) \geq \text{len}(x) - c$ for some small c, then we can say that the sequence x is random.

From regular Kolmogorov complexity to prefix Kolmogorov complexity. The above definition of randomness can be made even more intuitive if we take into account that we can flip the coin as many times as we want, so the coin-flipping sequence can be extended to any length, the fact that we stopped at some specific point should not change our opinion on whether this sequence is random or not. From this viewpoint, if it is not so easy to generate the binary sequence that we have so far, but much easier to generate its extension, then the resulting sequence should still be marked as not random. Generating an extension can be easier than the sequence itself: e.g., to generate a sequence 001001...001001 is somewhat easier than its initial segment 001001...00100 when we need to take special care of the last two 0s. To take this into account, we need to replace the original Kolmogorov complexity $C(x)$ with *prefix Kolmogorov complexity* $K(x)$, which is defined as the shortest length of the program that generates either the binary sequence x itself or a binary sequence starting with x; see, e.g., [3].

Conditional Kolmogorov complexity and the notion of information. We are interested in describing how the possibility to use an auxiliary sequence y can potentially affect our computations. In principle, when we use the bits from the given sequence y, we do not need to compute them and thus, it may be possible to have a shorter program for computing x. An analogy is that if we already have a program for computing $\sin(x)$, then computing $\sin^2(x)$ is fast: all you need to do is access the value of the sine and square it. In contrast, if we needed to compute $\sin^2(x)$ "from scratch", by using only basic hardware-supported arithmetic operations, we would need to perform a dozen or so operations—which would necessitate a much longer program.

From this viewpoint, it is reasonable to consider program for which an access to the i-th bit from the sequence y can be done by simply writing something like $y[i]$. The shortest length of such a using-calls-to-y program for computing either x or its extension is known as the *conditional prefix Kolmogorov complexity* and is usually denoted by $K(x\,|\,y)$.

In these terms, the fact that the use of y should not make computing x easier means that the complexity $K(x\,|\,y)$ should not be much smaller than the complexity of computing it without using y, i.e., that we should not have $K(x\,|\,y) \ll K(x)$. Of course, possibly using y does not mean that we have to, so when we compute $K(x\,|\,y)$, we consider all possible programs for computing x—including programs that do not use y at all. Thus, the length $K(x\,|\,y)$ of the shortest y-using program for computing x must be either shorter or of the same length as the length $K(x)$ of the shortest program that computes x without using y: $K(x\,|\,y) \le K(x)$.

Thus, the difference $K(x) - K(x\,|\,y)$ must be non-negative and not too large. This difference is known as *information* in y about x; it is denoted by $I(x : y)$. Similarly to the definition of randomness, the fact that using y should not affect our ability to compute x can be described as $I(x : y) \le c$ for some small $c > 0$.

Levin's formalization of the above intuitive idea: motivations. The above description is exactly what Levin used in formalizing the commonsense idea that the results of physical experiments should not help us solve mathematical problems.

All mathematics can be (and have been) formulated in terms of set theory ST, e.g., in terms of the usual Zermelo-Frenkel axiomatics ZF. In these terms, solving a mathematical problem—i.e., checking whether a given mathematical statement is true or not—is equivalent to checking whether a given statement from ST is true or not. Some mathematical problems are computational, not about proving results, but such problems can also be described in these terms: namely, if we want to compute the value of a real number (such as π), this means that for each n, we must find out whether the n-th digit in the binary expansion of this number is 0 or 1—and the fact that the n-th digit is 1 is also a mathematical statement that can be, in principle, formalized in set theory.

Each statement from ST is a finite combination of symbols, so in a computer, it is represented as a sequence of 0s and 1s—which, in its turn, can be interpreted as a natural number. Moreover, it is easy to check whether a given natural number—i.e., in effect, a given sequence of 0s and 1s—is a syntactically correct statement of set theory. Thus, by trying all possible natural numbers 0, 1, 2, ..., and checking whether each of them is a syntactically correct statement of set theory, we can enumerate all possible such statements into a sequence S_1, S_2, ...In these terms, perfect mathematical knowledge can be represented by an infinite binary sequence $\alpha = \alpha_1\alpha_2\ldots$ where $\alpha_i = 1$ if and only if the i-th statement α_i is true.

Similarly, all possible observation and measurement results can also be placed in a single binary sequence. Indeed, each observation and measurement result has to be represented in a computer—whether it is a numerical value or a textual observation—and can, thus, be naturally represented as a sequence of 0s and 1s. Each measurement result can thus be described as a sequence of bits.

Descriptions of the corresponding experimental and observational settings—what is usually called *metadata*—can also be described in a computer. So, for each bit of each measurement result, we have a description, e.g., "the 5-th bit of measuring wind speed at UTEP campus at 12 pm on August 27, 2019". We can similarly sort such descriptions, and get a potentially infinite binary sequence $\omega = \omega_1\omega_1\ldots$ Of course, this sequence is only *potentially* infinite: at any given moment of time, we have only finitely many measurement and observation results, and thus, we only know a finite part of this potentially infinite sequence.

Levin's formalization of the above commonsense idea is that for every m and n, using the first m bits of the measurement-results sequence $\omega_{1:m} \stackrel{\text{def}}{=} \omega_1\ldots\omega_m$ does not help us compute the truth values $\alpha_{1:n} \stackrel{\text{def}}{=} \alpha_1\ldots\alpha_n$ of the first n statements of set theory.

We already know how to formalize this "inability to help". Thus, we get the following precise description.

Resulting Levin's formalization of the above intuitive idea. There exists a small integer $c > 0$ for which, for all m and n, we have $I(\alpha_{1:n} : \omega_{1:m}) \le c$.

Comment. Levin called this formalization the *Independence Postulate*.

This formalization is in perfect accordance with modern physics. According to modern (quantum) physics, we have deterministic equations describing the dynamics of the wave function (i.e., the system's state). This state, in its turn, determines the probability of different measurement results, and the actual sequence of measurement results is random with respect to the corresponding probability measure; see, e.g., [1, 5].

For such sequence ω—which are random with respect to some computable probability measure—Levin's Independence Postulate is indeed true; see, e.g., [3]. In this sense, Levin's Independence Postulate is in perfect accordance with modern physics.

2 Remaining Challenge and Our Proposed Solution to This Challenge

Physicists believe that every theory is approximate. It is good that modern quantum physics is in accordance with the physicists' intuition—which is formally described by Levin's Independence Principle.

But we should take into account that physicists have yet another intuition (see, e.g., [1, 5]): namely, many of them believe that no theory is final, that no matter what theory we formulate, no matter how well this theory describes the current experimental results, it will eventually turn out to be only a good approximation—there will be new experiments, new data that will require a modification of this theory.

This happened with Newton's mechanics—which needed to be modified to take into account relativistic and quantum effects, this will happen—many physicists believe—with modern relativistic quantum physics as well.

In other words, there will be no limit to progress of science: as science progresses, we will get more and more accurate models of reality.

How can we describe this belief in precise terms. In general, the above belief means that whatever physical law we come up with which is consistent with all physical experiments and observations so far, eventually we will come up with experimental data that violates this law. In terms of our notations, currently available results of experiments and observations simply form an initial fragment $\omega_{1:n}$ of the potentially infinite sequence ω of all such results. From the mathematical viewpoint, a physical law is simply a property $P(\omega_{1:n})$ that limits possible values of such fragments to those that satisfy this property.

Thus, the above physicists' belief is that for each such property P, there exists an integer M—corresponding to some future moment of time—at which the fragment $\omega_{1:M}$ will *not* satisfy the corresponding property P.

Resulting challenge. In particular, the physicists' belief means that no matter what constant c we select in our description of Levin's Independence Principle, there will be a value M for which this principle will be violated, i.e., for which we will have

$$I(\alpha_{1:n}, \omega_{1:M}) > c.$$

So, contrary to the physicists' intuition (and to modern physics), under this belief, a sequence of physical observations cannot be random—in the above-described precise algorithmic sense of this randomness.

In other words, the two physicists' intuitions—the intuition about randomness and the intuition about infinite progress of physics—are not fully compatible. How can we reconcile these two intuitions?

How to reconcile the two intuitions: suggestion and challenges. Due to the second (progress-of-science) intuition, we cannot require—as Levine did—that all the values of the information $I(\alpha_{1:n}, \omega_{1:m})$ are bounded by a constant. However, intuitively, the first (randomness) intuition tell us that we cannot expect too much information above complex statements by simply looking at nature. We do not expect that we can find a solution to a complex mathematical problem by simply measuring, the sizes of the tree leaves.

In other words, since we cannot require that the amount of information $I(\alpha_{1:n}, \omega_{1:m})$ should be bounded, we can require that it be *small*—i.e., it should not grow too fast with m.

This idea is informal, can we formalize it? Unfortunately, not really: if we select some slowly growing function $c(m)$ and require that

$$I(\alpha_{1:n}, \omega_{1:m}) \le c(m),$$

we will have the same problem as with the original Levin's Independence Postulate: that, according to the progress-of-science intuition, there will be some M for which this inequality will be violated.

Thus, the only way to reconcile the two intuitions is to make an informal statement.

Thus, *there are fundamental reasons why informal knowledge is needed for describing the real world.*

Acknowledgements This work was supported in part by the National Science Foundation grants 1623190 (A Model of Change for Preparing a New Generation for Professional Practice in Computer Science) and HRD-1242122 (Cyber-ShARE Center of Excellence).

References

1. R. Feynman, R. Leighton, M. Sands, *The Feynman Lectures on Physics* (Addison Wesley, Boston, MA, 2005)
2. L.A. Levin, Randomness conservation inequalities: information and independence in mathematical theories. Inf. Control **61**, 15–37 (1984)
3. M. Li, P. Vitanyi, *An Introduction to Kolmogorov Complexity and Its Applications* (Springer, New York, 2008)
4. C. Papadimitriou, *Computational Complexity* (Addison-Wesley, Reading, MA, 1994)
5. K.S. Thorne, R.D. Blandford, *Modern Classical Physics: Optics, Fluids, Plasmas, Elasticity, Relativity, and Statistical Physics* (Princeton University Press, Princeton, NJ, 2017)

Scale-Invariance and Fuzzy Techniques Explain the Empirical Success of Inverse Distance Weighting and of Dual Inverse Distance Weighting in Geosciences

Laxman Bokati, Aaron Velasco, and Vladik Kreinovich

Abstract Once we measure the values of a physical quantity at certain spatial locations, we need to interpolate these values to estimate the value of this quantity at other locations x. In geosciences, one of the most widely used interpolation techniques is inverse distance weighting, when we combine the available measurement results with the weights inverse proportional to some power of the distance from x to the measurement location. This empirical formula works well when measurement locations are uniformly distributed, but it leads to biased estimates otherwise. To decrease this bias, researchers recently proposed a more complex dual inverse distance weighting technique. In this paper, we provide a theoretical explanation both for the inverse distance weighting and for the dual inverse distance weighting. Specifically, we show that if we use the general fuzzy ideas to formally describe the desired property of the interpolation procedure, then physically natural scale-invariance requirements select only these two distance weighting techniques.

1 Formulation of the Problem

Need for interpolation of spatial data. In many practical situations, we are interested in the value of a certain physical quantity at different spatial locations. For example, in geosciences, we may be interested in how elevation and depths of different geological layers depend of the spatial location. In environmental sciences, we may be interested in the concentration of different substances in the atmosphere at different locations. etc.

L. Bokati · A. Velasco · V. Kreinovich (✉)
University of Texas at El Paso, El Paso, TX 79968, USA
e-mail: vladik@utep.edu

L. Bokati
e-mail: lbokati@miners.utep.edu

A. Velasco
e-mail: aavelasco@utep.edu

B. Bede et al. (eds.), *Fuzzy Information Processing 2020*,
Advances in Intelligent Systems and Computing 1337,
https://doi.org/10.1007/978-3-030-81561-5_32

In principle, at each location, we can measure—directly or indirectly—the value of the corresponding quantity. However, we can only perform the measurement at a finite number of locations. Since we are interested in the values of the quantity at all possible locations, we need to estimate these values based on the measurement results—i.e., we need to *interpolate* and *extrapolate* the spatial data.

In precise terms: we know the values $q_i = q(x_i)$ of the quantity of interest q at several locations x_i, $i = 1, 2, \ldots, n$. Based on this information, we would like to estimate the value $q(x)$ of this quantity at a given location x.

Inverse distance weighting. A reasonable estimate q for $q(x)$ is a weighted average of the known values $q(x_i)$: $q = \sum\limits_{i=1}^{n} w_i \cdot q_i$, with $\sum\limits_{i=1}^{n} w_i = 1$. Naturally, the closer is the point x to the point x_i, the larger should be the weight w_i—and if the distance $d(x, x_i)$ is large, then the value $q(x_i)$ should not affect our estimate at all. So, the weight w_i with which we take the value q_i should decrease with the distance.

Empirically, it turns out that the best interpolation is attained when we take the weight proportional to some negative power of the distance: $w_i \sim (d(x, x_i))^{-p}$ for some $p > 0$. Since the weights have to add up to 1, we thus get

$$w_i = \frac{(d(x, x_i))^{-p}}{\sum\limits_{j=1}^{n} (d(x, x_j))^{-p}}.$$

This method—known as *inverse distance weighting*—is one of most widely used spatial interpolation methods; see, e.g., [3–5, 8, 9, 14].

First challenge: why inverse distance weighting? In general, the fact that some algorithm is empirically the best means that we tried many other algorithms, and this particular algorithm worked better than everything else we tried. In practice, we cannot try all possible algorithms, we can only try finitely many different algorithms. So, in principle, there could be an algorithm that we did not try and that will work better than the one which is currently empirically the best.

To be absolutely sure that the empirically found algorithm is the best, it is thus not enough to perform more testing: we need to have some theoretical explanation of this algorithm's superiority. Because of this, every time we have some empirically best alternative, it is desirable to come up with a theoretical explanation of why this alternative is indeed the best—and if such an explanation cannot be found, maybe this alternative is actually not the best?

Thus, the empirical success of inverse distance weighting prompts a natural question: is this indeed the best method? This is the first challenge that we will deal with in this paper.

Limitations of inverse distance weighting. While the inverse distance weighting method is empirically the best among different distance-dependence interpolation techniques, it has limitations; see, e.g., [7].

Specifically, it works well when we have a reasonably uniformly distributed spatial data. The problem is that in many practical cases, we have more measurements in some areas and fewer in others. For example, when we measure meteorological quantities such as temperature, humidity, wind speed, we usually have plenty of sensors (and thus, plenty of measurement results) in cities and other densely populated areas, but much fewer measurements in not so densely populated areas—e.g., in the deserts.

Let us provide a simple example explaining why this may lead to a problem. Suppose that we have two locations A and B at which we perform measurements:

- Location A is densely populated, so we have two measurement results q_A and $q_{A'}$ from this area.
- Location B is a desert, so we have only one measurement result q_B from this location.

Since locations A and A' are very close, the corresponding values are also very close, so we can safely assume that they are equal: $q_A = q_{A'}$. Suppose that we want to use these three measurement results to predict the value of the quantity x at a midpoint C between the locations A and B.

Since C is exactly in the middle between A and B, when estimating q_C, intuitively, we should combine the values q_A and q_B with equal weights, i.e., take $q_C = \dfrac{q_A + q_B}{2}$. From the commonsense viewpoint, it should not matter whether we made a single measurement at the location A or we made two different measurements.

However, this is *not* what we get if we apply the inverse distance weighting. Indeed, in this case, since all the distance are equal $d(A, C) = d(A', C) = d(B, C)$, the inverse distance weighting leads to

$$q_C = \frac{q_A + q_{A'} + q_B}{3} = \frac{2}{3} \cdot q_A + \frac{1}{3} \cdot q_B.$$

Dual inverse distance weighting: an empirically efficient way to overcome this limitation. To overcome the above limitation, a recent paper [7] proposed a new method called *dual inverse distance weighting*, a method that is empirically better than all previously proposed attempts to overcome this limitation.

In this method, instead of simply using the weight $w_i \sim (d(x, x_i))^{-p}$ depending on the distance, we also give more weight to the points which are more distant from others—and less weight to points which are close to others, by using a formula $w_i \sim (d(x, x_i))^{-p} \cdot \left(\sum\limits_{j \neq i} (d(x_i, x_j))^{p_2}\right)$, for some $p_2 > 0$.

Let us show, on an example, that this idea indeed helps overcome the above limitation. Indeed, in the above example of extrapolating from the three points $A \approx A'$ and B to the midpoint C between A and B (for which $d(A, C) = d(B, C)$), we have $d(A, A') \approx 0$ and $d(A, B) \approx d(A', B)$. Thus, we get the following expressions for the additional factors $f_i = \sum\limits_{j \neq i} (d(x_i, x_j))^{p_2}$:

$$f_A = (d(A, A'))^{p_2} + (d(A, B))^{p_2} \approx (d(A, B))^{p_2},$$

$$f_{A'} = (d(A', A))^{p_2} + (d(A', B))^{p_2} \approx (d(A, B))^{p_2},$$

and

$$f_B = (d(B, A))^{p_2} + (d(B, A'))^{p_2} \approx 2(d(A, B))^{p_2}.$$

So, the weights w_A and $w_{A'}$ with which we take the values q_A and $q_{A'}$ are proportional to

$$w_A \approx w_{A'} \sim (d(A, C))^{-p} \cdot f_A \approx (d(A, C))^{-p} \cdot (d(A, B))^{p_2},$$

while

$$w_B \approx w_B \sim (d(B, C))^{-p} \cdot f_2 \approx (d(A, C))^{-p} \cdot 2(d(A, B))^{p_2}.$$

The weight w_B is thus twice larger than the weights w_A and $w_{A'}$: $w_B = 2w_A = 2w_{A'}$. So the interpolated value of q_C is equal to

$$q_C = \frac{w_A \cdot q_A + w_{A'} \cdot q_{A'} + w_B \cdot q_B}{w_A + w_{A'} + w_B} = \frac{w_A \cdot q_A + w_A \cdot q_{A'} + 2w_A \cdot q_A}{w_A + w_{A'} + 2w_A}.$$

Dividing both numerator and denominator by $2w_A$ and taking into account that $q_{A'} = q_A$, we conclude that $q_C = \frac{q_A + q_B}{2}$, i.e., exactly the value that we wanted.

Second challenge: why dual inverse distance weighting? In view of the above, it is also desirable to come up with a theoretical explanation for the dual inverse weighting method as well. This is the second challenge that we take on in this paper.

2 What Is Scale Invariance and How It Explains the Empirical Success of Inverse Distance Weighting

What is scale invariance. When we process the values of physical quantities, we process real numbers. It is important to take into account, however, that the numerical value of each quantity depends on the measuring unit. For example, suppose that we measure the distance in kilometers and get a numerical value d such as 2 km. Alternatively, we could use meters instead of kilometers. In this case, the exact same distance will be described by a different number: 2000 m.

In general, if we replace the original measuring unit with a new one which is λ times smaller, all numerical values will be multiplied by λ, i.e., instead of the original numerical value x, we will get a new numerical value $\lambda \cdot x$.

Scale-invariance means, in our case, that the result of interpolation should not change if we simply change the measuring unit. Let us analyze how this natural requirement affects interpolation.

General case of distance-dependent interpolation. Let us consider the general case, when the further the point, the smaller the weight, i.e., in precise terms, when the weight w_i is proportional to $f(d(x, x_i))$ for some decreasing function $f(z)$: $w_i \sim f(d(x, x_i))$. Since the weights should add up to 1, we conclude that

$$w_i = \frac{f(d(x, x_i))}{\sum_j f(d(x, x_j))}, \tag{1}$$

and thus, our estimate q for $q(x)$ should take the form

$$q = \sum_{i=1}^{n} \frac{f(d(x, x_i))}{\sum_j f(d(x, x_j))} \cdot q_i. \tag{2}$$

In this case, scale-invariance means that for each $\lambda > 0$, if we replace all the numerical distance values $d(x, x_i)$ with "re-scaled" values $\lambda \cdot d(x, x_i)$, then we should get the exact same interpolation result, i.e., that for all possible values of q_i and $d(x, x_i)$, we should have

$$\sum_{i=1}^{n} \frac{f(\lambda \cdot d(x, x_i))}{\sum_j f(\lambda \cdot d(x, x_j))} \cdot q_i = \sum_{i=1}^{n} \frac{f(d(x, x_i))}{\sum_j f(d(x, x_j))} \cdot q_i. \tag{3}$$

Scale-invariance leads to inverse distance weighting. Let us show that the requirement (3) indeed leads to inverse distance weighting.

Indeed, let us consider the case when we have only two measurement results:

- at the point x_1, we got the value $q_1 = 1$, and
- at point x_2, we got the value $q_2 = 0$.

Then, for any point x, if we use the original distance values $d_1 \stackrel{\text{def}}{=} d(x, x_1)$ and $d_2 \stackrel{\text{def}}{=} d(x, x_2)$, the interpolated value q at this point will have the form

$$q = \frac{f(d_1)}{f(d_1) + f(d_2)}.$$

On the other hand, if we use a λ times smaller measuring unit, then the extrapolation formula leads to the values

$$\frac{f(\lambda \cdot d_1)}{f(\lambda \cdot d_1) + f(\lambda \cdot d_2)}.$$

The requirement that the interpolation value does not change if we simply change the measuring unit implies that these two expression must coincide, i.e., that we must have:

$$\frac{f(\lambda \cdot d_1)}{f(\lambda \cdot d_1) + f(\lambda \cdot d_2)} = \frac{f(d_1)}{f(d_1) + f(d_2)}. \quad (4)$$

If we take the inverse of both sides of this formula, i.e., flip the numerator and denominator in both sides, we get

$$\frac{f(\lambda \cdot d_1) + f(\lambda \cdot d_2)}{f(\lambda \cdot d_1)} = \frac{f(d_1) + f(d_2)}{f(d_1)}. \quad (5)$$

Subtracting number 1 from both sides, we get a simplified expression

$$\frac{f(\lambda \cdot d_2)}{f(\lambda \cdot d_1)} = \frac{f(d_2)}{f(d_1)}. \quad (6)$$

If we divide both sides by $f(d_2)$ and multiply by $f(\lambda \cdot d_1)$, we get the equivalent equality in which variables d_1 and d_2 are separated:

$$\frac{f(\lambda \cdot d_2)}{f(d_2)} = \frac{f(\lambda \cdot d_1)}{f(d_1)}. \quad (7)$$

The left-hand side of this formula does not depend on d_1; thus, the right-hand side does not depend on d_1 either, it must thus depend only on λ. Let us denote this right-hand side by $c(\lambda)$. Then, from $\dfrac{f(\lambda \cdot d_1)}{f(d_1)} = c(\lambda)$, we conclude that

$$f(\lambda \cdot d_1) = c(\lambda) \cdot f(d_1) \quad (8)$$

for all possible values of $\lambda > 0$ and d_1.

It is known that for decreasing functions $f(z)$, the only solutions to the functional equation (8) are functions $f(z) = c \cdot z^{-p}$ for some $p > 0$; see, e.g., [1]. For this function $f(z)$, the extrapolated value has the form $\sum w'_i \cdot q_i$, with

$$w'_i = \frac{c \cdot (d(x, x_i))^{-p}}{\sum_{j=1}^{n} c \cdot (d(x, x_j))^{-p}}.$$

If we divide both numerator and denominator by c, we get exactly the inverse distance weighting formula.

Thus, scale-invariance indeed leads to inverse distance weighting.

Comment. For smooth function $f(x)$, the above result about solutions of the functional equation can be easily derived. Indeed, differentiating both sides of the equality (8) by λ and taking $\lambda = 1$, we get

$$f'(d_1) \cdot d_1 = \alpha \cdot f(d_1),$$

where we denoted $\alpha \stackrel{\text{def}}{=} c'(1)$, i.e., we have

$$\frac{df}{dd_1} = \alpha \cdot f.$$

If we divide both sides by f and multiply by dd_1, we separate d_1 and f:

$$\frac{df}{f} = \alpha \cdot \frac{dd_1}{d_1}.$$

Integrating both sides, we get $\ln(f) = \alpha \cdot \ln(d_1) + C$, where C is the integration constant. Applying $\exp(z)$ to both sides and taking into account that $\exp(\ln(f)) = f$ and

$$\exp(\alpha \cdot \ln(d_1) + C) = \exp(\alpha \cdot \ln(d_1)) \cdot \exp(C) = \exp(C) \cdot (\exp(\ln(d_1))^\alpha = \exp(C) \cdot d_1^\alpha,$$

we get $f(d_1) = c \cdot d_1^\alpha$, where we denoted $c \stackrel{\text{def}}{=} \exp(C)$. Since the function $f(z)$ is decreasing, we should have $\alpha < 0$, i.e., $\alpha = -p$ for some $p > 0$. The statement is proven.

3 Scale Invariance and Fuzzy Techniques Explain Dual Inverse Weighting

What we want: informal description. In the previous section, when computing the estimate q for the value $q(x)$ of the desired quantity at a location x, we used, in effect, the weighted average of the measurements results q_i, with the weights decreasing as the distance $d(x, x_i)$ decreases—i.e., in more precise terms, with weights proportional to $f(d(x, x_i))$ for some decreasing function $f(z)$. In this case, scale-invariance implies that $f(z) = z^{-p}$ for some $p > 0$.

As we have mentioned in Section 1, we need to also give more weight to measurements at locations x_i which are far away from other location—and, correspondingly, less weight to measurements at locations which are close to other locations. In terms of weights, we would like to multiply the previous weights $f(d(x, x_i)) = (d(x, x_i))^{-p}$ by an additional factor f_i depending on how far away is location x_i from other locations. The further away the location x_i from other locations, the higher the factor f_i shall be. In other words, the factor f_i should be larger or smaller depending on our degree of confidence in the following statement:

$d(x_i, x_1)$ is large and $d(x_i, x_2)$ is large and $\ldots d(x_i, x_n)$ is large.

Let us use fuzzy techniques to translate this informal statements into precise terms. To translate the above informal statement into precise terms, a reasonable idea is to use fuzzy techniques—techniques specifically designed for such a translation; see, e.g., [2, 6, 10, 12, 13, 15]. In this technique, to each basic statement—like "d is large"—we assign a degree to which, according to the expert, this statement is true. This degree is usually denoted by $\mu(d)$. In terms of these notations:

- the degree to which $d(x_i, x_1)$ is large is equal to $\mu(d(x_i, x_1))$;
- the degree to which $d(x_i, x_2)$ is large is equal to $\mu(d(x_i, x_2))$; etc.

To estimate the degree to which the above "and"-statement is satisfied, fuzzy techniques suggest that we combine the above degrees by using an appropriate "and"-operation (= t-norm) $f_\&(a, b)$). Thus, we get the following degree:

$$f_\&(\mu(d(x_i, x_1)), \mu(d(x_i, x_2)), \ldots, \mu(d(x_i, x_{i-1})), \mu(d(x_i, x_{i+1})), \ldots, \mu(d(x_i, x_n))).$$

It is known—see, e.g., [11]—that for any "and"-operation and for any $\varepsilon > 0$, there exists an ε-close "and"-operation of the type $f_\&(a, b) = g^{-1}(g(a) + g(b))$ for some monotonic function $g(a)$, where $g^{-1}(a)$ denotes the inverse function (i.e., the function for which $g^{-1}(a) = b$ if and only if $g(b) = a$). Since the approximation error ε can be arbitrarily small, for all practical purposes, we can safely assume that the actual "and"-operation has this g-based form. Substituting this expression for the "and"-operation into the above formula, we conclude that f_i should monotonically depend on the expression

$$g^{-1}(g(\mu(d(x_i, x_1))) + \cdots + g(\mu(d(x_i, x_n)))).$$

Since the function g^{-1} is monotonic, this means that f_i is a monotonic function of the expression

$$G(d(x_i, x_1)) + \cdots + G(d(x_i, x_n))),$$

where we denoted $G(d) \stackrel{\text{def}}{=} g(\mu(d))$. In other words, we conclude that

$$f_i = F(G(d(x_i, x_1)) + \cdots + G(d(x_i, x_n))) \tag{9}$$

for some monotonic function $F(z)$.

So, we get an estimate

$$q = \sum_{i=1}^{n} \frac{f_i \cdot (d(x, x_i))^{-p} \cdot q_i}{\sum\limits_{j=1}^{n} f_j \cdot (d(x, x_j))^{-p}}, \tag{10}$$

where the factors f_i are described by the formula (9).

Let us recall the motivation for the factors f_i. As we have mentioned earlier, the main motivation for introducing the factors f_i is to make sure that for the midpoint C between A and B, we will have the estimate $\frac{q_A + q_B}{2}$, even if we perform two (or more) measurements at the point A. Let us analyze for which functions $F(z)$ and $G(z)$ this requirement is satisfied.

For the purpose of this analysis, let us consider the case when we have m measurement locations $A_1, \ldots, A_m$ in the close vicinity of the location A and one measurement result at location B. Let d denote the distance $d(A, B)$ between the locations A and B. For all the measurement locations $A_1, \ldots, A_m$, and B, the distance to the point C is the same—equal to $d/2$. Thus, in this case, the factors $(d(x, x_i))^{-p}$ in the formula (10) are all equal to each other. So, we can divide both the numerator and the denominator by the formula (10) by this common factor, and get a simplified expression

$$q = \sum_{i=1}^{n} \frac{f_i \cdot q_i}{\sum_{j=1}^{n} f_j}.$$

Since for the points $A_1, \ldots, A_m$ we have the same measurement results q_i (we will denote them by q_A), and the same factors f_i (we will denote them by f_A), we get

$$q = \frac{m \cdot f_A \cdot q_A + f_B \cdot q_B}{m \cdot f_A + f_B}. \tag{11}$$

We want to make sure that this value is equal to the arithmetic average $\frac{q_A + q_B}{2}$. Thus, the coefficient at q_A in the formula (11) should be equal to $1/2$:

$$\frac{m \cdot f_A}{m \cdot f_A + f_B} = \frac{1}{2}.$$

If we multiply both side by their denominators and subtract $m \cdot f_A$ from both sides, we get $m \cdot f_A = f_B$. Due to the formula (9), this means

$$m \cdot F(G(d) + (m - 1) \cdot G(0)) = F(m \cdot G(d)). \tag{12}$$

In the limit $d = 0$, this formula becomes $m \cdot F(m \cdot G(0)) = F(m \cdot G(0))$, thus $F(m \cdot G(0)) = 0$. Since the function $F(z)$ is monotonic, we cannot have $G(0) \neq 0$, since then we would have $F(z) = 0$ for all z. Thus, $G(0) = 0$, $F(G(0)) = F(0) = 0$, and the formula (12) takes the form $F(m \cdot G(d)) = m \cdot F(G(d))$. This is true for any value $z = G(d)$, so we have $F(m \cdot z) = m \cdot F(z)$ for all m and z.

- In particular, for $z = 1$, we get $F(m) = c \cdot m$, where $c \stackrel{\text{def}}{=} F(1)$.
- For $z = 1/m$, we then have $F(1) = c = m \cdot F(1/m)$, hence

$$F(1/m) = c \cdot (1/m).$$

- Similarly, we get $F(p/q) = F(p \cdot (1/q)) = p \cdot F(1/q) = p \cdot (c \cdot (1/q)) = c \cdot (p/q)$. So, for all rational values $z = p/q$, we get $F(z) = c \cdot z$.

Since the function $F(z)$ is monotonic, the formula $F(z) = c \cdot z$ is true for all values z.

Dividing both the numerator and the denominator by the coefficient c, we conclude that

$$q = \sum_{i=1}^{n} \frac{F_i \cdot (d(x, x_i))^{-p} \cdot q_i}{\sum\limits_{j=1}^{n} F_j \cdot (d(x, x_j))^{-p}}, \tag{13}$$

where we denoted

$$F_i \stackrel{\text{def}}{=} G(d(x_i, x_1)) + \cdots + G(d(x_i, x_n)). \tag{14}$$

Let us now use scale-invariance. We want to make sure that the estimate (13) does not change after re-scaling $d(x, y) \to d'(x, y) = \lambda \cdot d(x, y)$, i.e., that the same value q should be also equal to

$$q = \sum_{i=1}^{n} \frac{F_i' \cdot (d'(x, x_i))^{-p} \cdot q_i}{\sum\limits_{j=1}^{n} F_j' \cdot (d'(x, x_j))^{-p}}, \tag{15}$$

where

$$F_i' = G(d'(x_i, x_1)) + \cdots + G(d'(x_i, x_n)). \tag{16}$$

Here, $(d'(x, x_i))^{-p} = \lambda^{-p} \cdot (d(x, x_i))^{-p}$. Dividing both the numerator and the denominator of the right-hand side of the formula (15) by λ^{-p}, we get a simplified expression

$$q = \sum_{i=1}^{n} \frac{F_i' \cdot (d(x, x_i))^{-p} \cdot q_i}{\sum\limits_{j=1}^{n} F_j' \cdot (d(x, x_j))^{-p}}. \tag{17}$$

The two expressions (13) and (17) are linear in q_i. Thus, their equality implies that coefficients at each q_i must be the same. In particular, this means that the ratios of the coefficients at q_1 and q_2 must be equal, i.e., we must have

$$\frac{F_1 \cdot (d(x, x_1))^{-p}}{F_2 \cdot (d(x, x_2))^{-p}} = \frac{F_1' \cdot (d(x, x_1))^{-p}}{F_2' \cdot (d(x, x_2))^{-p}},$$

i.e.,

$$\frac{F_1}{F_2} = \frac{F_1'}{F_2'}.$$

For the case when we have three points with $d(x_1, x_2) = d(x_1, x_3) = d$ and $d(x_2, x_3) = D$, due to the formula (14), this means that

$$\frac{2G(d)}{G(d) + G(D)} = \frac{2G(\lambda \cdot d)}{G(\lambda \cdot d) + G(\lambda \cdot D)}.$$

Inverting both sides, multiplying both sides by 2 and subtracting 1 from both sides, we conclude that

$$\frac{G(D)}{G(d)} = \frac{G(\lambda \cdot D)}{G(\lambda \cdot d)}$$

for all λ, d, and D. We already know—from the first proof—that this implies that $G(d) = c \cdot d^{p_2}$ for some c and p_2, and that, by dividing both numerator and denominator by c, we can get $c = 1$.

Thus, we indeed get a justification for the dual inverse distance weighting.

Acknowledgements This work was supported in part by the National Science Foundation grants 1623190 (A Model of Change for Preparing a New Generation for Professional Practice in Computer Science) and HRD-1242122 (Cyber-ShARE Center of Excellence).

References

1. J. Aczel, J. Dhombres, *Functional Equations in Several Variables* (Cambridge University Press, Cambridge, UK, 1989)
2. R. Belohlavek, J.W. Dauben, G.J. Klir, *Fuzzy Logic and Mathematics: A Historical Perspective* (Oxford University Press, New York, 2017)
3. Q. Chen, G. Liu, X. Ma, G. Marietoz, Z. He, Y. Tian, Z. Weng, Local curvature entropy-based 3D terrain representation using a comprehensive quadtree. ISPRS J. Photogramm. Remote Sens. **139**, 130–145 (2018)
4. K.C. Clarke, *Analytical and Computer Cartography* (Pnetice Hall, Englewood Cliffs, New Jersey, 1990)
5. N. Henderson, L. Pena, The inverse distance weighting interpolation applied to a particular form of the path rubes method: theory and computation for advection in uncompressible flow. Appl. Math. Comput. **304**, 114–135 (2017)
6. G. Klir, B. Yuan, *Fuzzy Sets and Fuzzy Logic* (Prentice Hall, Upper Saddle River, New Jersey, 1995)
7. Z. Li, X. Zhang, R. Zhu, Z. Zhiang, Z. Weng, Integrating data-to-data correlation into inverse distance weighting. Comput. Geosci. (2019). https://doi.org/10.1007/s10596-019-09913-9
8. Q. Liang, S. Nittel, J.C. Whittier, S. Bruin, Real-time inverse distance weighting interpolation for streaming sensor data. Trans. GIS **22**(5), 1179–1204 (2018)
9. I. Loghmari, Y. Timoumi, A. Messadi, Performance comparison of two global solar radiation models for spatial interpolation purposes. Renew. Sustain. Energy Rev. **82**, 837–844 (2018)
10. J.M. Mendel, *Uncertain Rule-Based Fuzzy Systems: Introduction and New Directions* (Springer, Cham, Switzerland, 2017)
11. H.T. Nguyen, V. Kreinovich, P. Wojciechowski, Strict Archimedean t-norms and t-conorms as universal approximators. Int. J. Approx. Reason. **18**(3–4), 239–249 (1998)
12. H.T. Nguyen, C.L. Walker, E.A. Walker, *A First Course in Fuzzy Logic* (Chapman and Hall/CRC, Boca Raton, Florida, 2019)

13. V. Novák, I. Perfilieva, J. Močkoř, *Mathematical Principles of Fuzzy Logic* (Kluwer, Boston, Dordrecht, 1999)
14. D. Shepard, A two-dimensional interpolation function for irregularly-spaced data, in *Proceedings of the 1968 23rd ACM National Conference* (1968), pp. 517–524
15. L.A. Zadeh, Fuzzy sets. Inf. Control **8**, 338–353 (1965)

Is There a Contradiction Between Statistics and Fairness: From Intelligent Control to Explainable AI

Christian Servin and Vladik Kreinovich

Abstract At first glance, there seems to be a contradiction between statistics and fairness: statistics-based AI techniques lead to unfair discrimination based on gender, race, and socio-economical status. This is not just a fault of probability techniques: similar problems can happen if we use fuzzy or other techniques for processing uncertainty. To attain fairness, several authors proposed not to rely on statistics and instead, explicitly add fairness constraints into decision making. In this paper, we show that the seeming contradiction between statistics and fairness is caused mostly by the fact that the existing systems use simplified models; contradictions disappear if we replace them with more adequate (and thus more complex) statistical models.

1 Formulation of the Problem

Social applications of AI. Recent AI techniques like deep learning have led to many successful applications. For example, we can apply deep learning to decide whose loan applications should be approved and whose applications should be rejected—and if approved, what interest should we charge. We can apply deep learning to decide which candidates for graduate program to accept—and for those accepted what financial benefits to offer as an enticement.

In all such cases, we feed the system with numerous past examples of successes and failures. Based on these example, the systems tries its best to predict whether a given loan or a given potential student will be a success or not. Statistically, these systems

C. Servin
Computer Science and Information Technology Systems Department, El Paso Community College (EPCC), 919 Hunter Dr., El Paso, TX 79915-1908, USA
e-mail: cservin1@epcc.edu

V. Kreinovich (✉)
Department of Computer Science, University of Texas at El Paso, 500 W. University, El Paso, TX 79968, USA
e-mail: vladik@utep.edu

B. Bede et al. (eds.), *Fuzzy Information Processing 2020*,
Advances in Intelligent Systems and Computing 1337,
https://doi.org/10.1007/978-3-030-81561-5_33

seem to work well: they predict success or failure better than human decision makers. However, the results are often not satisfactory; see, e.g., [1, 9]. Let us explain why.

Many current social applications of AI are unsatisfactory. On average, loan applications from poorer geographic areas have a higher default rate. This is a known fact, and statistical methods underlying machine learning find this out. As a result, the system naturally recommends rejection of all loans from these areas. This is not fair to people with good credit record who happen to live in the not-so-good areas. Moveover, it is also detrimental to the bank since the bank will miss on profiting from such potentially successful loans.

Similarly, it is known that in many disciplines women has a lower success rates in getting their PhDs than men—and take longer when they succeed. One of the main reasons for this is that raising children requires much more efforts from women than from men. A statistical system, crudely speaking, does not care about the reasons, it just takes this statistical fact into account and preferably selects males. Not only this is not fair, this way the universities miss a lot of talent—and nowadays, with not much need for routine boring work, talent and creativity are extremely important, they should be nurtured, not rejected.

So is there a contradiction between statistics and fairness? At first glance, it may seem that there is a contradiction between statistical methods and our ideas of fairness. In other words, it seems that if we want the systems to be fair, we cannot rely on statistics only, we need to supplement statistics with additional fairness constraints.

The need for such constraints is usually formulated—in our opinion, not fully accurately—as the need for *explainable AI*; see, e.g., [6] and references therein. The main idea behind explainable AI is that instead of relying on a machine learning system as a black box, we extract some rules from this system—and if these rules are not fair, we replace them with fairer rules.

What we show in this paper. In this paper, we show that the seeming inconsistency comes from the fact that we use simplified statistical models. We show that a more detailed description of the corresponding uncertainty—be it probabilistic or fuzzy uncertainty—eliminates this seeming contradiction, and enables the system to come up with fair decisions without any need for additional constraints.

2 How Current Techniques Lead to Unfair Decisions: Simplified Examples

Let us give some examples. In other to explain why the existing techniques can lead to unfair solutions, let us give some detailed simplified examples. We will start with statistical examples. Then, we will show that mathematically similar examples—this time not related to fairness—can be found in applications of fuzzy techniques as well—namely, when we apply the usual intelligent control techniques.

A simplified statistical example. Let us consider a statistical version of a classical AI example:

- birds normally fly,
- penguins are birds,
- penguins normally do not fly, and
- Sam is a penguin.

The question is: does Sam fly? To make it into a statistical example, let us add some probabilities. Let us assume that 90% of the birds fly, and that 99% of the penguins do not fly (of course, in reality, 100% of the penguins do not fly, but let us keep it under 100% since in most real-life situations, we are never 100% sure about anything).

From the viewpoint of common sense, the information about birds flying in general is rather irrelevant for our situation—since we know that Sam is not just any bird, it is a penguin, a very specific type of bird for which we know the probability of flying. So, to find the probability of Sam flying, we should only take into account information about penguins and thus, conclude that the probability of Sam flying is $100 - 99 = 1\%$.

However, this is not what we would get if we use the standard statistical techniques. Indeed, from the purely statistical viewpoint, here we have two rules that lead us to two different conclusions:

- since Sam is a bird, we can make a conclusion A that Sam flies, with probability $a = 90\%$; and
- since Sam is a penguin, we can make a conclusion B that Sam does not fly, with probability $b = 99\%$.

These two conclusions cannot be both right, since the probabilities of Sam flying and not flying should add up to 1, and here we have $0.9 + 0.99 = 1.89 > 1$. This means that these conclusions are inconsistent.

From the purely logical viewpoint, if we have two statements A and B each of which may be true or false, we can have four possible situations:

- both A and B are true, i.e., $A \,\&\, B$;
- A is true but B is false, i.e., $A \,\&\, \neg B$;
- A is false but B is true, i.e., $\neg A \,\&\, B$; and
- both A and B are false, i.e., $\neg A \,\&\, \neg B$.

In general, the probabilities $P(.)$ of all four situations can be obtained by using the Maximum Entropy Principle—a natural extension of the Laplace Indeterminacy Principle—according to which, if we do not know the dependence between two random variables, then we should assume that they are independent; see, e.g., [3]. For independent events, probabilities multiply, so $P(A \,\&\, B) = P(A) \cdot P(B) = a \cdot b$, $P(A \,\&\, \neg B) = a \cdot (1 - b)$, $P(\neg A \,\&\, B) = (1 - a) \cdot b$, and $P(\neg A \,\&\, \neg B) = (1 - a) \cdot (1 - b)$.

In our case, the statements A and B are inconsistent, so we cannot have $A \,\&\, B$ and we cannot have $\neg A \,\&\, \neg B$. The only two consistent options are $A \,\&\, \neg B$ and

$\neg A \,\&\, B$. Thus, the true probabilities $P(A)$ and $P(B)$ of A and B can be found if we restrict ourselves to consistent situations:

$$P(A) = P(A \mid \text{consistent}) = \frac{P(A \,\&\, \text{consistent})}{P(\text{consistent})} =$$

$$\frac{P(A \,\&\, \neg B)}{P(A \,\&\, \neg B) + P(\neg A \,\&\, B)} = \frac{a \cdot (1-b)}{a \cdot (1-b) + (1-a) \cdot b}$$

and, of course, $P(B) = 1 - P(A)$.

In our example, with $a = 0.9$ and $b = 0.99$, we get

$$P(A) = \frac{0.9 \cdot 0.01}{0.9 \cdot 0.01 + 0.1 \cdot 0.99} = \frac{0.009}{0.009 + 0.099} = \frac{1}{12} \approx 8\%.$$

So, instead the desired 1%, we get a much larger 8% probability—clearly affected by the general rule that birds normally fly.

This is a simplified example, but it explains why recommendation systems based on usual statistical rules becomes biased: if a person with a perfect credit history happens to live in a poor neighborhood, in which the overall loan success rate is small, this person's chances of getting a loan will be decreased. Similarly, if a female student with perfect credentials applies for a graduate program, the system would be treating her less favorably—since in general, in computer science, female students succeed with lower frequency.

In both cases, we have clearly unfair situations—the system designers may honestly give female students a better chance to succeed, but instead, their inference system perpetrates the inequality.

A simplified fuzzy example. A fuzzy-related reader may view the above example as one more example of why statistical methods are not always applicable, and why alternative methods—such as fuzzy methods—are needed. Alas, we will show that a very similar example is possible if we use the usual fuzzy techniques.

This problem may not be well known for fuzzy recommendation systems—since there are not too many of them actively used—but it is exactly the same problem that is well known in fuzzy control, the traditional application area of fuzzy techniques; see, e.g., [2, 4, 5, 7, 8, 10].

Indeed, suppose that we have two rules that describe how the control u should depend on the value x of the measured quantity:

- if x is small, then u is small; and
- if $x = 0.2$, then $u = 0.3$.

Suppose also that the notion "small" is described by a triangular membership function $\mu_{\text{small}}(x)$ which is equal to $\max(1 - |x|, 0)$.

From the common sense viewpoint, the first rule is more general, while the second rule—which is actually in full agreement with the first one—describes a specific knowledge that we have about control corresponding to the value $x = 0.2$. Such situations can happen, e.g., when we combine the general expert knowledge (the first rule) with the results of specific calculations (second rule).

In this case, if we actually have the value $x = 0.2$ for which we know the exact control value $u = 0.3$, we should return this control value.

One of the usual ways of dealing with a system of fuzzy rules "if $A_i(x)$ then $B_i(u)$" ($i = 1, \ldots, n$) is to take into account that a control u is reasonable for given value x if:

- either the first rule is applicable, i.e., this rule's condition $A_1(x)$ is satisfied *and* thus, its conclusion $B_1(u)$ is also satisfied,
- or the second rule is applicable, i.e., this rule's condition $A_2(x)$ is satisfied and thus, its conclusion $B_2(u)$ is also satisfied, etc.

If we denote this property "u is reasonable for x" by $R(x, u)$, and use the usual notations & for "and" and $\vee$ for "or", then the above text will become the following formula:

$$R(x, u) \leftrightarrow (A_1(x) \,\&\, B_1(u)) \vee (A_2(x) \,\&\, B_2(u)) \vee \ldots$$

In line with the general fuzzy methodology, for situations in which we are not 100% sure about the properties A_i and B_j, we can apply the corresponding fuzzy versions $f_\&(a, b)$ and $f_\vee(a, b)$ of usual "and" and "or"—known as "and"- and "or"-operations (or, alternatively, t-norm and t-conorm)—to the degrees $\mu_{A_i}(x)$ and $\mu_{B_i}(u)$ to which these properties are satisfied. Then, for the degree $\mu_r(x, u)$ to which u is reasonable for x, we get the following formula:

$$\mu_r(x, u) = f_\vee(f_\&(\mu_{A_1}(x), \mu_{B_1}(u)), f_\&(\mu_{A_2}(x), \mu_{B_2}(u)), \ldots).$$

In particular, for the simplest possible "and"- and "or"-operations $f_\&(a, b) = \min(a, b)$ and $f_\vee(a, b) = \max(a, b)$, we get

$$\mu_r(x, u) = \max(\min(\mu_{A_1}(x), \mu_{B_1}(u)), \min(\mu_{A_2}(x), \mu_{B_2}(u)), \ldots).$$

Once we have this degree for each u, we can find the control $\overline{u}$ corresponding to x by requiring that its mean square deviation from the actual value u—weighted by this degree—is the smallest possible. In precise terms, for a given x, we minimize the expression $\int \mu_r(x, u) \cdot (u - \overline{u})^2$. Differentiating this expression with respect to $\overline{u}$ and equating the derivative to 0, we get the formula

$$\overline{u} = \frac{\int \mu_r(x, u) \cdot u\, du}{\int \mu_r(x, u)\, du}$$

known as *centroid defuzzification.*

Let us apply this technique to our two rules, for the case when $x = 0.2$ and thus, $\mu_{\text{small}}(x) = 0.8$. In the second rule, both the condition and the conclusion are crisp:

- we have $\mu_{A_2}(0.2) = 1$ and $\mu_{A_2}(x) = 0$ for all other values x, and

- we have $\mu_{B_2}(0.3) = 1$ and $\mu_{B_2}(u) = 0$ for all other values u.

Thus, for all $u \neq 0.2$, we have $\mu_r(x, u) = \min(\mu_{\text{small}}(u), 0.8)$ and for $u = 0.2$, we have $\mu_r(x, u) = 1$.

According to the centroid formula, the resulting control is the above ratio of two integrals. The single-point change in the function $\mu_r(x, u)$ does not affect its integral, so the numerator is simply equal to the integral of the product $\min(\mu_{\text{small}}(u), 0.8) \cdot u = \min(\max(1 - |u|), 0), 0.8) \cdot u$. This product is an odd function of u: the first factor does not change if we replace u with $-u$, and the second changes sign. Thus, its integral is 0, and so, the usual fuzzy methodology leads to the control $u = 0$—while from the viewpoint of common sense, we should get 0.3.

3 Using More Detailed Models Helps

General description of the problem. In all previous examples, we considered the case of situations when we have two rules describing a given situation. For example, in the case of loans:

- the first rule is that loans recipients from poor areas often default on a loan, and
- the second rule is that people with a good credit record usually pay back their loans.

From the common sense viewpoint, for a person with a good credit record living in a poor area, we should go with the second rule, but the above-described naive statistical approach—implemented in current machine learning systems—pays an unnecessarily high attention to the first rule as well.

Similarly, for Sam the penguin:

- we have a general rule applicable to all the birds—that they usually fly; and
- we have a second specific rule, applicable only to penguins—that they do not fly.

From the common sense viewpoint, since Sam in a penguin, we should go with the second rule, but the naive statistical approach gives too much weight to the first rule.

Idea. From the statistical viewpoint—or, more generally, from the viewpoint of data processing—how can we distinguish between a more general rule and a more specific rule? One important difference between a more general case is that this case describes a larger sample, while a more specific case describes a sub-sample of this sample, a sub-sample in which all the objects are, in some reasonable sense, similar and thus, differ from each other less than in the general sample. As a result, for many quantities characterizing the objects, the standard deviation σ corresponding to the larger sample is much larger than for a smaller sub-sample.

This is simple and reasonable, and—as we show—it helps put more weight on a more general rule and thus, helps avoid the contradiction between statistics and fairness.

How to combine two statistical rules with different means and standard deviations: reminder. To illustrate our point, let us consider the simplest situation when we have two statistical rules—coming from two independent sets of arguments or observation—that predict the value of a quantity x, and we are absolutely confident in both of these rules. Since these are statistical rules, they do not predict the exact value of the quantity, they only predict the probabilities of different possible values of this quantity. These probabilities can be described by the corresponding probability density functions $\rho_1(x)$ and $\rho_2(x)$.

If these were rules predicting two different quantities x_1 and x_2, then, due to the fact that these rules are assumed to be independent, the probability to have values x_1 and x_2 should be equal to the product $\rho_1(x_1) \cdot \rho_2(x_2)$. However, in our case, we know that these distributions describe the exact same quantity, i.e., that we have the additional condition $x_1 = x_2$. Thus, instead of the above 2-D probability density, we need to consider the *conditional* probability density, under the condition that $x_1 = x_2$. It is known that, in general, for $A \subseteq B$, the conditional probability $P(A \mid B)$ can be obtained from the probability $P(A)$ by diving it by the probability $P(B)$ that the condition is satisfied—i.e., in effect, by dividing the probability $P(A)$ by a constant. Thus, in our case, the resulting probability density is equal to $\rho(x) = c \cdot \rho_1(x) \cdot \rho_2(x)$, where c is a constant that can be determined from the condition that $\int \rho(x)\,dx = 1$, so that

$$\rho(x) = \frac{\rho_1(x) \cdot \rho_2(x)}{\int \rho_1(y) \cdot \rho_2(y)\,dy}.$$

In particular, if both probability distributions $\rho_1(x)$ and $\rho_2(x)$ are Gaussian, i.e., have the form $\rho_i(x) = \text{const} \exp\left(-\frac{(x - a_i)^2}{2\sigma_i^2}\right)$ for some means a_i and standard deviations σ_i, then, as one can easily check, the resulting distribution is also Gaussian, with mean a and standard deviation σ determined by the formulas $a = \frac{a_1 \cdot \sigma_1^{-2} + a_2 \cdot \sigma_2^{-2}}{\sigma_1^{-2} + \sigma_2^{-2}}$ and $\sigma^{-2} = \sigma_1{}^{2} + \sigma_2{}^{2}$.

How is this applicable to our examples. Let us consider the case of a loan. Here, we have two pieces of information about a loan applicant:

- the first piece of information is that this person has a good credit history;
- the second piece of information is that this person lives in a poor area.

To combine these two pieces of information, let us estimate the corresponding means and standard deviations.

Let us start with the estimates corresponding to people with good credit history. In most cases, people with good credit history return their loans—and return them on time. So, the mean value a_1 of the returned percentage of the loan x is close to 100, and the corresponding standard deviation is σ_1 is close to 0.

On the other hand, in general, for people living in a poor area, the returned percentages vary:

- some people living in the poor area struggle, but return their loans,

- some fail and become unable to return their loans.

Here, the average a_2 is clearly less that 100, and the standard deviation σ_2 is clearly much larger than σ_1:

$$\sigma_2 \gg \sigma_1.$$

If we multiply both the numerator and the denominator of the above formula for the combined value a by σ_1^2, we conclude that $a = \dfrac{a_1 + a_2 \cdot (\sigma_1^2/\sigma_2^2)}{1 + \sigma_1^2/\sigma_2^2}$. Since here $\sigma_1 \ll \sigma_2$, we get $a \approx a_1$. So, we conclude that the resulting estimate is fully determined by the fact that the applicant has a good credit history—and this estimate is practically *not* affected by the fact that the applicant happens to live in a poor area. This is exactly what we wanted the system to conclude.

Similar arguments help resolve the bird-fly puzzle. As a measure of a flying ability, we can take, e.g., the time that a bird can stay in the air.

- No penguin can really fly, so for penguins, this time is always small, and the standard deviation of this time is close to 0: $\sigma_1 \approx 0$.
- On the other hand, if we consider the population of all the birds, then on this general population, there is a large variance: some birds can barely fly for a few minutes, while others can fly for days and cross the oceans. For this piece of knowledge, the variance is huge and thus, the standard deviation σ_2 is also huge.

Here too, $\sigma_1 \ll \sigma_2$ and thus, our conclusion about Sam's ability to fly will be determined practically exclusively by the fact that Sam is a penguin—in full agreement with common sense.

4 How Is This Idea Applicable to Fuzzy

Usual relation between fuzzy and probability. As Lotfi Zadeh mentioned several times, from the mathematical viewpoint, the main difference between a probability density function $\rho(x)$ and a membership function $\mu(x)$ is in their normalization:

- for a probability density function, we have $\int \rho(x)\,dx = 1$, while
- for a membership function, we have $\max\limits_x \mu(x) = 1$.

As a result:

- if we have a probability density function $\rho(x)$, then we can normalize it as membership function, by taking

$$\mu(x) = \frac{\rho(x)}{\max\limits_y \rho(y)};$$

- if we have a membership function $\mu(x)$, then we can normalize it as a probability density function, by taking

$$\rho(x) = \frac{\mu(x)}{\int \mu(y)\,dy}.$$

Let us use this relation to combine fuzzy knowledge. We know how to combine probabilistic knowledge. So, if we have two membership functions $\mu_1(x)$ and $\mu_2(x)$, we can combine the corresponding pieces of knowledge as follows:

- first, we use the above relation to transform the given membership functions into probability density functions $\rho_i(x) = c_i \cdot \mu_i(x)$, for some constants c_i;
- second, we use the procedure described in the previous section to combine the probability density functions $\rho_1(x)$ and $\rho_2(x)$ into a single probability density function $\rho(x) = \text{const} \cdot \rho_1(x) \cdot \rho_2(x)$—which, due to the above relation between probability and fuzzy, takes the form $\rho(x) = c_3 \cdot \mu_1(x) \cdot \mu_2(x)$ for some constant c_3;
- finally, we transform the resulting probability function $\rho(x)$ back into a membership function, thus getting $\mu(x) = c_4 \cdot \rho(x)$ for some constant c_4, i.e., $\mu(x) = c \cdot \mu_1(x) \cdot \mu_2(x)$ for an appropriate constant c.

This idea allows us to avoid the problem of traditional defuzzification. Let us show that this combination rule enables us to avoid the above-described problem of traditional defuzzification. Indeed, if we have two rules:

- one rule corresponding to a very narrow membership function (i.e., in probabilistic terms, very small σ), and
- another rule with a very wide membership function (i.e., with large σ),

then, as we have mentioned in the previous section, in the combined function, the contribution of the wide rule will be largely ignored, and the conclusion will be practically identical with what the narrow rule recommends—exactly as we want.

What if we are only partly confident about some piece of knowledge? The above combination formula describes how to combine two rules about which we are fully confident. But what if we have some rules about which we are only partly confident?

One way to interpret degree of confidence in a statement is:

- to have a poll of N experts and,
- if M out of N experts confirm this statement, to take the corresponding proportion M/N as the desired degree of confidence.

Let us describe the membership function corresponding to the situation when only one expert confirms the statement by $\mu_1(x)$. In this case, according to the above combination formula, the case when M experts confirm the statement is described by a membership function proportional to $\mu_1^M(x)$. In particular, the case of full confidence, when all N experts confirm the statement, is described by the membership function

$\mu(x)$ which is proportional to $\mu_1^N(x)$: $\mu(x) \sim \mu_1^N(x)$. Thus, $\mu_1(x) \sim (\mu(x))^{1/N}$ and therefore, the membership function $\sim \mu_1^M(x)$ corresponding to degree of confidence $d = M/N$ is proportional to $(\mu(x))^{M/N} = \mu^d(x)$.

In general, if we have a rule like $A(x) \to B(u)$ relating the property $A(x)$ of the input (with membership function $\mu_A(x)$) and the property $B(u)$ of the desired control u (with membership function $\mu_B(u)$), then for each input x, our degree of confidence in the conclusion $B(u)$ is equal to $d = \mu_A(x)$. Thus, the resulting membership function about u should be proportional to $(\mu_B(u))^{\mu_A(x)}$. In we have several such rules $A_1(x) \to B_1(u)$, $A_2(x) \to B_2(u)$, etc., then the resulting membership function should be proportional to the product of membership functions corresponding to individual rules, i.e., to the product

$$(\mu_{B_1}(u))^{\mu_{A_1}(x)} \cdot (\mu_{B_2}(u))^{\mu_{A_2}(x)} \cdot \ldots$$

Acknowledgements This work was supported in part by the US National Science Foundation grants 1623190 (A Model of Change for Preparing a New Generation for Professional Practice in Computer Science) and HRD-1242122 (Cyber-ShARE Center of Excellence).

References

1. R.K.E. Bellamy et al., Think your Artifical Intelligence software is fair? Thing again. Comput. Edge, pp. 14–18 (2020)
2. R. Belohlavek, J.W. Dauben, G.J. Klir, *Fuzzy Logic and Mathematics: A Historical Perspective* (Oxford University Press, New York, 2017)
3. E.T. Jaynes, G.L. Bretthorst, *Probability Theory: The Logic of Science* (Cambridge University Press, Cambridge, UK, 2003)
4. G. Klir, B. Yuan, *Fuzzy Sets and Fuzzy Logic* (Prentice Hall, Upper Saddle River, New Jersey, 1995)
5. J.M. Mendel, *Uncertain Rule-Based Fuzzy Systems: Introduction and New Directions* (Springer, Cham, Switzerland, 2017)
6. B. Mittelstadt, C. Russell, S. Wachter, Explaining explanations in AI, in *Proceedings of the 2019 ACM Fairness, Accounatbility, and Transparency Conference FAT'2019*, Atlanta, Georgia, January 29–31 (2019), pp. 279–288
7. H.T. Nguyen, C.L. Walker, E.A. Walker, *A First Course in Fuzzy Logic* (Chapman and Hall/CRC, Boca Raton, Florida, 2019)
8. V. Novák, I. Perfilieva, J. Močkoř, *Mathematical Principles of Fuzzy Logic* (Kluwer, Boston, Dordrecht, 1999)
9. C. O'Neil, *Weapons of Math Destruction: How Big Data Increases Inequality and Threatens Democracy* (Crown Books, New York, 2016)
10. L.A. Zadeh, Fuzzy sets. Inf. Control **8**, 338–353 (1965)

Which Algorithms Are Feasible and Which Are Not: Fuzzy Techniques Can Help in Formalizing the Notion of Feasibility

Olga Kosheleva and Vladik Kreinovich

Abstract Some algorithms are practically feasible, in the sense that for all inputs of reasonable length they provide the result in reasonable time. Other algorithms are not practically feasible, in the sense that they may work well for small-size inputs, but for slightly larger – but still reasonable-size – inputs, the computation time becomes astronomical (and not practically possible). How can we describe practical feasibility in precise terms? The usual formalization of the notion of feasibility states that an algorithm is feasible if its computation time is bounded by a polynomial of the size of the input. In most cases, this definition works well, but sometimes, it does not: e.g., according to this definition, every algorithm requiring a constant number of computational steps is feasible, even when this number of steps is larger than the number of particles in the Universe. In this paper, we show that by using fuzzy logic, we can naturally come up with a more adequate description of practical feasibility.

1 Formulation of the Problem

Some algorithm are feasible and some are not. Computer scientists have invented many different algorithms. Some of these algorithm are practically feasible, in the sense that for inputs of reasonable size, they require reasonable (and practically implementable) time. Examples of such algorithms include different algorithms for search, for sorting, for solving systems of linear equations, etc.; see, e.g., [2].

On the other hand, there are algorithms which always produce the correct results but which, in practice, only work for small size inputs – otherwise, they require an unrealistic amount of computation time. A good example is an exhaustive search algorithm for solving the propositional satisfiability problem – given a propositional

O. Kosheleva · V. Kreinovich (✉)
University of Texas at El Paso, 500 W. University, El Paso, TX 79968, USA
e-mail: vladik@utep.edu

O. Kosheleva
e-mail: olgak@utep.edu

B. Bede et al. (eds.), *Fuzzy Information Processing 2020*,
Advances in Intelligent Systems and Computing 1337,
https://doi.org/10.1007/978-3-030-81561-5_34

formula (i.e., an expression obtained from Boolean (yes-no) variables $v_1, \ldots, v_n$ by using "and", "or", and "not"), find the values of these variables that make the formula true. In principle, we can solve this problem by trying all 2^n possible tuples of values $(v_1, \ldots, v_n)$ – each variable has two possible values (true of false), so the tuple has 2^n possible values.

- It works for $n = 10$, when we need $2^{10} \approx 10^3$ computational steps.
- It works for $n = 20$, when we need $2^{20} \approx 10^6$ steps.
- It works for $n = 30$, when we need $2^{30} \approx 10^9$ computational steps, one second or less on a usual GigaHertz computer.

However, already for a very reasonable size input $n = 300$, we will need $2^{300} \approx 10^{100}$ computational steps – which would require time which is much much longer than the lifetime of the Universe. So this algorithm is clearly not practically feasible.

It is desirable to have a precise definition of feasibility. It would be nice to know which algorithm is practically feasible and which is not. It is not easy to make such a conclusion based on the above description of practical feasibility, since this description uses imprecise works like "reasonable". To make the corresponding conclusion, it is desirable to have a precise definition of what is feasible.

How is the notion of feasibility described now. The existing formal definition of feasibility is based on the following fact:

- for the vast majority of practically feasible algorithms – including search, sorting, solving systems of linear equations – the worst-case computation time $t(n)$ on inputs of size n is bounded by some polynomial of n, while
- for the vast majority of not practically feasible algorithms – like the above-described exhaustive search algorithm – the worst-case computation time is exponential – or at least grows faster than any polynomial.

Because of this fact, formally, an algorithm is called *feasible* if its worst-case computation time $t(n)$ is bounded by some polynomial – i.e., if there exists a polynomial $P(n)$ for which $t(n) \le P(n)$ for all n; see, e.g., [4, 8].

The current formal definition is not fully adequate. In many cases, the above formal definition correctly describes what is feasible and what is not feasible. However, there are cases when this definition does not adequately describe practical feasibility. Let us give two examples:

- when $t(n) = 10^{100} \cdot n$, this expression is a polynomial – so it is feasible according to the current formal definition – but it is clearly *not* practically feasible, since even for inputs of length 1, this algorithm requires impossible 10^{100} steps to finish; similar arguments can be given if $t(n)$ is a large constant – e.g., if $t(n) = 10^{100}$ for all input sizes n;
- on the other hand, when $t(n) = \lceil \exp(10^{-20} \cdot n) \rceil$, then, strictly speaking, it is an exponential function, so it grows faster than any polynomial (and is, thus, not feasible in the sense of the formal definition), but even when we input the whole body of current knowledge, with $n = 10^{18}$, this algorithm will work really fast – in $\lceil \exp(10^{-20} \cdot 10^{18}) \rceil = \lceil \exp(0.01) \rceil = 2$ steps.

So, we arrive at a natural question.

A natural question, and what we do in this paper. Can we come up with an alternative precise definition of feasibility that would be more adequate? In this paper, we show that fuzzy techniques (see, e.g., [1, 3, 5–7, 9]) can help in providing such a definition.

2 Analysis of the Problem and Possible Solution

Natural idea: using fuzzy techniques. The informal description of practical feasibility uses the natural-language word "reasonable". Like many other natural-language words – like "small", "large", etc. – this word is not precise. Different people may disagree on what is reasonable, and for large but not too large sizes n, even a single person can be unsure whether this size is reasonable or not.

It is precisely to deal with such imprecise ("fuzzy") words from natural language that Lotfi Zadeh invented fuzzy techniques. So, a natural idea is to use fuzzy techniques to formalize the notion of practical feasibility.

Let us apply fuzzy techniques. To use fuzzy techniques, let us first re-formulate the above description of practical feasibility in more precise terms. Practical feasibility means that for all possible lengths n, if n is reasonable, then $t(n)$ should be reasonable too. If we denote "n is reasonable" by $r(n)$, then the definition of practical feasibility takes the following form $\forall n\,(r(n) \to r(t(n)))$, or, equivalently,

$$(r(1) \to r(t(1)))\,\&\,(r(2) \to r(t(2)))\,\&\,\ldots \qquad (1)$$

In fuzzy logic, our degree of confidence in each statement S is described by a number from the interval [0, 1]:

- the value 1 means that we are absolutely confident that the statement S is true;
- the value 0 means that we are absolutely confident that the statement S is false; and
- values between 0 and 1 indicate intermediate situations, when we are confident only to some extent.

For each imprecise property like $r(n)$, we can describe, for each n, the degree $R(n)$ that this property is true (i.e., in our case, that n is reasonable). The mapping that assigns this degree to each n is known as the *membership function* describing the corresponding notion.

Clearly, if the value n is reasonable, then all smaller values are reasonable as well. Thus, the degree $R(n)$ should be non-strictly decreasing, from $R(1) = 1$ to $R(n) \to 0$ as n increases.

To come up with estimates of composite statements – obtained by using logical connectives like "and" and "if ... then" from the elementary statements – we can use fuzzy analogues of these connectives, i.e., appropriate extensions of the usual logical connectives from the two-valued set $\{0, 1\} = \{\text{false, true}\}$ to the whole interval $[0, 1]$.

The simplest possible "and"-operation is $\min(a, b)$, the simplest possible "or"-operation is $\max(a, b)$, and the simplest possible negation operation is $1 - a$. Implication $A \to B$ is, in classical logic, equivalent to $B \vee \neg A$. Thus, if we know the truth values a and b of (= degrees of confidence in) statement A and B, then the truth value of the implication $A \to B$ can be estimated as $\max(b, 1 - a)$. Thus, the truth value of the formula (2) – i.e., the degree $D(t)$ to which an algorithm with worst-case time complexity $t(n)$ is practically feasible – takes the following form:

$$D(t) = \min(\max(R(t(1)), 1 - R(1)), \max(R(t(2)), 1 - R(2)), \ldots) =$$

$$\min_n \max(R(t(n)), 1 - R(n)). \quad (2)$$

If we use a general "and"-operation $f_\&(a, b)$ and a general implication operation $f_\&(a, b)$, we get the following formula:

$$D(t) = f_\&(f_\to(R(1), R(t(1))), f_\to(R(2), R(t(2))), \ldots) \quad (3)$$

This is our precise definition of practical feasibility.

The proposed new precise definition of practical feasibility is indeed more adequate than the existing one. Let us show that already for the simplest possible operations $f_\&(a, b) = \min(a, b)$ and $f_\to(a, b) = \max(b, 1 - a)$, the above definition is more adequate that the existing formal definition.

Indeed, for example, according to the formal definition, any function with constant time $t(n) = t = \text{const}$ is feasible. What will happen is we use our definition (2) – or, to be precise, its simplest-case version (1)? When n increases, the value $R(n)$ decreases, thus the value $1 - R(n)$ increases and the value

$$\max(R(t(n)), 1 - R(n)) = \max(R(t), 1 - R(n))$$

also increases. So, the minimum $D(t)$ is attained when the size n is the smallest, i.e., when $n = 1$:

$$D(t) = \max(R(t), 1 - R(1)).$$

When the constant value t is small, this degree is reasonable and the degree $D(t)$ that this computation time corresponds to a practically feasible algorithm is also reasonable. However, as the constant t increases, the value $R(t)$ tends to 0 and thus, $D(t)$ tends to a very small (practically 0) degree of confidence $1 - R(1)$ that 1 is not feasible – i.e., as desired, such an algorithm stops being feasible for large t. Actually, here $D(t) \le R(t)$, so if the constant t is not reasonable, the corresponding time complexity is not practically feasible.

Similarly, for a function like $t(n) = \exp(10^{-20} \cdot n)$, the value $R(t(n))$ becomes very small for large n – but for large n, $R(n)$ is also close to 0 and thus, $1 - R(n)$

is close to 1 and hence, the maximum $\max(R(t(n)), 1 - R(n)) \geq 1 - R(n)$ is also close to 1. Thus, the fact that the value $R(t(n))$ is small for such huge n does not affect the minimum $D(t)$, and the degree of confidence that this computation time is practically feasible remains high.

How to actually compute the newly defined degree of feasibility: towards an algorithm. OK, the definition is reasonable, but how can we actually compute the corresponding degree (2)? Even in its simplest form (1), it is defined as the minimum of infinitely many terms!

It turns out that to come up with the degree $D(t)$, there is no need to actually compute all these infinitely many terms. Indeed, we can use the fact that:

- the function $R(n)$ is decreasing and tending to 0,
- thus $1 - R(n)$ is increasing and tending to 1,
- while $R(t(n))$ is decreasing and tending to 0.

So, for large n, we thus have $R(t(n)) \leq 1 - R(n)$.

If this inequality holds for some n, then for $n \geq n'$, due to the above-described monotonicity, we have

$$R(t(n')) \leq R(t(n)) \leq 1 - R(n) \leq 1 - R(n')$$

thus $R(t(n')) \leq 1 - R(n')$. So, if this inequality holds for some n, it holds for all larger values n as well. Hence, there exists the smallest value n_0 for which this inequality is true.

For all values $n \geq n_0$, we have

$$\max(R(t(n)), 1 - R(n)) = 1 - R(n).$$

This term increases with n, thus the smallest possible value of this term is attained when n is the smallest, i.e., when $n = n_0$. For this value n, we have

$$\max(R(t(n_0)), 1 - R(n_0)) = 1 - R(n_0).$$

For values $n < n_0$, we have $R(t(n)) > 1 - R(n)$ and thus,

$$\max(R(t(n)), 1 - R(n)) = R(t(n)).$$

This term decreases with n, thus the smallest possible value of this term is attained when n is the largest, i.e., when $n = n_0 - 1$. For this value n, we have

$$\max(R(t(n_0 - 1)), 1 - R(n_0 - 1)) = R(t(n_0 - 1)).$$

Thus, to find the smallest possible value of the maximum-expression

$$\max(R(t(n)), 1 - R(n)),$$

there is no need to consider all infinitely many values of this expression corresponding to all possible natural numbers n: it is sufficient to consider only two values of this expression, corresponding to $n = n_0$ and to $n = n_0 - 1$. So, we arrive at the following algorithm.

How to actually compute the newly defined degree of feasibility: algorithm. Find the first value n_0 for which $R(t(n)) \le 1 - R(n)$. This value can be found, e.g., by bisection (see, e.g., [2]). Then, for $n_0 > 1$, we have

$$D(t) = \min(R(t(n_0 - 1)), 1 - R(n_0)).$$

Comment. For $n_0 = 1$, we similarly get $D(t) = 1 - R(1)$.

Acknowledgements This work was supported in part by the US National Science Foundation grants 1623190 (A Model of Change for Preparing a New Generation for Professional Practice in Computer Science) and HRD-1242122 (Cyber-ShARE Center of Excellence).

References

1. R. Belohlavek, J.W. Dauben, G.J. Klir, *Fuzzy Logic and Mathematics: A Historical Perspective* (Oxford University Press, New York, 2017)
2. Th.H. Cormen, C.E. Leiserson, R.L. Rivest, C. Stein, *Introduction to Algorithms* (MIT Press, Cambridge, Massachusetts, 2009)
3. G. Klir, B. Yuan, *Fuzzy Sets and Fuzzy Logic* (Prentice Hall, Upper Saddle River, New Jersey, 1995)
4. V. Kreinovich, A. Lakeyev, J. Rohn, P. Kahl, *Computational Complexity and Feasibility of Data Processing and Interval Computations* (Kluwer, Dordrecht, 1998)
5. J.M. Mendel, *Uncertain Rule-Based Fuzzy Systems: Introduction and New Directions* (Springer, Cham, Switzerland, 2017)
6. H.T. Nguyen, C.L. Walker, E.A. Walker, *A First Course in Fuzzy Logic* (Chapman and Hall/CRC, Boca Raton, Florida, 2019)
7. V. Novák, I. Perfilieva, J. Močkoř, *Mathematical Principles of Fuzzy Logic* (Kluwer, Boston, Dordrecht, 1999)
8. C. Papadimitriou, *Computational Complexity* (Addison-Wesley, Reading, Massachusetts, 1994)
9. L.A. Zadeh, Fuzzy sets. Inf. Control **8**, 338–353 (1965)

Centroids Beyond Defuzzification

Juan Carlos Figueroa García, Christian Servin, and Vladik Kreinovich

Abstract In general, expert rules expressed by imprecise (fuzzy) words of natural language like "small" lead to imprecise (fuzzy) control recommendations. If we want to design an automatic controller, we need, based on these fuzzy recommendations, to generate a single control value. A procedure for such generation is known as defuzzification. The most widely used defuzzification procedure is centroid defuzzification, in which, as the desired control value, we use one of the coordinates of the center of mass ("centroid") of an appropriate 2-D set. A natural question is: what is the meaning of the second coordinate of this center of mass? In this paper, we show that this second coordinate describes the overall measure of fuzziness of the resulting recommendation.

1 Formulation of the Problem

Centroid defuzzification: a brief reminder. In fuzzy control (see, e.g., [1, 3–7]):

- we start with the expert rules formulated in terms in imprecise ("fuzzy") words from natural language, and
- we end up with a strategy that, given the current values of the inputs, provides recommendations on what control value u to use.

J. C. Figueroa García
Universidad Distrital Francisco José de Caldas, Bogotá, Colombia
e-mail: jcfigueroag@udistrital.edu.co

C. Servin
Computer Science and Information Technology Systems Department, El Paso Community College (EPCC), 919 Hunter Dr., El Paso, TX 79915-1908, USA
e-mail: cservin1@epcc.edu

V. Kreinovich (✉)
University of Texas at El Paso, 500 W. University, El Paso, TX 79968, USA
e-mail: vladik@utep.edu

B. Bede et al. (eds.), *Fuzzy Information Processing 2020*,
Advances in Intelligent Systems and Computing 1337,
https://doi.org/10.1007/978-3-030-81561-5_35

This recommendation is also fuzzy: for each possible value u, the system provides a degree $\mu(u) \in [0, 1]$ indicating to what extent this particular control value is reasonable for the given input.

- Such a fuzzy outcome is perfect if the main objective of the system is to advise a human controller.
- In many practical situations, however, we want this system to actually control.

In such situations, it is important to transform the fuzzy recommendation – as expressed by the function $\mu(u)$ (known as the *membership function*) – into a precise control value $\overline{u}$ that this system will apply. Such a transformation is known as *defuzzification*.

The most widely used defuzzification procedure is *centroid defuzzification*

$$\overline{u} = \frac{\int u \cdot \mu(u)\, du}{\int \mu(u)\, du}. \tag{1}$$

Geometric meaning of centroid defuzzification. The name for this defuzzification procedure comes from the fact that:

- if we take the subgraph of the function $\mu(u)$, i.e., the 2-D set

$$S \stackrel{\text{def}}{=} \{(u, y) : 0 \le y \le \mu(u)\},$$

- then the value (1) is actually the u-coordinate of this set's center of mass ("centroid") $(\overline{u}, \overline{y})$.

Natural question. In fuzzy technique, we only use the u-coordinate of the center of mass. A natural question is: is there a fuzzy-related meaning of the y-coordinate $\overline{y}$?

What we do in this paper. In this paper, we propose such a meaning.

Comment. This paper follows our preliminary results published in [2].

2 Fuzzy-Related Meaning of the "Other" Component of the Centroid

Mathematical formula for the y-component. In general, the y-component of the center of mass of a 2-D body S has the form

$$\overline{y} = \frac{\int_S y\, du\, dy}{\int_S du\, dy}.$$

The denominator is the same as for the u-component: it is equal to $\int \mu(u)\,du$. The numerator can also be easily computed as

$$\int_S y\,du\,dy = \int_u du \cdot \int_0^{\mu(u)} y\,dy = \frac{1}{2} \cdot \int_u \mu^2(u)\,du.$$

Thus, we have

$$\overline{y} = \frac{1}{2} \cdot \frac{\int \mu^2(u)\,du}{\int \mu(u)\,du}. \tag{2}$$

First meaning of this formula. The u-component of the centroid is the weighted average value of u, with weights proportional to $\mu(u)$. Similarly, the expression (2) is the weighted average value of $\mu(u)$.

Each value $\mu(u)$ is the degree of fuzziness of the system's recommendation about the control value u. Thus, the value $\overline{y}$ can be viewed with the weighted average value of the degree of fuzziness.

Let us show that this interpretation makes some sense.

Proposition 1

- *The value $\overline{y}$ is always between 0 and 1/2.*
- *For a measurable function $\mu(u)$, the value $\overline{y}$ is equal to 1/2 if and only if the value $\mu(u)$ is almost everywhere equal either to 0 or to 1.*

Comments.

- In other words, if we ignore sets of measure 0, the value $\overline{y}$ is equal to 1/2 if and only if the corresponding fuzzy set is actually crisp. For all non-crisp fuzzy sets, we have $\overline{y} < 1/2$.
- For a triangular membership function, one can check that we always have $\overline{y} = 1/3$. For trapezoid membership functions, $\overline{y}$ can take any possible value between 1/3 and 1/2: the larger the value-1 part, the larger $\overline{y}$.

Proof of Proposition 1 Since $\mu(u) \in [0, 1]$, we always have $\mu^2(u) \le \mu(u)$, thus $\int \mu^2(u)\,du \le \int \mu(u)\,du$, hence

$$\frac{\int \mu^2(u)\,du}{\int \mu(u)\,du} \le 1$$

and $\overline{y} \le 1/2$.

Vice versa, if $\overline{y} = 1/2$, this means that

$$\frac{\int \mu^2(u)\,du}{\int \mu(u)\,du} = 1.$$

Multiplying both sides of this equality by the denominator, we conclude that $\int \mu^2(u)\,du = \int \mu(u)\,du$, i.e., that $\int \left(\mu(u) - \mu^2(u)\right)\,du = 0$. As we have mentioned, the difference $\mu(u) - \mu^2(u)$ is always non-negative. Since its integral is 0, this means that this difference is almost always equal to 0 – and the equality

$$\mu(u) - \mu^2(u) = 0$$

means that either $\mu(u) = 0$ or $\mu(u) = 1$.

The proposition is proven.

A version of the first meaning. In general, in the fuzzy case, we have different values of the degree of confidence $\mu(u)$ for different possible control values u. A natural way to find the "average" degree of fuzziness is to find a single degree μ_0 which best represents all these values. This is natural to interpret as requiring that the mean square difference weighted by $\mu(u)$ – i.e., the value

$$\int (\mu(u) - \mu_0)^2 \cdot \mu(u)\,du$$

attains its smallest possible value. Differentiating the minimized expression with respect to μ_0 and equating the derivative to 0, we conclude that

$$\int 2(\mu_0 - \mu(u)) \cdot \mu(u)\,du = 0,$$

hence

$$\mu_0 = \frac{\int \mu^2(u)\,du}{\int \mu(u)\,du}.$$

Thus, $y_0 = (1/2) \cdot \mu_0$.

Second meaning. It is known that from the mathematical viewpoint, membership functions $\mu(u)$ and probability density functions $\rho(u)$ differ by their normalization:

- for a membership function $\mu(u)$, we require that $\max\limits_u \mu(u) = 1$, while
- for a probability density function $\rho(u)$, we require that $\int \rho(u)\,du = 1$.

For every non-negative function $f(u)$, we can divide it by an appropriate constant c and get an example of either a membership function or a probability density function:

- if we divide $f(u)$ by $c = \max\limits_v f(v)$, then we get a membership function

$$\mu(u) = \frac{f(u)}{\max\limits_v f(v)};$$

- if we divide $f(u)$ by $c = \int \mu(v)\,dv$, we get a probability density function

$$\rho(u) = \frac{f(u)}{\int f(v)\,dv}.$$

In particular, for each membership function $\mu(u)$, we can construct the corresponding probability density function

$$\rho(u) = \frac{\mu(u)}{\int \mu(v)\,dv}.$$

In terms of this expression $\rho(u)$, the formulas for both components of the center of mass take a simplified form:

- the result $\overline{u}$ of centroid defuzzification takes the form $\overline{u} = \int u \cdot \rho(u)\,du$, i.e., is simply equal to the expected value of control under this probability distribution;
- similarly, the value $\mu_0 = 2\overline{y}$ takes the form $\mu_0 = \int \mu(u) \cdot \rho(u)\,du$, i.e., is equal to the expected value of the membership function.

It should be mentioned that the formula $\int \mu(u) \cdot \rho(u)\,du$ was first proposed by Zadeh himself to describe the probability of the fuzzy event characterized by the membership function $\mu(u)$. Since this membership function characterizes to what extent a control value u is reasonable, the value μ_0 thus describes the probability that a control value selected by fuzzy control will be reasonable.

This interpretation is in good accordance with Proposition 1:

- if we are absolutely confident in our recommendations, i.e., if $\mu(u)$ is a crisp set, then this probability μ_0 is equal to 1 – and thus, $\overline{y} = (1/2) \cdot \mu_0$ is equal to 1/2;
- on the other hand, if we are not confident in our recommendations, then the probability μ_0 is smaller than 1 and thus, its half $\overline{y}$ is smaller than 1/2.

Acknowledgements This work was supported in part by the US National Science Foundation grants 1623190 (A Model of Change for Preparing a New Generation for Professional Practice in Computer Science) and HRD-1242122 (Cyber-ShARE Center of Excellence).
The authors are thankful to all the participants of the World Congress of the International Fuzzy Systems Association and the Annual Conference of the North American Fuzzy Information Processing Society IFSA/NAFIPS'2019 (Lafayette, Louisiana, June 18–21, 2019) for valuable discussions.

References

1. R. Belohlavek, J.W. Dauben, G.J. Klir, *Fuzzy Logic and Mathematics: A Historical Perspective* (Oxford University Press, New York, 2017)
2. J.C. Figueroa-García, E.R. Lopez, C. Franco-Franco, A note about the (x, y) coordinates of the centroid of a fuzzy set, in *Proceedings of the 5th Workshop on Engineering Applications WEA'2018, Medellin, Colombia, October 17–19, 2018*. ed. by J.C. Figueroa-García, E.R. Lopez-Santana, J.I. Rodriguez-Molano (Springer, 2018), pp. 78–88
3. G. Klir, B. Yuan, *Fuzzy Sets and Fuzzy Logic* (Prentice Hall, Upper Saddle River, New Jersey, 1995)
4. J.M. Mendel, *Uncertain Rule-Based Fuzzy Systems: Introduction and New Directions* (Springer, Cham, Switzerland, 2017)

5. H.T. Nguyen, C.L. Walker, E.A. Walker, *A First Course in Fuzzy Logic* (Chapman and Hall/CRC, Boca Raton, Florida, 2019)
6. V. Novák, I. Perfilieva, J. Močkoř, *Mathematical Principles of Fuzzy Logic* (Kluwer, Boston, Dordrecht, 1999)
7. L.A. Zadeh, Fuzzy sets. Inf. Control **8**, 338–353 (1965)

Equations for Which Newton's Method Never Works: Pedagogical Examples

Leobardo Valera, Martine Ceberio, Olga Kosheleva, and Vladik Kreinovich

Abstract One of the most widely used methods for solving equations is the classical Newton's method. While this method often works – and is used in computers for computations ranging from square root to division – sometimes, this method does not work. Usual textbook examples describe situations when Newton's method works for some initial values but not for others. A natural question that students often ask is whether there exist functions for which Newton's method never works – unless, of course, the initial approximation is already the desired solution. In this paper, we provide simple examples of such functions.

1 Formulation of the Problem

Newton's method: a brief reminder. One of the most widely used methods for finding a solution to a non-linear equation $f(x) = 0$ is a method designed many centuries ago by Newton himself; see, e.g., [1]. This method is based on the fact that the derivative $f'(x)$ is defined as the limit of the ratio $\frac{f(x+h) - f(x)}{h}$ when h tends to 0. This means that for small h, the derivative is approximately equal to this ratio. In this approximation, $f'(x) \approx \frac{f(x+h) - f(x)}{h}$. Multiplying both sides by h, we get $f'(x) \cdot h \approx f(x+h) - f(x)$. Thus, adding $f(x)$ to both sides, we get

L. Valera
The University of Tennessee Knoxville, Knoxville, TN, USA
e-mail: leobardovalera@gmail.com

M. Ceberio · O. Kosheleva · V. Kreinovich (✉)
University of Texas at El Paso, 500 W. University, El Paso, TX 79968, USA
e-mail: vladik@utep.edu

M. Ceberio
e-mail: mceberio@utep.edu

O. Kosheleva
e-mail: olgak@utep.edu

B. Bede et al. (eds.), *Fuzzy Information Processing 2020*,
Advances in Intelligent Systems and Computing 1337,
https://doi.org/10.1007/978-3-030-81561-5_36

$$f(x+h) \approx f(x) + h \cdot f'(x). \tag{1}$$

Suppose that we know some approximation x_k to the desired value x. For this approximation, $f(x_k)$ is not exactly equal to 0. To make the value $f(x)$ closer to 0, it is therefore reasonable to make a small modification of the current approximation, i.e., take $x_{k+1} = x_k + h$. For this new value, according to the formula (1), we have $f(x_{k+1}) \approx f(x_k) + h \cdot f'(x_k)$. To get the value $f(x_{k+1})$ as close to 0 as possible, it is therefore reasonable to take h for which

$$f(x_k) + h \cdot f'(x_k) = 0,$$

i.e., to take $h = -\dfrac{f(x_k)}{f'(x_k)}$. Thus, for the next approximation $x_{k+1} = x_k + h$, we get the following formula:

$$x_{k+1} = x_k - \frac{f(x_k)}{f'(x_k)}. \tag{2}$$

This is exactly what Newton has proposed.

If this method converges precisely – in the sense that we have $x_{k+1} = x_k$, then, from the formula (2), we conclude that $f(x_k) = 0$, i.e., that x_k is the desired solution. If this method converges approximately, i.e., if the difference $x_{k+1} - x_k$ is very small, then, from the formula (2), we conclude that the value $f(x_k)$ is also very small, and thus, we have a good approximation to the desired solution.

This method is still actively used to solve equations. In spite of this method being centuries old, it is still used to solve many practical problems. For example, this is how most computers compute the square root of a given number a, i.e., how computers compute the solution to the equation $f(x) = 0$ with $f(x) = x^2 - a$. For this function $f(x)$, we have $f'(x) = 2x$, thus Newton's formula (2) takes the form

$$x_{k+1} = x_k - \frac{x_k^2 - a}{2x_k}. \tag{3}$$

This formula can be further simplified if we take into account that $\dfrac{x_k^2}{2x_k} = \dfrac{x_k}{2}$ and thus, the formula (3) can be transformed into the following simplified form:

$$x_{k+1} = \frac{1}{2} \cdot \left(x_k + \frac{a}{x_k}\right). \tag{4}$$

The formula (4) is indeed faster to compute than the formula (3): both formulas require one division, but (3) also requires one multiplication (to compute x_k^2) and two subtractions, while (4) needs only one addition. (Both formulas need multiplication or division by 2, but for binary numbers, this is trivial – just shifting by 1 bit to the left or to the right.)

The resulting iterative process (4) converges fast. For example, to compute the square root of $a = 2$, we can start with $x_0 = 1$ and get

$$x_1 = \frac{1}{2} \cdot \left(1 + \frac{2}{1}\right) = 1.5$$

and

$$x_2 = \frac{1}{2} \cdot \left(1.5 + \frac{2}{1.5}\right) = 1.4166\ldots,$$

i.e., in only two iterations, we already have the first three digits of the correct answer $\sqrt{2} = 1.414\ldots$

Newton's method also lies behind the way computers divide. To be more precise, computes compute the ratio $\frac{a}{b}$ by first computing the inverse $\frac{1}{b}$, and then multiplying a by this inverse. To compute the inverse, computers contain a table of pre-computed values of the inverse for several fixed values B_i, and, then, for $b \approx B_i$, use the recorded inverse $\frac{1}{B_i}$ as the first approximation x_0 in the Newton's method. In this case, the desired equation has the form $b \cdot x - 1 = 0$, i.e., here $f(x) = b \cdot x - 1$. The actual derivative $f'(x)$ is equal to b, i.e., ideally we should have

$$x_{k+1} = x_k - \frac{1}{b} \cdot (b \cdot x_k - 1). \tag{5}$$

This may sound reasonable, but since the whole purpose of this algorithm is to compute the inverse value $\frac{1}{b}$, we do not know it yet and thus, we cannot use the above formula directly. What we *do* know, at this stage, is the current approximation x_k to the desired inverse value $\frac{1}{b}$. So, a natural idea is to use x_k instead of the inverse value in the formula (5). Then, we get exactly the form of Newton's method that computers use to compute the inverse:

$$x_{k+1} = x_k - (b \cdot x_k - 1) \cdot x_k. \tag{6}$$

It should be mentioned that, similar to the expression (3), this expression can also be further simplified, e.g., to

$$x_{k+1} = x_k \cdot (2 - b \cdot x_k). \tag{7}$$

Both formulas (6) and (7) require two multiplications, but (7) is slightly faster to compute since this formula requires only one subtraction, while the formula (6) requires two subtractions.

Sometimes, Newton's method does not work. While Newton's method is efficient, there are examples when it does not work – such examples are usually given in textbooks, explaining the need for alternative techniques.

Sometimes, this happens because the values x_k diverge – i.e., become larger and larger with each iteration, never converging to anything. Sometime, this happens because the values x_k from a loop: we get $x_0, \ldots, x_{k-1}$, and then we again get $x_k = x_0$, $x_{k+1} = x_1$, etc. – and the process also never converges.

A natural question. The textbook examples usually show that whether Newton's method is successful depends on how close is the initial approximation x_0 to the actual solution x:

- if x_0 is close to x, then usually, Newton's method converges, while
- if the initial approximation x_0 is far away from the actual solution x, Newton's method starts diverging.

A natural question – that students sometimes ask – is whether this is always the case, or whether there are examples when Newton's method never converges.

What we do in this paper. In this paper, we provide examples when Newton's method never converges, no matter what initial approximation x_0 we take – unless, of course, we happen to take exactly the desired solution x as the first approximation, i.e., unless $x_0 = x$.

2 First Example

Let us look for a simple example. Let us first look for examples in which the equation $f(x) = 0$ has only one solution. For simplicity, let us assume that the desired solution is $x = 0$.

Again, for simplicity, let us consider odd functions $f(x)$, i.e. functions for which $f(-x) = -f(x)$. Let us also consider the simplest possible case when the Newton's method does not converge: when the iterations x_k form a loop, and let us consider the simplest possible loop, when we have x_0, $x_1 \neq x_0$, and then again $x_2 = x_0$, etc.

How to come up with such a simple example. In general, the closer x_0 to the solution, the closer x_1 will be. If x_1 was on the same side of the solution as x_0, then:

- if $x_1 < x_0$, we would eventually have convergence, and
- if $x_1 > x_0$, we would have divergence,

but we want a loop. Thus, x_1 should be on the other side of x_0.

Since the function $f(x)$ is odd, the dependence of x_2 on x_1 is exactly the same as the dependence of x_1 on x_0. So:

- if $|x_1| < |x_0|$, we would have convergence, and
- if $|x_1| > |x_0|$, we would have divergence.

The only way to get a loop is thus to have $|x_1| = |x_0|$.

Since the values x_0 and x_1 are on the other solution of the solution $x = 0$, this means that we must have $x_1 = -x_0$. According to the formula (2), we have $x_1 = x_0 - \dfrac{f(x_0)}{f'(x_0)}$. Thus, the desired equality $x_1 = -x_0$ means that $-x_0 = x_0 - \dfrac{f(x_0)}{f'(x_0)}$. We want to have an example in which the Newton's process will loop for all possible initial values x_0 – except, of course, for the case $x_0 = 0$. Thus, the above equality must hold for all real numbers $x \neq 0$:

$$-x = x - \frac{f(x)}{f'(x)}. \tag{8}$$

Let us solve this equation. By moving the ratio the left-hand side and $-x$ to the right-hand side, we get

$$2x = \frac{f}{\dfrac{df}{dx}},$$

i.e., $2x = \dfrac{f \cdot dx}{df}$. We can now separate the variables x and f if we multiply both sides by df and divide both sides by f and by $2x$. As a result, we get $\dfrac{df}{f} = \dfrac{dx}{2x}$. Integrating both sides, we get $\ln(f) = \dfrac{1}{2} \cdot \ln(x) + C$, where C is the integration constant. Applying $\exp(z)$ to both sides of this equality, we get $f(x) = c \cdot \sqrt{x}$, where $c \stackrel{\text{def}}{=} \exp(C)$. Since we want an odd function, we thus get

$$f(x) = c \cdot \operatorname{sign}(x) \cdot \sqrt{|x|}, \tag{9}$$

where $\operatorname{sign}(x) = 1$ for $x > 0$ and $\operatorname{sign}(x) = -1$ for $x < 0$.

Of course, if we shift the function by some value a, we get a similar behavior. Thus, in general, we have a 2-parametric family of functions for which the Newton's method always loops:

$$f(x) = c \cdot \operatorname{sign}(x) \cdot \sqrt{|x - a|}. \tag{10}$$

Comment. Interestingly, the simplest example on which Newton's method never works – the example of a square root function $f(x)$ – is exactly inverse to the simplest example of a function $f(x) = x^2$ for which the Newton's method works perfectly.

3 Other Examples

Can we have other examples? Can we have similar always-looping examples for other functions, not just for the square root? Indeed, suppose that we have a non-negative function $f(x)$ defined for non-negative x, for which $f(0) = 0$ and for which, for each $x_0 > 0$, the next step of the Newton's method leads to the value $x_1 < 0$ – i.e., for which always

$$x - \frac{f(x)}{f'(x)} < 0. \quad (11)$$

This inequality can be reformulated as $f'/f < x$, i.e., as $\frac{\ln(f)}{\ln(x)} < 1$, i.e., the requirement that in the log-log scale, the slope is always smaller than 1.

We will also assume that the difference $x - \frac{f(x)}{f'(x)}$ monotonically depends on x.

How to design such looping examples. We would like to extend the function $f(x)$ to negative values x in such a way that the Newton's process will always loop. For convenience, let us denote, for each $x > 0$, $F(x) \stackrel{\text{def}}{=} -f(-x)$, where $f(-x)$ is the desired extension. Then, for $x < 0$, we have $f(x) = -F(-x)$.

When we start with the initial value $x > 0$, the next iteration is $-y$, where we denoted

$$y = \frac{f(x)}{f'(x)} - x. \quad (12)$$

Then, if we want the simplest loop, on the next iteration, we should get back the value x, i.e., we should have

$$x = (-y) - \frac{f'(-y)}{f'(-y)}.$$

Substituting $f(x) = -F(-x)$ into this equality, we get

$$x = \frac{F(y)}{F'(y)} - y,$$

i.e., equivalently,

$$\frac{F'(y)}{F(y)} = \frac{1}{x + y}$$

and thus,

$$F'(y) = \frac{F(y)}{x + y}, \quad (13)$$

where $y(x)$ is determined by the formula (12).

We thus have a differential equation that enables us to reconstruct, step-by-step, the desired function $F(y)$ and thus, the desired extension of $f(x)$ to negative values.

Specific examples. For example, when $f(x) = x^a$ for some $a > 0$, the inequality (11) implies that $a < 1$. One can check that in this case, we can take $F(y) = y^{1-a}$, i.e., extend this function to negative values x as $f(x) = -|x|^{1-a}$.

In particular, for $a = 1/2$, we get the above square root example.

Acknowledgements This work was supported in part by the US National Science Foundation grants 1623190 (A Model of Change for Preparing a New Generation for Professional Practice in Computer Science) and HRD-1242122 (Cyber-ShARE Center of Excellence).

Reference

1. Th.H. Cormen, C.E. Leiserson, R.L. Rivest, C. Stein, *Introduction to Algorithms* (MIT Press, Cambridge, Massachusetts, 2009)

Optimal Search Under Constraints

Martine Ceberio, Olga Kosheleva, and Vladik Kreinovich

Abstract In general, if we know the values a and b at which a continuous function has different signs—and the function is given as a black box—the fastest possible way to find the root x for which $f(x) = 0$ is by using bisection (also known as binary search). In some applications, however—e.g., in finding the optimal dose of a medicine—we sometimes cannot use this algorithm since, to avoid negative side effects, we can only try values which exceed the optimal dose by no more than some small value $\delta > 0$. In this paper, we show how to modify bisection to get an optimal algorithm for search under such constraint.

1 Where This Problem Came From

Need to select optimal dose of a medicine. This research started with a simple observation about how medical doctors decide on the dosage. For many chronic health conditions like high cholesterol, high blood pressure, etc., there are medicines that bring the corresponding numbers back to normal. An important question is how to select the correct dosage:

- on the one hand, if the dosage is too small, the medicine will not have the full desired effect;
- on the other hand, we do not want the dosage to be higher than needed: every medicine has negative side effects, side effects that increase with the increase in dosage, and we want to keep these side effects as small as possible.

M. Ceberio · O. Kosheleva · V. Kreinovich (✉)
University of Texas at El Paso, 500 W. University, El Paso, TX 79968, USA
e-mail: vladik@utep.edu

M. Ceberio
e-mail: mceberio@utep.edu

O. Kosheleva
e-mail: olgak@utep.edu

B. Bede et al. (eds.), *Fuzzy Information Processing 2020*,
Advances in Intelligent Systems and Computing 1337,
https://doi.org/10.1007/978-3-030-81561-5_37

In most such cases, there are general recommendations that provide a range of possible doses depending on the patient's age, weight, etc., but a specific dosage within this range has to be selected individually, based on how this patient's organism reacts to this medicine.

How the first doctor selected the dose. It so happened that two people having similar conditions ended up with the same daily dosage of 137 units of medicine, but interestingly, their doctors followed a different path to this value.

For the first patient, the doctor seems to have followed the usual bisection algorithm:

- this doctor started with the dose of 200—and it worked,
- so, the doctor tried 100—it did not work,
- the doctor tried 150—it worked,
- the doctor tried 125—it did not work,
- so, the doctor tried 137—and it worked.

The doctor could have probably continued further, but the pharmacy already had trouble with maintaining the exact dose of 137, so this became the final arrangement.

This procedure indeed follows the usual bisection (= binary search) algorithm (see, e.g., [1])—which is usually described as a way to solve the equation $f(x) = 0$ when we have an interval $[a, b]$ for which $f(a) < 0 < f(b)$. In our problem, $f(a)$ is the difference between the effect of the dose a and the desired effect:

- if the dose is not sufficient, this difference is negative, and
- if the dose is sufficient, this difference is non-negative (positive or 0).

In the bisection algorithm, at each iteration, we have a range $[\underline{x}, \overline{x}]$ for which $f(\underline{x}) < 0$ and $f(\overline{x}) > 0$. In the beginning, we have $[\underline{x}, \overline{x}] = [a, b]$. At each iteration, we take a midpoint $m = \dfrac{\underline{x} + \overline{x}}{2}$ and compute $f(m)$. Depending on the sign of $f(m)$, we make the following changes:

- if $f(m) < 0$, we replace $\underline{x}$ with m and thus, get a new interval $[m, \overline{x}]$;
- if $f(m) > 0$, we replace $\overline{x}$ with m and thus, get a new interval $[\underline{x}, m]$.

In both cases, we decrease the width of the interval $[\underline{x}, \overline{x}]$ by half. We stop when this width becomes smaller than some given value $\varepsilon > 0$; this value represents the accuracy with which we want to find the solution.

In the above example, based on the first experiment, we know that the desired dose is within the interval [0, 200]. So:

- we try $m = 100$ and, after finding that $f(m) < 0$ (i.e., that the dose $m = 100$ is not sufficient), we come up with the narrower interval [100, 200];
- then, we try the new midpoint $m = 150$, and, based on the testing result, we come up with the narrower interval [100, 150];
- then, we try the new midpoint $m = 125$, and, based on the testing result, we come up with the narrower interval [125, 150];

- in the last step, we try the new midpoint $m = 137$ (strictly speaking, it should be 137.5, but, as we have mentioned, the pharmacy cannot provide such an accuracy); now we know that the desired value is within the narrower interval

$$[125, 137].$$

Out of all possible values from the interval [125, 137], the only value about which we know that this value is sufficient is the value 137, so this value has been prescribed to the first patient.

The second doctor selected the same dose differently. Interestingly, for the second patient, the process was completely different:

- the doctor started with 25 units;
- then—since this dose was not sufficient—the dose was increased to 50 units;
- then the dose was increased to 75, 100, 125 units, and, finally, to 150 units.

The 150 units dose turned out to be sufficient, so the doctor knew that the optimal dose is between 125 and 150. Thus, this doctor tried 137, and it worked.

Comment. Interestingly, in contrast to the first doctor, this doctor could not convince the pharmacy to produce a 137 units dose. So this doctor's prescription of this dose consists of taking 125 units and 150 units in turn.

Why the difference? Why did the two doctors use different procedures?

Clearly, the second doctor needed more steps—and longer time—to come up with the same optimal dose: this doctor used 7 steps (25, 50, 75, 100, 125, 150, 137) instead of only 5 steps used by the first doctor (200, 100, 150, 125, 137). Why did this doctor not use a faster bisection procedure?

At first glance, it may seem that the second doctor was not familiar with bisection—but clearly this doctor *was* familiar with it, since, after realizing that the optimal dose is within the interval [125, 150], he/she checked the midpoint dose of 137.

The real explanation of why the second doctor did not use the faster procedure is that the second doctor was more cautious about possible side effects—probably, in this doctor's opinion, the second patient was vulnerable to possible side effects. Thus, this doctor decided not to increase the dose too much beyond the optimal value, so as to minimize possible side effects—while the first doctor, based on the overall health of the first patient, was less worried about possible side effects.

Natural general question. A natural next question is: under such restriction on possible tested values x, what is the optimal way to find the desired solution (i.e., to be more precise, the desired ε-approximation to the solution)?

It is known that if we do not have any constraints, then bisection is the optimal way to find the solution to the equation $f(x) = 0$; see, e.g., [1]. So, the question is—how to optimally modify bisection under such constraints?

2 Towards Formulating the Problem in Precise Terms

The larger the dose of the medicine, the larger the effect. There is a certain threshold x_0 after which the medicine has the full desired curing effect.

Every time we test a certain does x of the medicine of a patient:

- we either get the full desired effect, which would mean that $x_0 \le x$,
- or we do not yet get the full desired effect, which means that $x < x_0$.

We want to find the curing dose as soon as possible, i.e., after as few tests as possible.

If the only objective was to cure the disease, then, in principle, we could use any dose larger than or equal to x_0. However, the larger the dose, the larger the undesired side effects. So, we would like to prescribe a value which is as close to x_0 as possible. Of course, in real life, we can only maintain the dose with some accuracy $\varepsilon > 0$. So, we want to prescribe a value x_r which is ε-close to x_0, i.e., for which $x_0 \le x_r \le x_0 + \varepsilon$.

The only way to find the optimal dose x_r is to test different doses on a given patient. If, during this testing, we assign too large a dose, we may seriously harm the patient. So, it is desirable not to exceed x_0 too much when testing. Let us denote the largest allowed excess by δ. This means that we can only test values $x \le x_0 + \delta$.

Now, we can formulate the problem in precise terms.

3 Precise Formulation of the Problem and the Optimal Algorithm

Definition 1 By search under constraints, we mean the following problem:

- Given:
  - rational numbers $\varepsilon > 0$ and $\delta > 0$, and
  - an algorithm c that, for some fixed (unknown) value $x_0 > 0$, given a rational number $x \in [0, x_0 + \delta]$, checks whether $x < x_0$ or $x \ge x_0$; this algorithm will be called a *checking* algorithm.
- Find: a real number x_r for which $x_0 \le x_r \le x_0 + \varepsilon$.

Comment. We want to find the fastest possible algorithm for solving this problem. To gauge the speed of this algorithm, we will count the number of calls to the checking algorithm c.

Definition 2

- For every algorithm A for solving the search under constraints problem, let us denote the number of calls to the checking algorithm c corresponding to each instance $(\varepsilon, \delta, x_0)$ by $N_{A,\varepsilon,\delta}(x_0)$.

- We say that the algorithm A_0 for solving the search under constraint problem is *optimal* if for each ε and δ, the function $N_{A_0,\varepsilon,\delta}(x_0)$ is asymptotically optimal, i.e., that for every other algorithm A for solving the search under constraints problem, we have

$$N_{A_0,\varepsilon,\delta}(x_0) \le N_{A,\varepsilon,\delta}(x_0) + \text{const}$$

for some constant depending on ε, δ, and A.

Proposition *The following algorithm $\mathscr{A}$ is optimal:*

- *First, we apply the algorithm c to values δ, 2δ, ..., until we find a value i for which $i \cdot \delta < x_0 \le (i+1) \cdot \delta$.*
- *Then, we apply bisection process to the interval $[i \cdot \delta, (i+1) \cdot \delta]$ to find x_r:*
  - *In this process, at each moment of time, we have an interval $[\underline{x}, \overline{x}]$ for which*

$$\underline{x} < x_0 \le \overline{x}.$$

  - *We start with $[\underline{x}, \overline{x}] = [i \cdot \delta, (i+1) \cdot \delta]$.*
  - *At each iteration step, we apply the checking algorithm c to the midpoint*

$$m = \frac{\underline{x} + \overline{x}}{2}.$$

  - *If it turns out that $m < x_0$, we replace $[\underline{x}, \overline{x}]$ with $[m, \overline{x}]$.*
  - *If it turns out that $x_0 \le m$, we replace $[\underline{x}, \overline{x}]$ with $[\underline{x}, m]$.*
  - *In both cases, we decrease the width of the interval by 2.*
  - *We stop when this width becomes smaller than or equal to ε, i.e., when*

$$\overline{x} - \underline{x} \le \varepsilon.$$

  - *Then, we take $\overline{x}$ as the desired output x_r.*

Proof It is easy to prove that the algorithm $\mathscr{A}$ indeed solves the search under constraints problem. Indeed, increasing the previously tested value $x \le x_0$ is legitimate: since then $x + \delta \le x_0 + \delta$. By this increase, for each x_0, we will eventually find the value i for which $x_0 \le (i+1) \cdot \delta$—namely, $i = \left\lceil \frac{x_0}{\delta} \right\rceil - 1$. Then, by induction, we can prove that on each step of the bisection process, we indeed have $\underline{x} < x_0 \le \overline{x}$. And if $\underline{x} < x_0 \le \overline{x}$ and $\overline{x} - \underline{x} \le \varepsilon$, then indeed $x_0 \le x_r = \overline{x} \le \underline{x} + \delta < x_0 + \varepsilon$.

Optimality if also easy to prove: indeed, the algorithm $\mathscr{A}$ takes $\frac{x_0}{\delta}$ + const steps, where the constant—approximately equal to $\log_2\left(\frac{\delta}{\varepsilon}\right)$—covers the bisection part. Let us show that other algorithms A cannot use fewer steps.

Indeed, if v is the largest value for which we have already checked that $v < x_0$, then, at the next test, we cannot use the value $x > v + \delta$. Indeed, in this case, we have $v < x - \delta$ so for any x_0 from the interval $(v, x - \delta)$, we have $v < x_0 < x - \delta$ and thus, $x > x_0 + \delta$. So, for this x_0, the checking algorithm c is not applicable.

Thus, at each step, we cannot increase the tested value x by more than δ in comparison with the previously tested value. So, to get to a value $x \geq x_0$—which is our goal—we need to make at least $\frac{x_0}{\delta}$ calls to the checking algorithm c.

The proposition is proven.

Comment. This is exactly what the both doctors did, the difference is that:

- the first doctor used $\delta = 200$, while
- the second doctor used a much smaller value $\delta = 25$.

Acknowledgements This work was supported in part by the US National Science Foundation grants 1623190 (A Model of Change for Preparing a New Generation for Professional Practice in Computer Science) and HRD-1242122 (Cyber-ShARE Center of Excellence).

Reference

1. Th.H. Cormen, C.E. Leiserson, R.L. Rivest, C. Stein, *Introduction to Algorithms* (MIT Press, Cambridge, Massachusetts, 2009)

How User Ratings Change with Time: Theoretical Explanation of an Empirical Formula

Julio C. Urenda, Manuel Hernandez, Natalia Villanueva-Rosales, and Vladik Kreinovich

Abstract In many application areas, it is important to predict the user's reaction to new products. In general, this reaction changes with time. Empirical analysis of this dependence has shown that it can be reasonably accurately described by a power law. In this paper, we provide a theoretical explanation for this empirical formula.

1 Formulation of the Problem

For many industries, it is important to predict the user's reaction to new products. To make this prediction, it is reasonable to use the past records of the user's degree of satisfaction with different similar products.

One of the problems with such a prediction is that the user's degree of satisfaction may change with time: the first time you see an exciting movie or read an exciting book, you feel very happy, when you see this movie the second time, you may notice holes in the plot or outdated (and thus, somewhat clumsy) computer simulation.

Because of this phenomenon, for each user, the ratings of the same product decrease with time. In other words, instead of the simplified formula $r = r(u, p)$ that describes how the ratings depend on the user u and on the product p, a more accurate estimates can be obtained if we take this dependence into account and use a more complex formula

$$r = r(u, p) + c_u(t),$$

J. C. Urenda · M. Hernandez · N. Villanueva-Rosales · V. Kreinovich (✉)
Departments of Mathematics and Computer Science, University of Texas at El Paso, El Paso, TX 79968, USA
e-mail: vladik@utep.edu

J. C. Urenda
e-mail: jcurenda@utep.edu

M. Hernandez
e-mail: mahernandez46@miners.utep.edu

N. Villanueva-Rosales
e-mail: nvillanuevarosales@utep.edu

B. Bede et al. (eds.), *Fuzzy Information Processing 2020*,
Advances in Intelligent Systems and Computing 1337,
https://doi.org/10.1007/978-3-030-81561-5_38

where a decreasing function $c_u(t)$—which is, in general, different for different users—change with time.

To test different ratings models, in the 2000s, Netflix had a competition in which different formulas were compared. The winning model (see, e.g., [3]) used an empirical formula

$$c_u(t) = \alpha_u \cdot \text{sign}(t - t_u) \cdot |t - t_u|^{\beta_u}, \quad (1)$$

where t_u is the mean date of rating, and α_u and β_u are parameters depending on the user. (Actually, it turned out that the value β_u is approximately the same for all the users.)

The question is: why this formula works well, while possible other dependencies on time do not work so well?

2 Our Explanation

2.1 *First Technical Comment*

The formula (1) does not uniquely determine the functions $r(u, p)$ and $c_u(t)$: we can add a constant to all the ratings $r(u, p)$ and subtract the same constant from all the values $c_u(t)$ and still get the same overall ratings r.

To avoid this non-uniqueness, we can, e.g., select $c_u(t)$ in such a way that $c_u(t_u) = 0$; this is, by the way, exactly what is done in the formula (1). This equality is easy to achieve: if we have a function $c_u(t)$ for which $c_u(t_u) \neq 0$, then we can consider new functions $\widetilde{c}_u(t_u) \stackrel{\text{def}}{=} c_u(t) - c_u(t_u)$ and $\widetilde{r}(u, p) \stackrel{\text{def}}{=} r(u, p) + c_u(t_u)$. Then, as one can easily see, we have

$$\widetilde{r}(u, p) + \widetilde{c}_u(t) = r(u, p) + c_u(t),$$

i.e., all the predicted ratings remain the same.

In view of this possibility, in the following text, we will assume that

$$c_u(t_u) = 0.$$

2.2 *First Idea: The Description Should Not Depend on the Unit for Measuring Time*

We are interested in finding out how the change in ratings depends on time. In a computer model, time is represented by a number, but the numerical value of time depends on what starting point we choose and what unit we use for measuring time. In our situation, there is a fixed moment t_u, so it is reasonable to use t_u as the starting point and thus use $T \stackrel{\text{def}}{=} t - t_u$ to measure time instead of the original t.

In the new scale, the formula describing how ratings change with time takes the form $C_u(T)$, so that $c_u(t) = C_u(t - t_u)$. The condition $c_u(t_u) = 0$ takes the form $C_u(0) = 0$.

In terms of the new time scale, the empirically best formula (1) leads to the following expression for $C_u(T)$:

$$C_u(T) = \alpha_u \cdot \text{sign}(T) \cdot |T|^{\beta_u}. \tag{2}$$

While there is a reasonable starting point for measuring time, there is no fixed unit of time. We could use years, months, weeks, days, whatever units make sense. If we replace the original measuring unit with a new unit which is λ times smaller, then all numerical values of time are multiplied by λ. So, instead of the original value T, we get a new value $\widetilde{T} = \lambda \cdot T$.

Since there is nothing special in selecting a measuring unit for time, it makes sense to require that the corresponding formulas not change when we make a different selection.

2.3 *This Has to Be Related to a Change in How We Measure Ratings*

Of course, we cannot simply require that the formula for $C_u(T)$ be invariant under the change $T \to \lambda \cdot T$. Indeed, in this case, we would have $C_u(\lambda \cdot T) = C_u(T)$ for all λ and T. Thus, for each $T_0 > 0$, by taking $T = 1$ and $\lambda = T_0$, we would be able to conclude that $C_u(T_0) = C_u(1)$—i.e., instead of the desired decreasing function, we would have a constant function $C_u(T) = \text{const}$.

This seeming problem can be easily explained if we take into account how similar scale invariance works in physics. For example, the formula $v = d/t$ that describes how the average velocity v depends on the distance d and time t clearly does not depend on what measuring unit we use for measuring distance: we could use meters, we could use centimeters, we could use inches. However, this does not mean that if we simply change the measuring unit for distance and thus replace the original value d with the new value $\lambda \cdot d$, the formula remains the same: for the formula to remain valid, we also need to accordingly change the unit for measuring velocity, e.g., from meters per second to centimeters per second.

Similarly in our case, when we describe the dependence of rating on time, we cannot just change the unit for time, we also need to change another unit—which, in this case, is the unit for ratings.

But does this change make sense? At first glance, it may seem that it does not: we ask the user to mark the quality of a product (e.g., of a movie) on a certain fixed scale (e.g., 0–5), so how can we change this scale? Actually, we can. Users are different. Some users use all the scale, and mark the worst of the movies by 0, and the best by 5. What happens when a new movie comes which is much better than anything that the user has been before? In this case, the user has no choice but to give a 5 to this movie as well—wishing that the scale had 6 or 7 or even more. Similarly, if a movie has a very negative experience with a movie, a much worse one than anything that he or she has seen before, this user places 0 and wishes that there was a possibility to give -1 or even -2.

Other users recognize this problem and thus, use only, e.g., grades from 1–4, reserving 0 and 5 for future very bad and very good products. Some professors grade the student papers the same way, using, e.g., only values up to 70 or 80 out of 100, and leaving higher grades for possible future geniuses.

In other words, while the general scale—from 0 to 5 or from 0 to 100—is indeed fixed, the way we use it changes from one user to another one. Some users use the whole scale, some "shrink" their ratings to fit into a smaller sub-scale. A natural way to describe this shrinking is by an appropriate linear transformations—this is how, e.g., we estimate the grade of a student who for legitimate reasons had to skip a test worth 20 points out of 100: if overall, the student earned 72 points out of 80, we mark it as $\frac{72}{80} \cdot 100 = 90$ points on a 0–100 scale.

Depending on what scale we use for ratings, the corresponding rating values change by a linear formula: $r \to \widetilde{r} = a \cdot r + b$. In particular, for the difference between the ratings, we get

$$r_1 - r_2 \to (a \cdot r_1 + b) - (a \cdot r_2 + b) = a \cdot (r_1 - r_2).$$

So, when we change the unit for measuring time by a λ times smaller one, we may need to according re-scale our difference $C(T)$ between the ratings. Thus, we arrive at the following precise formulation of the desired invariance.

2.4 Formal Description of Unit-Independence

We want to select a function $C_u(T)$ for which, for each $\lambda > 0$, there exists a value $a(\lambda)$ for which

$$C_u(\lambda \cdot T) = a(\lambda) \cdot C_u(T). \quad (2)$$

It is also reasonable to assume that the function $C_u(T)$ continuously change with time—or at least change with time in a measurable way.

2.5 What Can We Conclude Based on This Independence.

It is known (see, e.g., [1]), that every measurable (in particular, every continuous) solution to the Eq. (2) for $T > 0$ has the form

$$C_u(T) = \alpha_u^+ \cdot T^{\beta_u^+}, \tag{3}$$

for some α_u^+ and β_u^+.

Similarly, for $T < 0$, we get

$$C_u(T) = \alpha_u^- \cdot |T|^{\beta_u^-}, \tag{4}$$

for some α_u^- and β_u^-.

These formulas are similar to the desired formula (1), but we still have too many parameters: four instead of the desired two. To get the exact form (1), we need one more idea.

2.6 *Second Idea: The Change in Rating Should Be the Same Before and After t_u*

It is reasonable to require that for each time interval $T > 0$, the change of rating should be the same before and after t_u, i.e., the change of ratings between the moments $t_u - T$ and t_u should be the same as the change of ratings between the moments t_u and $t_u + T$.

The change of ratings between the moments $t_u - T$ and t_u is equal to

$$c_u(t_u) - c_u(t_u - T) = -(c_u(t_u - T) - c_u(T)) = -C_u(-T).$$

The change of ratings between the moments $t_u + T$ and t_u is simply equal to

$$c_u(t_u + T) - c_u(t_u) = C_u(T).$$

Thus, the above requirement means that for every $T > 0$, we should have

$$-C_u(-T) = C_u(T).$$

Substituting the expressions (3) and (4) into this formula, and taking into account that $|-T| = T$, we conclude that for each $T > 0$, we have

$$\alpha_u^+ \cdot T^{\beta_u^+} = -\alpha_u^- \cdot T^{\beta_u^-}.$$

Since this must be true for all T, we must have $\alpha_u^+ = -\alpha_u^-$ and $\beta_u^+ = \beta_u^-$.

Thus, for both $T > 0$ and $T < 0$, we indeed have the formula (1), with $\alpha_u = \alpha_u^+$ and $\beta_u = \beta_u^+$. The formula (1) is explained.

Acknowledgements This work was supported in part by the US National Science Foundation grants 1623190 (A Model of Change for Preparing a New Generation for Professional Practice in Computer Science) and HRD-1242122 (Cyber-ShARE Center of Excellence).

References

1. J. Aczel, J. Dhombres, *Functional Equations in Several Variables* (Cambridge University Press, Cambridge, 1989)
2. Y. Koren, *The BellKor Solution to the Netflix Trand Prize* (2009). https://www.netflixprize.com/assets/GrandPrize2009$_$BPC$_$BellKor.pdf
3. Y. Koren, R. Bell, C. Volinsky, Matrix factorization techniques for recommender systems. Computer **42**(8), 30–37 (2019)

Why a Classification Based on Linear Approximation to Dynamical Systems Often Works Well in Nonlinear Cases

Julio C. Urenda and Vladik Kreinovich

Abstract It can be proven that linear dynamical systems exhibit either stable behavior, or unstable behavior, or oscillatory behavior, or transitional behavior. Interesting, the same classification often applies to nonlinear dynamical systems as well. In this paper, we provide a possible explanation for this phenomenon, i.e., we explain why a classification based on linear approximation to dynamical systems often works well in nonlinear cases.

1 Formulation of the Problem

1.1 Dynamical Systems Are Ubiquitous

To describe the state of a real-life system at any given moment of time, we need to know the values $x = (x_1, \ldots, x_n)$ of all the quantities that characterize this system. For example, to describe the state of a mechanical system consisting of several pointwise objects, we need to know the position and velocities of all these objects. To describe the state of an electric circuit, we need to know the currents and voltages, etc.

In many real-life situation, the corresponding systems are deterministic—in the sense that the future states of the system are uniquely determined by its current state. Sometimes, to make the system deterministic, we need to enlarge its description so that it incorporates all the objects that affect its dynamics. For example, the system consisting of Earth and Moon is not deterministic in its original form—since the Sun affects its dynamics, but once we add the Sun, we get a system with a deterministic behavior.

J. C. Urenda · V. Kreinovich (✉)
Departments of Mathematics and Computer Science, University of Texas at El Paso, El Paso, TX 79968, USA
e-mail: vladik@utep.edu

J. C. Urenda
e-mail: jcurenda@utep.edu

B. Bede et al. (eds.), *Fuzzy Information Processing 2020*,
Advances in Intelligent Systems and Computing 1337,
https://doi.org/10.1007/978-3-030-81561-5_39

The fact that the future dynamics of the system is uniquely determined by its current state means, in particular, that the rate $\dot{x}$ with which the system changes is also uniquely determined by its current state, i.e., that we have $\dot{x} = f(x)$, for some function $f(x)$. This equation can be described coordinate-wise, as

$$\dot{x}_i = f_i(x_1, \ldots, x_n). \tag{1}$$

Systems that satisfy such equations are known as *dynamical systems*; see, e.g., [1].

1.2 Simplest Case: Linear Systems

The simplest case is when the rate of change $f_i(x_1, \ldots, x_n)$ of each variables is a linear function, i.e., when

$$\dot{x}_i = a_{i0} + \sum_{j=1}^{n} a_{ij} \cdot x_j. \tag{2}$$

In almost all such cases—namely, in all the cases when the matrix a_{ij} is non-degenerate—we can select constants s_i so that for the correspondingly shifted variables $y_i = x_i + s_i$, the system gets an even simpler form

$$\dot{y}_i = \sum_{j=1}^{n} a_{ij} \cdot y_j. \tag{3}$$

Indeed, substituting $x_i = y_i - s_i$ into the formula (2), and taking into account that $\dot{y}_i = \dot{x}_i$, we conclude that

$$\dot{x}_i = a_{i0} + \sum_{j=1}^{n} a_{ij} \cdot (y_j - s_j) = a_{i0} + \sum_{j=1}^{n} a_{ij} \cdot y_j - \sum_{j=1}^{n} a_{ij} \cdot s_j.$$

Thus, if we select the value s_j for which $a_{i0} = \sum\limits_{j=1}^{n} a_{ij} \cdot s_j$ for each i, we will indeed get the formula (2).

For the Eq. (2), the general solution is well known: it is a linear combination of expressions of the type $t^k \cdot \exp(\lambda \cdot t)$, where λ is an eigenvalue of the matrix $\|a_{ij}\|$—which is, in general, a complex number $\lambda = a + \mathrm{i} \cdot b$, and k is a natural number which does not exceed the multiplicity of this eigenvalue. In real-number terms, we get a linear combination of the expressions $t^k \cdot \exp(a \cdot t) \cdot \sin(b \cdot t + \varphi)$.

Depending on the values of λ, we have the following types of behavior:

- when $a < 0$ for all the eigenvalues, then the system is *stable*: no matter what state we start with, it asymptotically tends to the state $y_1 = \ldots = y_n = 0$;

- when $a > 0$ for at least one eigenvalue, then the system is *unstable*: the deviation from the 0 state exponentially grows with time;
- when $a = 0$ and $b \neq 0$, we get an *oscillatory behavior*; and
- when $a = b = 0$, we get a *transitional behavior*, when a system linearly (or quadratically etc.) moves from one state to another.

1.3 A Similar Classification Works Well in Non-linear Cases, But Why?

Interestingly, a similar classification works well for nonlinear dynamical systems as well, but why? In this paper, we will try to explain this fact.

2 Our Explanation

2.1 We Need Finite-Dimensional Approximations

We want to describe how the state $x(t) = (x_1(t), \ldots, x_n(t))$ of a dynamical system changes with time t. In general, the set of all possible smooth functions $x_i(t)$ is infinite-dimensional, i.e., we need infinitely many parameters to describe it. However, in practice, at any given moment, we can only have finitely many parameters. Thus, it is reasonable to look for finite-parametric approximations. A natural idea is to fix some smooth functions $e_k(t) = (e_{k1}(t), \ldots, e_{kn}(t))$, $1 \leq k \leq K$, and consider linear combinations

$$x(t) = \sum_{k=1}^{K} c_k \cdot e_k(t). \tag{4}$$

2.2 Shift-Invariance

For dynamical systems, there is no fixed moment of time. The equations remain the same if we change the starting point for measuring time, i.e., if we replace the original temporal variable t with the new variable $t' = t + t_0$.

It is therefore reasonable to require that the approximating family (4) be invariant with respect to the same transformation, i.e., in other words, that all shifted functions $e_k(t + t_0)$ can also be represented in the same form (4). Let us show that this reasonable requirement explains the above phenomenon.

Comment. This derivation will be similar to the one given in [2].

2.3 Towards the Explanation

The formula (4) means that for each component i, we have

$$x_i(t) = \sum_{k=1}^{K} c_k \cdot e_{ki}(t). \quad (5)$$

The fact that shifted functions can be represented in this form means that for each k, i, and t_0, we have

$$e_{ki}(t + t_0) = \sum_{\ell=1}^{K} c_{k\ell i}(t_0) \cdot e_{\ell i}(t), \quad (6)$$

for some coefficients $c_{k\ell i}(t_0)$ depending on k, ℓ, i, and t_0.

Let us fix i and k and select K different moments of time t_m, $m = 1, \ldots, K$. For these moments of time, (6) takes the form

$$e_{ki}(t_m + t_0) = \sum_{\ell=1}^{K} c_{k\ell i}(t_0) \cdot e_{\ell i}(t_m). \quad (7)$$

Thus, we get K linear equations for determining K unknowns $c_{k1i}(t_0), \ldots, c_{kKi}(t_0)$. Cramer's formula describes the solution to a system of linear equations as a rational (and thus, smooth) function of its coefficients and right-hand sides. Thus, each coefficient $c_{k\ell i}(t_0)$ is a smooth function of the values $e_{ki}(t_m + t_0)$ and $e_{\ell i}(t_m)$. Since the functions $e_{ki}(t)$ are smooth, the dependence of the coefficients $c_{k\ell i}(t_0)$ on t_0 is also differentiable.

Since all the functions involved in the formula (6) are differentiable, we can differentiate this formula with respect to t_0 and get

$$\dot{e}_{ki}(t + t_0) = \sum_{\ell=1}^{K} \dot{c}_{k\ell i}(t_0) \cdot e_{\ell i}(t). \quad (8)$$

In particular, for $t_0 = 0$, we get

$$\dot{e}_{ki}(t) = \sum_{\ell=1}^{K} a_{k\ell i} \cdot e_{\ell i}(t), \quad (9)$$

where we denoted $a_{k\ell i} \stackrel{\text{def}}{=} \dot{c}_{k\ell i}(t_0)$.

So, we conclude that the functions $e_{ki}(t)$ satisfy the system of linear differential equations with constant coefficients—and we have already mentioned that the solutions to such systems are exactly the functions leading to a known classification of linear dynamical system behaviors.

This explains why for nonlinear systems, we also naturally observe similar types of behavior.

Acknowledgements This work was supported in part by the US National Science Foundation grants 1623190 (A Model of Change for Preparing a New Generation for Professional Practice in Computer Science) and HRD-1242122 (Cyber-ShARE Center of Excellence).

References

1. M.W. Hirsch, S. Smale, R.L. Devaney, *Differential Equations, Dynamical Systems, and an Introduction to Chaos* (Academic, Waltham, Massachisetts, 2013)
2. H.T. Nguyen, V. Kreinovich, *Applications of Continuous Mathematics to Computer Science* (Kluwer, Dordrecht, 1997)

How Mathematics and Computing Can Help Fight the Pandemic: Two Pedagogical Examples

Julio C. Urenda, Olga Kosheleva, Martine Ceberio, and Vladik Kreinovich

Abstract With the 2020 pandemic came unexpected mathematical and computational problems. In this paper, we provide two examples of such problems—examples that we present in simplified pedagogical form. The problems are related to the need for social distancing and to the need for fast testing. We hope that these examples will help students better understand the importance of mathematical models.

1 First Example: Need for Social Distancing

Formulation of the problem. This problem is related to the pandemic-related need to observe a social distance of at least 2 m (6 ft) from each other.

Two persons are on two sides of a narrow-walkway street, waiting for the green light. They start walking from both sides simultaneously. For simplicity, let us assume that they walk with the same speed.

If they follow the shortest distance path—i.e., a straight line connecting their initial locations A and B—they will meet in the middle, which is not good. So one of them should move somewhat to the left, another somewhat to the right. At all moments of time, they should be at least 2 m away from each other. What is the fastest way for them to do it?

J. C. Urenda (✉) · O. Kosheleva · M. Ceberio · V. Kreinovich
University of Texas at El Paso, El Paso, TX 79968, USA
e-mail: jcurenda@utep.edu

O. Kosheleva
e-mail: olgak@utep.edu

M. Ceberio
e-mail: mceberio@utep.edu

V. Kreinovich
e-mail: vladik@utep.edu

B. Bede et al. (eds.), *Fuzzy Information Processing 2020*,
Advances in Intelligent Systems and Computing 1337,
https://doi.org/10.1007/978-3-030-81561-5_40

Towards formulating this problem in precise terms. The situation is absolutely symmetric with respect to the reflection in the midpoint M of the segment AB. So, it is reasonable to require that the trajectory of the second person can be obtained from the trajectory of the first person by this reflection. Thus, at any given moment of time, the midpoint M is the midpoint between the two persons. In these terms, the requirement that they are separated by at least 2 m means that each of them should always be at a distance at least 1 m from the midpoint M. In other words, both trajectories should avoid the disk of radius 1 m with a center at the midpoint M.

We want the fastest possible trajectory. Since the speed is assumed to be constant, this means that they should follow the shortest possible trajectory. In other words, we need to find the shortest possible trajectory going from point A to point B that avoids the disk centered at the midpoint M of the segment AB.

Solution. To get the shortest path, outside the disk, the trajectory should be straight, and where it touches the circle, it should be smooth. Thus, the solution is as follows:

- first, we follow a straight line until it touches the circle as a tangent,
- then, we follow the circle,
- and finally, we follow the straight line again—which again starts as a tangent to the circle.

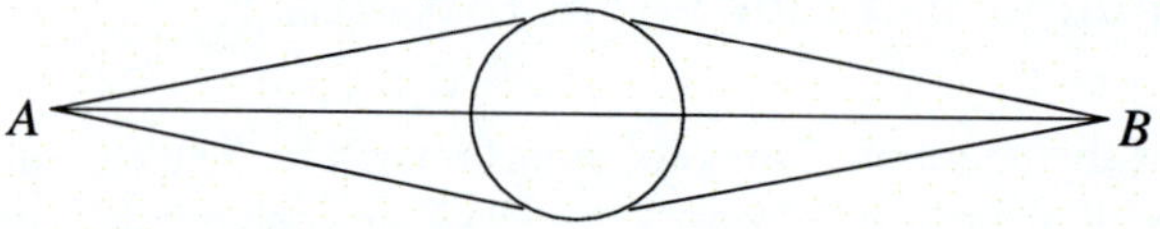

2 Second Example: Need for Fast Testing

Formulation of the problem. One of the challenges related to the COVID-19 pandemic is that this disease has an unusually long incubation period—about 2 weeks. As a result, people with no symptoms may be carrying the virus and infecting others. As of now, the only way to prevent such infection is to perform massive testing of the population. The problem is that there is not enough test kits to test everyone.

What was proposed. To solve this problem, researchers proposed the following idea [1, 2]: instead of testing everyone individually, why not combine material from a group of several people and test each combined sample by using a single test kit. If no viruses are detected in the combined sample, this means that all the people from the corresponding group are virus-free, so there is no need to test them again. After this, we need to individually test only folks from the groups that showed the presence of the virus.

Resulting problem. Suppose that we need to test a large population of N people. Based on the previous testing, we know the proportion p of those who have the virus. In accordance with the above idea, we divide N people into groups. The question is: what should be the size s of each group?

If the size is too small, we are still using too many test kits. If the size is too big, every group, with a high probability, has a sick person, so we are not dismissing any people after such testing, and thus, we are not saving any testing kits at all. So what is the optimal size of the group?

Comment. Of course, this is a simplified formulation, it does not take into account that for large group sizes s, when each individual testing material is diluted too much, tests may not be able to detect infected individuals.

Let us formulate this problem in precise terms. If we divide N people into groups of s persons each, we thus get N/s groups.

The probability that a person is virus-free is $1 - p$. Thus, the probability that all s people from a group are virus-free is $(1 - p)^s$. So, out of N/s groups, the number of virus-free groups is $(1 - p)^s \cdot (N/s)$. Each of these groups has s people, so the overall number of tested people can be obtained by multiplying the number of virus-free groups by s, resulting in $(1 - p)^s \cdot N$. For the remaining $N - (1 - p)^s \cdot N$ folks, we need individual testing. So, the overall number of needed test kits is

$$N_t = \frac{N}{s} + N - (1 - p)^s \cdot N. \tag{1}$$

We want to minimize the number of test kits, i.e., we want to find the group size s for which the number (1) is the smallest possible.

Solution. Differentiating the expression (1) with respect to s, equating the derivative to 0, and dividing both sides of the resulting equality by N, we get

$$-\frac{1}{s^2} - (1 - p)^s \cdot \ln(1 - p) = 0. \tag{2}$$

For small p, we have $(1 - p)^s \approx 1$ and $\ln(1 - p) \approx -p$, so the formula (2) takes the form $-\frac{1}{s^2} + p \approx 0$, i.e.,

$$s \approx \frac{1}{\sqrt{p}}. \tag{3}$$

For example, for $p = 1\%$, we have $s \approx 10$; for $p = 0.1\%$, we get $s \approx 30$; and for $p = 0.01\%$, we get $s \approx 100$.

The resulting number of tests (1) can also be approximately estimated. When the group size s is described by the approximate formula (3), we have $\frac{N}{s} \approx \sqrt{p} \cdot N$. If we take into account that $(1 - p)^s \approx 1 - p \cdot s$, then

$$N - (1 - p)^s \cdot N \approx p \cdot s \cdot N \approx \sqrt{p} \cdot N.$$

Thus, we get

$$N_t \approx \sqrt{p} \cdot N. \tag{4}$$

For example, for $p = 1\%$, we need 10 times fewer test kits than for individual testing; for $p = 0.1\%$, we need 30 times fewer test kits; and for $p = 0.01\%$, we need 100 times fewer test kits.

Acknowledgements This work was supported in part by the US National Science Foundation grants 1623190 (A Model of Change for Preparing a New Generation for Professional Practice in Computer Science) and HRD-1242122 (Cyber-ShARE Center of Excellence).

References

1. T.S. Perry, Researchers are using algorithms to tackle the coronavirus test shortage: the scramble to develop new test kits that deliver faster results. IEEE Spectr **57**(6), 4 (2020)
2. N. Shental, S. Levy, V. Wuvshet, S. Skorniakov, Y. Shemer-Avni, A. Porgador, and T. Hertz, Efficient High Throughput SARS-CoV-2 Testing to Detect Asymptomatic Carriers, medRxiv preprint, Accessed 20 April 2020. https://doi.org/10.1101/2020.04.14.20064618

Natural Invariance Explains Empirical Success of Specific Membership Functions, Hedge Operations, and Negation Operations

Julio C. Urenda, Orsolya Csiszár, Gábor Csiszár, József Dombi, György Eigner, and Vladik Kreinovich

Abstract Empirical studies have shown that in many practical problems, out of all symmetric membership functions, special *distending* functions work best, and out of all hedge operations and negation operations, fractional linear ones work the best. In this paper, we show that these empirical successes can be explained by natural invariance requirements.

1 Formulation of the Problem

Fuzzy techniques: a brief reminder. In many applications, we have knowledge formulated in terms of imprecise ("fuzzy") terms from natural language, like "small", "somewhat small", etc. To translate this knowledge into computer-understandable form, Lotfi Zadeh proposes *fuzzy techniques*; see, e.g., [1, 11, 12, 14, 15, 17]. According to these techniques, each imprecise property like "small" can be described

J. C. Urenda (✉) · V. Kreinovich
University of Texas at El Paso, El Paso, TX 79968, USA
e-mail: jcurenda@utep.edu

V. Kreinovich
e-mail: vladik@utep.edu

O. Csiszár
Faculty of Basic Sciences, University of Applied Sciences Esslingen, Esslingen, Germany
e-mail: orsolya.csiszar@nik.uni-obuda.hu

O. Csiszár · G. Eigner
Institute of Applied Mathematics, Óbuda University, Budapest, Hungary
e-mail: eigner.gyorgy@nik.uni-obuda.hu

G. Csiszár
Institute of Materials Physics, University of Stuttgart, Stuttgart, Germany
e-mail: gabor.csiszar@mp.imw.uni-stuttgart.de

J. Dombi
Institute of Informatics, University of Szeged, Szeged, Hungary
e-mail: dombi@inf.u-szeged.hu

B. Bede et al. (eds.), *Fuzzy Information Processing 2020*,
Advances in Intelligent Systems and Computing 1337,
https://doi.org/10.1007/978-3-030-81561-5_41

by assigning, to each value x of the corresponding quantity, a degree $\mu(x)$ to which, according to the expert, this property is true. These degrees are usually selected from the interval [0, 1], so that 1 corresponds to full confidence, 0 to complete lack of confidence, and values between 0 and 1 describe intermediate degrees of confidence. The resulting function $\mu(x)$ is known as a *membership function*.

In practice, we can only ask finitely many questions to the expert, so we only elicit a few values $\mu(x_1)$, $\mu(x_2)$, etc. Based on these values, we need to estimate the values $\mu(x)$ for all other values x. For this purpose, usually, we select a family of membership functions—e.g., triangular, trapezoidal, etc.—and select a function from this family which best fits the known values.

For terms like "somewhat small", "very small", the situation is more complicated. We can add different "hedges" like "somewhat", "very", etc., to each property. As a result, we get a large number of possible terms, and it is not realistically possible to ask the expert about each such term. Instead, practitioners estimate the degree to which, e.g., "somewhat small" is true based on the degree to which "small" is true. In other words, with each linguistic hedge, we associate a function h from [0, 1] to [0, 1] that transforms the degree to which a property is true into an estimate for the degree to which the hedged property is true.

Similarly to the membership functions, we can elicit a few values $h(x_i)$ of the hedge operation from the experts, and then we extrapolate and/or interpolate to get all the other values of $h(x)$. Usually, a family of hedge operations is pre-selected, and then we select a specific operation from this family which best fits the elicited values $h(x_i)$.

Similarly, instead of asking experts for their degrees of confidence in statements containing negation, such as "not small", we estimate the expert's degree of confidence in these statements based on their degrees of confidence in the positive statements. The corresponding operation $n(x)$ is known as the *negation operation*.

1.1 Need to Select Proper Membership Functions, Proper Hedge Operations, and Proper Negation Operations

Fuzzy techniques have been successfully applied to many application areas. However, this does not necessarily mean that every time we try to use fuzzy techniques, we get a success story. The success (or not) often depends on which membership functions and which hedge and negation operations we select: for some selections, we get good results (e.g., good control), for other selections, the results are not so good.

1.2 *What We Do in This Paper*

There is a lot of empirical data about which selections work better. In this paper, we provide a general explanation for several of these empirically best selections, an explanation based on the natural concepts of invariance.

Specifically, we explain the following empirically successful selections:

- for symmetric membership functions that describe properties like "small", for which $\mu(x) = \mu(-x)$ and the degree $\mu(|x|)$ decreases with $|x|$, in many practical situations, the most empirically successful are so-called *distending* membership functions, i.e., functions of the type

$$\mu(x) = \frac{1}{1 + a \cdot |x|^b} \tag{1}$$

  for some a and b; see, e.g., [7–9];
- among hedge and negation operations, in many practical situations, the most efficient are fractional linear functions

$$h(x) = \frac{a + b \cdot x}{1 + c \cdot x} \tag{2}$$

  for some a, b, and c; see, e.g., [2–5].

2 Analysis of the Problem

2.1 *Re-scaling*

The variable x describes the value of some physical quantity, such a distance, height, difference in temperatures, etc. When we process these values, we deal with numbers, but numbers depend on the selection of the measuring unit: if we replace the original measuring unit with a new one which is λ times smaller, then all the numerical values will be multiplied by λ: $x \to X = \lambda \cdot x$. For example, 2 m become $2 \cdot 100 = 200$ cm. This transformation from one measuring scale to another is known as *re-scaling*.

2.2 *Scale-Invariance: Idea*

In many physical situations, the choice of a measuring unit is rather arbitrary. In such situations, all the formulas remain the same no matter what unit we use.

For example, the formula $y = x^2$ for the area of the square with side x remains valid if we replace the unit for measuring sides from meters with centimeters—of course, we then need to appropriately change the unit for y, from square meters to square centimeters. In general, invariance of the formula $y = f(x)$ means that for each re-scaling $x \to X = \lambda \cdot x$, there exists an appropriate re-scaling $y \to Y$ for which the same formula $Y = f(X)$ will be true for the correspondingly re-scaled variables X and Y.

2.3 *Let Us Apply This Idea to the Membership Function*

It is reasonable to require that the selection of the best membership functions should also not depend on the choice of the unit for measuring the corresponding quantity x. In other words, it is reasonable to require that for each $\lambda > 0$, there should exist some reasonable transformation $y \to Y = T(y)$ of the degree of confidence for which $y = \mu(x)$ implies $Y = \mu(X)$.

2.4 *So, What Are Reasonable Transformations of the Degree of Confidence?*

One way to measure the degree of confidence is to have a poll: ask N experts how many of them believe that a given value x is, e.g., small, count the number M of whose who believe in this, and take the ratio M/N as the desired degree $y = \mu(x)$.

As usual with polls, the more people we ask, the more adequately we describe the general opinion. So, to get a more accurate estimate for $\mu(x)$, it is reasonable to ask more people. When we have a limited number of people to ask, it is reasonable to ask top experts in the field. When we start asking more people, we are thus adding people who are less experienced—and who may therefore be somewhat intimidated by the opinions of the top experts. This intimidation can be expressed in different ways:

- some new people may be too shy to express their own opinion, so they will keep quiet; as a result, if we add A people to the original N, we will still have the same number M of people voting "yes", and the new ratio will be equal to $Y = \frac{M}{N+A}$, i.e., to $Y = a \cdot y$, where $a \stackrel{\text{def}}{=} \frac{N}{N+A}$;
- some new people will be too shy to think on their own and will vote with the majority; so for the case when $M > N/2$, we will have

$$Y = \frac{M+A}{N+A},$$

i.e., since $M = y \cdot N$, we will have

$$Y = \frac{y \cdot N + A}{N + A} = a \cdot y + b,$$

where a is the same as before and $b = \frac{A}{N + A}$;

- we may also have a situation in which a certain proportion c of the new people keep quiet while the others vote with the majority; in this case, we have

$$Y = \frac{M + (1 - c) \cdot A}{N + A} = a \cdot y + b,$$

where $a = (1 - c) \cdot \frac{A}{N + A}$.

In all these cases, we have a linear transformation $Y = a \cdot y + b$. So, it seems reasonable to identify reasonable transformations with linear ones. We will call the corresponding scale-invariance L-scale-invariance (L for Linear).

2.5 What Membership Functions We Consider

We consider symmetric properties, for which $\mu(-x) = \mu(x)$, so it is sufficient to consider only positive values x. Specifically, we consider properties like "small" for which the degree of confidence decreases with x, going all the way to 0 as x increases. We will call such membership functions s-membership functions (s for small). Thus, we arrive at the following definition.

Definition 1 By an $s - membership\ function$, we means a function $\mu : (0, \infty) \to [0, 1]$ that, starting with $\mu(0) = 1$, decreases with x (i.e., for which $x_1 > x_2$ implies $\mu(x_1) \geq \mu(x_2)$) and for which $\lim\limits_{x \to \infty} = 0$.

Definition 2 We say that an s-membership function $\mu(x)$ is $L - scale - invariant$ if for every $\lambda > 0$, there exist values $a(\lambda)$ and $b(\lambda)$ for which $y = \mu(x)$ implies $Y = \mu(X)$, where $X = \lambda \cdot x$ and $Y = a(\lambda) \cdot y + b(\lambda)$.

Unfortunately, this does not solve our problem: as the following result shows, the only L-scale-invariant s-membership functions are constants;

Proposition 1 *The only L-scale-invariant s-membership functions are constant functions* $\mu(x) = \text{const}$.

2.6 Discussion

What does this result mean? We considered two possible types of reasonable transformations of the degrees of confidence—which both turned out to be linear, and this was not enough. So probably there are other reasonable transformations of degrees of confidence. How can we describe such transformations?

Clearly, if we have a reasonable transformation, then its inverse is also reasonable. Also, a composition of two reasonable transformations should be a reasonable transformation too. So, in mathematical terms, reasonable transformations should form a *group*.

This group should be finite-dimensional, in the sense that different transformations should be uniquely determined by a finite number of parameters—since in the computer, we can store only finitely many parameters. We also know that linear transformations are reasonable. So, we are looking for a finite-dimensional group of transformations from real numbers to real numbers that contains all linear transformations. It is known (see, e.g., [10, 13, 16]) that all such transformations are piece-wise linear, i.e., have the form

$$\mu \to \frac{a \cdot \mu + b}{1 + c \cdot \mu}.$$

Thus, we arrive at the following definitions.

3 Which Symmetric Membership Functions Should We Select: Definitions and the Main Result

Definition 3 We say that an s-membership function $\mu(x)$ *is scale − invariant* if for every $\lambda > 0$, there exist values $a(\lambda)$, $b(\lambda)$, and $c(\lambda)$ for which $y = \mu(x)$ implies $Y = \mu(X)$, where $X = \lambda \cdot x$ and

$$Y = \frac{a(\lambda) \cdot y + b(\lambda)}{1 + c(\lambda) \cdot y}.$$

Proposition 2 *The only scale-invariant s-membership functions are distending membership functions (1).*

3.1 Discussion

This result explains the empirical success of distending functions.

4 Which Hedge Operations and Negation Operations Should We Select

4.1 Discussion

We would like hedging and negation operations $y = h(x)$ to be also invariant, i.e., that for each natural transformation $X = T(x)$, there should be a transformation $Y = S(y)$ for which $y = h(x)$ implies $Y = h(X)$. Now that we know what are natural transformations of membership degrees—they are fractional-linear functions—we can describe this requirement in precise terms.

Definition 4 We say that a monotonic function $y = h(x)$ from an open (finite or infinite) interval D to real numbers is $h - scale - invariant$ if for every fractional-linear transformation $X = T(x)$, there exists a fractional-linear transformation $Y = S(y)$ for which $y = h(x)$ implies $Y = h(X)$.

Proposition 3 *The only h-scale-invariant functions are fractionally linear ones.*

4.2 Discussion

- This result explains the empirical success of fractional-linear hedge operations and negation operations.
- As we show in the proof, it is sufficient to require that a fractional linear transformation S exist only for all *linear* transformations T.

5 Proofs

5.1 Proof of Proposition 1

We will prove this result by contradiction. Let us assume that the function $\mu(x)$ is not a constant, and let us derive a contradiction.

Substituting the expressions for X, Y, and $y = \mu(x)$ into the formula $Y = \mu(X)$ describing L-scale-invariance, we conclude that for every x and for every λ, we have

$$\mu(\lambda \cdot x) = a(\lambda) \cdot \mu(x) + b(\lambda). \tag{3}$$

It is known that monotonic functions are almost everywhere differentiable. Due to the formula (3), if a function $\mu(x)$ is differentiable at some point $x = x_0$, it is also differentiable at any point of the type $\lambda \cdot x_0$ for every $\lambda > 0$—and thus, that it is differentiable for all $x > 0$.

Since the function $\mu(x)$ is not constant, there exist values $x_1 \neq x_2$ for which $\mu(x_1) \neq \mu(x_2)$. For these values, the formula (3) has the form

$$\mu(\lambda \cdot x_1) = a(\lambda) \cdot \mu(x_1) + b(\lambda); \quad \mu(\lambda \cdot x_2) = a(\lambda) \cdot \mu(x_2) + b(\lambda).$$

Subtracting the two equations, we get

$$\mu(\lambda \cdot x_1) - \mu(\lambda \cdot x_2) = a(\lambda) \cdot (\mu(x_1) - \mu(x_2)),$$

thus

$$a(\lambda) = \frac{\mu(\lambda \cdot x_1) - \mu(\lambda \cdot x_2)}{\mu(x_1) - \mu(x_2)}.$$

Since the function $\mu(x)$ is differentiable, we can conclude that the function $a(\lambda)$ is also differentiable. Thus, the function $b(\lambda) = \mu(\lambda \cdot x) - a(\lambda) \cdot \mu(x)$ is differentiable too.

Since all three functions $\mu(x)$, $a(\lambda)$, and $b(\lambda)$ are differentiable, we can differentiate both sides of the equality (3) with respect to λ. If we substitute $\lambda = 1$, we get $x \cdot \mu'(x) = A \cdot \mu(x) + B$, where we denoted $A \stackrel{\text{def}}{=} a'(1)$, $B \stackrel{\text{def}}{=} b'(1)$, and $\mu'(x)$, as usual, indicates the derivative. Thus, $x \cdot \dfrac{d\mu}{dx} = A \cdot \mu + B$. We cannot have $A = 0$ and $B = 0$, since then $\mu'(x) = 0$ and $\mu(x)$ would be a constant. Thus, in general, the expression $A \cdot \mu + B$ is not 0, so

$$\frac{d\mu}{A \cdot \mu + B} = \frac{dx}{x}.$$

If $A = 0$, then integration leads to $\dfrac{1}{B} \cdot \mu(x) = \ln(x) + c_0$, where c_0 is the integration constant. Thus, $\mu(x) = B \cdot \ln(x) + B \cdot c_0$. This expression has negative values for some x, while all the values $\mu(x)$ are in the interval [0, 1]. So, this case is impossible.

If $A \neq 0$, then we have $d(A \cdot \mu + B) = A \cdot d\mu$, hence

$$\frac{d(A \cdot \mu + B)}{A \cdot \mu + B} = A \cdot \frac{dx}{x}.$$

Integration leads to $\ln(A \cdot \mu(x) + B) = A \cdot \ln(x) + c_0$. By applying $\exp(z)$ to both sides, we get $A \cdot \mu(x) + B = \exp(c_0) \cdot x^A$, i.e., $\mu(x) = A^{-1} \cdot \exp(c_0) \cdot x^A - B/A$. This expression tends to infinity either for $x \to \infty$ (if $A > 0$) or for $x \to 0$ (if $A < 0$). In both cases, we get a contradiction with our assumption that $\mu(x)$ is always within the interval [0, 1].

The proposition is proven.

5.2 Proof of Proposition 2

Substituting the expressions for X, Y, and $y = \mu(x)$ into the formula $Y = \mu(X)$ describing scale-invariance, we conclude that for every x and for every λ, we have

$$\mu(\lambda \cdot x) = \frac{a(\lambda) \cdot \mu(x) + b(\lambda)}{1 + c(\lambda) \cdot \mu(x)}. \tag{4}$$

Similarly to the proof of Proposition 1, we can conclude that the function $\mu(x)$ is differentiable for all $x > 0$.

Multiplying both sides of the equality (4) by the denominator, we conclude that

$$\mu(\lambda \cdot x) + c(\lambda) \cdot \mu(x) \cdot \mu(\lambda \cdot x) = a(\lambda) \cdot \mu(x) + b(\lambda).$$

So, for three different values x_i, we have the following three equations:

$$\mu(\lambda \cdot x_i) + c(\lambda) \cdot \mu(x_i) \cdot \mu(\lambda \cdot x_i) = a(\lambda) \cdot \mu(x_i) + b(\lambda), \quad i = 1, 2, 3.$$

We thus have a system of three linear equations for three unknowns $a(\lambda)$, $b(\lambda)$, and $c(\lambda)$. By Cramer's rule, the solution to such a system is a rational (hence differentiable) function of the coefficients and the right-hand sides. So, since the function $\mu(x)$ is differentiable, we can conclude that the functions $a(\lambda)$, $b(\lambda)$, and $c(\lambda)$ are differentiable as well.

Since all the functions $\mu(x)$, $a(\lambda)$, $b(\lambda)$, and $c(\lambda)$ are differentiable, we can differentiate both sides of the formula (4) with respect to λ. If we substitute $\lambda = 1$ and take into account that for $\lambda = 1$, we have $a(1) = 1$ and $b(1) = c(1) = 0$, we get

$$x \cdot \frac{d\mu}{dx} = B \cdot \mu + A - C \cdot \mu^2,$$

where A and B are the same as in the previous proof and $C \stackrel{\text{def}}{=} c'(1)$.

For $x \to \infty$, we have $\mu(x) \to 0$, so $\mu'(x) \to 0$, and thus $A = 0$ and

$$x \cdot \frac{d\mu}{dx} = B \cdot \mu - C \cdot \mu^2,$$

i.e.,

$$\frac{d\mu}{B \cdot \mu - C \cdot \mu^2} = \frac{dx}{x}. \tag{5}$$

As we have shown in the proof of Proposition 1, we cannot have $C = 0$, so $C \neq 0$. One can easily see that

$$\frac{1}{\mu - \frac{B}{C}} - \frac{1}{\mu} = \frac{\frac{B}{C}}{\mu \cdot \left(\mu - \frac{B}{C}\right)} = \frac{-B}{B \cdot \mu - C \cdot \mu^2}.$$

Thus, by multiplying both sides of equality (5) by $-B$, we get

$$\frac{d\mu}{\mu - \frac{B}{C}} - \frac{d\mu}{\mu} = -B \cdot \frac{dx}{x}.$$

Integrating both sides, we get

$$\ln\left(\mu(x) - \frac{B}{C}\right) - \ln(\mu) = -B \cdot \ln(x) + c_0.$$

By applying $\exp(z)$ to both sides, we get

$$\frac{\mu(x) - \frac{B}{C}}{\mu(x)} = C_0 \cdot x^{-B}$$

for some constant C_0, i.e.,

$$1 - \frac{B/C}{\mu} = C_0 \cdot x^{-B},$$

hence

$$\frac{B/C}{\mu} = 1 - C_0 \cdot x^{-B}$$

and

$$\mu(x) = \frac{B/C}{1 - C_0 \cdot x^{-B}}.$$

From the condition that $\mu(0) = 1$, we conclude that $B < 0$ and $B/C = 1$. From the condition that $\mu(x) \leq 1$, we conclude that $C_0 < 0$. Thus, we get the desired formula

$$\mu(x) = \frac{1}{1 + |C_0| \cdot x^{|B|}}.$$

The proposition is proven.

5.3 Proof of Proposition 3

For constant functions the statement is trivial, since every constant function is fractional-linear. Therefore, it is sufficient to prove for non-constant functions $h(x)$.

Similarly to the proof of Proposition 2, we can prove that the function $h(x)$ is differentiable. Let $x \in D$, and let λ and x_0 from an open neighborhood of 1 and 0 respectively be such that $\lambda \cdot x \in D$ and $x + x_0 \in D$. Since the function $h(x)$ is h-scale-invariant, there exist fractional-linear transformations for which

$$h(x + x_0) = \frac{a(x_0) \cdot h(x) + b(x_0)}{1 + c(x_0) \cdot h(x)} \tag{6}$$

and

$$h(\lambda \cdot x) = \frac{d(\lambda) \cdot h(x) + e(\lambda)}{1 + f(\lambda) \cdot h(x)}. \tag{7}$$

Similarly to the proof of Proposition 2, we can prove that the functions $a(x_0)$, ..., are differentiable. Similar to the proof of Proposition 2, we can differentiate the formula (7) with respect to λ and take $\lambda = 1$, then we get:

$$x \cdot h' = D \cdot h + E - F \cdot h^2. \tag{8}$$

Similarly, differentiating the formula (6) with respect to x_0 and taking $x_0 = 0$, we get:

$$h' = A \cdot h + B - C \cdot h^2. \tag{9}$$

Let us consider two cases: $C \neq 0$ and $C = 0$.

Let us first consider the case when $C \neq 0$. By completing the square, we get $h' = A \cdot h + B - C \cdot h^2 = \widehat{A} - C \cdot (h - h_0)^2$ for some $\widehat{A}$ and h_0, i.e.,

$$h' = \widehat{A} - C \cdot H^2, \tag{10}$$

where $H \stackrel{\text{def}}{=} h - h_0$. Substituting $h = H + h_0$ into the right-hand of the formula (8), we conclude that

$$x \cdot h' = \widehat{D} \cdot H + \widehat{E} - F \cdot H^2 \tag{11}$$

for some constants $\widehat{D}$ and $\widehat{E}$. Dividing (11) by (10), we get

$$x = \frac{\widehat{D} \cdot H + \widehat{E} - F \cdot H^2}{\widehat{A} - C \cdot H^2}, \tag{12}$$

so

$$\frac{dx}{dH} = \frac{(\widehat{D} - 2F \cdot H) \cdot (\widehat{A} - C \cdot H^2) - (\widehat{D} \cdot H + \widehat{E} - F \cdot H^2) \cdot (-2C \cdot H)}{(\widehat{A} - C \cdot H^2)^2} =$$

$$\frac{\widehat{A} \cdot \widehat{D} - 2(\widehat{A} \cdot F - C \cdot \widehat{E}) \cdot H + C \cdot \widehat{D} \cdot H^2}{(\widehat{A} - C \cdot H^2)^2}. \tag{13}$$

On the other hand,

$$\frac{dx}{dH} = \frac{1}{\frac{dH}{dx}} = \frac{1}{\widehat{A} - C \cdot H^2}. \tag{14}$$

The right-hand sides of the formulas (13) and (14) must be equal, so for all H, we have

$$\widehat{A} \cdot \widehat{D} - 2(\widehat{A} \cdot F - C \cdot \widehat{E}) \cdot H + C \cdot \widehat{D} \cdot H^2 = \widehat{A} - C \cdot H^2.$$

Since the two polynomials of H are equal, the coefficients at 1, H, and H^2 must coincide.

Comparing the coefficients at H^2, we get $C \cdot \widehat{D} = -C$. Since $C \neq 0$, we conclude that $\widehat{D} = -1$. Comparing the coefficients at 1, we get $\widehat{A} \cdot \widehat{D} = \widehat{A}$, i.e., $-\widehat{A} = \widehat{A}$ and thus $\widehat{A} = 0$. Comparing the coefficients at H and taking into account that $\widehat{A} = 0$, we get $0 = \widehat{A} \cdot F - C \cdot \widehat{E} = -C \cdot \widehat{E}$. Since $C \neq 0$, this implies $\widehat{E} = 0$. So, the formula (12) takes the form

$$x = \frac{\widehat{D} \cdot H - F \cdot H^2}{-C \cdot H^2} = \frac{\widehat{D} - F \cdot H}{-C \cdot H}.$$

Thus x is a fractional linear function of H, hence H (and therefore $h = H + h_0$) is also a fractional linear function of x.

Let us now consider the case when $C = 0$. In this case, $h' = A \cdot h + B$ and $x \cdot h' = D \cdot h + E - F \cdot h^2$, thus

$$x = \frac{x \cdot h'}{h'} = \frac{D \cdot h + E - F \cdot h^2}{A \cdot h + B}.$$

If $F = 0$, then x is a fractional linear function of $h(x)$ and hence, h is also a fractional-linear function of x.

So, it is sufficient to consider the case when $F \neq 0$. In this case, by completing the square, we can find constants $\widetilde{D}$, h_0, and $\widehat{B}$ for which, for $H = h - h_0$, we have

$$x \cdot h' = D \cdot h + E - F \cdot h^2 = \widetilde{D} - F \cdot H^2 \tag{15}$$

and

$$h' = A \cdot h + B = A \cdot H + \widehat{B}. \tag{16}$$

Dividing (15) by (16), we have

$$x = \frac{\widetilde{D} - F \cdot H^2}{A \cdot H + \widehat{B}}. \tag{17}$$

Thus,

$$\frac{dx}{dH} = \frac{(-2F \cdot H) \cdot (A \cdot H + \widehat{B}) - (\widehat{D} - F \cdot H^2) \cdot A}{(A \cdot H + \widehat{B})^2}$$

$$= \frac{-A \cdot \widehat{D} - 2\widehat{B} \cdot F \cdot H - A \cdot F \cdot H^2}{(A \cdot H + \widehat{B})^2}.$$

On the other hand,

$$\frac{dx}{dH} = \frac{1}{\frac{dH}{dx}} = \frac{1}{A \cdot H + \widehat{B}}.$$

By equating the two expressions for the derivative and multiplying both sides by $(A \cdot H + \widehat{B})^2$, we conclude that

$$-A \cdot \widehat{D} - 2\widehat{B} \cdot F \cdot H - A \cdot F \cdot H^2 = A \cdot H + \widehat{B},$$

thus $A \cdot F = 0$, $A = -2\widehat{B} \cdot F$, and $-A \cdot \widehat{D} = \widehat{B}$. If $A = 0$, then we have $\widehat{B} = 0$, so $h' = 0$ and h is a constant—but we consider the case when the function $h(x)$ is not a constant. Thus, $A \neq 0$, hence $F = 0$, and the formula (17) describes x as a fractional-linear function of H.

In both cases $C \neq 0$ and $C = 0$, we obtain an expression of x in terms of H (hence h) that is fractional linear. Since the inverse of a fractional linear is fractional linear, the function $h(x)$ is also fractional linear.

The proposition is proven.

Acknowledgements This work was supported in part by the grant TUDFO/47138-1/2019-ITM from the Ministry of Technology and Innovation, Hungary, and by the US National Science Foundation grants 1623190 (A Model of Change for Preparing a New Generation for Professional Practice in Computer Science) and HRD-1242122 (Cyber-ShARE Center of Excellence).

References

1. R. Belohlavek, J.W. Dauben, G.J. Klir, *Fuzzy Logic and Mathematics: A Historical Perspective* (Oxford University Press, New York, 2017)
2. O. Csiszar, J. Dombi, in *Generator-Based Modifiers and Membership Functions in Nilpotent Operator Systems*, vol. 3–5, (Budapest, Hungary, 2019), pp. 99–105
3. J. Dombi, O. Csiszar, Implications in bounded systems. Inf. Sci. **283**, 229–240 (2014)
4. J. Dombi, O. Csiszar, The general nilpotent operator system. Fuzzy Sets Syst. **261**, 1–19 (2015)
5. J. Dombi, O. Csiszar, Equivalence operators in nilpotent systems. Fuzzy Sets Syst. **299**, 113–129 (2016)
6. J. Dombi, T. Szépe, Arithmetic-based fuzzy control. Iran. J. Fuzzy Syst. **14**(4), 51–66 (2017)
7. J. Dombi, A. Hussain, in *Interval Type-2 Fuzzy Control Using the Distending Function*, vol. 18–21, ed. by A.J. Tallón-Ballesteros (Kitakyushu, Japan, 2019), pp. 705–714
8. J. Dombi, A. Hussain, in *Data-Driven Arithmetic Fuzzy Control Using the Distending Function*, vol. 22–24, (Nice, France, 2019), pp. 215–221

9. J. Dombi, A. Hussain, A new approach to fuzzy control using the distending function. J. Process Control **86**, 16–29 (2020)
10. V.M. Guillemin, S. Sternberg, An algebraic model of transitive differential geometry. Bull. Am. Math. Soc. **70**(1), 16–47 (1964)
11. G. Klir, B. Yuan, *Fuzzy Sets and Fuzzy Logic* (Prentice Hall, Upper Saddle River, New Jersey, 1995)
12. J.M. Mendel, *Uncertain Rule-Based Fuzzy Systems: Introduction and New Directions* (Springer, Cham, Switzerland, 2017)
13. H.T. Nguyen, V. Kreinovich, *Applications of Continuous Mathematics to Computer Science* (Kluwer, Dordrecht, 1997)
14. H.T. Nguyen, C.L. Walker, E.A. Walker, *A First Course in Fuzzy Logic* (Chapman and Hall/CRC, Boca Raton, Florida, 2019)
15. V. Novák, I. Perfilieva, J. Močkoř, *Mathematical Principles of Fuzzy Logic* (Kluwer, Boston, Dordrecht, 1999)
16. I.M. Singer, S. Sternberg, Infinite groupsof Lie and Cartan, Part I. J. d'Analyse Math. **15**, 1–113 (1965)
17. L.A. Zadeh, Fuzzy sets. Inf. Control **8**, 338–353 (1965)